THE

Illustrated AutoCAD® 2008

QUICK REFERENCE

Ralph Grabowski

Autodesk·

THOMSON

DELMAR LEARNING

Australia · Canada · Mexico · Singapore · Spain · United Kingdom · United States

THOMSON
™
DELMAR LEARNING

Autodesk

The Illustrated AutoCAD® 2008 Quick Reference
Ralph Grabowski

Vice President, Technology and Trades SBU:
Dave Garza

Senior Acquisitions Editor:
Jim Gish

Marketing Director:
Debbie Yarnell

Channel Manager:
Kevin Rivenburg

Marketing Coordinator:
Mark Pierro

Production Director:
Patty Stephan

Content Project Manager:
Betsy Hough

Editorial Assistant:
Sarah Timm

Senior Project Manager:
John Fisher

Production Manager:
Stacy Masucci

Book Design and Typesetting:
Ralph Grabowski

Cover Images:
Getty Images

Library of Congress Cataloging-in-Publication Data:

ISBN: 1-4283-1160-2

NOTICE TO THE READER

Function Keys

Key	Description
F1	Displays online help
F2	Toggles text & graphics windows
F3	Toggles object snap
F4	Toggles tablet mode
F5	Cycles through isoplanes
F6	Toggles dynamic UCS
F7	Toggles grid display
F8	Toggles ortho mode
F9	Toggles snap mode
F10	Toggles polar tracking
F11	Toggles object snap tracking
F12	Toggles dynamic input
Ctrl F4	Closes drawing
Ctrl F6	Switches to next drawing
Alt F4	Closes AutoCAD
Alt F8	Runs VBA
Alt F11	Opens VBA IDE

Acad.exe Startup Switches

Switch	Description
/b	Runs .scr script file
/c	Specifies path to alt .cfg hardware
/layout	Displays named layout
/ld	Loads ARX or DBX app
/nohardware	Disables graphics board hardware acceleration 🗗
/nologo	Suppresses AutoCAD logo
/nossm	Suppresses Sheet Set Manager
/p	Loads user-defined profile
/pl	Background plot/publish
/r	Restores default pointer
/s	Sets paths to support folders
/set	Loads .dst sheetset file
/t	Loads .dwt template file
/v	Displays named view
/w	Shows workspace at startup

Mouse Buttons

Button	Description
①	Picks objects
②	Displays cursor menu
③ (Wheel)	Zooms or pans
Ctrl ②	Displays osnap cursor menu

Control Keys

Key	Description
Ctrl A	Selects all objects
Ctrl Shift A	Toggles group mode
Ctrl B	Toggles snap mode
Ctrl C	Copies to the Clipboard
Ctrl Shift C	Copies with base point
Ctrl D	Cycles coordinate display
Ctrl E	Cycles through isoplanes
Ctrl F	Toggles object snap mode
Ctrl G	Toggles grid display
Ctrl H	Toggles pick style
Ctrl K	Displays Hyperlinks dialog box
Ctrl L	Toggles ortho mode
Ctrl N	Opens new drawing
Ctrl O	Opens drawing files
Ctrl P	Displays Plot dialog box
Ctrl Q	Toggles log file
Ctrl R	Cycles through viewports
Ctrl S	Saves drawing
Ctrl Shift S	Displays Save As dialog box
Ctrl T	Toggles tablet mode
Ctrl U	Toggles polar tracking
Ctrl V	Pastes from the Clipboard
Ctrl Shift V	Pastes with insertion point
Ctrl X	Cuts to the Clipboard
Ctrl Y	Redoes last undo (MRedo)
Ctrl Z	Undoes last command
Ctrl 0	Toggles cleanscreen mode
Ctrl 1	Toggles Properties window
Ctrl 2	Toggles DesignCenter window
Ctrl 3	Toggles Tool Palettes window
Ctrl 4	Toggles SheetSet Manager
Ctrl 6	Toggles dbConnect Manager
Ctrl 7	Toggles Markup Set Manager
Ctrl 8	Toggles QuickCalc window
Ctrl 9	Toggles the command line

Other Keys

Key	Description
Esc	Cancels commands and grips
Delete	Deletes selected objects
Enter	Executes and repeats commands

About This Book

The Illustrated AutoCAD 2008 Quick Reference presents concise facts about all commands found in AutoCAD 2008 and earlier. The clear format of this reference book is illustrates with hundreds of figures. Each command starts on its own page, and includes one or more of the following: command line options, dialog box and palette options, toolbar icons, shortcut menus, related commands, and related system variables. Plus:

- Variations of commands, such as the **View** and **-View** commands.
- Dozens of AutoCAD 2008 commands and system variables not documented by Autodesk.
- Icons, such as ☑ and ⊙, that indicate default settings of check boxes, radio buttons, and other contols in dialog boxes and palettes.
- Over 100 definitions of acronyms and hard-to-understand terms.
- Nearly 1,000 context-sensitive tips.
- Express Tool command names and descriptions, listed in Appendix A.
- Obsolete commands that no longer work in AutoCAD, listed in Appendix B.
- All system variables, including those not listed by the SetVar command, listed in Appendix C.

The name of each command shown in mixed upper and lower case to help you understand the name, which is often condensed. For example, the **VpClip** command is short for "ViewPort CLIP." Each command includes all alternative methods of command input:

- Alternate command spelling, such as **Donut** and **Doughnut**.
- **'** (the apostrophe prefix) indicating transparent commands, such as **'Blipmode**.
- All aliases, such as **L** for the Line command.
- Pull-down menu picks, such as **Draw ⮰ Construction Line** for the XLine command.
- Control-key combinations, such as **Ctrl+E** for the Isoplane toggle.
- Function keys, such as **F1** for the Help command.
- Alt-key combinations, such as **Alt+TE** for the Spell command.
- Dashboard palette names, such as **Block Attributes** for the AttDef command.

The version or release number indicates when the command first appeared in AutoCAD, such as **Ver. 1.0**, **Rel. 9**, or **2008** — useful when working with older versions of AutoCAD.

A special thank you to Stephen Dunning for editing the text, and to Bill Fane for his technical editing of this book. *Soli Deo Gloria!*

Ralph Grabowski
Abbotsford, British Columbia, Canada
March 23, 2007

Table of Contents

▣ Indicates commandd new to AutoCAD 2008

'About

Displays the version number and other information about AutoCAD.

Command	Alias	Ctrl+	F-key	Alt+	Menu Bar	Dashboard
about	...	...	...	HA	Help ↳About	...

Command: about

Displays dialog box:

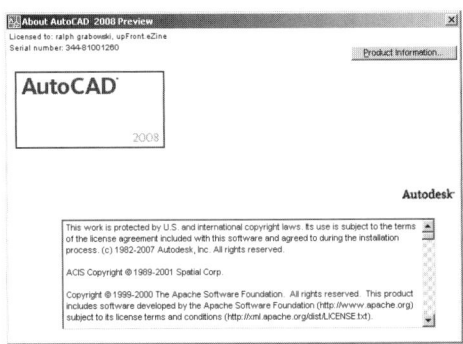

ABOUT AUTOCAD DIALOG BOX

☒ dismisses dialog box; alternatively, press ESC.

Product Information displays the Product Information dialog box:

Update displays a dialog box for changing the software serial number (*new to 2008*).

License Agreement opens the computer's default word processor, and then displays the Autodesk Software License Agreement document (*license.rtf*).

Activate runs the Product Activation wizard.

Save As displays the Save As dialog box, which records the information displayed above.

Close returns to AutoCAD.

RELATED COMMANDS

Properties displays information about selected objects.

Status displays information about the drawing and environment.

RELATED SYSTEM VARIABLES

_PkSer displays the AutoCAD software serial number.

_Server displays the network authorization code.

AcisIn, AcisOut

Rel. 13   AcisIn imports *.sat* files into drawings to create 3D solids, 2D regions, and bodies; AcisOut exports 3D solids, 2D regions, and bodies in SAT format.

Command	Alias	Ctrl+	F-key	Alt+	Menu Bar	Dashboard
acisin	...	...	...	IA	Insert	...
					⍜ACIS File	

Command: acisin

Displays Select ACIS File dialog box. Select a .sat file, and then click Open.

DIALOG BOX OPTIONS

Cancel dismisses the dialog box.

Open opens the selected *.sat* file.

Command: acisout
Select objects: *(Select one or more objects.)*
Select objects: *(Press Enter to end object selection.)*

Displays Create ACIS File dialog box. Enter a name for the .sat file, and then click Save.

DIALOG BOX OPTIONS

Cancel dismisses the dialog box.

Save saves the selected object(s) in a *.sat* file.

RELATED COMMAND

AmeConvert converts AME v2.0 and v2.1 solid models and regions into solids.

RELATED SYSTEM VARIABLE

AcisOutVer specifies the ACIS version number for exporting models with the AcisOut command.

RELATED FILE

.sat is the ASCII format of ACIS model files; short for "save as text."

TIPS

- **AcisOut** exports 3D solids, 2D regions, and bodies only.

- Other ACIS-based software can read *.sat* files.

- "ACIS" comes from the first names of the original developers, "Andy, Charles, and Ian's System." In Greek mythology, Acis was the lover of the goddess Galatea; when Acis was killed by the jealous Cyclops, Galatea turned the blood of Acis into a river.

- ACIS is the name of solids modeling technology from the Spatial Technologies division of Dassault Systemes, and is used by numerous 3D CAD packages. As of AutoCAD 2004, Autodesk uses its own ACIS-derived solids modeler, called ShapeManager.

'AdCenter, 'AdcClose

<u>**2000**</u> Opens and closes the Design Center palette (short for Autocad Design CENTER).

Command	Aliases	Ctrl+	F-key	Alt+	Menu Bar	Dashboard
adcenter	adc	2	...	TPG	Tools	...
	content				⅋Palettes	
	dc, dcenter				⅋DesignCenter	
adcclose	...	2	...	TPG	Tools	...
					⅋Palettes	
					⅋DesignCenter	

Command: adcenter

Displays DesignCenter palette:

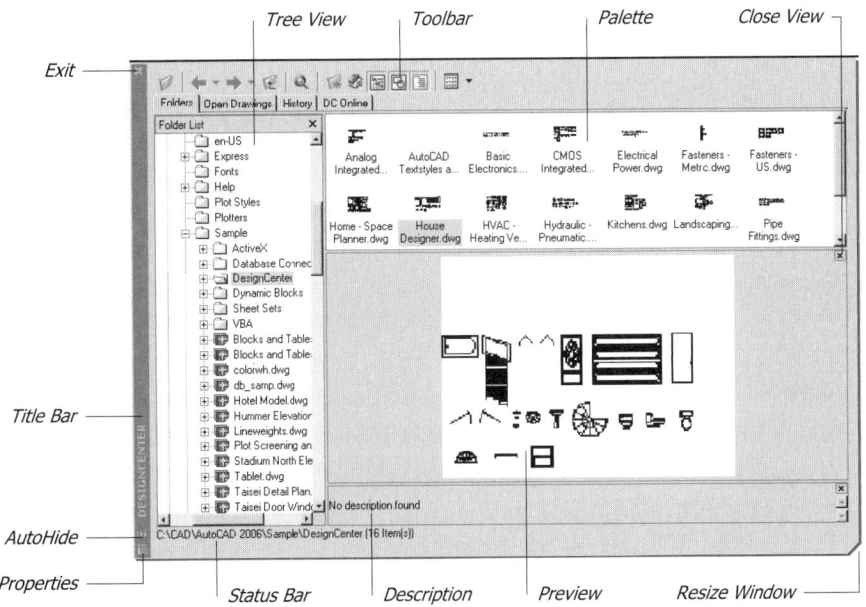

Command: adcclose

Closes the DesignCenter palette.

TABS

Folders displays the content of the local computer, as well as networked computers.

Open Drawings displays the content of drawings currently open in AutoCAD.

History displays the drawings previously opened.

DC Online displays the content available from Autodesk via the Internet.

Palette toolbar

Load displays the Open dialog box to open the following vector and raster file types: *.dwg, .dws, .dwf, .dxf, .bil, .bmp, .cal, .cg4, .dib, .flc, .fli, .gif, .gp4, .ig4, .igs, .jpg, .jpeg, .mil, .pat, .pcx, .png, .rlc, .rle, .rst, .tga,* and *.tif.*

Back returns to the previous view.

Forward goes to the next view.

Up moves up one folder level.

Search opens the Search dialog box to search for AutoCAD files (on the computer) and the following objects (in drawings): blocks, dimstyles, drawings, hatch patterns and *.pat* files, layers, layouts, linetypes, text styles, table styles, and xrefs.

Favorites displays the files in the *\documents and settings\<username>\favorites\autodesk* folder.

Home displays the files in the *\autocad 2008\sample\designcenter* folder.

Tree View Toggle hides and displays the Folders and Open Drawings tree views.

Preview toggles the display of the preview image of *.dwg* and raster files.

Description toggles the display of the description area.

View changes the display format of the palette area.

RELATED COMMAND

AdcNavigate specifies the initial path for DesignCenter.

RELATED SYSTEM VARIABLE

AdsState reports whether DesignCenter is open or not.

TIPS

- Use Design Center to track drawings and parts of drawings, such as block libraries.

- You can drag blocks, table styles, and other drawing parts from the Design Center palette into the drawing, and blocks onto Tool palette.

- Design Center allows you to share text and dimension styles between drawings.

- The Design Center palette can switch between floating and docked modes by right-clicking and selecting the option from the menu.

- Ctrl+2 opens and closes the Design Center palette each time you press the keys.

- This command cannot be used transparently (within another command) until it has been used at least once non-transparently.

AdcNavigate, AdcCustomNavigate

<u>2000</u> Specifies the initial path for Design Center to access content.

Command	Alias	Ctrl+	F-key	Alt+	Menu Bar	Dashboard
adcnavigate	...	...	...	...	...	...
adccustomnavigate						

Command: adcnavigate

Opens DesignCenter, if not already open.

Enter pathname <>: *(Enter a path, and then press Enter.)*

COMMAND LINE OPTION

Enter pathname specifies the path, such as *c:\program files\autocad2008\sample.*

. . . .

AdCustomNavigate command
Command: adccustomnavigate

Points to the folder containing custom objects, which are displayed in Design Center's Custom tab. The Custom tab is displayed only by when custom applications are registered with AutoCAD, such as AutoCAD Architectural.

Enter pathname <>: *(Enter a path, and then press Enter.)*

RELATED COMMAND

AdCenter opens the Design Center palette.

TIPS

- AutoCAD uses the path specified by this command to locate content displayed by DesignCenter's Desktop option.

- You can enter paths to files, folders, or network locations:

Example of folder path:	*c:\program files\autocad 2008\sample*
Example of file path:	*c:\design center\welding.dwg*
Example of network path:	*\\downstairs\c\project*

. . . .

AecToAcad

<u>**2006**</u> Converts AutoCAD Architectural custom objects to AutoCAD objects; short for "Architecture Engineering Construction TO AutoCAD."

Command	Alias	Alt+	Menu Bar	Dashboard
aectoacad	-exporttoautocad	...	...	...

Command: aectoacad
File format: 2007
Bind xrefs: Yes
Bind type: Insert
Filename prefix:
Filename suffix:
Export options [Format/Bind/bind Type/Maintain/Prefix/Suffix/?] <Enter for filename>: *(Enter an option.)*
Export drawing name <d:\path\filename.dwg>: *(Press Enter.)*

COMMAND OPTIONS

Format specifies the DWG version.

Bind determines whether xrefs are bound.

bind Type specifies the type of xref binding.

Maintain specifies how to deal with blocks in ADT custom objects.

Prefix specifies a prefix for file names.

Suffix specifies a suffix for file names.

? lists the setting for each option.

Format options
Enter file format [r14/2000/2004/2007] <2007>:

r14 saves drawing in Release 14 DWG format.

2000 saves drawing in 2000/2000i/2001 DWG format.

2004 saves drawing in 2004/2005/2006 DWG format.

2007 saves drawing in 2007/8 DWG format.

Bind option
Bind xrefs [Yes/No] <Yes>:

Yes binds externally-referenced files to the drawing.

No stores links to xref files.

bind Type option
Bind type [Bind/Insert] <Insert>:

Bind makes xrefs part of the drawing.

Insert inserts xrefs into the drawing as if they were blocks.

Maintain option
Maintain resolved properties [Yes/No] <Yes>:

Yes explodes blocks in AEC custom objects.

No keeps blocks as unexploded xrefs.

Prefix and Suffix options
Filename prefix <>:
Filename suffix <>:

Filename Prefix and **Filename Suffix** respectively add text in front of and behind the current drawing's file name.

RELATED SYSTEM VARIABLE

ProxyNotice controls the display of custom objects in drawings:

- **0** — Custom objects are not displayed.
- **1** — Custom objects are displayed as proxy objects.
- **2** — Custom objects are displayed by a bounding box.

TIPS

- AutoCAD and related software have the ability to create *custom objects* which have special properties not otherwise found in AutoCAD. These objects are controlled by programming code written by in ARX. Autodesk products, such as Architectural Desktop and Mechanical Desktop, use custom objects. When drawings made with these programs are opened in AutoCAD, custom objects are displayed as *proxy objects*, which cannot be edited by AutoCAD.

- This command explodes custom objects created in ADT so that they can be edited by AutoCAD, and then saved to file by another name in order to preserve the original drawing.

- Exploded custom objects lose their intelligence.

- The following commands are also available for dealing with ADT custom objects:

  - **-AecDwgUnits** sets the units for inserting ADT objects into AutoCAD drawings.

  - **AecDisplayManagerConfigsSelection** changes the display configuration.

  - **AecFileOpenMessage** toggles display of dialog box warning of opening drawings from older releases of ADT.

  - **AecFileSaveMessage** toggles display of warning message when saving to previous versions.

  - **AecObjectCopyMessage** toggles display of warning message when saving to newer versions.

  - **AecObjRelDump, AecObjrelDumpEnhancedReferences**, and **AecObjRelShowEnhancedReferences** perform AEC custom object database dumps to the text screen.

  - **AecObjRelShow** indicates which ADT objects are related to each other.

  - **AecObjRelUpdate** updates ADT objects.

  - **AecPostDxfinFix** fixes problems with imported ADT drawings that cannot be repaired by AutoCAD's Audit command.

  - **AecSetXrefConfig** allows the same xref to be displayed several times.

  - **AecVersion** reports the version number of the ADT object enabler.

Ai_Box/Cone/Dish/Dome/Mesh/ Pyramid/Sphere/Torus/Wedge

Draws 3D boxes as mesh objects (undocumented commands); they do not draw surfaces.

Command	Alias	Ctrl+	F-key	Alt+	Menu Bar	Dashboard
ai_box	...	...	...	...	...	...

Command: ai_box
Specify corner point of box: *(Pick a point.)*
Specify length of box: *(Pick a point.)*
Specify width of box or [Cube]: *(Pick a point, or type **C**.)*
Specify height of box: *(Pick a point.)*
Specify rotation angle of box about the Z axis or [Reference]: *(Specify the rotation angle, or type **R**.)*

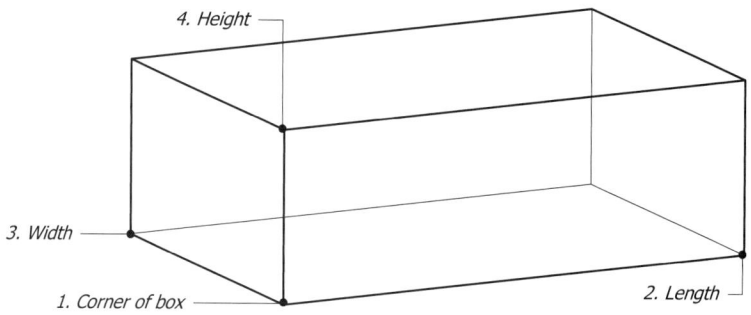

COMMAND LINE OPTIONS

Corner of box specifies the initial corner of the box.

Length specifies the length of one side of the box along the x-axis.

Cube creates a cube based on Length.

Width specifies the width of the box, along the y-axis.

Height specifies the height of the box, along the z-axis.

Rotation angle specifies the angle the box rotates about the z-axis.

Reference prompts you to pick two points that represent the new angle.

 Ai_Cone command

Command: ai_cone
Specify center point for base of cone: *(Pick a point.)*
Specify radius for base of cone or [Diameter]: *(Pick a point, or type D.)*
Specify radius for top of cone or [Diameter] <0>: *(Specify the radius, or type D; enter 0 for a pointy top.)*

Specify height of cone: *(Pick a point.)*
Enter number of segments for surface of cone <16>: *(Enter a number between 3 and 255, or press Enter.)*

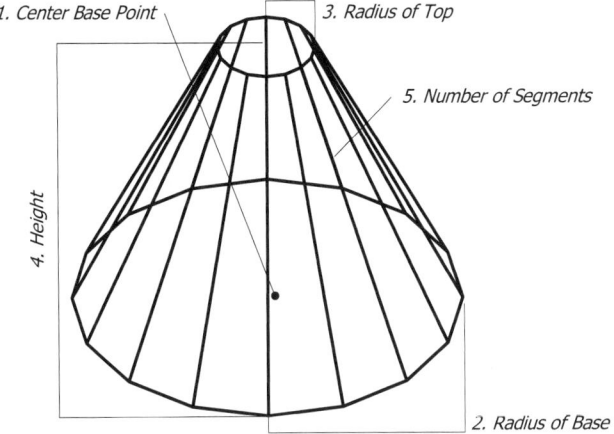

1. Center Base Point
3. Radius of Top
5. Number of Segments
4. Height
2. Radius of Base

COMMAND LINE OPTIONS

Base center point specifies the center point of the base of the cone.

Diameter of base specifies the diameter of the base.

Radius of base specifies the radius of the base.

Diameter of top specifies the diameter of the top of the cone.

Radius of top specifies the radius of the top of the cone; 0 = cone with a point.

Height specifies the height of the cone.

Number of segments specifies the number of "lines" defining curved surfaces; default = 16.

  Ai_Dish and Ai_Dome commands

Command: ai_dish
Specify center point of dish: *(Pick a point.)*
Specify radius of dish or [Diameter]: *(Pick a point, or type D.)*
Enter number of longitudinal segments for surface of dish <16>: *(Enter a value between 3 and 255, or press Enter.)*
Enter number of latitudinal segments for surface of dish <8>: *(Enter a value between 3 and 255, or press Enter.)*

COMMAND LINE OPTIONS

Center of dish specifies the center of the dish's base.

Diameter specifies the diameter of the dish.

Radius specifies the radius of the dish.

Number of longitudinal segments specifies the number of "lines" that define the curved surface in the vertical direction; default = 16.

Number of latitudinal segments specifies the number of "lines" that define the curved surface in the horizontal direction; default = 8.

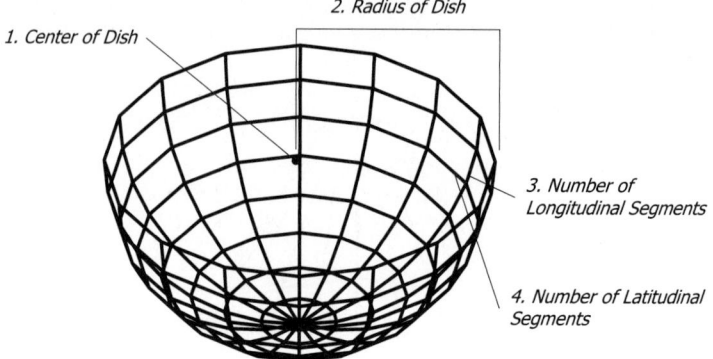

1. Center of Dish

2. Radius of Dish

3. Number of Longitudinal Segments

4. Number of Latitudinal Segments

Ai_Mesh command

Command: ai_mesh
Specify first corner point of mesh: *(Pick point 1.)*
Specify second corner point of mesh: *(Pick point 2.)*
Specify third corner point of mesh: *(Pick point 3.)*
Specify fourth corner point of mesh: *(Pick point 4.)*
Enter mesh size in the M direction: *(Specify a number.)*
Enter mesh size in the N direction: *(Specify a number.)*

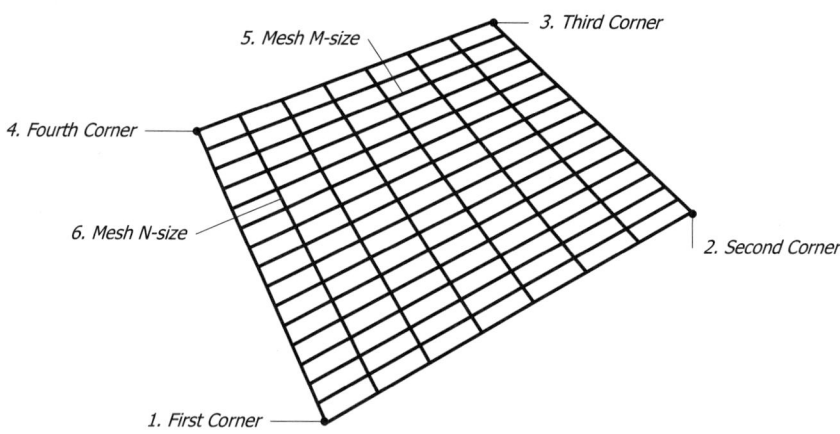

5. Mesh M-size

3. Third Corner

4. Fourth Corner

6. Mesh N-size

2. Second Corner

1. First Corner

COMMAND LINE OPTIONS

First corner specifies the mesh's first corner.

Second corner specifies the mesh's second corner.

Third corner specifies the mesh's third corner.

Fourth corner specifies the mesh's last corner.

Mesh size M direction specifies the number of horizontal "lines" that define the mesh's surface.

Mesh size N direction specifies the number of vertical "lines" that define the mesh's surface.

Ai_Pyramid command

Command: ai_pyramid
Specify first corner point for base of pyramid: *(Pick point 1.)*
Specify second corner point for base of pyramid: *(Pick point 2.)*
Specify third corner point for base of pyramid: *(Pick point 3.)*
Specify fourth corner point for base of pyramid or [Tetrahedron]: *(Pick point 4, or type T.)*

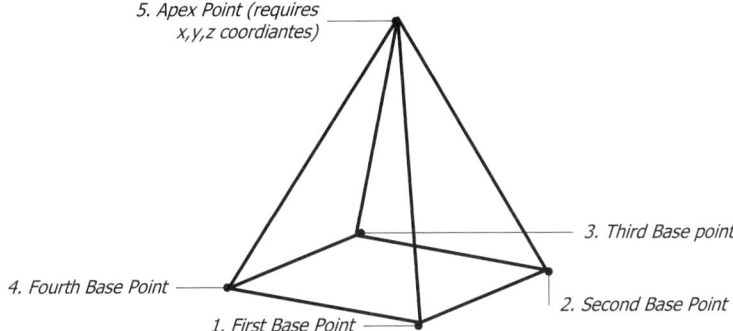

5. Apex Point (requires x,y,z coordiantes)

3. Third Base point

4. Fourth Base Point

2. Second Base Point

1. First Base Point

COMMAND LINE OPTIONS

First base point specifies the pyramid's first base point.

Second base point specifies pyramid's second base point.

Third base point specifies the pyramid's third base point.

Fourth base point specifies the pyramid's last base point.

Tetrahedron draws a pyramid with triangular sides.

Ridge specifies a ridge-top for the pyramid; see figure below.

Top specifies a flat-top for the pyramid; see figure below.

Apex point specifies a point for the pyramid's top

 Ai_Sphere command

Command: ai_sphere
Specify center point of sphere: *(Pick point 1.)*
Specify radius of sphere or [Diameter]: *(Pick point 2, enter a radius, or type D.)*
Enter number of longitudinal segments for surface of sphere <16>: *(Enter a value.)*
Enter number of latitudinal segments for surface of sphere <16>: *(Enter a value.)*

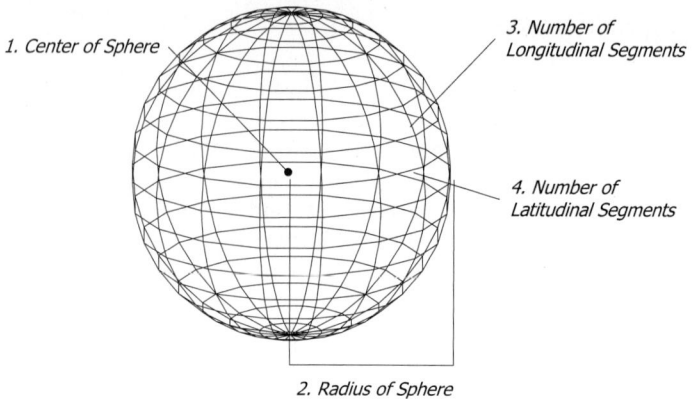

1. Center of Sphere

3. Number of
Longitudinal Segments

4. Number of
Latitudinal Segments

2. Radius of Sphere

COMMAND LINE OPTIONS

Center of sphere specifies the center point of the sphere.

Diameter specifies the diameter of the sphere.

Radius specifies the radius of the sphere.

Number of longitudinal segments specifies the number of "lines" that define the curved surface in the vertical direction; default = 16.

Number of latitudinal segments specifies the number of "lines" that define the curved surface in the horizontal direction; default = 16.

 Ai_Torus command

Command: ai_torus
Specify center point of torus: *(Pick point 1.)*
Specify radius of torus or [Diameter]: *(Pick point 2, specify the radius, or type D.)*
Specify radius of tube or [Diameter]: *(Pick point 3, specify the radius, or type D.)*
Enter number of segments around tube circumference <16>: *(Press Enter.)*
Enter number of segments around torus circumference <16>: *(Press Enter.)*

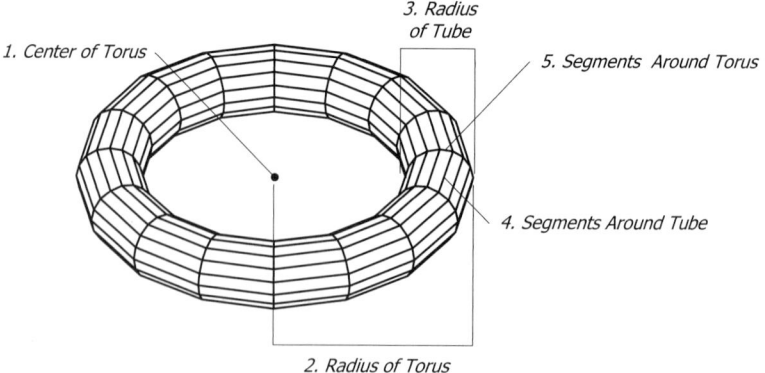

3. Radius
of Tube

1. Center of Torus

5. Segments Around Torus

4. Segments Around Tube

2. Radius of Torus

COMMAND LINE OPTIONS

Center of torus specifies the center point of the torus.

Diameter of torus specifies the diameter of the torus as measured across the center of the tube.

Radius of torus specifies the radius of the torus, as measured from the center of the torus to the center of the tube.

Diameter of tube specifies the diameter of the tube cross-section.

Radius of tube specifies the radius of the tube cross-section.

Segments around tube circumference <16> specifies the number of faces defining the curved surface; default = 16, range = 3 to 255.

Segments around torus circumference <16> specifies the number of faces defining the curved surface; default = 16, range = 3 to 255.

 Ai_Wedge Command

Command: ai_wedge
Specify corner point of wedge: *(Pick point 1.)*
Specify length of wedge: *(Pick point 2.)*
Specify width of wedge: *(Pick point 3.)*
Specify height of wedge: *(Pick point 4.)*
Specify rotation angle of wedge about the Z axis: *(Enter a rotation angle, or type 0 for no rotation.)*

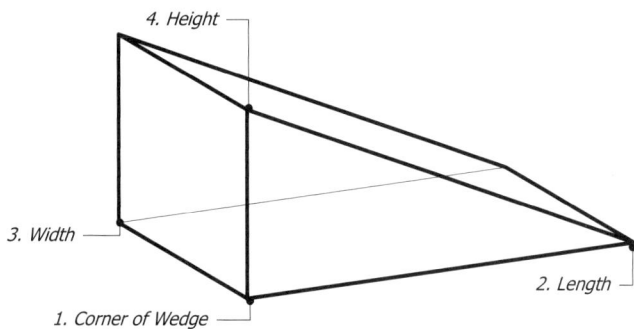

COMMAND LINE OPTIONS

Corner of wedge specifies the corner of the wedge's base.

Length specifies the length of the wedge's base.

Width specifies the width of the wedge's base.

Height specifies the height of the wedge.

Rotation angle specifies the angle the box rotates about the z-axis.

The following items apply to the Ai_ surface model commands:

RELATED SYSTEM VARIABLES

SurfU specifies the surface mesh density in the m-direction.

SurfV specifies the surface mesh density in the n-direction.

TIPS

- As of AutoCAD 2007, Autodesk changed the name of these objects from "surface" to "mesh" objects.

- Types of 3D objects drawn by the Ai_ series of commands:

  Ai_Box — Rectangular boxes, and cubes.
  Ai_Dish — Bottom half of a sphere.
  Ai_Dome — Top half of a sphere.
  Ai_Cone — Pointy cones, and truncated cones.
  AI_Mesh — Non-planar polyface meshes.
  Ai_Pyramid — Pyramids, truncated pyramids, tetrahedrons, truncated tetrahedrons, and roof shapes.
  Ai_Sphere — Spheres.
  Ai_Toru — Tori (donuts).
  Ai_Wedge — Wedges.

- This command creates 3D mesh (hollow) objects; to draw solid models, use commands such as Box and Sphere.

- When specifying the Width and Height in Ai_Box and Ai_Wedge, move the cursor back to the Corner of box point; otherwise the box may have a different size than you expect.

- You *cannot* perform Boolean operations (such as intersect, subtract, and union) on 3D mesh objects. These mesh objects are made of a single polymesh; Explode converts each side to an independent 3D face object.

- If no z coordinate is specified, then the base of mesh objects is drawn at the current setting of the Elevation system variable; the centers of spheres and tori are at the current elevation, unless you specify the z coordinate.

- Mesh *m* and *n* sizes (SurfU and SurfV) are limited to values between 2 and 256. The meaning of these varies with the type of mesh model created.

- You can use the Hide, VisualStyle, and Render commands on these mesh model shapes.

- To draw a 2D pyramid, enter no z-coordinate for the Ridge, Top, and Apex point options; use the .xy filter to specify the z-coordinate for these options.

- The point you pick for the Corner of wedge option determines the taller end of the wedge.

- The Ai_Torus command draws donut shapes.; the tube diameter cannot exceed the torus radius.

- Use the .xy filter first to specify the x,y coordinate, followed by the z coordinate, as follows:

  First corner:.xy
  of *(Pick a point.)*
  need Z: *(Specify z.)*

- The Ai_Dish command draws the bottom half of a sphere; Ai_Dome draws the top half.

'Ai_CircTan

Draws circles tangent to three points (an undocumented command).

Command	Alias	Ctrl+	F-key	Alt+	Menu Bar	Dashboard
ai_circtan	...	...	...	...	...	...

Command: ai_circtan
Enter Tangent spec: *(Pick an object.)*
Enter second Tangent spec: *(Pick an object.)*
Enter third Tangent spec: *(Pick an object.)*

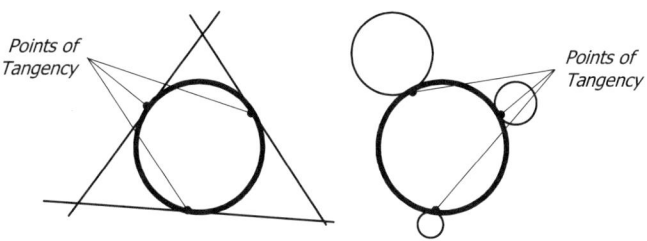

Left: Circle drawn tangent to three lines.
Right: Circle drawn tangent to three circles.

COMMAND LINE OPTION

Enter Tangent spec picks the objects to which the circle will be made tangent.

TIPS

- If a circle cannot be drawn between the three tangent points, AutoCAD complains, "Circle does not exist."

- This command is meant for use in toolbar and menu macros. ; it is an alternative to the Circle command's 3P option with the TANgent object snap.

Ai_Custom_Safe/Product_Support _ Safe/Training_Safe/Access_SubCtr

Accesses Autodesk support Web pages (undocumented commands).

Commands	Alias	Ctrl+	F-key	Alt+	Menu Bar	Dashboard
ai_custom_safe	...	...	...	HRD	Help	...
					⮑ Additional Resources	
					⮑ Online Development Center	
ai_product_support_safe				HRK	Help	
					⮑ Online Resources	
					⮑ Support Knowledge Base	
ai_training_safe				HRT	Help	
					⮑ Online Resources	
					⮑ Online Training Resources	
access_subctr						

Command: ai_custom_safe
_.browser Enter Web location (URL) <http://www.autodesk.com> http:// www.www.autodesk.com/developautocad

Opens Web browser, and accesses Autodesk's Web site.

Command: access_subctr
Enter subscription resource request ID (1-4): *(Enter a digit.)*

Opens Web browser, and accesses Autodesk's developer center Web site.

COMMAND LINE OPTIONS
None.

TIPS

- The ai_-commands access the following Web pages:

    ai_custom_safe — AutoCAD Developer Center
    ai_product_support_safe — AutoCAD Support
    ai_training_safe — AutoCAD Training

- The options of the Access_SubCtr command access the following Autodesk Subscription Web pages:

    1 — e-Learning Catalog
    2 — Create Support Request
    3 — View Support Requests
    4 — Edit Subscription Center Profile

- These commands are meant for use in menu and toolbar macros.

AiDimFlipArrow

Reverses the direction of selected arrowheads (undocumented command).

Command	Alias	Ctrl+	F-key	Alt+	Menu Bar	Dashboard
aidimfliparrow	...	...	...	...	...	...

Command: aidimfliparrow
Select objects: *(Select one or more arrowheads.)*
Select objects: *(Press Enter.)*

Top: Dimension and arrowheads as drawn by AutoCAD's DimLinear command.
Above: Both arrowheads flipped by the AiDimFlipArrow command.

COMMAND LINE OPTIONS

Select objects selects one or more arrowheads.

TIPS

- In some cases, AutoCAD places arrowheads differently from your preference. This command allows you retroactively to change the direction arrowheads point.

- Selecting one arrowhead allows you to flip it independently of the companion arrowhead.

- This command has no effect on arrowheads in leaders.

AiDimPrec

Changes the displayed precision of existing dimensions (undocumented command).

Command	Alias	Ctrl+	F-key	Alt+	Menu Bar	Dashboard
aidimprec	...	...	...	...	...	...

Command: aidimprec
Enter option [0/1/2/3/4/5/6] <4>: *(Enter a digit.)*
Select objects: *(Select one or more dimensions; non-dimensions are ignored.)*
Select objects: *(Press Enter.)*

Before (left) and after (right) applying AiDimPrec = 1 to a decimal dimension.

Before (left) and after (right) applying AiDimPrec = 1 to a fractional dimension.

COMMAND LINE OPTIONS

Enter option specifies the precision (number of decimal places, or fractional equivalent); enter a number between 0 and 6.

Select objects selects one or more dimensions.

TIPS

■ This command allows you retroactively to change the displayed precision of selected dimensions.

■ Alternatively, you can select a dimension, right click, and then select Precision from the shortcut menu.

■ Zero to six decimal places can be specified; fractional units are rounded to the nearest fraction:

       0 — Rounded to the nearest unit.
       1 — 1/2"
       2 — 1/4"
       3 — 1/8"
       4 — 1/16"
       5 — 1/32"
       6 — 1/64"

■ *Caution!* Because AiDimPrec rounds off dimensions, it can display false values. The dimension line below measures 3.4375", but setting AiDimPrec to 0 rounds it down to 3".

Applying AiDimPrec = 0 to a 3 $^{7}/_{16}$" dimension.

AiDimStyle

Saves and applies preset dimension styles (undocumented command).

Command	Alias	Ctrl+	F-key	Alt+	Menu Bar	Dashboard
aidimstyle	...	...	...	...	...	...

Command: aidimstyle
Enter option [1/2/3/4/5/6/Other/Save] <1>: *(Specify option, then press Enter.)*
Select objects: *(Select one or more dimensions.)*
Select objects: *(Press Enter.)*

COMMAND LINE OPTIONS

Enter option specifies a predefined dimension style, numbered 1 through 6.

Other applies a named dimension style to selected dimension(s).

Save saves the style of the selected dimension(s).

Select objects selects one or more dimensions.

Other option
Enter option [1/2/3/4/5/6/Other/Save] <1>: o

Displays dialog box after selecting objects:

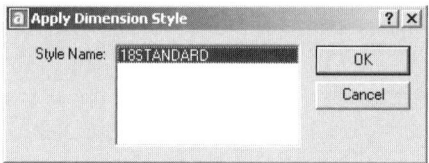

Save option
Enter option [1/2/3/4/5/6/Other/Save] <1>: s

Displays dialog box after selecting exactly one dimension:

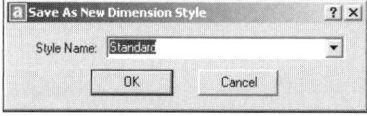

OK saves style to the selected dimension style name.

TIPS

- This command quickly applies and saves dimensions styles. It is used by the right-click shortcut menu: left-click a dimension to select it, right click, and then select Dim Style from the shortcut menu.

- *Caution!* The Save option overwrites existing dimstyles; it does not create new style names. It does warn you, and asks for permission to make the change.

Ai_Dim_TextAbove/Center/Home

Moves dimension text relative to dimension lines (undocumented commands).

Command	Alias	Ctrl+	F-key	Alt+	Menu Bar	Dashboard
ai_dim_textabove	...	...	...	...	...	...
ai_dim_textcenter						
ai_dim_texthome						

Command: ai_dim_textabove
Select objects: *(Select one or more dimensions.)*
Select objects: *(Press Enter.)*

Before (left) and after (right) applying Ai_Dim_TextAbove .

Before (left) and after (right) applying Ai_Dim_TextCenter.

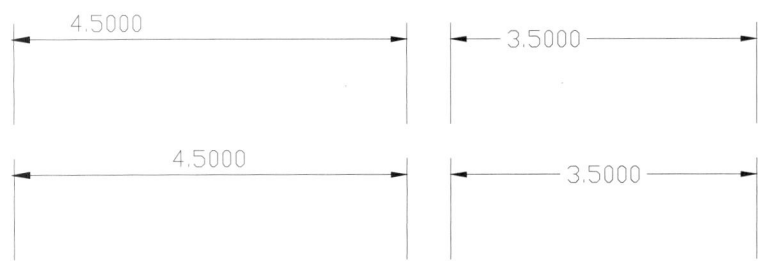

Before (top) and after (bottom) applying Ai_Dim_TextHome.

COMMAND LINE OPTION

Select objects selects one or more dimensions.

TIPS

- Select a dimension, right-click, and then select Dim Text Position from the menu:
  - **Ai_Dim_TextAbove** makes dimensions compliant with JIS dimensioning.
  - **Ai_Dim_TextCenter** centers text vertically on the dimension line, but not horizontally.
  - **Ai_Dim_TextHome** centers text horizontally on the dimension line, but not vertically.

- Use the DimTEdit command to align text to the left, center, or right on horizontal dimensions.

AiDimTextMove

Moves dimension text (undocumented command).

Command	Alias	Ctrl+	F-key	Alt+	Menu Bar	Dashboard
aidimtextmove	...	...	...	...	...	...

Command: aidimtextmove
Enter option [0/1/2] <2>: *(Enter an option, and then press Enter.)*
Select objects: *(Select one dimension.)*
Select objects: *(Press Enter.)*

Before (left) and after (right) applying AiDimTextMove = 0 to dimension text.

Before (left) and after (right) applying AiDimTextMove = 1 to dimension text.

Before (left) and after (right) applying AiDimTextMove = 2 to dimension text.

COMMAND LINE OPTIONS

Enter option specifies the nature of text movement:

    0 — Text moves with the dimension line.
    1 — Adds a leader to the moved text.
    2 — Text moves independently of dimension line and leader (default).

Select objects selects one or more dimensions.

TIPS

- This command allows you retroactively to change the position of dimension text. It is used by the right-click shortcut menu: select a dimension, right-click, and then select Dim Text Position from the shortcut menu.

- Although the command allows you to select more than one dimension, it operates on the first-selected dimension only.

Ai_Fms, Ai_PSpace

Ai_PSpace switches to layout mode; Ai_Fms does the same, and then goes to floating model space (short for Floating Model Space; both are undocumented commands).

Command	Alias	Ctrl+	F-key	Alt+	Menu Bar	Dashboard
ai_fms	...	...	...	...	...	...
ai_pspace						

Command: ai_pspace

Switches to the last active layout.

Command: ai_fms

Switches to the last active layout, and then to the first floating model viewport.

The heavy border indicates the currently-active floating viewport in model space.

COMMAND LINE OPTIONS

None.

TIPS

- The Ai_Fms command sets system variable Tilemode to 0, and then runs the MSpace command.

- These commands are meant for use with menu and toolbar macros.

Ai_Molc

Changes the current layer to the one on which the selected object is located (short for Make Object Layer Current; undocumented command).

Command	Alias	Ctrl+	F-key	Alt+	Menu Bar	Dashboard
ai_molc	...	...	...	...	...	Layers

Command: ai_molc
Select object whose layer will become current: *(Pick an object.)*

COMMAND LINE OPTION

Select object selects a single object.

RELATED COMMANDS

Layer displays the Layer Properties Manager dialog box.

LayerP reverts to the previous layer.

MatchProp applies the properties of one object to other objects.

RELATED SYSTEM VARIABLE

CLayer holds the name of the current layer.

TIPS

- This command is activated by the Make Object's Layer Current button on the Layers toolbar.

- Click the button, and then select an object: the object's layers becomes the current layer.

- Merely selecting an object displays its layer on the Object Properties toolbar without changing the current layer.

AiObjectScaleAdd/Remove

Adds (and removes) the current annotative scale to (and from) annotative objects (an undocumented command).

Command	Alias	Ctrl+	Alt+	Menu Bar	Dashboard
aiobjectscaleadd	...	...	...	...	Annotation Scaling
aiobjectscaleremove					Annotation Scaling

Command: aiobjectscaleadd
Select annotative objects: *(Pick one or more objects.)*
Select annotative objects: *(Press Enter to continue.)*
n objects updated to support annotation scale *<n:n>*.

. . . .

AiObjectsScaleRemove Command
Command: aiobjectscaleremove
Select annotative objects: *(Pick one or more objects.)*
Select annotative objects: *(Press Enter to continue.)*
1 object scale removed.

COMMAND LINE OPTION
Select objects selects one or more objects; AutoCAD filters out non-annotative objects.

RELATED COMMANDS
ObjectScale lists all annotative objects associated with the selected object(s).

AnnoReset resets scale and location of selected annotative objects.

AnnoUpdate updates annotative objects to their styles.

Properties toggles the annotative property.

RELATED SYSTEM VARIABLE
CAnnoScale holds the name of the current annotation scale for the current viewport.

TIPS
- The AiObjectScaleAdd command adds the current annotative scale, as displayed by the Annotation Scale droplist.

- When the pickbox is over an annotative object, AutoCAD displays this icon:

  When the object has two or more scale factors, AutoCAD displays this icon:

- To see all the scale factors applied to an annotative object, use the ObjectScale command.

Ai_SelAll

Selects all objects in drawings in the current space, including those off-screen (undocumented command).

Command	Alias	Ctrl+	F-key	Alt+	Menu Bar	Dashboard
'ai_selall	...	a	...	...	...	...

Command: ai_selall
Selecting objects...done.

COMMAND LINE OPTIONS

None.

TIPS

- This command is meant for use in menu macros and toolbars.

- Use the Ctrl+A shortcut to select all objects in the drawing, other than those on frozen layers.

- To deselect all objects in AutoLISP, use the undocumented **(ai_deselect)** function.

Align

Rel.12 Moves, transforms, and rotates objects in three dimensions.

Command	Alias	Ctrl+	F-key	Alt+	Menu Bar	Dashboard
align	al	...	...	M3L	Modify	...
					↳3D Operation	
					↳Align	

Command: align
Select objects: *(Select one or more objects to be moved.)*
Select objects: *(Press Enter.)*
Specify first source point: *(Pick a point.)*
Specify first destination point: *(Pick a point.)*
Specify second source point: *(Pick a point.)*
Specify second destination point: *(Pick a point, or press Enter.)*
Specify third source point or <continue>: *(Pick a point, or press Enter.)*
Specify third destination point: *(Pick a point.)*

COMMAND LINE OPTIONS

First point moves object in 2D or 3D when one source and destination point are picked.

Second point moves, rotates, and scales object in 2D or 3D when only two source and destination points are picked.

Third point moves objects in 3D when three source and destination points are picked.

Continue option
Scale objects based on alignment points? [Yes/No] <N>: *(Type Y or N.)*

Specifies that the distance between the first and second source points is to be scaled to the distance between the first and second destination points.

RELATED COMMANDS

Mirror3d mirrors objects in three dimensions.

Rotate3d rotates objects in three dimensions.

TIPS

■ Enter the first pair of points to define the move vector (distance and direction):

> **Specify first source point:** *(Pick a point.)*
> **Specify first destination point:** *(Pick a point.)*
> **Specify second source point:** *(Press Enter.)*

■ Enter two pairs of points to define a 2D (or 3D) transformation, scaling, and rotation:

> **First** — Base point for alignment.
> **Second** — Rotation angle.
> **Third** — Planes aligned by source and destination points.

■ The third pair defines the 3D transformation.

AmeConvert

<u>Rel.13</u> Converts PADL solid models and regions created by AME v2.0 and v2.1 (AutoCAD Releases 11 and 12) to ShapeManager solid models.

Command	Alias	Ctrl+	F-key	Alt+	Menu Bar	Dashboard
ameconvert	...	...	...	...	...	...

Command: ameconvert
Select objects: *(Select one or more objects.)*
Processing Boolean operations.

COMMAND LINE OPTION

Select objects selects AME objects to convert; ignores non-AME objects, such as the ACIS solids produced by AutoCAD Release 13 through 2002, and ShapeManager solids from AutoCAD 2004-2008.

RELATED COMMAND

AcisIn imports ACIS models from a *.sat* file.

TIPS

- After the conversion process, the AME model remains in the drawing in the same location as the solid model. Erase, if necessary.

- AME holes may become blind holes (solid cylinders) in the solid model.

- AME fillets and chamfers may be placed higher or lower in the solid model.

- Once Release 12 PADL drawings are converted to AutoCAD 2008 solid models, they cannot be converted back to PADL format.

- This command ignores objects that are neither AME solids nor regions.

- Old AME models are stored in AutoCAD as anonymous block references.

DEFINITIONS

ACIS — solids modeling technology used by AutoCAD in Release 13 through 2002.

AME — short for "Advanced Modeling Extension," the solids modeling module used by AutoCAD in Releases 10 through 12.

PADL — short for "Parts and Description Language," the solids modeling technology used by AutoCAD Releases 10 through 12.

ShapeManager — the solids modeling technology used by AutoCAD 2004 through 2008.

AniPath

2007 Specifies animation paths along which cameras move; can be saved as movie files; short for "ANImation PATH."

Command	Alias	Ctrl+	Alt+	Menu Bar	Dashboard
anipath	...	...	VM	View	3D Navigate
				⮡ Motion Path Animation	

Command: anipath

Displays the Motion Path Animation dialog box:

DIALOG BOX OPTIONS

Camera

⦿ **Point** places the camera in the drawing.

◯ **Path** guides the camera along an object in the drawing: a line, arc, elliptical arc, circle, polyline, spline, or 3D polyline.

🔲 **Pick** picks the camera point or selects the path in the drawing.

List lists previously selected points and paths.

Target

◯ **Point** points the camera at a target in the drawing.

⦿ **Path** points the camera along a path defined by an object (line, circle, etc.) in the drawing.

🔲 **Pick** picks the target point or selects the path in the drawing.

List lists previously selected target points and paths.

Animation Settings

Frame Rate (FPS) specifies the speed of the animation; ranges from 1 to 60 frames per second.

Number of Frames specifies the total number of frames to be captured for the animation.

Duration (seconds) specifies duration of the animation in seconds; linked to Number of Frames.

Visual Style selects the preset visual style or rendering quality for the animation: as displayed, rendered, 3D hidden, 3D wireframe, conceptual, realistic, draft, low, medium, high, presentation.

Format specifies the animation file format: AVI, MOV, MPG, or WMV.

Resolution provides a list of resolutions, ranging from 160x120 to 1024x768.

☑ **Corner Deceleration** reduces the camera's speed around curves.

☐ **Reverse** reverses the animation direction.

☑ **When Previewing Show Camera Preview** displays the Animation Preview dialog box.

Preview previews the animation.

Animation Preview dialog box

Animation Preview controls

▶ **Play** plays the animation once.

● **Record** records the animation beginning with the current frame (non-operational).

❚❚ **Pause** pauses the animation at the current frame; click Play to continue.

■ **Save** displays the Save As dialog box; saves the animation in AVI, MOV, MPG, or WMV format.

 Visual Styles selects the visual style for the animation.

 Slider moves through the animation by frames.

RELATED COMMAND

 VisualStyles creates and edits custom visual styles.

TIPS

■ Cameras are created automatically when the motion path is specified.

■ Deleting objects that define motion paths also deletes the path.

■ Targets must be linked to paths when cameras are linked to points.

■ Changing the frame rate determines the duration, and vice versa.

■ MOV format requires QuickTime player; WMV format requires Media Player 9.

AnnoReset, AnnoUpdate

__2008__ AnnoReset resets the location of all scale representations to the current one.

AnnoUpdate updates objects when annotative styles change.

Command	Alias	Ctrl+	F-key	Alt+	Menu Bar	Dashboard
annoreset	...	...	...	...	...	...
annoupdate						

Command: annoreset
Reset alternative scale representations to current position.
n **found,** *n* **was not annotative.**

. . . .

AnnoUpdate Command
Command: annoupdate
Select objects: *(Pick one or more objects.)*
Select objects: *(Press Enter.)*
n **found,** *n* **was updated.**

COMMAND LINE OPTION

Select objects selects annotative objects to be updated.

RELATED COMMANDS

ObjectScale assigns and edits scale factors to annotative objects.

AiObjectScaleAdd adds scale factors to annotative objects.

AiObjectScaleRemove removes scale factors from annotative objects.

RELATED SYSTEM VARIABLES

AnnoAllVisible toggles the display of annotative objects not at the current annotation scale.

AnnoAutoScale updates annotative objects when the annotation scale is changed.

AnnotativeDwg toggles whether the drawing acts like an annotative block when inserted into other drawings.

CAnnoScale reports the name of the current annotation scale for the current space; a separate scale is stored for model space and for each paperspace viewport.

CAnnoScaleValue reports the current value of the annotation scale.

TIPS

- Annotatively-scaled objects are displayed only when their assigned scale factor matches the model space's scale factor, whether in Model tab or in a layout tab's viewport.

- These objects can have annotative scale factors: text and mtext, attributes and blocks, leaders and multileaders, dimensions and tolerances, and hatch patterns. As well, linetype scaling can match the model space scale factor with the MsLtScale system variable.

- See the ObjectScale command for more on annotative scaling.

. . . .

'Aperture

V. 1.3 Sets the size (in pixels) of the object snap target height, or box cursor.

Command	Alias	Ctrl+	F-key	Alt+	Menu Bar	Dashboard
aperture	...	...	...	...	...	...

Command: aperture
Object snap target height (1-50 pixels) <10>: *(Enter a value.)*

Aperture size = 1 (left), 10 (center), and 50 pixels (right).

COMMAND LINE OPTION

Height specifies the height of the object snap cursor's target; range is 1 to 50 pixels.

RELATED COMMAND

Options allows you to set the aperture size interactively (Drafting tab).

RELATED SYSTEM VARIABLES

ApBox toggles the display of the aperture box cursor (*undocumented*).

Aperture contains the current target height, in pixels:

    1 — Minimum size.
    10 — Default size, in pixels.
    50 — Maximum size.

TIPS

- Besides the aperture cursor, AutoCAD has two other similar-looking cursors: the *osnap* cursor appears only during object snap selection; the *pick* cursor appears anytime AutoCAD expects object selection. The size of both cursors can be changed; to change the size of the pick cursor, use the Pickbox command.

- By default, the box cursor does not appear. Nevertheless, it determines how close you have to be to an object for AutoCAD to "snap" to it.

- To display the box cursor, use the Options command, select the Drafting tab, and then enable the Display AutoSnap Aperture box option. Alternatively, use the undocumented ApBox system variable.

- Use the Options command to change the size of the aperture visually: select the Drafting tab, and then move the Aperture Size slider.

AppLoad

<u>Rel.12</u>   Creates a list of LISP, VBA, ObjectARx, and other applications to load
into AutoCAD (short for APPlication LOADer).

Command	Alias	Ctrl+	F-key	Alt+	Menu Bar	Dashboard
appload	ap	...	...	TL	Tools	...
					⤷Load Applications	

Command: appload

Displays dialog box:

LOAD/UNLOAD APPLICATIONS DIALOG BOX

Look in lists the names of drives and folders available to this computer.

File name specifies the name of the file to load.

Files of type displays a list of file types:

Filetype	Meaning
ARX	objectARX.
DVB	Visual Basic for Applications (VBA).
DBX	objectDBX.
FAS	FASt load autolisp.
LSP	autoLiSP.
VLX	Visual Lisp eXecutable.

Load loads all or selected files into AutoCAD.

Loaded Applications displays the names of applications already loaded into AutoCAD.

History List displays the names of applications previously saved to this list.

☐ **Add to History** adds the file to the History List tab.

Unload unloads all or selected files out of AutoCAD.

Close exits the dialog box.

Startup Suite

Contents displays dialog box:

List of applications lists the file names and paths of applications to be loaded automatically each time AutoCAD starts. AutoLISP related files (*.lsp*, *.fas*, and *.vlx*) are loaded whenever a drawing is loaded or a new drawing is created. All others are loaded when AutoCAD starts.

Add displays the Add File to Startup Suite dialog box; allows you to select one or more application files.

Remove removes the application from the list.

Close returns to the Load/Unload Applications dialog box.

RELATED COMMANDS

Arx lists ObjectARX programs currently loaded in AutoCAD.

VbaLoad loads VBA applications.

RELATED AUTOLISP FUNCTIONS

(load) loads an AutoLISP program.

(autoload) predefines commands to load related AutoLISP programs.

STARTUP SWITCH

/ld loads an ARX or DBX application when AutoCAD starts up.

TIPS

- Use **AppLoad** when AutoCAD does not automatically load a command; alternatively, you can drag files from Windows Explorer into the Loaded Applications list.

- ObjectARX, VBA, and DBX applications are loaded immediately; FAS, LSP, and VLX files are loaded after this dialog box closes.

- This command is limited to loading 50 applications at a time.

- Load the *asdkTPCtest.arx* application for undocumented Tablet PC support.

- The *acad2008doc.lsp* file establishes autoloader and other utility functions, and is loaded automatically each time a drawing is opened; the *acad2008.lsp* file is loaded only once per AutoCAD session. Use the AcadLspAsDoc system variable to control whether these files are loaded with AutoCAD.

Arc

V. 1.0   Draws 2D arcs of less than 360 degrees, by eleven methods.

Command	Alias	Ctrl+	F-key	Alt+	Menu Bar	Dashboard
arc	a	...	...	DA	Draw	2D Draw
					⤷Arc	

Command: arc
Specify start point of arc or [CEnter]: *(Pick a point, or enter the CE option.)*
Specify second point of arc or [CEnter/ENd]: *(Pick a point, or enter an option.)*
Specify end point of arc: *(Pick a point.)*

COMMAND LINE OPTIONS

SSE (start, second, end) arc options

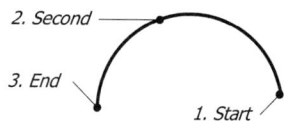

Start point indicates the start point of a three-point arc (point 1, above).
Second point indicates a second point anywhere along the arc (2).
Endpoint indicates the end point of the arc (3).

SCE (start, center, end), SCA (start, center, angle), and SCL (start, center, length) options

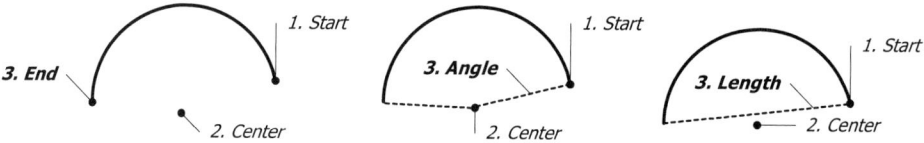

Start point indicates the start point of a two-point arc (1).
Center indicates the arc's center point (2).
Endpoint indicates the arc's end point (2).
Angle indicates the arc's included angle (3).
Length of chord indicates the length of the arc's chord (3).

SEA (start, end, angle), SED (start, end, direction), SER (start, end, radius), and SEC (start, end, center) options

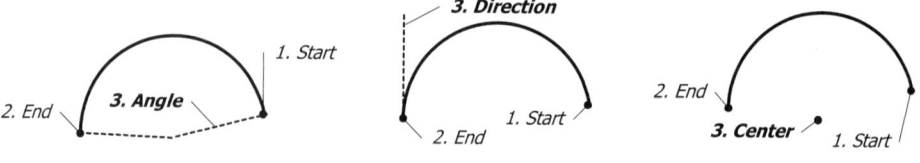

Start point indicates the start point of a two-point arc (1).

End indicates the arc's end point (2).

Angle indicates the arc's included angle (3).

Direction indicates the tangent direction from the arc's start point (3).

Center point indicates the arc's center point (3).

Radius indicates the arc's radius.

CSE (center, start, end), CSA (center, start, angle), and CSL (center, start, length) options

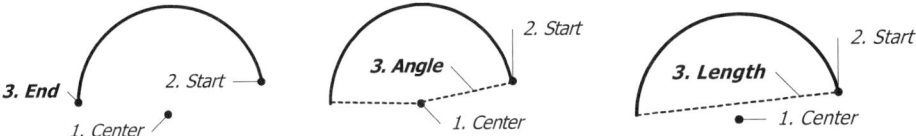

Center indicates the center point of a two-point arc (1).

Start point indicates the arc's start point (2).

Endpoint indicates the arc's end point (3).

Angle indicates the arc's included angle (3).

Length of chord indicates the length of the arc's chord (4).

Continued Arc option

Arc continues from last point.

Enter continues the arc tangent from the end point of last-drawn line or arc.

RELATED COMMANDS

Circle draws an "arc" of 360 degrees.

DimArc dimensions the length of the arc.

DimCenter places center marks at the arc's center.

Ellipse draws elliptical arcs.

Polyline draws connected polyline arcs.

ViewRes controls the roundness of arcs.

RELATED SYSTEM VARIABLES

LastAngle saves the included angle of the last-drawn arc (read-only).

WhipArc controls the smoothness in display of circles and arcs.

TIPS

- Arcs are drawn counterclockwise.

- In most cases, it is easier to draw a circle, and then use the Trim command to convert the circle into an arc.

- It can be easier to use the Fillet command to create arcs tangent to lines.

- To start an arc precisely tangent to the end point of the last line or arc, press Enter at the 'Specify start point of arc or [CEnter]:' prompt.

- You can drag the arc only during the last-entered option.

- Specifying an x,y,z-coordinate as the starting point of the arc draws the arc at the z-elevation.

- The components of AutoCAD arcs:

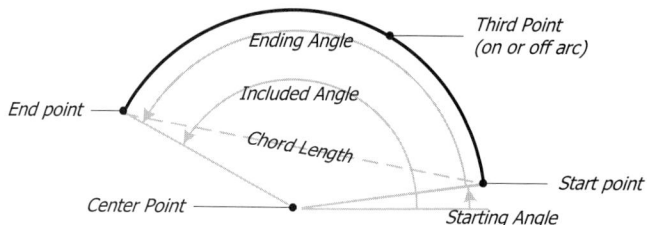

- AutoCAD 2006 adds three more stretch handles (shown as triangles) to arcs:

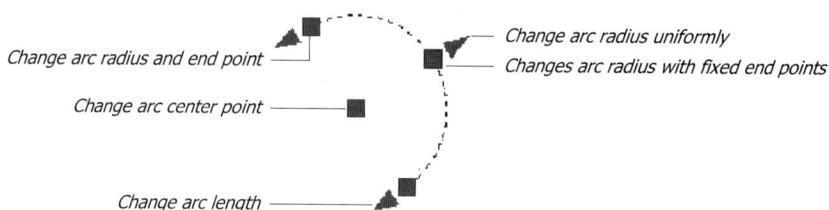

- When the chord length is positive, the minor arc is drawn counterclockwise from the start point; when negative, the major arc is drawn counterclockwise.

Archive

<u>**2005**</u>   Packages all files related to the current sheet set.

Command	Alias	Ctrl+	F-key	Alt+	Menu Bar	Dashboard
archive	...	...	...	...	...	...
-archive						

Command: archive

When no sheet sets are open, AutoCAD complains, "No Sheet Set is Open," and terminates the command. (To open a sheet set, use the OpenSheetset command.)

When at least one sheet set is open, AutoCAD displays the following dialog box:

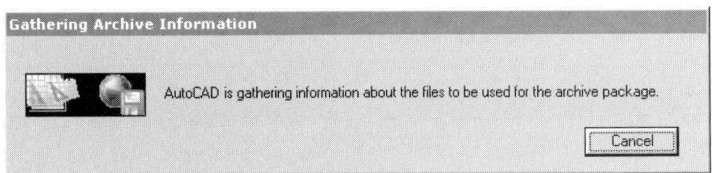

After a moment, AutoCAD displays the next dialog box:

ARCHIVE A SHEET SET DIALOG BOX

Sheets tab displays sheets included with the archive.

Files Tree tab displays names of drawing and support files, grouped by category.

Files Table tab displays file names in alphabetical order.

Enter notes to include with this archive provides space for entering notes.

View Report displays the View Archive Report dialog box.

Modify Archive Setup displays the Modify Archive Setup dialog box.

Files Tree tab

☑ File included in archive.

☐ File excluded from archive.

Add File displays the Add File to Archive dialog box, which adds files to the archive set.

Files Table tab

VIEW REPORT DIALOG BOX

Save As saves the report in *.txt* format (plain ASCII text).

Close closes the dialog box, and then returns to the previous dialog box.

MODIFY ARCHIVE SETUP DIALOG BOX

Archive Type and Location

Folder (set of files) archives uncompressed files in new and existing folders. This is the best option when archiving to CD or FTP.

Self-Extracting Executable (*.exe) archives files in a compressed, self-extracting file. Uncompress by double-clicking the file.

Zip (*.zip) (*default*) archives files as a compressed ZIP file. Uncompress the file using PkZip or WinZip. This is the best option when sending by email.

File Format

Keep existing drawing file formats archives files in their native format.

AutoCAD 2007/LT 2007 Drawing Format archives files in AutoCAD 2007/8 formats.

AutoCAD 2004/LT 2004 Drawing Format archives files in AutoCAD 2004/5/6 formats.

AutoCAD 2000/LT 2000 Drawing Format archives files in AutoCAD 2000/i/1 formats.

Warning! Some objects and properties not found in earlier AutoCAD versions may be changed or lost.

Archive File Folder specifies the location in which to archive the files. When no location is specified, the archive is created in the same folder as the *.dst* file.

Archive File Name

Prompt for a File Name prompts the user for the file name.

Overwrite if Necessary uses the specified file name, and overwrites the existing file of the same name.

Increment File Name if Necessary uses the specified file name, and appends a digit to avoid overwriting the existing file of the same name.

Archive

⊙ **Use Organized Folder Structure** preserves the folder structure in the archive, but makes the changes listed below; allows you to specify the name of the folder tree. The option is unavailable when saving archives to the Internet. Autodesk reminds you of the following:

- Relative paths remain unchanged.
- Absolute paths inside the folder tree are converted to relative paths; absolute paths outside the folder tree are converted to "No Path," and are moved to the folder tree.
- *Fonts* folder created when font files are included in the archive.
- *PlotCFG* folder created when plotter configuration files are included.
- *SheetSets* folder created for sheet set support files; *.dst* sheet set data file is in root folder.

○ **Place All Files in One Folder** locates all files in a single folder.

○ **Keep Files and Folders As Is** preserves the folder structure in the archive. The option is unavailable when saving archives to the Internet.

☐ **Include Fonts** includes all *.ttf* and *.shx* font files in the archive.

☑ **Include Textures from Materials** includes all material definitions used by 3D models (*new to AutoCAD 2008*).

☑ **Include Photometric Web files** includes IES files that define web light distributions (*new to AutoCAD 2008*).

☐ **Set Default Plotter to 'None'** resets the plotter to None for all drawings in the archive.

☐ **Prompt for Password** displays a dialog box for specifying a password for the archive; not available when the Folder archive type is selected.

☑ **Include Sheet Set Data (DST) File** includes the *.dst* sheet set data file with the archive.

. . . .

-Archive Command

Command: -archive
Sheet Set name or [?] <Sheet Set>: *(Enter a name or ?.)*
Enter an option [Create archive package/Report only] <Create>: *(Type C or R.)*
Gathering files ...

Sheet Set name specifies the name of the sheet set to archive.

? lists the names of sheet sets open in the current drawing.

Create archive package creates the archive in ZIP format.

Report only displays the Save Report File As dialog box for saving the archive report as a text file; also prompts you for a Transmittal Note.

RELATED COMMANDS

eTransmit creates an archive of only the current drawing and its support files.

SheetSet creates and controls sheet sets.

TIPS

- This command works only when at least one sheet set is open in the current drawing.
- Archive files can become large when they hold many drawing files.
- The -Archive is useful for creating scripts and macros that archive the same group of drawings and other files repeatedly.

. . . .

Area

V. 1.0 Calculates the area and perimeter of objects.

Command	Alias	Ctrl+	F-key	Alt+	Menu Bar	Dashboard
area	aa	...	...	TQA	Tools	...
					⬑Inquiry	
					⬑Area	

Command: area
Specify first corner point or [Object/Add/Subtract]: *(Pick a point, or enter option.)*
Specify next corner point or press ENTER for total: *(Pick a point, or press Enter.)*

Sample response:
Area = 1.8398, Perimeter = 6.5245

COMMAND LINE OPTIONS

First corner point specifies the first point to begin measurement.

Object selects the object to be measured.

Add switches to add-area mode.

Subtract switches to subtract-area mode.

Enter end the area outline.

RELATED COMMANDS

Dist returns the distance between two points.

Id lists the x,y,z coordinates of a selected point.

List reports on the area and other properties of objects.

Properties reports the area of ojbects, as well as of single and multi-hatched areas.

MassProp returns surface area, and so on, of solid models.

RELATED SYSTEM VARIABLES

Area contains the most recently-calculated area.

Perimeter contains the most recently-calculated perimeter.

TIPS

- Before subtracting, you must use the Add option. Picking points or using the Object option first prevents you from subtracting.

- At least three points must be picked to calculate an area; AutoCAD "closes the polygon" with a straight line before measuring the area.

- The Object option returns the following information:

    Circle, ellipse — area and circumference.
    Planar closed spline — area and circumference.
    Closed polyline, polygon — area and perimeter.
    Open objects — area and length.
    Region — net area of all objects in region.
    2D solid — area.

- Areas of wide polylines are measured along center lines; closed polylines must have only one closed area.

Array

V. 1.3 Creates 2D linear, rectangular, and polar arrays of objects.

Commands	Aliases Ctrl+	F-key	Alt+	Menu Bar	Dashboard
array	ar ...	...	MA	Modify ↳Array	2D Draw
-array	-ar				

Command: array

Displays dialog box:

ARRAY DIALOG BOX

⊙ **Rectangular Array** displays the options for creating a rectangular array.

○ **Polar Array** displays the options for creating a circular array.

🔲 **Select objects** dismisses the dialog box temporarily, so that you can select the objects in the drawing:

 Select objects: *(Select one or more objects.)*
 Select objects: *(Press Enter.)*
 Press Enter, or right-click to return to the dialog box.

Preview dismisses the dialog box temporarily, so that you can see what the array will look like.

Rectangular Array

Rows specifies the number of rows; minimum=1, maximum=32767.

Columns specifies the number of columns; minimum=1, maximum=32767.

Row offset specifies the distance between the center lines of the rows; use negative numbers to draw rows in the negative x-direction (to the left).

Column offset specifies the distance between the center lines of the columns; use negative numbers to draw rows in the negative y-direction (downward).

Angle of array specifies the angle of the array, which "tilts" the x and y axes of the array.

Polar Array

Center point specifies the center of the polar array.

Method specifies the method by which the array is constructed:

- Total number of items and angle to fill.
- Total number of items and angle between items.
- Angle to fill and angle between items.

Total number of items specifies the number of objects in the array; minimum=2.

Angle to fill specifies the angle of "arc" to construct the array; min=1 deg; max=360 deg.

Angle between items specifies the angle between each object in the array.

Rotate items as copied

☑ Objects are rotated so that they face the center of the array.

☐ Objects are not rotated.

More displays additional options for constructing a polar array.

Object Base Point

Set to object's default:

☑ The base point of the object is used, as listed below.

☐ The base point is specified by the user; if objects are not rotated, select a base point to avoid unexpected results.

Arc, circle, ellipse — center point.
Polygon, rectangle — first vertex.
Line, polyline, 3D polyline, ray, spline — start point.
Donut — start point.
Block, mtext, text — insertion point.
Xline — midpoint.
Region — grip point.

Base point specifies the x and y coordinates of the new base point.

-Array Command

Command: -array
Select objects: *(Select one or more objects.)*
Select objects: *(Press Enter.)*
Enter the type of array [Rectangular/Polar] <R>: *(Type R or P.)*

Rectangular options
Enter the number of rows (---) <1>: *(Enter a value, or press Enter.)*
Enter the number of columns (|||) <1>: *(Enter a value, or press Enter.)*
Enter the distance between rows or specify unit cell (---): *(Enter a value.)*
Specify the distance between columns (|||): *(Enter a value.)*

Polar options
Specify center point of array or [Base]: *(Select one or more objects, or type B.)*
Enter the number of items in the array: *(Enter a value.)*
Specify the angle to fill (+=ccw, -=cw) <360>: *(Enter a value, or press Enter.)*
Rotate arrayed objects? [Yes/No] <Y>: *(Type Y or N.)*

COMMAND LINE OPTIONS

R creates a rectangular array from the selected object.

P creates a polar array from the selected object.

B specifies the base point of objects, and the center point of the array.

Center point specifies the center point of the array.

Rows specifies the number of horizontal rows.

Columns specifies the number of vertical columns.

Unit cell specifies the vertical and horizontal spacing between objects.

RELATED COMMANDS

3dArray creates a rectangular or polar array in 3D space.

Copy creates one or more copies of the selected object.

MInsert creates a rectangular block array of blocks.

RELATED SYSTEM VARIABLE

SnapAng determines the default angle of a rectangular array.

TIPS

- To create a rectangular array at an angle, use the Rotation option of the Snap command.

- Rectangular arrays are drawn upward in the positive x-direction, and to the right in the positive y-direction; to draw the array in the opposite directions, specify negative row and column distances.

- Polar arrays are drawn counterclockwise; to draw the array clockwise, specify a negative angle.

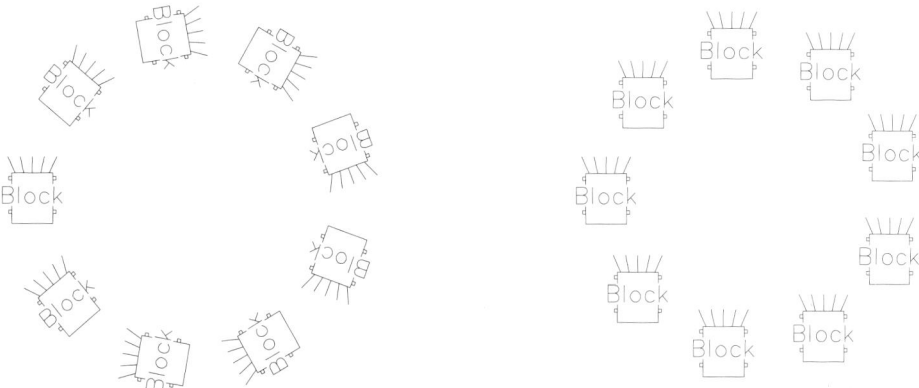

Nine-item polar arrays — rotated (left) and unrotated (right).

- For linear arrays, enter 1 for the number of rows or columns, or use Divide or Measure.

Arx

<u>Rel.13</u> Loads and displays information about ObjectARX programs.

Command	Alias	Ctrl+	F-key	Alt+	Menu Bar	Dashboard
arx	...	...	...	...	...	...

Command: arx
Enter an option [?/Load/Unload/Commands/Options]: *(Enter an option.)*

COMMAND LINE OPTIONS

? lists the names of currently loaded ObjectARX programs.

Load loads the ObjectARX program into AutoCAD.

Unload unloads the ObjectARX program out of memory.

Commands lists the names of commands associated with each ObjectARX program.

Options options

CLasses lists the class hierarchy for ObjectARX objects.

Groups lists the names of objects entered into the "system registry."

Services lists the names of services entered in the ObjectARX "service dictionary."

RELATED COMMAND

AppLoad loads LISP, VBA, ObjectDBX, and ObjectARX programs via a dialog box.

RELATED AUTOLISP FUNCTIONS

(arx) lists currently loaded ObjectARX programs.

(arxload) loads an ObjectARX application.

(autoarxload) predefines commands that load the ObjectARX program.

(arxunload) unloads an ObjectARX application.

RELATED FILE

***.arx** are objectARX program files.

TIPS

■ Use the Load option to load external commands that do not seem to work.

■ Use the Unload option of the Arx command to remove ObjectARX programs from AutoCAD to free up memory.

. . . .

Removed Commands

The following ASE (AutoCAD SQL Extension) commands were removed from AutoCAD 2000: **AseAdmin**, **AseExport**, **AseLinks**, **AseRows**, **AseSelect**, and **AseSqlEd**. They were replaced by **DbConnect**.

The **Assist** and **AssistClose** commands were removed from AutoCAD 2008, and replaced by the InfoCenter on the right end of the menu bar.

. . . .

AttachURL

Rel.14 Attaches hyperlinks to objects and areas.

Command	Alias	Ctrl+	F-key	Alt+	Menu Bar	Dashboard
attachurl	...	...	...	...	...	...

Command: attachurl
Enter hyperlink insert option [Area/Object] <Object>: *(Type A or O.)*
Select objects: *(Select one or more objects.)*
Select objects: *(Press Enter.)*
Enter hyperlink <current drawing>: *(Enter an address.)*

COMMAND LINE OPTIONS

Area creates rectangular hyperlinks by specifying two corners of a rectangle.

Object attaches hyperlinks after you select one or more objects.

Enter hyperlink requires you to enter a valid hyperlink.

Area Options
First corner picks the first corner of the rectangle.

Other corner picks the second corner of the rectangle.

RELATED COMMANDS

Hyperlink displays a dialog box for adding a hyperlink to an object.

SelectUrl selects all objects with attached hyperlinks.

TIPS

- The hyperlinks placed in the drawing can link to *any* other file: another AutoCAD drawing, an office document, or a file located on the Internet.

- Autodesk recommends that you use the following URL (uniform resource locator) formats:

File Location	Example URL
Web Site	**http://***servername/pathname/filename***.dwg**
FTP Site	**ftp://***servername/pathname/filename***.dwg**
Local File	**file:///***drive:/pathname/filename***.dwg**
	or **file:////***localPC/pathname/filename***.dwg**
Network File	**file://***localhost/drive:/pathname/filename***.dwg**

- The URL (hyperlink) is stored as follows:

    One object — stored as xdata (extended entity data).
    Multiple objects — stored as xdata in each object.
    Area — stored as xdata in a rectangular object on layer URLLAYER.

- The Area option creates a layer named URLLAYER with the default color of red, and places a rectangle object on this layer; do not delete the layer.

AttDef

V. 2.0 Defines attribute modes and prompts (short for ATTribute DEFinition).

Commands	Aliases	Ctrl+	F-key	Alt+	Menu Bar	Dashboard
attdef	att	...	...	DKD	Draw	Block Attributes
	ddattdef				⤷ Block	
					⤷ Define Attributes	
-attdef	-att					

Command: attdef

Displays dialog box (redesigned in AutoCAD 2008):

After you click OK, AutoCAD prompts:

Specify start point: *(Pick a point to locate the attribute text.)*

ATTRIBUTE DEFINITION DIALOG BOX

Mode options

☐ **Invisible** makes the attribute text invisible.

☐ **Constant** uses constant values for the attributes.

☐ **Verify** verifies the text after input.

☐ **Preset** presets the variable attribute text.

☑ **Lock Position** prevents attributes from moving, required for attributes in dynamic blocks *(new to AutoCAD 2008)*.

☐ **Multiple Lines** allows multi-line attribute values, when on *(new to AutoCAD 2008)*.

Insertion Point options

☑ **Specify On-screen** picks the insertion point with cursor.

X specifies the x coordinate insertion point.

Y specifies the y coordinate insertion point.

Z specifies the z coordinate insertion point.

Attribute options

Tag identifies the attribute.

Prompt prompts the user for input.

Default sets the default value for the attribute (formerly "Value").

 Insert Field displays the Field dialog box; select a field, and then click OK. See the Field command.

Open Multiline Eidtor displays the multiline attribute editor; available only when the Multiple Lines option is turned on in the Modes area (*new to AutoCAD 2008*). See AttIPedit command.

Text Settings options

Justification sets the text justification.

Text Style selects a text style.

Annotative toggles annoatative scaling for this attribute definition (*new to AutoCAD 2008*).

Text Height specifies the height of the text.

prompts 'Height:' at the command line; pick two points, or enter a value.

Rotation sets the rotation angle.

prompts 'Rotation angle:' at the command line; pick two points, or enter a value.

Boundary Width specifies the width of margins for multiline attribute values; available only when the Multiple Lines option is turned on (*new to AutoCAD 2008*).

prompts 'Multiline attribute width:' at the command line; pick two points, or enter a value.

☐ **Align below previous attribute definition** places the text automatically below the previous attribute.

. . . .

-ATTDEF Command

Command: -attdef
Current attribute modes: Invisible=N Constant=N Verify=N Preset=N Annotative=N Multiple line=N
Enter an option to change [Invisible/Constant/Verify/Preset/Lock position/ Annotative/Multiple lines] <done>: *(Enter an option.)*
Enter attribute tag: *(Enter text, and then press Enter.)*
Enter attribute prompt: *(Enter text, and then press Enter.)*
Enter default attribute value: *(Enter text, and then press Enter.)*
Specify start point of text or [Justify/Style]: *(Pick a point, or enter an option.)*
Specify height <0.200>: *(Enter a value.)*
Specify rotation angle of text <0>: *(Enter a value.)*

COMMAND LINE OPTIONS

Attribute mode selects the mode(s) for the attribute:

- **Invisible** toggles visibility of attribute text in drawing.
- **Constant** toggles fixed or variable value of attribute.
- **Verify** toggles confirmation prompt during input.
- **Preset** toggles automatic insertion of default values.

. . . .

- **Lock position** toggles locking of the attribute value's position.
- **Annotative** toggles annotation scaling of attribute values (*new to AutoCAD 2008*).
- **Multiple lines** toggles multi-line attribute values (*new to AutoCAD 2008*).

Start point indicates the start point of the attribute text.

Justify selects the justification mode for the attribute text.

Style selects the text style for the attribute text.

Height specifies the height of the attribute text; not displayed if the style specifies a height other than 0.

Rotation angle specifies the angle of the attribute text.

RELATED COMMANDS

AttDisp controls the visibility of attributes.

EAttEdit edits the values of attributes.

EAttExt extracts attributes and other data to tables or files on disc.

AttRedef redefines an attribute or block.

Block creates blocks with attributes.

Insert inserts blocks; if block has attributes, prompts for their values.

RELATED SYSTEM VARIABLES

AFlags holds the default value of modes in bit form.

AttMulti toggles the ability to create multi-line attributes (*new to AutoCAD 2008*).

TIPS

- Constant attributes cannot be edited.

- Attribute tags cannot be *null* (have no value); attribute values may be null.

- You can enter any characters for the attribute tag, except spaces and exclamation marks. All characters are converted to uppercase.

- When you press Enter at 'Attribute Prompt,' AutoCAD uses the attribute *tag* as the prompt.

Left: Block with attribute **value** ("attribute").
Right: Attribute **tags** ("sname" and so on).

- AutoCAD does not prevent you from using the same tag over and over again. But during attribute extraction, you may have difficulty distinquishing between tags.

'AttDisp

<u>V. 2.0</u>   Controls the display of all attributes in the drawing (short for ATTribute DISPlay).

Command	Alias	Ctrl+	F-key	Alt+	Menu Bar	Dashboard
attdisp	...	...	...	VLA	View	...
					↳Display	
					↳Attribute Display	

Command: attdisp
Enter attribute visibility setting [Normal/ON/OFF] <Normal>: *(Enter an option.)*
Regenerating drawing.

*Attribute display **Normal** (left), **Off** (center), and **On** (right).*

COMMAND LINE OPTIONS

Normal displays attributes according to AttDef setting.

ON displays all attributes, regardless of AttDef setting.

OFF displays no attributes, regardless of AttDef setting.

RELATED COMMAND

AttDef defines new attributes, including their default visibility.

RELATED SYSTEM VARIABLE

AttMode holds the current setting of AttDisp:

  0 — Off: no attributes are displayed.
  1 — Normal: invisible attributes are not displayed.
  2 — On: all attributes are displayed.

TIPS

- When RegenAuto is off, use Regen after AttDisp to change to attribute display.

- When you define invisible attributes, use AttDisp to view them.

- Use AttDisp to turn off the display of attributes, which increases display speed and reduces drawing clutter.

AttEdit

V. 2.0   Edits attributes in drawings (short for ATTribute EDIT).

Commands	Aliases	Ctrl+	F-key	Alt+	Menu Bar	Dashboard
attedit	ate	...	...	MOAS	Modify	...
	ddatte				⮡ Object	
					⮡ Attribute	
					⮡ Single	
-attedit	-ate			MOAG	Modify	
	atte				⮡ Object	
					⮡ Attribute	
					⮡ Global	

Command: attedit
Select block reference: *(Pick a block.)*
   Displays dialog box:

EDIT ATTRIBUTES DIALOG BOX

Block Name names the selected block.

Attribute-specific prompts allow you to change attribute values.

.... **Open Multiline Editor** displays the multiline attribute editor; available with multiline attribute values only (*new to AutoCAD 2008*). See the AttIPedit command.

Previous displays the previous list of attributes, if any.

Next displays the next list of attributes, if any.

-AttEdit Command
Command: -attedit

One-at-time Attribute Editing options
Edit attributes one at a time? [Yes/No] <Y>: *(Type Y.)*
Enter block name specification <*>: *(Press Enter to edit all.)*
Enter attribute tag specification <*>: *(Press Enter to edit all.)*
Enter attribute value specification <*>: *(Press Enter to edit all.)*
Select Attributes: *(Select one or more attributes.)*
Select Attributes: *(Press Enter.)*
Enter an option [Value/Position/Height/Angle/Style/Layer/Color/Next] <N>: *(Enter an option, and then press Enter.)*

During single attribute editing, AttEdit marks the current attribute with an 'X.'

Global Attribute Editing options
Edit attributes one at a time? [Yes/No] <Y>: *(Type Y.)*
Performing global editing of attribute values.

Edit only attributes visible on screen? [Yes/No] <Y>: *(Press Enter.)*
Enter block name specification <*>: *(Press Enter.)*
Enter attribute tag specification <*>: *(Press Enter.)*
Enter attribute value specification <*>: *(Press Enter.)*
Select Attributes: *(Select one or more attributes.)*
Select Attributes: *(Press Enter.)*

Enter string to change: *(Enter existing string, and then press Enter.)*
Enter new string: *(Enter new string, and then press Enter.)*

COMMAND LINE OPTIONS
Value changes or replaces the value of the attribute.

Position moves the text insertion point of the attribute.

Height changes the attribute text height.

Angle changes the attribute text angle.

Style changes the text style of the attribute text.

Layer moves the attribute to a different layer.

Color changes the color of the attribute text.

Next edits the next attribute.

RELATED SYSTEM VARIABLE

AttDia toggles use of AttEdit during the Insert command.

RELATED COMMANDS

AttDef defines an attribute's original value and parameter.

AttDisp toggles an attribute's visibility.

AttRedef redefines attributes and blocks.

EAttEdit edits all aspects of attributes.

AttIPedit edits attribute values in blocks (*new to AutoCAD 2008*).

Explode reduces attribute values to their tags.

TIPS

■ Constant attributes cannot be edited with this command; only attributes parallel to the current UCS can be edited.

■ The DdEdit command also displays this dialog box.

■ Unlike other text input to AutoCAD, attribute values are case-sensitive.

■ To edit null attribute values, use -AttEdit's global edit option, and enter \ (backslash) at the 'Enter attribute value specification' prompt.

■ The wildcard characters **?** and ***** are interpreted literally at the 'Enter string to change' and 'Enter new string' prompts.

■ To edit the different parts of an attribute, use the following commands:

Attedit — selects non-constant attribute *values* in one block.
-AttEdit — selects attribute *values* and *properties* (such as position, height, and style) in one block or in all attributes.

■ When selecting attributes for global editing, you may pick the attributes, or use the following selection modes: Window, Last, Crossing, BOX, Fence, WPolygon, and CPolygon.

■ You may use wildcards in the block name, tag, and value specifications:

#	matches any single numeric character.
@	matches any single alphabetic character.
.	matches any single non-alphabetic character.
*	matches any string.
?	matches any single character.
~	matches anything but the following pattern.
[ ]	matches any single character enclosed.
[~]	matches any single character not enclosed.
[-]	matches any single character in the enclosed range.
'	treats the next character as a non-wild-card character.

■ AttEdit does not trim leading and trailing spaces from attribute values. Be sure to avoid entering them to reduce unexpected results.

AttExt

V. 2.0 Extracts attribute data from drawings to files on disk (short for ATTribute EXTract).

Commands	Alias	Ctrl+	F-key	Alt+	Menu Bar	Dashboard
attext	ddattext	...	...	...	...	...
-attext						

Command: attext

Displays dialog box:

ATTRIBUTE EXTRACTION DIALOG BOX

File Format options

⊙ **Comma Delimited File (CDF)** creates a CDF text file, where commas separate fields.

○ **Space Delimited File (SDF)** creates an SDF text file, where spaces separate fields.

○ **DXF Format Extract File (DXX)** creates an ASCII DXF-format file.

Select Objects returns to the graphics screen to select attributes for export.

Template File specifies the name of the TXT template file for CDF and SDF files.

Output File specifies the name of the attribute output file, *.txt* for CDF and SDF formats, or *.dxx* for DXF format.

. . . .

-ATTEXT Command
Command: -attext
Enter extraction type or enable object selection [Cdf/Sdf/Dxf/Objects] <C>:
(Enter an option.)

Displays the Select Template File dialog box; select the template file.

Displays the Create Extract File dialog box.

. . . .

COMMAND LINE OPTIONS

Cdf outputs attributes in comma-delimited format.

Sdf outputs attributes in space-delimited format.

Dxf outputs attributes in DXF format.

Objects selects objects from which to extract attributes.

RELATED COMMANDS

AttDef defines attributes.

DataExtraction provides a wizard for extracting attributes and other drawing data to tables and external data files (*new to AutoCAD 2008*).

RELATED FILES

*.txt required extension for template file; extension for CDF and SDF files.

*.dxx extension for DXF extraction files.

TIPS

- It is easier to use the EAttExt command than this command for attribute extraction.

- **CDF** is short for "Comma Delimited File"; it has one record for each block reference; a comma separates each field; single quotation marks delimit text strings.

- **SDF** is short for "Space Delimited File"; it has one record for each block reference; fields have fixed width padded with spaces; string delimiters are not used.

- **DXF** is short for "Drawing Interchange File"; it contains only block reference, attribute, and end-of-sequence DXF objects; no template file is required.

- CDF files use the following conventions:

    Specified field widths are the maximum width.
    Positive number fields have a leading blank.
    Character fields are enclosed in ' ' (single quotation marks).
    Trailing blanks are deleted.
    Null strings are '' (two single quotation marks).
    Use spaces; do not use tabs.
    Use the C:DELIM and C:QUOTE records to change the field and string delimiters to another character.

- To output the attributes to the printer, specify:

Logical Filename	Meaning
CON	Displays on text screen.
PRN *or* **LPT1**	Prints to parallel port 1.
LPT2 *or* **LPT3**	Prints to parallel ports 2 or 3.

- Before you can specify the SDF or CDF option, you must create a template file.

AttIPedit

2008 Edits multiline attribute values (short for "ATTribute In Place EDITor").

Command	Alias	Ctrl+	F-key	Alt+	Menu Bar	Dashboard
attipedit	ati	...	...	...	...	...

Command: attipedit
Select attribute to edit: *(Pick an attribute value.)*

Select single-line attribute: displays simple inplace editor, like the Text command (no toolbar).

Select multi-line attribute: displays either a large or small toolbars.

COMMAND LINE OPTION
Select attribute to edit selects attribute value to be edited.

TOOLBAR AND SHORTCUT MENU OPTIONS
See the MText command.

RELATED COMMANDS
AttEdit edits all of a block's attributes through a dialog box.

AttDef defines new attributes.

EAttEdit edits all aspects of attributes.

RefEdit also edits attributes in-place.

RELATED SYSTEM VARIABLES
AttIpe toggles the size of the toolbar; see tips.

AttMulti toggles the creation of multi-line attributes.

TIP
- This command works with text in attribute definitions (not yet part of a block insertion) and attributes in blocks.

- The AttIpe system variable toggles the size of the multi-line attribute editors. When 1, displays the MText-like toolbar. When 0, displays this condensed version of the toolbar:

- Double-clicking attribute values runs the Enhanced Attribute Editor dialog box (EAttEdit command).

- There are now more attribute editing commands than you can shake a stick at: AttEdit, -AttEdit, EAttEdit, BAttMan, RefEdit, DdEdit, Properties, and now AttIPedit.

AttRedef

Rel.13 Redefines blocks and attributes (short for ATTribute REDEFinition).

Command	Alias	Ctrl+	F-key	Alt+	Menu Bar	Dashboard
attredef	at	...	...	...	...	...

Command: attredef
Name of Block you wish to redefine: *(Enter name of block.)*
Select objects for new Block...
Select objects: *(Select one or more objects.)*
Select objects: *(Press Enter.)*
Insertion base point of new block: *(Pick a point.)*

COMMAND LINE OPTIONS

Name of Block you wish to redefine specifies the name of the block to be redefined.

Select objects selects objects for the new block.

Insertion base point of new block picks the new insertion point.

RELATED COMMANDS

AttDef defines an attribute's original value and parameter.

AttDisp toggles an attribute's visibility.

EAttEdit edits the attribute's values.

Explode reduces attribute values to tags.

TIPS

- Existing attributes retain their values.

- Existing attributes not included in the new block are erased.

- New attributes added to an existing block take on default values.

AttSync

2002 Updates blocks with new attribute definitions (short for ATTribute SYNChronization).

Command	Alias	Ctrl+	F-key	Alt+	Menu Bar	Dashboard
attsync	...	...	...	...	...	**Block Attributes**

Command: attsync
Enter an option [?/Name/Select] <Select>: *(Specify an option.)*
Select a block: *(Pick a block.)*
ATTSYNC block name? [Yes/No] <Yes>: *(Type Y or N.)*

COMMAND LINE OPTIONS
   ? lists the names of all blocks in the drawing.

   Name enters the name of the block.

   Select selects a single block with the cursor.

RELATED COMMANDS
   AttDef defines an attribute.

   BattMan edits the attributes in a block definition.

   EAttEdit edits the attributes in block references.

TIPS
- This command is used together with other attributed-related commands, in the following order:

    1. AttDef and Block define attributes, and attach them to blocks.

    2. Insert inserts the block, and gives values to the attributes.

    3. EAttEdit changes the attributes in selected blocks, adding, deleting, or modifying the attribute definitions.

    4. BAttMan changes the attributes in the orignal blocks.

    5. AttSync updates the attributes to the new definition. (BAttMan also performs this task.)

- This command does not operate if the drawing lacks blocks with attributes. AutoCAD complains, "This drawing contains no attributed blocks."

Audit

<u>Rel.11</u> Examines drawing files for structural errors.

Command	Alias	Ctrl+	F-key	Alt+	Menu Bar	Dashboard
audit	...	...	...	FUA	File ✎Drawing Utilities ✎Audit	...

Command: audit
Fix any errors detected? [Yes/No] <N>: *(Type Y or N.)*

Sample output

```
AutoCAD Text Window - C:\CAD\Red Deer\Sample\8th floor furniture.dwg
Edit

Command: audit

Fix any errors detected? [Yes/No] <N>: y
  24      Blocks audited
Pass 1 2988     objects audited
Pass 2 2988     objects audited
Pass 3 4800     objects audited
Total errors found 0 fixed 0

Command: |
```

COMMAND LINE OPTIONS

N reports errors found in drawing files, but does not fix errors.

Y reports and fixes errors.

RELATED COMMANDS

DrawingRecover displays a palette listing recovered drawings.

Save saves recovered drawings to disk.

Recover recovers damaged drawing files.

RELATED SYSTEM VARIABLE

AuditCtl creates *.adt* audit log files when set to 1.

RELATED FILE

.adt is the audit log file, which records the auditing process.

TIPS

- The Audit command is a diagnostic tool for validating and repairing the contents of *.dwg* files.

- Objects with errors are placed in the Previous selection set. Use an editing command, such as Copy, to view the objects.

- If Audit cannot fix a drawing file, try the Recover command.

. . . .

Removed Commands

Background is replaced by the Background option of the **View** command, as of AutoCAD 2007.

. . . .

AutoPublish

2008 Automatically publishes the current drawing as a DWF file when saved or closed.

Command	Alias	Ctrl+	F-key	Alt+	Menu Bar	Dashboard
autopublish	...	...	...	...	...	...

Command: autopublish
AutoPublish DWF or specify override [Location] <AutoPublish>: *(Press Enter, or type L.)*

COMMAND LINE OPTIONS

AutoPublish immediately exports the drawing as a *.dwf* file.

Location specifies the folder for published files; displays the Select A Folder for Generated Files dialog box.

To access the options for this command: enter the Options command, choose the Plot and Publish tab, and then click the Automatic DWF Publishing Settings button; displays dialog box:

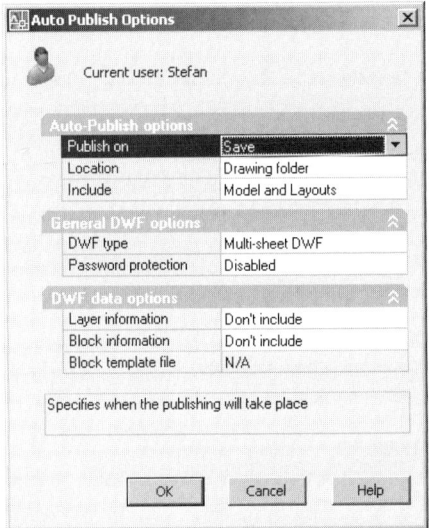

DIALOG BOX OPTIONS

Auto-Publish Options
Publish On specifies when the drawing is exported in DWF format:

- **Save** exports the drawing when it is saved.
- **Close** exports the drawing when it is closed, whether saved or not.
- **Prompt on Save** asks if you want the drawing exported in DWF format.
- **Prompt on Close** asks if you want the drawing exported in DWF format; see dialog box below. This option works only if you agree to save changes to the drawing.

Location specifies the drive and folder in which the *.dwf* files are saved. Preset options include the drawing's folder, a folder named *dwf* below the drawing's folder, *my documents*, and any folder you choose.

Include determines whether the following are included in the *.dwf* file:

- **Model** exports only the Model tab.
- **Layouts** exports all layout tabs.
- **Model and Layouts** exports all tabs.

General DWF Options

DWF Type chooses between multiple single-sheet and single multisheet DWF files.

Password Protection provides an optional password for DWF files.

DWF Data Options

Layer Information toggles the inclusion of layers in the DWF file (layers for 3D drawings are never included).

Block Information determines whether information about blocks and attributes are included.

Block Template File specifies the name of the BLK file to use for formatting block and attribute information:

- **Create** opens the Publish Block Template dialog box for creating new block templates.
- **Edit** opens the Select Block Template dialog box for selecting a BLK block template file.

RELATED COMMANDS

3dDwf exports 3D drawings in DWF format.

Archive creates multi-page DWF files from sheetsets.

Publish creates multi-page DWF files from one or more drawings.

Plot exports drawings in DWF format.

RELATED SYSTEM VARIABLES

AutoDwfPublish toggles the AutoPublish feature; default = 0 (off).

BackgroundPlot toggles whether plots are carried out in the background.

TIPS

- Use this command and its related system variable to have AutoCAD automatically generate *.dwf* files of the current drawing; this synchronizes DWG and DWF files, to ensure they are identical.

- This command does not turn on the AutoDwfPublish system variable, curiously enough.

'Base

<u>V. 1.0</u>   Changes the insertion point of the drawing, located by default at (0,0,0).

Command	Alias	Ctrl+	F-key	Alt+	Menu Bar	Dashboard
base	...	...	...	DKB	Draw	...
					↳ Block	
					↳ Base	

Command: base
Enter base point <0.0,0.0,0.0>: *(Pick a point.)*

COMMAND LINE OPTION

Enter base point specifies the x, y, z coordinates of the new insertion point.

RELATED COMMANDS

Block specifies the insertion point of new blocks.

Insert inserts another drawing into the current drawing.

Xref references other drawings.

RELATED SYSTEM VARIABLE

InsBase contains the current setting of the drawing's insertion point.

TIPS

- Use this command to shift the insertion point of the current drawing.

- This command does not affect the current drawing. Instead, the relocated base point comes into effect when you insert or xref the drawing into another drawing.

BAttMan

<u>2002</u>   Edits all aspects of attributes in blocks; works with one block at a time (short for Block ATTribute MANager).

Command	Alias	Ctrl+	Alt+	Menu Bar	Dashboard
battman	...	...	MO⤷AB	Modify	Block Attributes
				⤷Object	
				⤷Attribute	
				⤷Block Attribute Manager	

Command: battman

When the drawing contains no blocks with attributes, displays error message, "This drawing contains no attributed blocks."

When drawing contains at least one block with attributes, displays dialog box:

BLOCK ATTRIBUTE MANAGER DIALOG

Select block hides the dialog box, and then prompts, 'Select a block:'.

Block lists the names of blocks in the drawing, and displays the name of the selected block.

Sync changes the attributes in block insertions to match the changes made here.

Move Up moves the selected attribute tag up the list; constant attributes cannot be moved.

Move Down moves the attribute tag down the list.

Edit displays the Edit Attribute dialog box; see the EAttEdit command.

Remove removes the attribute tag and related data from the block; it does not operate when the block contains a single attribute.

Settings displays the Settings dialog box.

Apply applies the changes to the block definition.

SETTINGS DIALOG BOX

Display In List
  ☑ **Tag** toggles (turns on and off) the display of the column of tags.
  ☑ **Prompt** toggles display of the column of attribute prompts.
  ☑ **Default** toggles the display of the attribute's default value.
  ☑ **Modes** toggles the display of the attribute's modes: invisible, constant, verify, and/or preset.
  ☑ **Annotative** toggles annotative scaling of the attribute text *(new to AutoCAD 2008)*.
  ☐ **Style** toggles the display of the attribute's text style name.

  ☐ **Justification** toggles the display of the attribute's text justification.
  ☐ **Height** toggles the display of the attribute's text height.
  ☐ **Rotation** toggles the display of the attribute's text rotation angle.
  ☐ **Width Factor** toggles the display of the attribute's text width factor.
  ☐ **Oblique Angle** toggles the display of the attribute's text obliquing angle (slant).
  ☐ **Boundary Width** toggles the margin constraint of multi-line text *(new to AutoCAD 2008)*.

  ☐ **Layer** toggles the display of the attribute's layer.
  ☐ **Linetype** toggles the display of the attribute's linetype.
  ☐ **Color** toggles the display of the attribute's color.
  ☐ **Lineweight** toggles the display of the attribute's lineweight.
  ☐ **Plot style** toggles the display of the attribute's plot style name (available only when plot styles are turned on).
  Select All selects all display options.
  Clear All clears all display options, except tag name.

  Emphasize duplicate tags:
    ☑ Highlights duplicate attribute tags in red.
    ☐ Does not highlight duplicate tags.

Apply changes to existing references:

☑ applies changes to all block instances that reference this definition in the drawing.

☐ applies the new attribute definitions only to newly-inserted blocks.

RELATED COMMANDS

AttDef defines attributes.

Block binds attributes to symbols.

Insert inserts blocks, and then allows you to specify the attribute data.

TIPS

- Use this command to edit and remove attribute definitions, as well as to change the order in which attributes appear.

- The Sync option does not change the values you assigned to attributes.

- When an attribute has a mode of Constant, it cannot be moved up or down the list.

- Turning on all the options displays a lot of data. To see all the data columns, you can stretch the dialog box.

With the cursor, grab the edge of the dialog box to make it larger and smaller.

- An attribute definition cannot be changed to Constant via the Edit Attribute dialog box.

- The Remove option does not work when the block contains a single attribute.

BEdit

2006

Edits blocks and external references; assigns actions to blocks to create dynamic blocks (short for Block EDITor).

Command	Alias	Ctrl+	F-key	Alt+	Menu Bar	Dashboard
bedit	be	...	...	TB✎	Tools	...
					✎Block Editor	

-bedit

Command: bedit

Displays dialog box:

Enter unique name to create new block.

Select entire drawing or individual blocks.

Icon indicates dynamic block.

EDIT BLOCK DEFINITION DIALOG BOX

Block to create or edit

- To create a new block, enter a name.
- To edit an existing block, select its name from the list.
- To edit the drawing as a block, select "<Current Drawing>."

OK causes AutoCAD to display the Block Editor.

Cancel returns to the drawing editor.

. . . .

-BEDIT Command

Command: -bedit

Enter block name or [?]: *(Enter the name of a block or type ?.)*

Enter block name specifies the block to edit.

? lists the names of blocks in the drawing, if any.

Block Editor Toolbar

Block Editing Environment

Block Authoring Palettes

Grips Associated
wtih Parameters

Parameters

Actions Associated
with Parameters

Block Editor Toolbar

Left end of toolbar:

Edit or
Create Block
Definition

Block
Definition
Name

Authoring
Palettes

Define
Attribute

Update Parameter
and Action Text Size

Save

Save As

Parameter

Action

Close Block
Editor

Right end of toolbar:

Toggle
Visibility Modes

Manage Visibility States
Dialog Box

Make Visible

Make Invisible

Name of State

Block Editor: **BAction**

Adds actions to dynamic block definitions (alias: ac).

Command: baction
Select parameter: *(Select one parameter with which to associate an action.)*
Select action type [Array/Move/Scale/Stretch/Polar Stretch]: *(Enter an option.)*

Note: the actions available depend on the parameter selected.

Block Editor: **BActionSet**

Respecifies selection set of objects associated with actions.

Select action: *(Select an action from the dynamic block definition.)*
Specify selection set for action object [New/Modify] <New>: *(Enter an option.)*

Modify modifies the existing selection set by adding and subtracting objects.

New creates a new selection set.

Block Authoring Palette: **BActionTool**

Defines changes to geometry of dynamic block references by changing custom properties.

Select action type [Array/Lookup/Flip/Move/Rotate/Scale/sTretch/Polar stretch]: *(Enter an option.)*
Select parameter: *(Select a parameter to associate with the action.)*
Specify selection set for action
Select objects: *(Select one or more objects.)*

COMMAND OPTIONS

Array adds array actions to dynamic block definitions; selected objects are arrayed.

Flip adds flip actions to dynamic block definitions; objects are flipped about mirror line.

Lookup adds lookup actions to the dynamic block definitions; lookup actions display the shortcut menu with options.

Move adds move actions to dynamic block definitions; objects move in linear or polar directions.

Rotate adds rotate actions to dynamic block definitions; prompts "Specify action location or [Base type]".

Scale adds resize actions to dynamic blocks; objects resize in linear, polar, or x,y directions.

Stretch adds stretch actions to dynamic blocks; objects are stretched in point, linear, polar, or x,y directions.

 Polar Stretch adds stretch actions to dynamic blocks; stretched by distance and angle.

Several actions use these options:

Base Type specifies the type of base point to use for the action:

• **Dependent** rotates dynamic block about the associated parameter's base point.

• **Independent** rotates dynamic block about a specified base point.

Multiplier changes the associated parameter value by the specified factor; prompts "Enter distance multiplier <1.0>."

Offset increases and decreases angles by the specified number of degrees; prompts "Enter angle offset <0>."

XY specifies distances in the x-, y-, and xy-directions from the parameter base point; prompts "Enter XY distance type [X/Y/XY]."

· ·

Block Editor: **BAssociate**

Associates orphaned actions with parameters.

Select action object: *(Select one action not associated with parameters.)*
Select parameter to associate with action: *(Select the parameter to associate.)*

· ·

Block Editor: **BAttOrder**

Controls the order in which attributes are listed.

Displays dialog box:

Move Up moves the selected attribute up the list.

Move Down moves the selected attribute down the list.

OK exits the dialog box, and returns to the Block Editor.

Cancel cancels changes, and returns to the Block Editor.

· ·

Block Editor: **BAuthorPalette** and **BAuthorPaletteClose**

Toggles the display of the Block Authoring Palettes window.

Block Editor: **BClose**

Closes the Block Editor environment, and returns to the drawing editor (alias: bc)

Block Editor: **BCycleOrder**

Changes the cycle order of grips in dynamic block references.

Displays dialog box:

Move Up moves the selected parameter up the list.

Move Down moves the selected parameter down the list.

Cycling toggles cycling of the selected parameter.

Block Editor: **BGripSet**

Creates, resets, and deletes grips associated with parameters.

Select parameter: *(Select one parameter in the dynamic block definition.)*

Enter number of grip objects for parameter or reset position [0/1/2/4/ Reposition]: *(Enter number of grips to display, or type R.)*

0 - 4 specifies the number of grips to display; number varies, depending on the parameter. Enter **0** to hide the display of grips.

Reposition repositions grips to their default positions (a.k.a. reset).

Grip	Name	Grip Manipulation	Associated Parameters
∪	Alignment	Aligns with objects within 2D planes.	Alignment
⤡	Flip	Flips (mirrors).	Flip
◁	Linear	Moves along defined directions or axes.	Linear
▽	Lookup	Displays lists of items.	Visibility, Lookup
○	Rotation	Rotates about axes.	Rotation
□	Standard	Moves in any direction within 2D planes.	Base, Point, Polar, XY

Block Editor: **BLookupTable**

Creates lookup tables for dynamic block definitions.

If dynamic block definition contains a lookup action and at least one lookup parameter, displays dialog box:

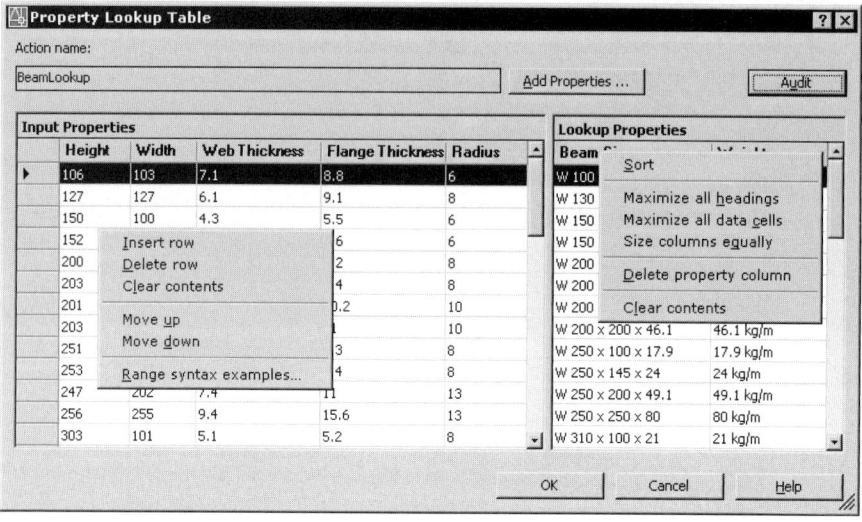

Add Properties displays the Add Parameter Properties dialog box for adding parameter properties to this table.

Audit checks data to ensure that rows are unique.

Right-click data or rows:

Insert Row adds a row above the selected row.

Delete Row erases the row and its data.

Clear Contents erases the data in the selected cell or row.

Move Up moves the row up.

Move Down moves the row down.

Range Syntax Examples displays help on writing ranges.

Right-click headings (columns):

Sort sorts data in the selected column.

Maximize All Headings makes each column wide enough to show the heading's wording.

Maximize All Data Cells makes each column wide enough to show the data's wording.

Size Columns Equally makes all columns the same width, fitted to the dialog box width.

Delete Property Column erases the selected column.

Clear Contents erases the data in the selected cell or row.

Block Editor: **BParameter**

Adds parameters with grips to dynamic block definitions (alias: **param**).

Enter parameter type [Alignment/Base/pOint/Linear/Polar/XY/Rotation/ Flip/Visibility/looKup] <last>: *(Enter an option.)*

Select a parameter, and then associate an action with the parameter:

Parameter	Options	Action(s)	Comments
Alignment	Perpendicular	...	Automatically rotates and aligns with other
	Base point		objects.
Base	Location	...	Defines base points for dynamic blocks.
Flip	Reflection Line	Flip	Flips objects about their reflection lines.
Linear	Start Point	Array	Constrains grip movement to preset angles.
	Endpoint	Move	
	Chain	Scale	
	Midpoint	Stretch	
	List		
	Increment		
Lookup	Action Name	Lookup	Defines custom properties that evaluate
	Properties		values from lists and tables you define.
	Audit		
	Input		
	Lookup		
Point	Location	Move	Defines x,y locations in drawings.
	Chain	Stretch	
Polar	Base Point	Array	Constrains grip movement to distances
	Endpoint	Move	and angles.
	Chain	Polar Stretch	
	List	Scale	
	Increment	Stretch	
Rotation	Base Point	Rotate	Defines angles.
	Radius		
	Base Angle		
	Default Angle		
	Chain		
	List		
	Increment		
Visibility	Display	...	Toggles the visibility states of blocks.
XY	Base Point	Array	Shows x,y distances from the base
	Endpoint	Move	points of blocks.
	Chain	Scale	
	List	Stretch	
	Increment		

Block Editor: **BSave** and **BSaveAs**
Saves block definitions.

BSave saves the changes to the block definition. Changes are also saved when exiting the Block Editor (alias: bs)

BSaveAs saves the block definition by another name; displays dialog box.

Blocks are saved in the drawing, not to disk.

Block Editor: **BvHide** and **BvShow**
Changes the visibility state of objects in dynamic block definitions.

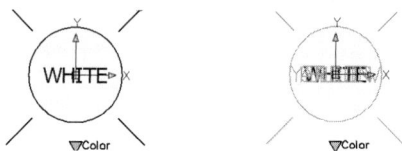

Left: *Visibility state set to* ***Current*** *(White).*

Right: *Visibility state set to* ***All*** *(White, Green, Red, Blue, Yellow, and Amber).*

Select objects to hide/show: *(Select objects to hide or show.)*

Hide/Show for current state or all visibility states [Current/All]: *(Enter an option.)*

Current sets the visibility of selected objects to the current visibility state.

All sets the visibility of selected objects to all visibility states, as defined by the block.

Block Editor: **BvState**
Creates, sets, renames, and deletes visibility states in dynamic blocks (alias: **vs**)

Displays dialog box:

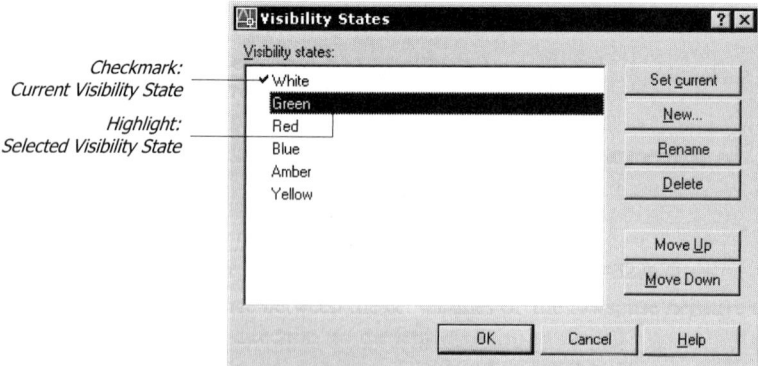

Set Current sets the selected visibility state as current in the Block Editor; does not affect the default visibility state of blocks in drawings.

Rename changes the name of the selected visibility state.

Delete erases the selected visibility state.

Move Up moves the selected visibility state up the list.

Move Down moves the selected visibility state down the list.

New displays the New Visibility State dialog box.

Visibility State Name specifies the name of the new visibility state.

○ **Hide All Existing Objects in New State** hides all objects in this visibility state.

○ **Show All Existing Objects in New State** shows all objects in this visibility state.

⊙ **Leave Visibility of Existing Objects Unchanged in New State** retains the visibility state of objects as the current visibility state.

RELATED COMMANDS

AttDef defines attributes for blocks.

Block makes blocks.

Insert inserts blocks into drawings.

RefEdit edits blocks and xrefs in-place.

WBlock writes blocks to individual files on disk.

RELATED SYSTEM VARIABLES

BlockEditLock toggles editing of dynamic blocks in the Block Editor.

BlockEditor reports whether the Block Editor is active.

The following system variables affect the Block Editor only:

BActionColor specifies the text color of actions.

BDependencyHighlight determines whether dependent objects are highlighted when parameters, actions, and grips are selected.

BGripObjColor specifies the color of grips.

BGripObjSize specifies the size of custom grips relative to the window size.

BParameterColor specifies the color of parameters.

BParameterFont specifies the font for parameters and actions.

BParameterSize specifies the size of parameter text and features relative to the window size.

BVMode toggles the display of hidden objects between invisible or dimmed.

TIPS

- Selecting a block before entering the BEdit or -BEdit command causes AutoCAD immediately to enter the Block Editor environment. Alternatively, you can double-click a dynamic block to open it in the Block Editor.

- Entering the name of a nonexistent block creates a new block for the Block Editor.

- Most of AutoCAD's drawing and editing commands are available in the Block Editor.

- Values in lookup tables are limited to 256 characters per cell, including distances and angles for points, linear, polar, XY, and rotation parameters; text string parameters; properties for flip; visibility parameter values; and architectural and mechanical units, such as 23'1/2". Invalid values are automatically reset to the last valid value.

- Commas are delimiters between values. Brackets [ ] specify inclusive ranges; parentheses ( ) specify exclusive ranges, as follows:

[8,31]	Any values between 8 and 31 inclusive.
(9,15)	Any values between 9 and 15, excluding 9 and 15.
[,25]	Any values up to and including 25.
(56,)	Any values larger than 56.

  When this icon appears, it means that no action is associated with the parameter.

- As of AutoCAD 2008, grips display tooltips describing their purpose, as illustrated below:

DEFINITIONS

Actions — determine how the geometry of dynamic blocks change when their custom properties are modified; actions are associated with parameters.

Parameters — define custom properties for dynamic block references; actions must be associated with parameters.

Orphaned Actions — arise when parameters are removed from block definitions, orphaning their associated actions.

Renamed Commands

The **BHatch** and **-BHatch** commands are found under the **Hatch** and **-Hatch** commands.

'Blipmode

V. 2.1 Turns the display of pick-point markers, known as "blips," on and off.

Command	Alias	Ctrl+	F-key	Alt+	Menu Bar	Dashboard
blipmode	...	...	...	...	...	...

Command: blipmode
Enter mode [ON/OFF] <OFF>: on

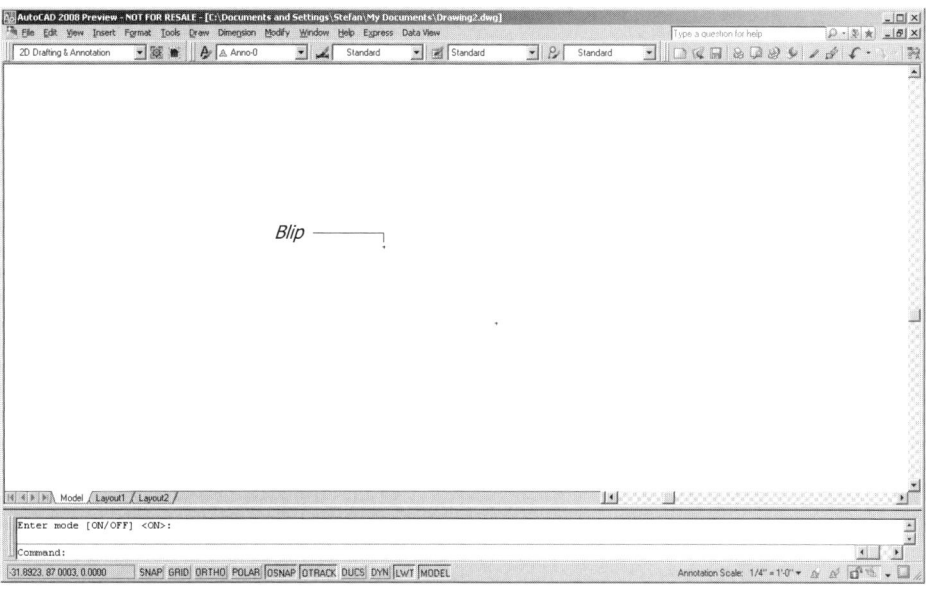

COMMAND LINE OPTIONS

ON turns on the display of pick-point markers.

OFF turns off the display of pick-point markers.

RELATED COMMANDS

Options allows blipmode toggling via a dialog box.

Redraw cleans blips off the screen.

RELATED SYSTEM VARIABLE

Blipmode contains the current blipmode setting.

TIPS

- You cannot change the size of the blip mark.

- Blip marks are erased by any command that redraws the view, such as Redraw, Regen, Zoom, and Vports.

Block

V. 1.0 Defines a group of objects as a single named object; creates symbols.

Commands	Aliases Ctrl+	F-key	Alt+	Menu Bar	Dashboard
block	b ...	...	DKM	Draw	Block Attributes
	bmake			⤷ Block	2D Draw
	bmod			⤷ Make	
	acadblockdialog				
-block	-b				

Command: block

Displays dialog box, redesigned in AutoCAD 2008:

BLOCK DEFINITION DIALOG BOX

Name names the block (*maximum = 255 characters, including spaces, numbers, and other special characters, except \ | / * ? : < or >*).

Base Point

Specify On-screen prompts you 'Specify insertion base point:' after you click OK (*new to AutoCAD 2008*).

Pick point dismisses the dialog box; AutoCAD prompts, 'Specify insertion base point:'. Pick a point that specifies the block's insertion point, usually the lower-left corner.

X, Y, and **Z** specify the x, y, z coordinates of the insertion point.

Objects

Specify On-screen prompts you 'Select objects:' after you click OK (*new to AutoCAD 2008*).

Select objects dismisses the dialog box; AutoCAD prompts, 'Select objects:'. Select one or more objects that make up the block:

○ **Retain** leaves objects in place after the block is created.

⊙ **Convert to Block** erases objects making up the block, and replaces them with the block.

○ **Delete** erases the objects making up the block; the block is stored in drawing.

✎ **Quick Select** displays the Quick Select dialog box; see the QSelect command.

Behavior
(New to AutoCAD 2008.)

☑ **Annotative** toggles annotative scaling for this block.

☐ **Match Block Orientation to Layout** ensures the block rotates to match the layout.

☑ **Scale Uniformly** prevents different scale factors being applied to the x, y, and z directions.

☑ **Allow Exploding** allows the block to be exploded.

Settings
Insert Units selects the units for the block when dragged from the Design Center.

Hyperlink displays the Insert Hyperlink dialog box; see the Hyperlink command.

Description
Description describes the block.

Open in Block Editor opens the block in the Block Editor.

(AutoCAD 2008 removes the Preview Icon and X, Y, Z Scale options from this dialog box.)

. . . .

-BLOCK Command
Command: -block
Enter block name or [?]: *(Enter a name, or type ?.)*
Specify insertion base point or [Annotative]: *(Pick a point, or type A.)*
Select objects: *(Select one or more objects.)*
Select objects: *(Press Enter.)*

COMMAND LINE OPTIONS
Block name allows you to name the block.

? lists the names of blocks stored in the drawing.

Insertion base point specifies the x, y coordinates of the block's insertion point.

Annotative applies annotative scaling to the block.

Select objects selects the objects and attributes that make up the block.

Annotative options
(New to AutoCAD 2008.)
Create annotative block [Yes/No] <N>: *(Type Y.)*
Match orientation to layout in paper space viewports [Yes/No] <N>:

Create annotative block determines whether the block takes annotative scale factors.

Match orientation ensures the block rotates to match the layout.

RELATED COMMANDS
BEdit opens blocks in the Block Editor environment; makes dynamic blocks.

BlockIcon generates icons of blocks defined in earlier releases of AutoCAD.

. . . .

Explode reduces blocks to their original objects.

Insert adds blocks or other drawings to the current drawing.

Oops returns objects to the screen after creating the block.

Purge removes unused blocks from drawings.

RefEdit edits blocks and xrefs in-place.

WBlock writes blocks as drawings to *.dwg* files on disk.

XRef displays another drawing in the current drawing.

RELATED FILES

All *.dwg* drawing files can be inserted as blocks.

RELATED SYSTEM VARIABLES

InsName default block name.

InsUnits drawing units for blocks dragged from the DesignCenter.

TIPS

■ Blocks consist of these parts:

Insertion Point   Attribute (Ext. #)

■ The names of blocks have up to 255 alphanumeric characters, including $, -, and _.

■ Use the INSertion object snap to select the insertion point of blocks.

■ Objects within a block definition take on the properties specified by their layers, with one exception: if the objects were created on layer 0, then upon insertion they take on the properties specified by the host layer.

■ AutoCAD has five types of blocks:

User blocks	Named blocks created by users.
Nested blocks	Blocks inside other blocks.
Unnamed blocks	Blocks created by AutoCAD.
Xrefs	Externally-referenced drawings.
Dependent blocks	Blocks in externally-referenced drawings.

■ AutoCAD sometimes creates unnamed blocks, also called "anonymous blocks":

*An	Groups.
*Dn	Associative dimensions.
*Un	Created by AutoLISP or ObjectARx apps.
*Xn	Hatch patterns.

■ AutoCAD automatically purges unreferenced anonymous blocks when drawings are first loaded.

BlockIcon

<u>2000</u>   Creates preview images for all blocks in drawings created with AutoCAD Release 14 or earlier.

Command	Alias	Ctrl+	F-key	Alt+	Menu Bar	Dashboard
blockicon	...	...	...	FUU	File	...
					⤷Drawing Utilities	
					⤷Update Block Icons	

Command: blockicon
Enter block names <*>: *(Enter a name, a wildcard pattern, or press Enter for all names.)*
n **blocks updated.**

COMMAND LINE OPTION

Enter block names specifies the blocks for which to create icons; press Enter to add icons to all blocks in the drawing.

RELATED COMMANDS

AdCenter displays the icons created by this command:

Block creates new blocks and their icons.

Insert inserts blocks; dialog box displays icons generated by this command.

TIP

■ AutoCAD creates icons of blocks automatically; this command is necessary only for drawings older than AutoCAD 2000 containing blocks.

BmpOut

<u>Rel.13</u> Exports the current viewport as a raster image in BMP bitmap format.

Command	Alias	Ctrl+	F-key	Alt+	Menu Bar	Dashboard
bmpout	...	...	...	...	...	...

Command: bmpout

*Displays Create BMP File dialog box. Enter a file name, and then click **Save**.*

Select objects or <all objects and viewports>: *(Press Enter to select all objects, or select individual objects.)*

DIALOG BOX OPTION

Save saves drawings as BMP format raster files.

RELATED COMMANDS

JpgOut exports objects and viewports in JPEG format.

PngOut exports objects and viewports in PNG format.

TifOut exports objects and viewports in TIFF format.

WmfOut exports selected objects in WMF format.

RELATED WINDOWS COMMANDS

PRT SCR saves screen to the Clipboard.

ALT+PRT SCR saves the topmost window to the Clipboard.

TIPS

■ The *.bmp* extension is short for "bitmap," a raster file standard for Windows.

■ This command creates uncompressed *.bmp* files.

Boundary

Rel.12 Creates boundaries as polylines or 2D regions.

Command	Aliases Ctrl+	F-key	Alt+	Menu Bar	Dashboard
boundary	bo ...	...	DB	Draw	Q9
	bpoly			↳Boundary	
-boundary	-bo				

Command: boundary

Displays dialog box, redesigned in AutoCAD 2006:

BOUNDARY CREATION DIALOG BOX

🔲 **Pick Points** picks points inside closed areas.

Island detection:

☑ Island detection is turned on.

☐ Island detection is turned off.

Boundary Retention

Retain Boundaries is grayed out (unavailable), because the point of this command is to create boundaries, not discard them!

Object Type constructs the boundary from:

- **Polylines** forms the boundary from a polyline.
- **Region objects** forms the boundary from a 2D region.

Boundary Set

Boundary Set defines how objects are analyzed for defining boundaries (default = current viewport).

🔲 **New** creates new boundary sets; AutoCAD dismisses the dialog box, and then prompts you to select objects.

-BOUNDARY Command

Command: -boundary
Specify internal point or [Advanced options]: *(Pick a point, or type A.)*

COMMAND LINE OPTIONS

Specify internal point creates a boundary based on the point you pick.

Advanced options options
Enter an option [Boundary set/Island detection/Object type]: *(Enter an option.)*

Boundary set defines the objects **-Boundary** analyzes when defining a boundary from a specified pick point. It chooses from a new set of objects, or all objects visible in the current viewport.

Island detection:

- **On** uses objects within the outermost boundary as boundary objects.
- **Off** fills objects within the outermost boundary.

Object type specifies polyline or region as the boundary object.

RELATED COMMANDS

Hatch creates boundaries for hatches and fills.

PLine draws polylines.

PEdit edits polylines.

Region creates 2D regions from a collection of objects.

RELATED SYSTEM VARIABLE

HpBound specifies the default object used to create boundary:

   0 — Draw as region.
   1 — Draw as polyline (default).

TIPS

- Use this command to measure irregular areas:

   1. Apply the Boundary command to an irregular area.
   2. Use the Area command to find the area and perimeter of the boundary.

- Use Boundary together with the Offset command to help create *poching*, areas partially covered by hatching.

Box

Rel.13 Draws a 3D box as a solid model.

Command	Alias	Ctrl+	F-key	Alt+	Menu Bar	Dashboard
box	...	...	...	DMB	Draw	3D Make
					⤷ Modeling	
					⤷ Box	

Command: box
Specify first corner or [Center]: *(Pick point 1, or type the C option.)*
Specify other corner or [Cube/Length]: *(Pick point 2, or enter an option.)*
Specify height or [2Point] <-1.2757>: *(Pick point 3, or type the 2 option.)*

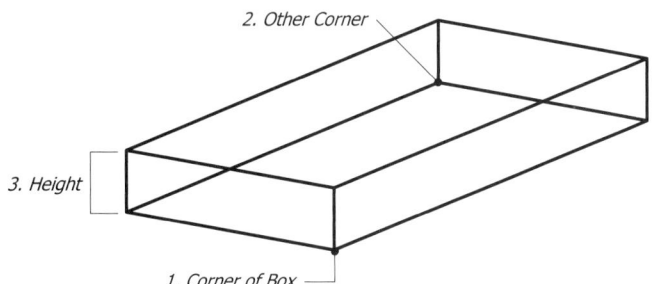

2. Other Corner

3. Height

1. Corner of Box

COMMAND LINE OPTIONS

First corner specifies one corner for the base of box.

Center draws the box about a center point.

Other corner specifies the second corner for the base of box.

Cube draws a cube box — all sides have the same length.

Length specifies the length along x axis, width along y axis, and height along z axis.

Height specifies the height of the box.

2point specifies height as the distance between two points.

RELATED COMMANDS

Ai_Box draws 3D surface model boxes.

Cone draws 3D solid cones.

Cylinder draws 3D solid tubes.

Sphere draws 3D solid balls.

Torus draws 3D solid donuts.

Wedge draws 3D solid wedges.

RELATED SYSTEM VARIABLES

DispSilh displays 3D objects as silhouettes after hidden-line removal and shading.

DragVs specifies the visual style during construction of 3D solid primitives..

TIP

- Once the box is placed in the drawing, you can edit it interactively. Select the box, and then notice the grips. Select a grip, which turns red, and then move it. Press Esc to exit direct editing.

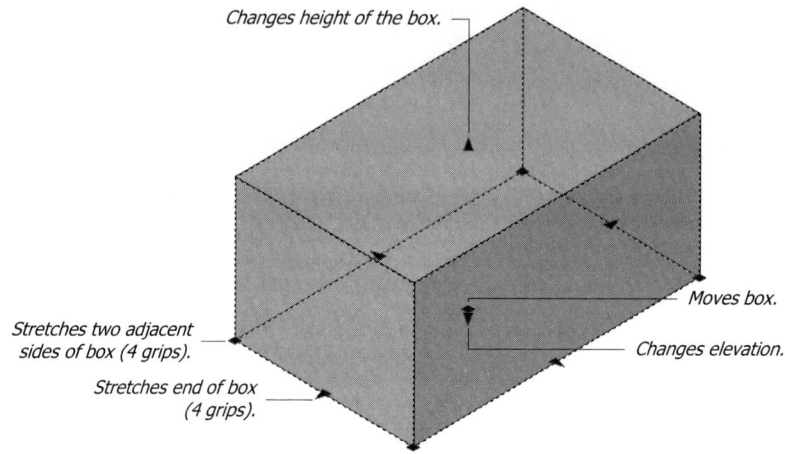

Changes height of the box.

Stretches two adjacent sides of box (4 grips).

Stretches end of box (4 grips).

Moves box.

Changes elevation.

Break

V. 1.4 Removes portions of objects.

Command	Alias	Ctrl+	F-key	Alt+	Menu Bar	Dashboard
break	br	...	...	MK	Modify	2D Draw
					⌐Break	

Command: break
Select object: *(Select one object — point 1.)*
Specify second break point or [First point]: *(Pick point 2, or type F.)*

1. First Point   *2. Second Point*

Breaking a polyline at two points.

COMMAND LINE OPTIONS

Select object selects one object to break; the pick point becomes the first break point, unless the F option is used at the next prompt.

@ uses the first break point's coordinates for the second break point.

First Point options
Enter first point: *(Pick a point.)*
Enter second point: *(Pick a point.)*

First point specifies the first break point.

Second point specifies the second break point.

RELATED COMMANDS

Change changes the length of lines.

PEdit removes and relocates vertices of polylines.

Trim shortens the length of open objects.

TIPS

- Use this command to convert circles into arcs; pick the break points clockwise to keep the portion between pick points. Use the Join command to convert arcs back to circles.

- This command can erase a portion of an object (as shown in the figure above) or remove the end of an open object.

- The second point does not need to be on the object; AutoCAD breaks the object at the point nearest to the pick point.

- The Break command works on the following objects: lines, arcs, circles, polylines, ellipses, rays, xlines, and splines, as well as objects made of polylines, such as donuts and polygons.

BRep

2007 Removes construction history from 3D solids; short for "Boundary REPresentation."

Command	Alias	Ctrl+	F-key	Alt+	Menu Bar	Dashboard
brep	...	...	...	...	...	...

Command: brep
Select 3D solids: *(Select one or more solid models.)*
Select 3D solids: *(Press Enter to exit command.)*

Box

Cone

3D Solid Model

Left: *Before applying BRep, the model's original primitives can be selected.*
Right: *After applying BRep, the history of original primitives disappears.*

COMMAND LINE OPTION

Select 3D solids selects one or more objects from which to remove history.

RELATED SYSTEM VARIABLES

ShowHist toggles the preservation of history in solid models.

SolidHist toggles the display of primitive history entities.

Browser, Browser2

Rel.14 Prompts for a Web address, and then launches the Web browser.

Commands	Alias	Ctrl+	F-key	Alt+	Menu Bar	Dashboard
browser	...	...	...	...	...	Y8
browser2						

Command: browser
Enter Web location (URL) <http://www.autodesk.com>: *(Enter a Web address.)*
Launches browser:

COMMAND LINE OPTION

Enter Web location specifies the URL; see the AttachURL command for information about URLs.

RELATED SYSTEM VARIABLE

InetLocation contains the name of the default URL.

TIPS

- *URL* is short for "uniform (or universal) resource locator," the universal file-naming system used on the Internet; also called a link or hyperlink. An example of a URL is http://www.autodeskpress.com, the Autodesk Press Web site.

- Many file dialog boxes also give you access to the Web browser; see the Open command.

- The undocumented Browser2 command also opens URLs.

'Cal

Rel.12 Calculates algebraic and vector geometry at the command line (*short for CALculator*).

Command	Alias	Ctrl+	F-key	Alt+	Menu Bar	Dashboard
cal	...	...	...	...	...	...

Command: cal
>>Expression: (*Enter an expression, and then press Enter.*)

COMMAND LINE OPTIONS

( )	Grouping of expressions.	yzof	y,z coordinates of a point.
[ ]	Vector expressions.	xof	x coordinate of a point.
		yof	y coordinate of a point.
+	Addition.	zof	z coordinate of a point.
-	Subtraction.	rxof	Real x coordinate of a point.
*	Multiplication.	ryof	Real y coordinate of a point.
/	Division.	rzof	Real z coordinate of a point.
^	Exponentiation.	cur	x, y, z coordinates of a picked point.
&	Vector product of vectors.	rad	Radius of object.
sin	Sine.	pld	Point on line, distance from.
cos	Cosine.	plt	Point on line, using parameter *t*.
tang	Tangent.	rot	Rotated point through angle
asin	Arc sine.		about origin.
acos	Arc cosine.	ill	Intersection of two lines.
atan	Arc tangent.	ilp	Intersection of line and plane.
ln	Natural logarithm.	dist	Distance between two points.
log	Logarithm.	dpl	Distance between point and line.
exp	Natural exponent.	dpp	Distance between point and plane.
exp10	Exponent.	ang	Angle between lines.
sqr	Square.	nor	Unit vector normal.
sqrt	Square root.	vec	Vector translation between two points.
abs	Absolute value.	vec1	Unit vector direction.
round	Round off.	dee	Distance between two endpoints.
trunc	Truncate.	ille	Intersection of two lines defined
cvunit	Converts units using *acad.unt*.		by endpoints.
w2u	WCS to UCS conversion.	mee	Midpoint between two endpoints.
u2w	UCS to WCS conversion.	nee	Unit vector in the x,y-plane normal
r2d	Radians-to-degrees conversion.		to two endpoints
d2r	Degrees-to-radians conversion.	vee	Vector from two endpoints.
pi	The value PI (3.14159).	vee1	Unit vector from two endpoints.
xyof	x,y coordinates of a point.	end	Endpoint object snap.
xzof	x,z coordinates of a point.	ins	Insertion point object snap.

int	Intersection object snap.		qua	Quadrant object snap.
mid	Midpoint object snap.		per	Perpendicular object snap.
cen	Center object snap.		tan	Tangent object snap.
nea	Nearest object snap.			
nod	Node object snap.		ESC	Exits Cal mode.

RELATED COMMAND

QuickCal displays a graphical user interface for the Cal command.

RELATED SYSTEM VARIABLES

UserI1 — UserI5 store integers.

UserR1 — UserR5 store real numbers.

TIPS

- To use Cal, type expressions at the >> prompt.

  For example, to find the area of a circle (pi*r^2) with radius of 1.2 units, enter the following:

  Expression >> pi*(1.2^2)
  4.52389

- **Cal** recognizes these prefixes:

*	Scalar product of vectors.
&	Vector product of vectors.

- And the following suffixes:

r	Radian (degrees is the default).
g	Grad.
'	Feet (unitless distance is the default).
"	Inches.

- Because 'Cal is a transparent command, it can perform calculations in the middle of other commands, and then return the value to that command.

  For example, to set the offset distance to the radius of a circle:

  Command: offset
  Specify offset distance or [Through] <Through>: 'cal
  >> Expression: rad
  >> Select circle, arc or polyline segment for RAD function: *(Pick circle.)* 2.0
  Select object to offset or <exit>: *(And so on.)*

- This command works with real numbers and integers. The smallest and largest integers are -32768 and 32767.

Camera

2000 Creates 3D perspective views.

Command	Alias	Ctrl+	Alt+	Menu Bar	Dashboard
camera	cam	...	VT	View ↳Create Camera	3D Navigate

Command: camera
Current camera settings: Height=0'-0" Lens Length=50.0 mm
Specify camera location: *(Pick a point, or enter coordinates.)*
Specify target location: *(Pick another point, or enter coordinates.)*
**Enter an option [?/Name/LOcation/Height/Target/LEns/Clipping/View/
eXit]<eXit>:** *(Enter an option, or press Enter to exit the command.)*

Turn on the camera glyph with the CameraDisplay system variable.

COMMAND LINE OPTIONS

Specify camera location specifies the x, y, z coordinates for the camera ("look from").
Specify target location specifies the x, y, z coordinates for the target ("look at").

? option
Enter camera name(s) to list <*> lists the names of cameras stored in the drawing, which are accessed through the **View** command.

```
Camera Name
----------------
Camera1
Camera2
```

Name option
Enter name for new camera specifies the name for the current camera-target position, which it saves for later reuse.

LOcation option
Specify camera location locates the camera using x,y,z coordinates.

Height option
Specify camera height elevates the camera above the x,y-plane; saved in the CameraHeight system variable.

Target option
Specify target location locates the target point and determines the camera's direction.

LEns option
Specify lens length in mm specifies the angle of view; 50mm is normal; smaller numbers (such as 35) widen the view but also distort it, while larger numbers (such as 100) narrow the view and foreshorten it.

Clipping options
Enable front clipping plane? toggles the front clipping plane, which is used to cut off part of the view.

Specify front clipping plane offset from target plane specifies the location of the plane.

Enable back clipping plane? toggles the rear clipping plane.

Specify back clipping plane offset from target plane specifies the location of the plane.

View option
Switch to camera view? changes the viewpoint to that of the camera.

RELATED COMMANDS
DView has a CAmera option, which interactively sets a new 3D viewpoint.

View lists the names of saved camera views.

VPoint sets a 3D viewpoint through x, y, z coordinates or angles.

3dOrbit sets a new 3D viewpoint interactively.

RELATED SYSTEM VARIABLES
CameraDisplay toggles the display of the camera glyphs.

CameraHeight specifies the height of the camera above the x,y-plane.

TIPS
- Colors of the camera and field of view can be changed in the Options dialog box.

- Saved cameras are listed in the **View** command's dialog box.

- After placing a camera, select its glyph to edit the viewpoint and to display the Camera Preview window:

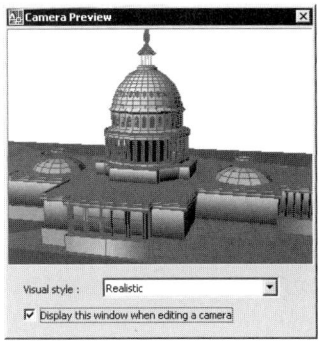

- Grips are used interactively to adjust the camera, target, and field of view. If clipping planes were specified, they can also be adjusted using grips.

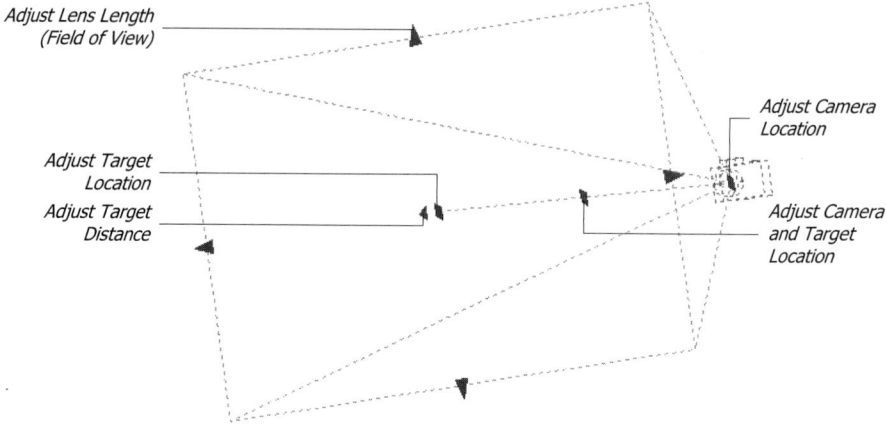

- Double-click a camera glyph to display the Properties:

Left: *Camera properties in the Properties palette.*
Right: *Camera properties in the View Manager dialog box.*

- The View command also lists the names of cameras and their properties, but provides access to many more options.

- The 3D Navigation toolbar contains a droplist with the names of stored camera views. The Camera Adjustment toolbar has nothing to do with this command.

- Entering the Camera command automatically turns on the perspective viewpoint.

Chamfer

V. 2.1 Bevels the intersection of two lines, all vertices of 2D polylines, and the faces of 3D solid models.

Command	Alias	Ctrl+	F-key	Alt+	Menu Bar	Dashboard
chamfer	cha	...	...	MCC⬏	Modify ⬏Chamfer	2D Draw

Command: chamfer
(TRIM mode) Current chamfer Dist1 = 0.0, Dist2 = 0.0
Select first line or [Undo/Polyline/Distance/Angle/Trim/mEthod/Multiple]: *(Pick an object, or enter an option.)*
Select second line: *(Pick an object.)*

 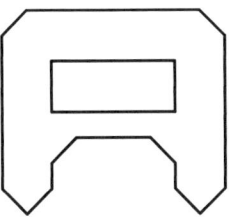

Left: Original object drawn with polylines.
Right: Chamfered with the Polyline option.

COMMAND LINE OPTIONS

Select first line selects the first line, arc, face, or edge.

Select second line selects the second line, arc, face, or edge.

Undo reverses the last chamfer operation within the command.

Multiple allows more than one pair of lines to be chamfered.

Polyline options
Select 2D polyline: *(Pick a polyline.)*
***n* lines were chamfered**

Select 2D polyline chamfers *all* segments of a 2D polyline; if the polyline is not closed with the **Close** option, the first and last segments are not chamfered.

Distance options
Specify first chamfer distance <0.5000>: *(Enter a value.)*
Specify second chamfer distance <0.5000>: *(Enter a value.)*

First distance specifies the chamfering distance along the line picked first.

Second distance specifies the chamfering distance along the line picked second.

Angle options
Specify chamfer length on the first line <1.0000>: *(Enter a distance.)*
Specify chamfer angle from the first line <0>: *(Enter an angle.)*

Chamfer length specifies the chamfering distance along the line picked first.

Chamfer angle specifies the chamfering angle by an angle from the line picked first.

Trim options
Enter Trim mode option [Trim/Notrim] <Trim>: *(Type T or N.)*

   Trim trims or extends lines, edges, and faces after chamfer.

   No trim does not trim lines, edges, and faces after chamfer.

Original Lines (before chamfering)   *Chamfer (with trim)*   *Chamfer (with no trim)*

Intersecting lines before and after chamfering.

Method options
Enter trim method [Distance/Angle] <Angle>: *(Type D or A.)*

   Distance determines chamfer by two specified distances.

   Angle determines chamfer by the specified angle and distance.

. . . .

Chamfering 3D Solids — Edge Mode Options
Command: chamfer
(TRIM mode) Current chamfer Dist1 = 0.0, Dist2 = 0.0
Polyline/.../<Select first line>: *(Pick a 3D solid model.)*
Select base surface: *(Pick surface edge on the solid model — see 1, below.)*
Next/<OK>: *(Type N, or select the next surface, or OK to end selection.)*
Enter base surface distance <0.0>: *(Enter a value.)*
Enter other surface distance <0.0>: *(Enter a value.)*
Loop/<Select edge>: *(Select an edge — see 2, below.)*
Loop/<Select edge>: *(Press Enter.)*

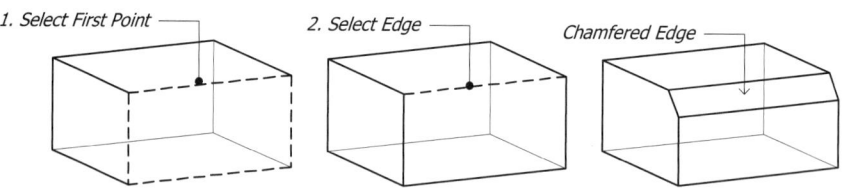

1. Select First Point   *2. Select Edge*   *Chamfered Edge*

. . . .

Chamfering 3D Solids — Loop Mode Options
Command: chamfer
(TRIM mode) Current chamfer Dist1 = 0.0, Dist2 = 0.0
Polyline/.../<Select first line>: *(Pick a 3D solid model.)*
Select base surface: *(Pick a surface edge — see 1, following.)*
Next/<OK>: *(Type N, or select the next surface, or OK to end selection.)*
Enter base surface distance <0.0>: *(Enter a value.)*
Enter other surface distance <0.0>: *(Enter a value.)*

. . . .

Loop/<Select edge>: *(Type L.)*
Edge/<Select edge loop>: *(Pick a surface edge — see 2, below.)*
Edge/<Select edge loop>: *(Press Enter.)*

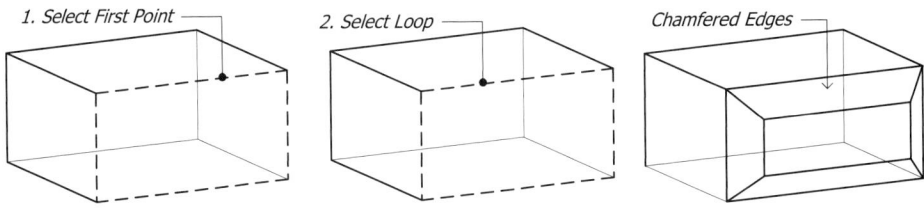

1. Select First Point   2. Select Loop   Chamfered Edges

COMMAND LINE OPTIONS

Select first line selects the 3D solid.

Next selects the adjacent face, or press Enter to accept face.

Enter base surface distance specifies the first chamfer distance.

Enter other surface distance specifies the second chamfer distance.

Loop selects all edges of the face.

Select edge selects a single edge of the face.

RELATED COMMANDS

Fillet rounds the intersection with a radius.

SolidEdit edits the faces and edges of solids.

RELATED SYSTEM VARIABLES

ChamferA is the first chamfer distance.

ChamferB is the second chamfer distance.

ChamferC is the length of chamfer.

ChamferD is the chamfer angle.

ChamMode toggles chamfer measurement:

   0 — Chamfer by two distances (default).
   1 — Chamfer by distance and angle.

TrimMode determines whether lines/edges are trimmed after chamfer:

   0 — Do not trim selected edges.
   1 — Trim selected edges (default).

TIPS

- When TrimMode is set to 1 and lines do not intersect, Chamfer extends or trims the lines to intersect before chamfering.

- When the two objects are not on the same layer, Chamfer places the chamfer line on the current layer; the chamfer line is placed on the current layer.

Change

V. 1.0   Modifies the color, elevation, layer, linetype, linetype scale, lineweight, plot style, material, annotative scaling, and thickness of most objects, and additional properties of lines, circles, blocks, text, and attributes.

Command	Alias	Ctrl+	F-key	Alt+	Menu Bar	Dashboard
change	-ch	...	...	...	...	...

Command: change
Select objects: *(Pick one or more objects.)*
Select objects: *(Press Enter.)*
Specify change point or [Properties]: *(Pick a point, an object, or type P.)*
Enter property to change [Color/Elev/LAyer/LType/ltScale/LWeight/Thickness/Material/Plotstyle/Annotative]: *(Enter an option.)*

COMMAND LINE OPTIONS

Specify change point selects the object to change:

(pick a line) indicates the new length of lines.

(pick a circle) indicates the new radius of circles.

(pick a block) indicates the new insertion point or rotation angle of blocks.

(pick text) indicates the new location of text.

(pick an attribute) indicates the attribute's new text insertion point, text style, height, rotation angle, text, tag, prompt, or default value.

ENTER changes the insertion point, style, height, rotation angle, and text of text strings.

Properties options

Color changes the color of objects.

Elev changes the elevation of objects.

LAyer moves the object to different layers.

LType changes the linetype of objects.

ltScale changes the scale of linetypes.

LWeight changes the lineweight of objects.

Thickness changes the thickness of objects, except blocks and 3D solids.

Material changes the material assigned to objects.

Plotstyle changes the plot style of objects (available only when plot styles are turned on).

Annotative sets the annotative property for the objects *(new to AutoCAD 2008)*.

RELATED COMMANDS

AttRedef changes blocks and attributes.

ChProp contains the properties portion of the Change command.

Color changes the current color setting.

Elev changes the working elevation and thickness.

LtScale changes the linetype scale.

Properties changes most aspects of all objects.

PlotStyle sets the plot style.

Materials assigns materials and textures to objects.

ObjectScale assigns annotative scale factors to objects.

RELATED SYSTEM VARIABLES

CAnnoScale contains the current annotative scale (*new to AutoCAD 2008*).

CeColor contains the current color setting.

CeLType contains the current linetype setting.

CelWeight contains the current lineweight.

CircleRad contains the current circle radius.

CLayer contains the name of the current layer.

CMaterial contains the name of the default material.

CPlotstyle contains the name of the current plot style.

Elevation contains the current elevation setting.

LtScale contains the current linetype scale.

TextSize contains the current height of text.

TextStyle contains the current text style.

Thickness contains the current thickness setting.

TIPS

- The Change command cannot change:

  The size of donuts.
  The radius and length of arcs.
  The length of polylines.
  The justification of text.

- Use this command to change the endpoints of groups of lines to a common vertex:

Left: Original lines.
Center: *Line endpoints changed with ortho mode turned off.*
Right: *Line endpoints changed with ortho on.*

- Turn on ortho mode to extend or trim a group of lines, without needing a cutting edge (as do the Extend and Trim commands).

- The PlotStyle option is not displayed when plot styles are not turned on.

'CheckStandards

2001 Checks drawings for adherence to standards previously specified by the
Standards command.

Command	Alias	Ctrl+	F-key	Alt+	Menu Bar	Dashboard
checkstandards	chk	...	...	TSK	Tools	...
					⤷CAD Standards	
					⤷Check	

Command: checkstandards

*When the **Standards** command has not set up standards for the drawing, AutoCAD displays this error message:*

*When standards have been set up for the drawing, this command checks the drawing against the CAD
standards, and then displays this dialog box:*

*Click **OK**.*

*To change settings in the Configure Standards dialog box, click **Settings**.*

CAD STANDARDS SETTINGS DIALOG BOX OPTIONS

Notification Settings options

- ○ Disables standards notifications.
- ○ Displays alert upon standards violation.

⊙ Displays standards status bar icon.

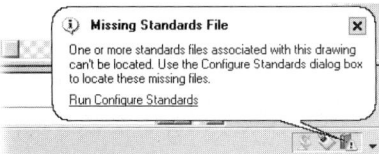

Check Standards Settings options

Automatically fix non-standard properties:

☑ Fixes properties not matching the CAD standard automatically.

☐ Steps manually through properties not matching the CAD standard.

Show ignored problems:

☑ Displays problems marked as ignored.

☐ Does not display ignored problems.

Preferred standards file to use for replacements selects the default *.dws* file.

CHECK STANDARDS DIALOG BOX

When an object does not match the standards, AutoCAD displays this dialog box:

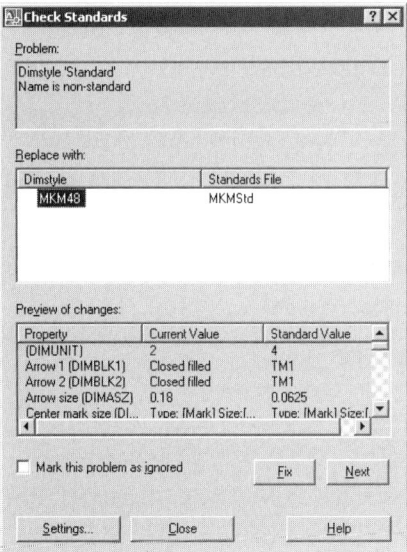

Problem describes properties in drawings that do not match the standard; this dialog box displays one problem at a time.

Replace With lists linetypes, text styles, etc. found in the *.dws* standards file.

Mark this problem as ignored:

☑ Ignores nonstandard properties, and marks them with the user's login name; some errors are always ignored by AutoCAD, such as settings for layer 0 and DefPoints.

☐ Does not ignore nonstandard properties.

Fix replaces the nonstandard property with the selected standard; the color check mark icon means fixes are available.

Next displays the next nonstandard property.

Settings displays the Check Standards Settings dialog box.

Close closes the dialog box; displays this warning dialog box if standards are not fully checked:

TRAY ICON

When drawings contain drawings standards, AutoCAD displays a book icon on the tray:

Associated Standards
File(s)

*Click the icon to start the **CheckStandards** command. Right-clicking the icon displays the shortcut menu:*

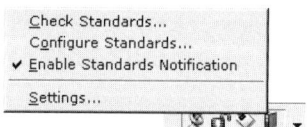

Check Standards runs the CheckStandards command.

Configure Standards runs the Standards command.

Enable Standards Notification toggles the display of the balloon.

Settings displays the CAD Standards Settings dialog box.

RELATED COMMANDS

Standards selects *.dws* standards files.

LayTrans translates layers between drawings.

RELATED SYSTEM VARIABLE

StandardsViolation determines whether alerts are displayed when a CAD standard is violated in the current drawing.

RELATED PROGRAM

DwgCheckStandards.Exe is the external Batch Standards Checker program that checks one or more drawings at a time. It checks drawings without needing AutoCAD.

RELATED FILES

.dws drawing standards file; stored in DWG format.

.chs standard check file; stored in XML format.

TIP

■ While this dialog box is open, you can use the following shortcut keys:

| F4 | Fix problem. |
| F5 | Next problem. |

ChProp

Rel.10 Modifies the color, layer, linetype, linetype scale, lineweight, plot style, annotative scale, and thickness of most objects (short for CHange PROPerties).

Command	Alias	Ctrl+	F-key	Alt+	Menu Bar	Dashboard
chprop	...	...	...	...	...	...

Command: chprop
Select objects: *(Select one or more objects.)*
Select objects: *(Press Enter.)*
Enter property to change [Color/LAyer/LType/ltScale/LWeight/Thickness/Plotstyle/Material/Annotative]: *(Enter an option.)*

COMMAND LINE OPTIONS

Color changes the color of objects.

LAyer moves objects to a different layer.

LType changes the linetype of objects.

ltScale changes the linetype scale.

LWeight changes the lineweight of objects.

Thickness changes the thickness of all objects, except blocks.

PlotStyle changes the plot style assigned to the objects.

Material changes the material assigned to objects.

Annotative sets the annotative property for the objects *(new to AutoCAD 2008)*.

RELATED COMMANDS

Change changes lines, circles, blocks, text and attributes.

Color changes the current color setting.

Elev changes the working elevation and thickness.

LtScale changes the linetype scale.

LWeight sets the lineweight options.

ObjectScale assigns annotative scale factors to objects.

Properties changes most aspects of all objects.

PlotStyle sets the plot style.

RELATED SYSTEM VARIABLES

See the Change command.

TIPS

- This command cannot change the elevation of object, so use the Change or Properties commands instead.

- The Plotstyle option is not displayed when plot styles are not turned on.

ChSpace

2007 Moves objects between paper and model space, scaling them appropriately; a former Express Tool (short for CHange SPACE).

Command	Alias	Ctrl+	F-key	Alt+	Menu Bar	Dashboard
chprop	...	...	...	...	...	...

Command: chprop

*In model tab, AutoCAD responds, "** Command not allowed in Model Tab **."*

In layout tabs, the command continues:

Select objects: *(Select one or more objects.)*
Select objects: *(Press Enter.)*
n **object(s) changed from PAPER space to MODEL space.**
Objects were scaled by a factor of 192.000000708072 to maintain visual appearance.

COMMAND LINE OPTION

Select objects selects the objects to be moved.

Warning! Objects are moved, not copied, which means they no longer appear in the other space.

RELATED COMMAND

SpaceTrans converts the height of text between model and paper space.

Circle

V. 1.0 Draws 2D circles by five different methods.

Command	Alias	Ctrl+	F-key	Alt+	Menu Bar	Dashboard
circle	c	...	...	DC	Draw	2D Draw
					↳Circle	

Command: circle
Specify center point for circle or [3P/2P/Ttr (tan tan radius)]: *(Pick a point, or enter an option.)*
Specify radius of circle or [Diameter]: *(Pick a point, or type D.)*

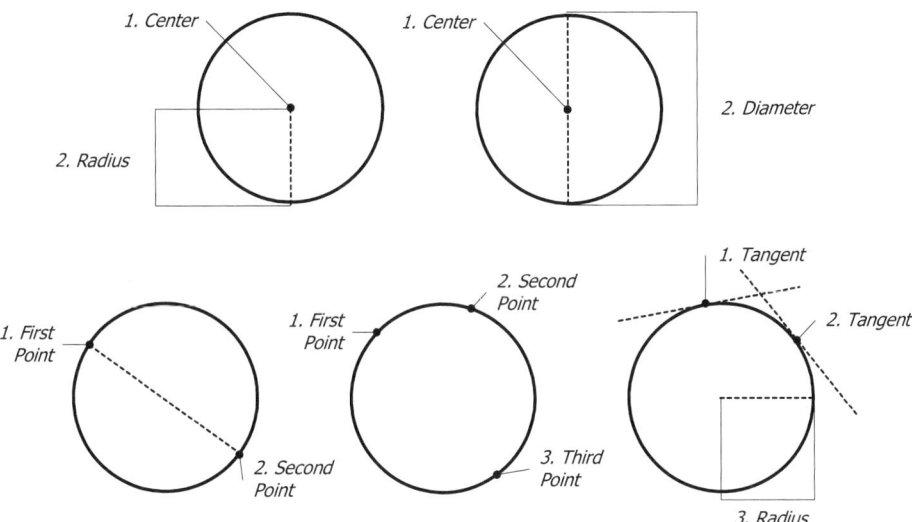

COMMAND LINE OPTIONS

Center and Radius or Diameter Options
Specify center point for circle or [3P/2P/Ttr (tan tan radius)]: *(Pick a point, or enter an option.)*
Specify radius of circle or [Diameter] <0.5>: *(Pick a point, or type D.)*
Specify diameter of circle <0.5>: *(Enter a value.)*

Center point indicates the circle's center point.

Radius indicates the circle's radius.

Diameter indicates the circle's diameter.

3P (three-point) Options
Specify first point on circle: *(Pick a point.)*
Specify second point on circle: *(Pick a point.)*
Specify third point on circle: *(Pick a point.)*

First point indicates the first point on the circle.

Second point indicates the second point on the circle.

Third point indicates the third point on the circle.

2P (two-point) Options
Specify first end point of circle's diameter: *(Pick a point.)*
Specify second end point of circle's diameter: *(Pick a point.)*

First end point indicates the first point on the circle.

Second end point indicates the second point on the circle.

TTR (tangent-tangent-radius) Options
Specify point on object for first tangent of circle: *(Pick a point.)*
Specify point on object for second tangent of circle: *(Pick a point.)*
Specify radius of circle <0.5>: *(Enter a radius.)*

First tangent indicates the first point of tangency.

Second tangent indicates the second point of tangency.

Radius indicates the first point of radius.

Tan Tan Tan *(menu-only option)*:
From the Draw menu, select Circle | Tan,Tan,Tan. AutoCAD prompts:
Command: _circle Specify center point for circle or [3P/2P/Ttr (tan tan radius)]: _3p Specify first point on circle: _tan to *(Pick an object.)*
Specify second point on circle: _tan to *(Pick an object.)*
Specify third point on circle: _tan to *(Pick an object.)*

AutoCAD draws a circle tangent to the three points, if possible.

RELATED COMMANDS

Ai_CircTan draws circles tangent to three objects.

Arc draws arcs.

Donut draws solid-filled circles or donuts.

Ellipse draws elliptical circles and arcs.

Sphere draws 3D solid balls.

RELATED SYSTEM VARIABLE

CircleRad specifies the default circle radius.

TIPS

- The 3P (three-point) circle defines three points on the circle's circumference.

- When drawing TTR (tangent, tangent, radius) circles, AutoCAD draws the circles with tangent points closest to the pick points; note that more than one circle placement is possible.

- Using the TTR option automatically turns on the TANgent object snap.

- Giving circles thickness turns them into hollow cylinders.

CleanScreenOn, CleanScreenOff

2004 Maximizes the drawing area.

Command	Alias	Ctrl+	F-key	Alt+	Menu Bar	Dashboard
cleanscreenon	...	**0** *(zero)*	...	**VCC**	**View**	...
					⤷**Clean Screen**	
cleanscreenoff	...	**0**	...	**VCC**	**View**	...
					⤷**Clean Screen**	

Command: cleanscreenon

AutoCAD turns off the title bar, toolbars, and window edges, and maximizes the AutoCAD window to the full size of your computer's screen:

Command: cleanscreenoff

AutoCAD returns to normal.

TIPS

- Alternatively, you can click the Cleanscreen icon in the tray on the status bar.

- For an even larger drawing area, turn off the scroll bars and layout tabs through the Options | Display dialog box, and then drag the command prompt area into a window.

- To toggle this command, press Ctrl+0 (zero).

Close, CloseAll

2000 Closes the current drawing, or all drawings; does not exit AutoCAD.

Command	Alias	Ctrl+	F-key	Alt+	Menu Bar	Dashboard
close	...	F4	...	FC	File ⤷Close	...
closeall	...	...	...	WL	Window ⤷Close All	...

Command: close

Closes the current drawing. When drawings are not saved since last change, displays dialog box:

Command: closeall

Closes all open drawings.

AutoCAD DIALOG BOX

Yes displays the Save Drawing As dialog box, or saves the drawing if it has been previously saved.

No exits the drawing without saving it.

Cancel returns to the drawing.

RELATED COMMANDS

Quit exits AutoCAD.

Open opens additional drawings, each in its own window.

AutoPublish exports drawing as DWF files when the drawing is closed.

RELATED SYSTEM VARIABLES

DbMod reports whether the drawing has been modified since the last save.

AutoDwfPublish exports drawing in DWF format, when set to 1.

TIPS

- As an alternative to this command, you can click the **x** button on the title bar:

- If AutoPublish is turned on, then AutoCAD automatically publishes the drawing as a DWF file when you close the drawing (*new to AutoCAD 2008*). See the AutoPublish command.

'Color

V. 2.5 Sets the new working color.

Commands	Aliases	Ctrl+	F-key	Alt+	Menu Bar	Dashboard
color	ddcolor	...	...	OC	Format	Object Properties
	col				↳Color	
	colour					
-color						

Command: color

Displays dialog box:

SELECT COLOR DIALOG BOX

Index Color tab

AutoCAD Color Index (ACI) selects one of AutoCAD's colors (short for "AutoCAD Color Index").

Bylayer sets the color to BYLAYER (color 256); color 257 is ByEntity in some situations.

Byblock sets the color to BYBLOCK.

Color sets the color by number or name.

True Color tab

Hue selects the hue (color), ranging from 0 (red) to 360 (violet).

Saturation selects the saturation (intensity of color), ranging from 0 (gray) to 100 (color).

Luminance selects the luminance (brightness of color), ranging from 0 (white) to 100 (black).

Color specifies the color number as hue, saturation, luminance.

Color model selects the type of color model:

• **HSL** is hue, saturation, luminance.

- **RGB** is red, green, blue.

Left: *HSL color picker.*
Right: *RGB color picker.*

Red selects the range of red from 0 to 255.

Green selects the range of green from 0 to 255.

Blue selects the range of blue from 0 to 255.

Color specifies the color number as red, green, blue.

Color Book tab

Color book selects a predefined collection of colors.

Color specifies the name of the Pantone, RAL, and DIC colors.

Historical notes: The **RAL** color system was designed in 1927 to standardize colors by limiting the number of color gradations, at first to just 30, but now to over 1,600. RAL (Reichs Ausschuß für Lieferbedingungen – German for "Imperial Committee for Supply Conditions") is administered by the German Institute for Quality Assurance and Labeling <www.ral.de>.

The **Pantone Color System** was designed in 1963 to specify color for graphic arts, textiles, and plastics, based on the assumption that colors are seen differently by different individuals. Designers typically work with Pantone's fan-format book of standardized colors <www.pantone.com>.

The **DIC Color Guide** is the Japanese standard for colors, developed by Dainippon Ink and Chemicals <www.dic.co.jp/eng/index.html>.

-COLOR Command

Command: -color

Enter default object color [Truecolor/COlorbook] <BYLAYER>: *(Enter a color number or name, or enter an option.)*

COMMAND LINE OPTIONS

BYLAYER sets the working color to the color of the current layer.

BYBLOCK sets the working color of inserted blocks.

color number sets the working color using number (1 through 255), name, or abbreviation:

Color Number	Color Name	Abbreviation	Comments
1	Red	R	
2	Yellow	Y	
3	Green	G	
4	Cyan	C	
5	Blue	B	
6	Magenta	M	
7	White	W	Or black.
8 - 249			Additional colors.
250 - 255			Shades of gray.

Truecolor Options

Red, Green, Blue: *(Specify color values separated by commas.)*

Red, Green, Blue specifies color by red, green, and blue in the range from 0 to 255:

Red, Green, Blue: 255,128,0

COlorbook Options

Enter Color Book name: *(Specify a name, such as Pantone.)*
Enter color name: *(Enter a color name, such as 11-0103TC.)*

Enter Color Book name specifies the name of a color book.

Enter color name specifies the name of the color.

RELATED COMMANDS

ChProp changes the color via the command line in fewer keystrokes.

Properties changes the color of objects via a dialog box.

RELATED SYSTEM VARIABLE

CeColor is the current object color setting.

TIPS

- 'BYLAYER' means that objects take the color assigned to that layer. When BYLAYER objects on layer 0 are part of a block definition, they take on the properties of whichever layer the block was inserted on, or the layer to which the block was moved.

- 'BYBLOCK' objects in a block definition take on the color assigned to the block. This may be the color in effect at the time the block was inserted, or the color to which the block was changed.

- The Layer command's VP Color overrides allow different colors in each viewport *(new to AutoCAD 2008)*.

- White objects display as black when the background color is white. Color "0" cannot be specified; AutoCAD uses it internally as the background color.

- The *colorwh.dwg* drawing has 255- and 16.7 million-color wheels.

CommandLine, CommandLineHide

<u>**2006**</u>   Toggles the display of the command line window.

Command	Alias	Ctrl+	F-key	Alt+	Menu Bar	Dashboard
commandline	cli	9	...	T	Tools	...
					⮑ Command Line	
commandlinehide		9				

Command: commandline

Displays the command line window:

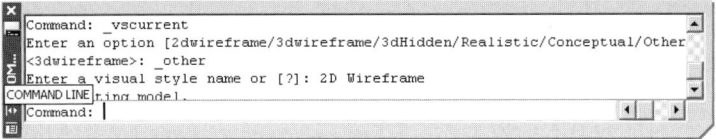

Command: commandlinehide

Hides the command line window.

RELATED SYSTEM VARIABLES

CmdInputHistory specifies the maximum number of previous commands remembered by AutoCAD.

CmdNames reports the name of the current command.

TIPS

▪ When the command line window is missing, you can type commands and see them in the drawing area when the DYN button is depressed on the status bar, and system variable DynMode is set to 1 or 3.

▪ The command line window normally shows three lines of history; you can stretch the window to show a greater or lesser number of lines.

▪ The window can be made translucent: right-click the title bar, and then select Transparency from the shortcut menu. This does not work when AutoCAD is using hardware acceleration.

▪ Press F2 to display the text window, which displays the last 400 lines of command history.

Compile

Rel.12 Compiles *.shp* shape and font files and *.pfb* font files into *.shx* files.

Command	Alias	Ctrl+	F-key	Alt+	Menu Bar	Dashboard
compile	...	...	...	...	...	...

Command: compile

Displays Select Shape or Font File dialog box. Select a .shp or .pfb file, and then click Open.

DIALOG BOX OPTIONS

Open opens the *.shp* or *.pfb* file for compiling.

Cancel closes the dialog box without loading the file.

RELATED COMMANDS

Load loads compiled *.shx* shape files into the current drawing.

Style loads *.shx* and *.ttf* font files into the current drawing.

RELATED SYSTEM VARIABLE

ShpName is the current *.shp* file name.

RELATED FILES

*.shp are AutoCAD font and shape source files.

*.shx are AutoCAD compiled font and shape files.

*.pfb are PostScript Type B font files.

TIPS

- As of AutoCAD Release 12, Style converts *.shp* font files on-the-fly; it is only necessary to use the Compile command to obtain *.shx* font files.

- Prior to Release 12, this command was an option on the number menu system that appeared when AutoCAD started up.

- As of Release 14, AutoCAD no longer directly supports PostScript font files. Instead, use the Compile command to convert *.pfb* files to *.shx* format.

- TrueType fonts are not compiled.

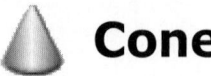

Cone

Rel.11 Draws 3D solid cones with circular or elliptical bases.

Command	Alias	Ctrl+	F-key	Alt+	Menu Bar	Dashboard
cone	...	...	...	DMO	Draw ↳Modeling ↳Cone	3D Make

Command: cone
Specify center point of base or [3P/2P/Ttr/Elliptical]: *(Pick center point 1.)*
Specify base radius or [Diameter] <2">: *(Specify radius 2, or type D.)*
Specify height or [2Point/Axis endpoint/Top radius] <2">: *(Specify height 3, or type A.)*

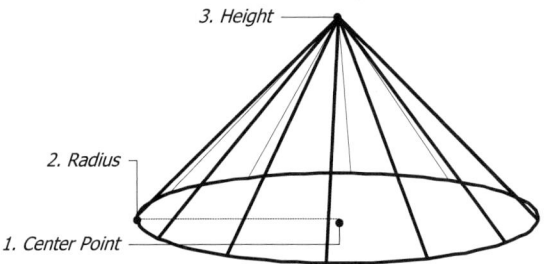

3. Height

2. Radius

1. Center Point

COMMAND LINE OPTIONS

Center point of base specifies the x,y,z coordinates of the center point of the cone's base.

3P picks three points that specify the base's circumference.

2P specifies two points on the base's circumference.

Ttr specifies two points of tangency (with other objects) and the radius.

Elliptical creates cone with an elliptical base.

Base radius specifies the cone's radius.

Diameter specifies the cone's diameter.

Height specifies the cone's height.

2Points picks two points that specify the z orientation of the cone.

Axis endpoint picks the other end of an axis formed from the center point.

Top radius creates a cone with a flat top.

RELATED COMMAND

Ai_Cone draws 3D surface model cones.

RELATED SYSTEM VARIABLES

DragVS specifies the visual style during cone creation.

IsoLines specifies the number of isolines on solid surfaces.

TIPS

- You define the elliptical base in two ways: by the length of the major and minor axes, or by the center point and two radii.

- To draw a cone at an angle, use the **Axis** option.

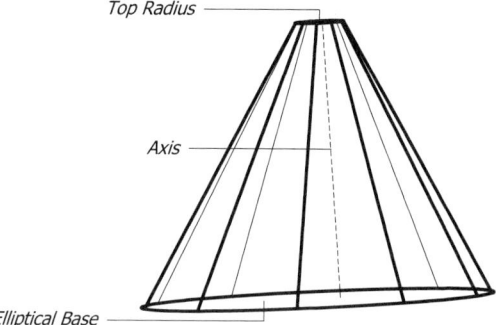

- Once the cone is placed in the drawing, you can edit it interactively. Select the cone, and then select a grip, which turns red, meaning it can be dragged. Press Esc to exit direct editing.

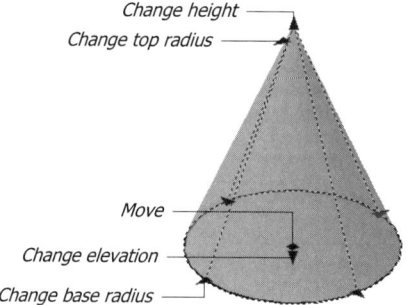

- Silhouette lines are displayed when IsoLines is set to 0 and DispSihl is set to 1. There is no need to hide or shade the cone to see them.

Changed Command

The **Config** command now displays the Options dialog box; see the Options command.

Convert

<u>Rel.14</u> Converts 2D polylines and associative hatches (created in R13 and earlier) to an optimized "lightweight" format.

Command	Alias	Ctrl+	F-key	Alt+	Menu Bar	Dashboard
convert	...	...	...	...	...	...

Command: convert
Enter type of objects to convert [Hatch/Polyline/All] <All>: *(Enter an option.)*
Enter object selection preference [Select/All] <All>: *(Type S or A.)*

COMMAND LINE OPTIONS

Hatch converts associative hatch patterns from anonymous blocks to hatch objects; displays warning dialog box:

Yes converts hatch patterns to hatch objects.

No does not convert hatch patterns.

Polyline converts 2D polylines to Lwpolyline objects.

All converts all polylines and hatch patterns.

Select selects the hatch patterns and 2D polylines to convert.

RELATED COMMANDS

BHatch creates associative hatch patterns.

PLine draws 2D polylines.

RELATED SYSTEM VARIABLE

PLineType determines whether pre-Release 14 polylines are converted in AutoCAD.

   0 — Not converted; PLine creates old-format polylines.

   1 — Not converted; PLine creates lwpolylines.

   2 — Converted; PLine creates lwpolylines (default).

TIPS

- When a Release 13 or earlier drawing is opened, AutoCAD automatically converts most (not all) 2D polylines to lwpolylines; hatch patterns are not automatically updated.

- Hatch patterns are automatically updated the first time the HatchEdit command is applied, or when their boundaries are changed.

- Polylines are not converted when they contain curve fit segments, splined segments, extended object data in their vertices, or 3D polylines.

- PLineType affects the following commands: Boundary (polylines), Donut, Ellipse (PEllipse = 1), PEdit (when converting lines and arcs), Polygon, and Sketch (SkPoly = 1).

ConvertCTB

2001 Converts a plot style file from CTB color-dependent format to STB named format (short for CONVERT Color TaBle).

Command	Alias	Ctrl+	F-key	Alt+	Menu Bar	Dashboard
convertctb	...	...	...	...	...	...

Command: convertctb

Displays Select File dialog box.

1. Select a .ctb file, and then click Open. AutoCAD displays the Create File dialog box.

2. Specify the name of an .stb file, and then click Save. When AutoCAD completes the conversion, it displays this dialog box:

RELATED COMMANDS

ConvertPStyles converts a drawing between color-dependent and named plot styles.

PlotStyle sets the plot style for a drawing.

RELATED FILES

.ctb are color-dependent plot style table files.

.stb are named plot style tables.

TIPS

- "Color-dependent plot styles" are used by older versions of AutoCAD, where the color of the object controls the pen selection.

- "Named plot styles" is the alternative introduced with AutoCAD 2000, which allows plotter-specific information to be assigned to layers and objects.

ConvertOldLights, ConvertOldMaterials

__2007__ Converts lights and materials from drawings created in AutoCAD 2006 and earlier.

Command	Alias	Ctrl+	F-key	Alt+	Menu Bar	Dashboard
convertoldlights	...	...	...	...	...	...
convertoldmaterials						

Command: convertoldlights

Lights defined in drawings created by AutoCAD 2006 and earlier are converted to the new format.

Command: convertoldmaterials

Materials defined in drawings created by AutoCAD 2006 and earlier are converted to the new format.

RELATED COMMANDS

LightList displays the Light List palette, which lists and edits lights in the current drawing.

Materials displays the Materials palette, which modifies materials.

RELATED SYSTEM VARIABLE

3dConversionMode determines how lights and materials are converted in drawings from AutoCAD 2007 and earlier.

TIP

- Autodesk warns that the conversion may not be correct in all cases. For example, the intensity of lights may need to be corrected, and the mapping of materials may need to be adjusted.

ConvertPStyles

2001 Converts drawings to either color-dependent or named plot styles (short for Convert Plot STYLES).

Command	Alias	Ctrl+	F-key	Alt+	Menu Bar	Dashboard
convertpstyles	...	...	...	...	...	...

Command: convertpstyles

Displays warning dialog box:

When AutoCAD has completed the conversion, it displays the message, "Drawing converted from Named plot style mode to Color Dependent mode."

AUTOCAD DIALOG BOX

OK proceeds with the conversion.

Cancel prevents the conversion.

RELATED COMMANDS

ConvertCTB converts a plot style file from CTB color-dependent format to STB named format.

PlotStyle sets the plot style for a drawing.

TIPS

- "Color-dependent plot styles" was used by older versions of AutoCAD, where the color of the object controlled the pen selection.

- "Named plot styles" is the alternative introduced with AutoCAD 2000, which allows plotter-specific information to be assigned to layers and objects.

ConvToSolid,
 # ConvToSurface

2007 Converts certain objects to solids, and open objects to surfaces.

Command	Alias	Ctrl+	F-key	Alt+	Menu Bar	Dashboard
convtosolid	...	...	...	...	...	3D Make
convtosurface						3D Make

Command: convtosolid

Converts these objects to solids: closed, uniform-width polylines with thickness, closed zero-width polylines with thickness, and circles with thickness.

Select objects: *(Select one or more objects.)*
Select objects: *(Press Enter.)*

Command: convtosurface

Converts open objects to surfaces: 2D solids, regions, open zero-width polylines with thickness, lines with thickness, arcs with thickness, and planar 3D faces.

Select objects: *(Select one or more objects.)*
Select objects: *(Press Enter.)*

RELATED COMMANDS

Extrude extrudes 2D objects into 3D solids or surfaces.

Revolve rotates 2D objects into 3D solids or surfaces.

Thicken thickens 2D objects into 3D solids or surfaces.

Sweep rotates 2D objects into 3D solids or surfaces.

Loft converts 2D objects into lofted 3D solids or surfaces.

RELATED SYSTEM VARIABLE

DelObj determines whether to retain or erase objects used to create the surfaces and solids.

TIP

- Use the Explode command to convert 3D solids with curved surfaces into surfaces.

Copy

V. 1.0 Creates one or more copies of objects.

Command	Alias	Ctrl+	F-key	Alt+	Menu Bar	Dashboard
copy	co	...	...	MY	Modify	2D Draw
	cp				⬦ Copy	

Command: copy
Select objects: *(Select one or more objects — point 1.)*
Select objects: *(Press Enter.)*
Specify base point or [Displacement] <Displacement>: *(Pick point 2.)*
Specify second point of displacement or <use first point as displacement>: *(Point 3.)*
Specify second point or [Exit/Undo] <Exit>: *(Press Enter to end the command.)*

COMMAND LINE OPTIONS

Base point indicates the starting point; does not need to be on the object.

Second point indicates the point to which to copy.

Displacement prompts for two points to use as the displacement.

Undo undoes the last copy.

Exit exits the command.

RELATED COMMANDS

Array draws a rectangular or polar array of objects.

MInsert places an array of blocks.

Offset creates parallel copies of lines, polylines, circles, and arcs.

RELATED SYSTEM VARIABLE

CopyMode stops this command after making one copy *(new to AutoCAD 2008)*.

TIPS

- Turn on Ortho to copy objects in a precise horizontal and vertical direction, or Polar to achieve precise incremental angles; use OSnap to copy objects precisely to a geometric feature.

- Inserting a block multiple times is more efficient than placing multiple copies.

CopyBase

<u>2000</u>   Copies selected objects to the Clipboard with a specified base point (short for COPY with BASEpoint).

Command	Alias	Ctrl+	F-key	Alt+	Menu Bar	Dashboard
copybase	...	Shift+C	...	EB	Edit	...
					⤷Copy with Base Point	

Command: copybase
Specify base point: *(Pick a point.)*
Select objects: *(Select one or more objects.)*
Select objects: *(Press Enter.)*

COMMAND LINE OPTIONS

Specify base point specifies the base point.

Select objects selects the objects to copy to the Clipboard.

RELATED COMMANDS

CopyClip copies selected objects to the Clipboard with a base point equal to the lower-left extents of the selected objects.

PasteBlock pastes objects from the Clipboard into drawings as blocks.

PasteClip pastes objects from the Clipboard into drawings.

TIPS

■ When PasteBlock pastes objects previously selected with the CopyBase command, AutoCAD prompts you 'Specify insertion point:', and then pastes the objects as a block with a name similar to A$C7E1B27BE.

■ When specifying the All option at the 'Select objects:' prompt, CopyBase selects only objects visible in the current viewport.

■ As of AutoCAD 2004, you can use the Ctrl+Shift+C shortcut for this command.

■ For best results, the technical editor suggests using the WmfOut command to obtain scalable 2D drawings with lineweights.

CopyClip

<u>Rel.12</u> Copies selected objects from the drawing to the Clipboard (short for COPY to CLIPboard).

Command	Alias	Ctrl+	F-key	Alt+	Menu Bar	Dashboard
copyclip	...	C	...	EC	Edit ↘Copy	...

Command: copyclip
Select objects: *(Select one or more objects.)*
Select objects: *(Press Enter.)*

The Clipboard Viewer shows a drawing copied from AutoCAD to the Clipboard.

COMMAND LINE OPTION

Select objects selects the objects to copy to the Clipboard.

RELATED COMMANDS

CopyBase copies objects to the Clipboard with a specified base point.

CopyHist copies Text window text to the Clipboard.

CopyLink copies the current viewport to the Clipboard.

CutClip cuts selected objects to the Clipboard.

PasteBlock pastes objects from the Clipboard into the drawing as a block.

PasteClip pastes objects from the Clipboard into the drawing.

RELATED WINDOWS COMMANDS

PrtScr copies the entire screen to the Clipboard

Alt+PrtScr copies the topmost window to the Clipboard.

TIPS

- Contrary to the AutoCAD *Command Reference*, text is *not* copied to the Clipboard in text format; instead, text is copied as AutoCAD objects. (Text copied from the command line is copied as plain text.)

- When the All option is specified at the 'Select objects' prompt, CopyClip selects only objects visible in the current viewport.

CopyHist

__Rel.13__ Copies all of the Text window text to the Clipboard (short for COPY HISTory).

Command	Alias	Ctrl+	F-key	Alt+	Menu Bar	Dashboard
copyhist	...	...	...	...	...	...

Command: copyhist

AutoCAD copies all text in the history window to the Clipboard.

COMMAND LINE OPTIONS
None.

RELATED COMMAND
CopyClip copies selected text from the drawing to the Clipboard.

RELATED WINDOWS COMMAND
ALT+PRT SCR copies the Text window to the Clipboard in graphics format.

TIPS
- To copy a selected portion of Text window text to the Clipboard, highlight the text first, then select Copy from the Text window's Edit menu bar.

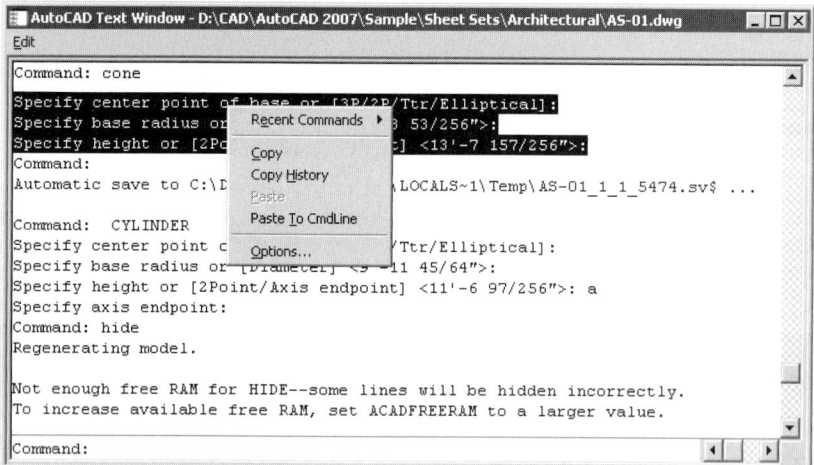

- To paste text to the command line, select Paste to Cmdline from the Edit menu. However, this only works when the Clipboard contains text — not graphics.

- As an alternative, right-click in the Text window to bring up the cursor menu.

CopyLink

Rel.13 Copies the current viewport to the Clipboard; optionally allows you to link the drawing to AutoCAD.

Command	Alias	Ctrl+	F-key	Alt+	Menu Bar	Dashboard
copylink	...	...		EL	Edit	...
					↳Copy Link	

Command: copylink

COMMAND LINE OPTIONS
None.

RELATED COMMANDS
CopyClip copies selected objects to the Clipboard.

CopyEmbed copies selected objects to the Clipboard.

CopyHist copies Text window text to the Clipboard.

CutClip cuts selected objects to the Clipboard.

PasteClip pastes objects from the Clipboard into the drawing.

RELATED WINDOWS COMMANDS
PRT SCR copies the entire screen to the Clipboard

ALT+PRT SCR copies the topmost window to the Clipboard.

TIPS
- In the other application, select Paste Special from the Edit menu to paste the AutoCAD image into the document; to link the drawing to AutoCAD, select the Paste Link option.

- AutoCAD does not let you link a drawing to itself.

- This command copies everything in the current viewport (if in model space) or the entire drawing (if in paper space).

- CopyEmbed is identical to CopyLink, except that CopyEmbed prompts you to select objects.

- If you use Paste Special | AutoCAD Drawing when pasting the drawing into a document, you can then double-click the drawing, which launches AutoCAD so that you can edit the drawing.

CopyToLayer

<u>2007</u>   Copies selected objects to the layer of another object; formerly an Express Tool.

Command	Alias	Ctrl+	F-key	Alt+	Menu Bar	Dashboard
copytolayer	...	...		OOPP⃰	Format	...
					⃰Layer Tools	
					⃰Copy Objects to New Layer	
-copytolayer						

Command: copytolayer
Select objects to copy: *(Pick one or more objects.)*
Select objects to copy: *(Press Enter.)*

Select object on destination layer or [Name] <Name>: *(Pick an object, or type N.)*
n **object(s) copied and placed on layer "***layername***".**
Specify base point or [Displacement/eXit] <eXit>: *(Enter an option or press Enter.)*

COMMAND LINE OPTIONS

Select objects to copy selects the objects to be copied.

Select object on destination layer specifies the layer through the selection of another object.

Name displays the Copy to Layer dialog box with names of all layers in drawing.

Specify base point specifies the starting point of the copy.

Displacement specifies the displacement distance, from original to copy.

eXit exits the command.

Name dialog box

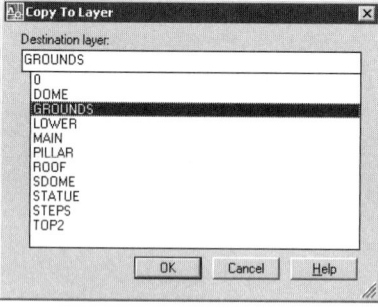

Select a layer name, and then click OK.

-COPYTOLAYER COMMAND

Command: -copytolayer
Select objects to copy: *(Pick one or more objects.)*
Select objects to copy: *(Press Enter.)*

Specify the destination layer name or [?/= (select object)] <0>: *(Enter the name of a layer, or enter an option.)*
n **object(s) copied and placed on layer** *"layername".*
Specify base point or [Displacement/eXit] <eXit>: *(Enter an option or press Enter.)*
Specify second point of displacement or <use first point as displacement>:

COMMAND LINE OPTIONS

Select objects to copy selects the objects to be copied.

Select object on destination layer specifies the layer through the selection of another object.

? lists the names of layers in the drawing.

= prompts to select an object whose layer should be used; AutoCAD prompts, "Select an object with the desired layer name."

Specify base point specifies the starting point of the copy.

Displacement specifies the displacement distance, from original to copy.

eXit exits the command.

RELATED COMMANDS

Copy copies objects onto the same layer.

Ai_Molc makes the layer of the selected object current.

ChSpace moves objects from paper to model space.

CUI

<u>2006</u>   Customizes many aspects of the user interface; replaces the Menu, Toolbar, and Customize commands (short for Customize User Interface).

Command Dashboard	Alias	Ctrl+	F-key	Alt+	Menu Bar	
cui	toolbar	...	...	TCI	Tools	...
	tbconfig				⤷Customize	
	to				⤷Interface	

Command: cui

Displays the Customize tab of the dialog box (redesigned in AutoCAD 2008):

The dialog box consists of panes; a "pane" is an area of the dialog box.

Customizations In (All CUI Files) pane

Right-click items to access shortcut menus, described later. Added to AutoCAD 2008: Dashboard Panels.

▼ **Categories** lists categories of CUI files.

📥 **Load Partial Customization File** displays the Open dialog box; see the CuiLoad command.

💾 **Save** saves all current customization files.

Command List pane

Right-click items to access shortcut menus.

▼ **Categories** droplist lists categories of commands.

⭐ **Find Command or Text** displays Find and Replace dialog box.

⭐ **Create a New Command** creates new commands.

🔍 **Search Command List** finds command description by name *(new to AutoCAD 2008)*.

Properties pane

Content varies, depending on the item being edited:

Name specifies the name of the command or macro.

Description specifies the help text displayed on the status bar.

Macro executes the code when the command, menu item, toolbar button, or shortcut keystroke is invoked; click the **...** button to display the Long String Editor dialog box:

ElementID reports the internal identification assigned to the command by AutoCAD.

Small image shows the 16x16-pixel bitmap image displayed on toolbars and so on; click the **...** button to display the Select Image File dialog box for loading *.bmp* image files.

Large image shows the 24x24-pixel bitmap image displayed on toolbars and so on.

Button Image pane

Small works with 16x16-pixel bitmap images.

Large works with 24x24-pixel bitmap images.

Both works with both sizes of images.

Edit displays the Button Editor dialog box.

Sorts properties according to categories.

Sorts properties alphabetically.

Toggles the display of the Tips box below the Properties grid.

SHORTCUT MENUS

Customizations In (All CUI Files) pane

Content of shortcut menus varies, depending on the item selected:

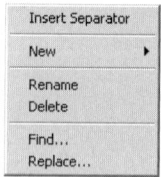

Insert Separator adds separation lines between items.

New adds an item specific to the section; examples include new toolbars and flyouts, new workspaces, and new menus and sub-menus.

Rename changes the name of the item.

Delete erases the item; AutoCAD displays a warning dialog box.

Find displays the Find tab of the Find and Replace dialog box.

Replace displays the Replace tab of the Find and Replace dialog box.

Command List pane

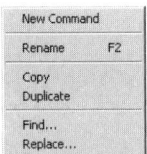

New Command adds new "commands" (macros, actually) to AutoCAD.

Rename displays the Rename Command dialog box for renaming items.

Copy copies item to the Clipboard; can be pasted elsewhere in CUI with the Paste command *(new to AutoCAD 2008)*. Do not attempt to paste into other applications.

Duplicate makes a copy of the command *(new to AutoCAD 2008)*.

Delete erases the item (available only for user-created commands).

BUTTON EDITOR DIALOG BOX

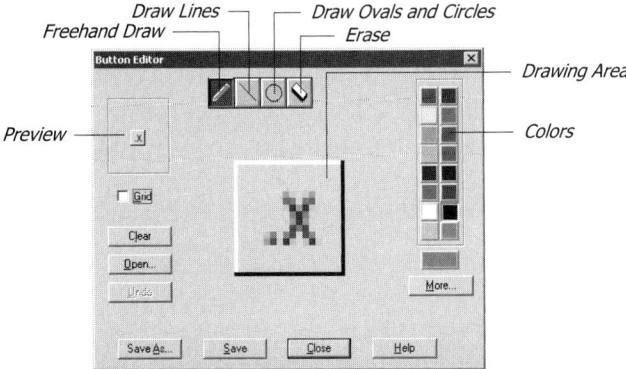

Grid:
- ☑ Displays a grid in the icon drawing area.
- ☐ Hides the grid.

Clear erases the icon.

Open opens a *.bmp* (bitmap) file, which can be used as an icon.

Undo undoes the last operation.

Save As saves the icon as a *.bmp* file.

Save saves changes to the icon in AutoCAD's internal storage.

Close closes the dialog box.

More displays the Select Color dialog box.

RELATED COMMANDS

CuiExport and **CuiImport** transfer customization settings between the local *acad.cui* and enterprise and partial *cui* files.

CuiLoad and **CuiUnload** loads and unloads *.cui* files; replaces the MenuLoad command.

Customize rearranges tool palettes.

QuickCui opens the CUI dialog box partially collapsed *(new to AutoCAD 2008)*.

Tablet calibrates, configures, and toggles digitizing tablet menus.

Workspace creates, modifies, and saves workspaces.

RELATED SYSTEM VARIABLES

CuiState reports whether the Customize User Interface dialog box is open or not.

EnterpriseMenu specifies the path and name of the *.cui* file used for the enterprise.

LockUi prevents the windows and toolbars from being moved; hold down the Ctrl key to override the lock.

TempOverrides turns on and off temporary override keys.

Tooltips toggles the display of tooltips.

DEFINITIONS

CUI — customization user interface.

Enterprise CUI — customization data shared by all users in an office; typically controlled by the CAD manager.

TIPS

- CUI replaces the Menu, Toolbar, and Customize commands as of AutoCAD 2006. The *.cui* file replaces the *.mnu*, *.mns*, and *.mnc* files.

- The LockUi system variable prevents the windows and toolbars from being moved; to override the lock, hold down the Ctrl key while dragging them. Alternatively click the padlock icon in the tray.

CuiExport, CuiImport

2006 Transfers user interface customization data between *.cui* files.

Command	Alias	Ctrl+	F-key	Alt+	Menu Bar	Dashboard
cuiexport	...	...	...	TC	**Tools**	...
					⤷**Customize**	
					⤷**Export Customizations**	
cuiimport	...	...	...	TC	**Tools**	...
					⤷**Customize**	
					⤷**Import Customizations**	

Command: cuiexport *or* cuiimport

Both commands display the same Transfer tab of the dialog box:

CUSTOMIZE USER INTERFACE DIALOG BOX

To import and export items, follow these steps:

1. Open a *.cui* or *.mnu* file in one panel.

2. Open another *.cui* or *.mnu* file in the second panel.

3. Drag items between the two panels.

4. Click the Save 🖫 button.

Dialog Box Toolbar

🗐 Creates new customization files.

📂 Opens customization and menu files; displays the Open dialog box.

🖫 Saves the current customization file; does not display a dialog box.

RELATED COMMANDS

Cui customizes elements of the user interface.

CuiLoad adds *.cui* and *.mnu* files to the user interface.

CuiLoad, CuiUnload

2006 Loads (and unloads) "partial" *.cui* files.

Command	Alias	F-key	Alt+	Menu Bar	Dashboard
cuiload	menuload	...	...	...	...
cuiunload	menuunload				

Command: cuiload *or* cuiunload

Both commands display the same dialog box:

LOAD/UNLOAD CUSTOMIZATIONS DIALOG BOX

Loaded Customization Groups lists the names of loaded menu groups and files.

Unload unloads selected menu group.

Load loads the selected menu group into AutoCAD.

File Name displays the name of the *.cui* file.

Browse displays the Select Customization File dialog box.

RELATED COMMANDS

CUI manipulates the user interface customization.

Tablet configures digitizing tablets for use with overlay menus.

RELATED FILE

*.cui is the customization user interface file; stored in XML format.

TIPS

- The CUILoad command allows you to add *partial* user interfaces, without replacing the entire UI structure.

- The *custom.cui* file is meant customizing menus independently of *acad.cui*.

- As of AutoCAD 2006, these two commands replace the MenuLoad and MenuUnload commands of earlier releases of AutoCAD.

CustomerInvolvementProgram

2008   Invites you to join the Customer Involvement Program (undocumented command; at 26 characters, the longest command name).

Command Dashboard	Alt+	Menu Bar	
customerinvolvementprogram	HC	Help	...
		⇘Customer Involvement Program	

Command: customerinvolvementprogram

Displays dialog box:

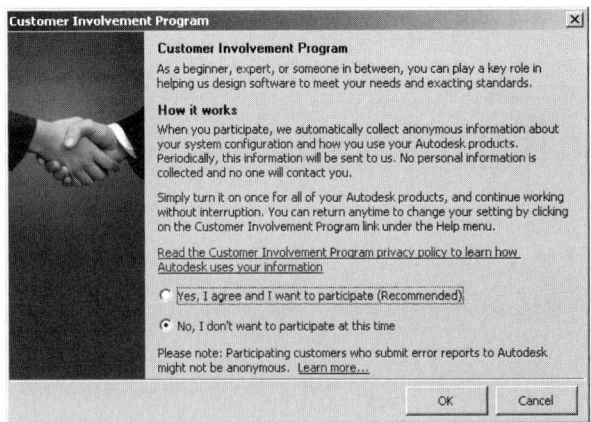

DIALOG BOX OPTIONS

Yes contacts Autodesk via the Internet to join the program.

No dismisses the dialog box; sometimes reminds you in seven days.

RELATED SYSTEM VARIABLE

CipMode provides command line access to the Customer Involvement Program.

Customize

<u>2000i</u> Manages tool palettes and creates palette groups.

Command Dashboard	Aliases	Ctrl+	F-key	Alt+	Menu Bar	
customize	...	...	...	TCP	Tools	...
					↳Customize	
					↳Tool Palettes	

Command: customize

Displays dialog box:

SHORTCUT MENUS

Palettes

Rename renames the selected palette.

New Tool Palette creates a new, blank palette.

Delete removes the selected palette.

Import imports palettes (*.xtp* files).

Export exports the selected palette in an XML-like *.xtp* file.

Palette Groups

Right-click a folder for the shortcut menu:

New Group creates new groups and subgroups.

Rename renames the selected group.

Delete removes the group.

Set Current makes the palette group current.

RELATED COMMANDS

ToolPalette toggles the display of the Tool Palette.

TpNavigate displays the specified tool palette or palette group (*new to AutoCAD 2008*).

CutClip

Rel.12 Cuts the selected objects from the drawing to the Clipboard (short for CUT to CLIPboard).

Command	Alias	Ctrl+	F-key	Alt+	Menu Bar	Dashboard
cutclip	...	X	...	ET	Edit	...
					↳Cut	

Command: cutclip
Select objects: *(Select one or more objects.)*
Select objects: *(Press Enter.)*

COMMAND LINE OPTION

Select objects selects the objects to cut to the Clipboard.

RELATED COMMANDS

BmpOut exports selected objects in the current view to a *.bmp* file.

CopyClip copies selected objects to the Clipboard.

CopyHist copies the Text window text to the Clipboard.

CopyLink copies the current viewport to the Clipboard.

PasteClip pastes objects from the Clipboard into the drawing.

RELATED WINDOWS COMMANDS

PRT SCR copies the entire screen to the Clipboard

ALT+PRT SCR copies the topmost window to the Clipboard.

TIPS

- When the All option is specified at the 'Select objects:' prompt, CutClip selects objects visible in the current viewport only.

- In the other application, use the Edit | Paste or Edit | Paste Special commands to paste the AutoCAD image into the document; the Paste Special command lets you specify the pasted format.

- You can use the Undo command to return the "cut" objects to the drawing.

Cylinder

Rel.12 Draws a 3D solid cylinder with a circular or elliptical cross-section.

Command	Alias	Ctrl+	F-key	Alt+	Menu Bar	Dashboard
cylinder	cyl	...	...	DMC	Draw	3D Make
					⬥Modeling	
					⬥Cylinder	

Command: cylinder
Specify center point of base or [3P/2P/Ttr/Elliptical]: *(Pick center point 1, or enter an option.)*
Specify base radius or [Diameter] <3>: *(Specify radius 2, or type D.)*
Specify height or [2Point/Axis endpoint] <1">: *(Specify height 3, or type C.)*

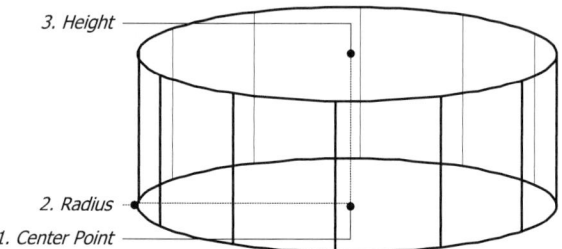

COMMAND OPTIONS

Center point of base specifies the x,y,z coordinates center point of the cylinder's base.

3P picks three points that specify the base's circumference.

2P specifies two points on the base's circumference.

Ttr specifies two points of tangency (with other objects) and the radius.

Elliptical creates elliptical cylinders.

Base radius specifies the cylinder's radius.

Diameter specifies the cylinder's diameter.

Height specifies the cylinder's height.

2Points picks two points that specify the z orientation of the cylinder.

Axis endpoint picks the other end of an axis formed from the center point.

RELATED COMMANDS

Ai_Cylinder creates cylinders made of 3D meshes.

Extrude creates cylinders and other extruded shapes with arbitrary cross-sections and sloped walls.

RELATED SYSTEM VARIABLES

DragVs determines the visual style while creating 3D solid objects.

IsoLines determines the number of isolines on curved solids.

TIP

- Once the cylinder is placed in the drawing, you can edit it interactively:

  1. Select the cylinder.
  2. Select one of the grips, which turns red.
  3. Drag the grip to change the cylinder's size or position.
  4. When done, press ESC to exit direct editing.

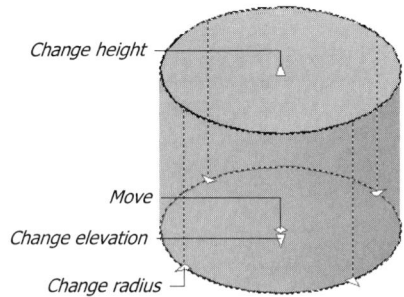

Change height

Move

Change elevation

Change radius

- The Axis endpoint option rotates the cylinder.

'Dashboard, 'DashboardClose

<u>2007</u> Opens and closes the Dashboard palette.

Command	Alias	Ctrl+	F-key	Alt+	Menu Bar	Dashboard
dashboard	...	...	...	TP	Tools	...
				⇖Enter	⇖Palettes	
					⇖Dashboard	
dashboardclose	...	...	...	TP	Tools	...
				⇖Enter	⇖Palettes	
					⇖Dashboard	

Command: dashboard

Displays one of two palettes. When workspace is set to "3D Modeling":

Command: dashboardclose

Closes the palette.

When workspace is set to "2D Drafting & Annotation" (new to AutoCAD 2008):

Layers Control Panel

2D Draw Control Panel

Flyout for more commands

Annotation Scaling Control Panel

Dimensions Control Panel

Text Control Panel

Multileaders Control Panel

Tables Control Panel

2D Navigate Control Panel

Object Properties Control Panel

Block Attributes Control Panel

Shortcut Menu

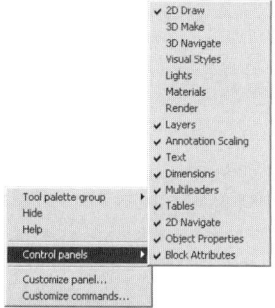

Tool Palette Group associates a tool palette group with the selected control panel.

Hide hides the selected control panel.

Help displays online help for the Dashboard.

Control Panels chooses the panels to display or hide.

Customize Panel displays the Customize User Interface dialog box; see the Cui command.

Customize Commands displays a collapsed CUI dialog box; see the QuickCui command.

RELATED SYSTEM VARIABLE

DashboardState reports whether the Dashboard palette is open or closed.

TIP

■ The workspace determines which Dashboard panels to display; these are defined by CUI.

DataExtraction

<u>**2008**</u> Runs a wizard that extracts data from the drawing, and then places them in a table or in an external file; replaces function of the EAttExt command.

Command	Aliases	Ctrl+	Alt+	Menu Bar	Dashboard
dataextraction	dx	...	TX	Tools	Block Attributes
	eattext			⇘Data Extraction	

-dataextraction

Command: dataextraction

Displays wizard:

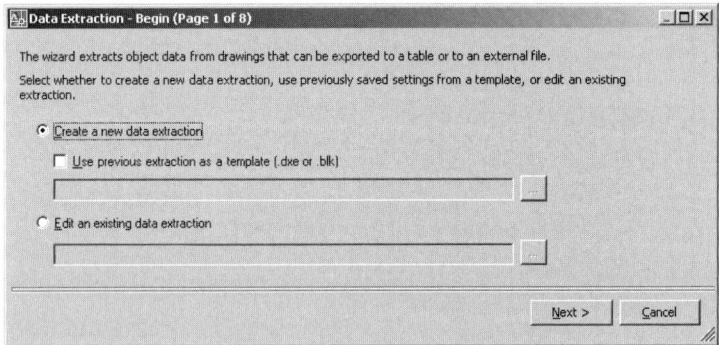

WIZARD OPTIONS

1. Begin page

   ⊙**Create a New Data Extraction** creates new templates for specifying which data to extract and how.

   ☐ **Use Previous Extraction as a Template** reuses existing *.dxe* and *.blk* template files.

   ... **Open Template** opens dialog box for selecting data extraction template files (*.dxe* or *.blk*).

   ○**Edit an Existing Data Extraction** modifies an earlier template.

   Next displays the Save Data Extraction As dialog box; enter a file name, and then click Save.

2. Define Data Source page

⊙ **Drawings/Sheet Set** prompts you to select any drawing and folder for data extraction.

   ☑ **Include Current Drawing** includes the current drawing in the data extraction. Turn off to ignore data in the current drawing.

○ **Select Objects in the Current Drawing** selects only the current drawing for data extraction.

   ▣ **Select Objects** selects specific objects in the current drawing.

Add Folder displays the Add Folder Options dialog box for selecting entire folders of drawing files to include in data extraction (detailed later).

Add Drawings displays the Select Files dialog box for selecting specific drawings files.

Remove removes the checked drawings or folders in the Drawing Files and Folders list from the data extraction.

Settings displays the Data Extraction - Additional Settings dialog box (detailed later).

3. Select Objects page

Click a header to change the sort order.

☑ **Display All Object Types** lists all objects, both blocks and non-blocks.

   ⊙ **Display Blocks Only** lists only blocks and dynamic blocks; includes xrefs.

   ○ **Display Non-Blocks Only** lists all objects, except for blocks and xrefs.

☑ **Display Blocks with Attributes Only** lists names of blocks with attributes and dynamic blocks with actions and parameters.

☑ **Display Only Objects Used in Drawings** lists employed objects.

Shortcut menu

   Check All selects all listed items.

   Uncheck All deselects all listed items.

   Invert Selection checks unchecked items and vice versa.

   Edit Display Name allows you to change names in the Display Name column.

4. Select Properties page

Click a header to change the sort order.

☑ **Category Filter** chooses subsets of object and drawing properties.

5. Refine Data page

Click a header to change the sort order.

☑ **Combine Identical Rows** combines rows with identical names.

☑ **Show Count Column** toggles display of the Count column.

☑ **Show Name Column** toggles the display of the Name column.

Link External Data displays the Link External Data dialog box; see the DataLink command.

Sort Columns Options displays the Sort Columns dialog box for sorting data by more than one column (detailed later); click headers to sort by one column.

Full Preview displays a window showing the final output.

Right-click any row for this shortcut menu with many options:

Sort Descending sorts data in selected column by descending order.

Sort Ascending sorts data in selected column by ascending order.

Sort Column Options displays the Sort Columns dialog box (detailed later).

Rename Column changes the name of the selected column's name.

Hide Column hides the selected column.

Show Hidden Columns shows a submenu of hidden column names.

Set Column Data Format displays the Set Cell Format dialog box (detailed later).

Insert Formula Column displays the Insert Formula Column dialog box (detailed later), and then inserts the formula in a new column to the right of the selected one.

Edit Formula Column displays the Edit Formula Column dialog box.

Remove Formula Column removes the selected formula column.

Combine Record Mode shows numeric data as separate values, or else combines identical numeric property rows into single rows.

Show Count Column toggles display of the Count column.

Show Name Column toggles display of the Name column.

Insert Totals Footer chooses from sum, max, min, or average for columns with numeric data; adds a row at the end of the table, inserting result of the arithmetic in the cell.

Remove Totals Footer removes the Totals footer.

Filter Options displays the Filter Column dialog box (detailed later).

Reset Filter removes the filter of the selected column.

Reset All Filters resets filters of all columns.

Copy to Clipboard copies the whole enchilada to the Clipboard; you can paste the data into other applications, such as word processors and spreadsheets.

6. Choose Output page

☐ **Insert Data Extraction Table into Drawing** automatically inserts a table with the extracted data, at the end of this wizard.

☐ **Output Data to External File** creates a data file in comma-separated (*.csv*), tab-separated (*.txt*), Access database (*.mdb*), or Excel spreadsheet (*.xls*) file format .

7. Table Style page

Table Style selects a named table style.

○ **Use Table in Table Style for Label Rows** (available when selected table style contains a template table) inserts a table with label cells in the top row and footer cells at the bottom.

⊙ **Manually Setup Table** enters a title and specification of the title, header, and data cell style.

Enter a Title for Your Table specifies the title for the table; this row is not overwritten when the table is updated.

Title Cell Style selects a cell style for the title row.

Header Cell Style selects a cell style for the header row.

Data Cell Style selects a cell style for the data rows.

☑ **Use Property Names as Additional Column Headers** uses the property name columns as the header row.

8. Finish page

Finish inserts the table in the drawing, or exports the data to the specified file.

DIALOG BOX OPTIONS

Add Folder Options dialog box

Folder reports the path to the folder.

☐ **Automatically Include New Drawings Added in this Folder to the Data Extraction** tracks when drawings are added to this folder, and includes them in the next data extraction process.

☑ **Include Subfolders** includes drawing files in subfolders.

☐ **Utilize a Wild-card Character to Select Drawings** reduces the number of drawings through the * (any number of characters) and ? (any single character) wildcards.

Data Extraction - Additional Settings dialog box

☑ **Extract Objects from Blocks** extracts data from objects inside of blocks.

☑ **Extract Objects from Xrefs** extracts data from objects and blocks in xrefs.

☑ **Include Xrefs in Block Counts** counts xrefs as blocks.

○ **Objects in Model Space** counts only objects in model space.

⊙ **All Objects in Drawing** counts objects in model and paper space.

Sort Columns dialog box

Add adds more column names to the Column list.

Remove removes the selected column name.

Move Up/Down moves the selected column up (or down) the list; higher columns have sort preference over lower columns.

Set Cell Format dialog box

Additional Format displays the Additional Format dialog box; see the Field command.

Insert / Edit Formula Column dialog box
*Drag an item from Columns into Formula, and then click a function button, such as + or *.*

+, -, *, / perform addition, subtraction, multiplication, or division on column values.

Validate checks that the formula will work.

Filter Column dialog box

Filter Based on the Following Conditions selects amongst greater than, greater than or equal to, less than, less than or equal to, equal to, not equal to, between, or outside of.

-DataExtraction Command

Command: -dataextraction
Enter the template file path for the extraction type: *(Enter path and file name for the .blk attribute extraction template .dxe or data extraction file.)*
Enter the output filetype [Csv/Txt/Xls/Mdb] <Csv>: *(Enter an option.)*
Enter output filepath: *(Enter the path and file name.)*
Specify insertion point: *(Pick a point to locate the table in the drawing.)*

RELATED COMMANDS
AttDef extracts attributes only.

DataLink links tables with external data files.

dbConnect links external database files with objects.

RELATED SYSTEM VARIABLE
DxEval determines when AutoCAD should check for changes in extracted data tables (sum of the following):

0 — No check is performed.
1 — Checked when drawing is opened.
2 — Checked when drawing is saved.
4 — Checked when drawing is plotted (default).
8 — Checked when drawing is published (default).
16 — Checked during the eTransmit and Archive commands.
32 — Checked before Save with Automatic Update.
64 — Checked before Plot with Automatic Update.
128 — Checked before Publish with Automatic Update.
256 — Checked before eTransmit or Archive with Automatic Update.

TIPS
- This command replaces and extends the EAttExt command by outputting object properties and drawing information.

- Formulas cannot be used for input to each other.

- Excel and Access files are limited to a maximum 255 columns.

DataLink, DataLinkUpdate

<u>2008</u> Links external spreadsheet files to tables in drawings.

Command	Alias	Ctrl+	F-key	Alt+	Menu Bar	Dashboard
datalink	dl	...	...	...	Tools ↳Data Links ↳Data Link Manager	Tables
datalinkupdate	dlu	...	...	...	Tools ↳Data Links ↳Update Data Links	Tables

Command: datalink

Displays dialog box:

DIALOG BOX OPTIONS

Links lists the links that exist between the drawing and external spreadsheet files, if any.

Preview toggles the display of the thumbnail image of spreadsheet cells.

Create a New A Excel Data Link displays the following dialog box for naming new links:

Name names the new data link.

OK displays the following dialog box for selecting the spreadsheet file:

... selects a spreadsheet *.xls* or *.xls*, or comma-delimited *.csv* file; the files don't need to be produced by Excel, despite the exclusive reference to Excel in this command.

OK updates the dialog box to show the spreadsheet data:

▼ **Select the Excel Sheet to Link To** shows the path and file name of the data file.

⊙ **Link Whole Sheet** uses all data in the file (for eventual linking with a table in the drawing.)

○ **Link To a Named Range** uses a named range (subset) of data in the file.

○ **Link to Range** uses a range (subset) of data in the file, such as:

A3:E15	Part of the spreadsheet, from cell A3 to E15.
A:E	Entire columns, from A through E.
3:15	Entire rows, from 3 to 15.

Preview updates the thumbnail image.

⊙ expands the dialog box to show more options:

☑ **Convert Data Types to Text** converts formulas to results, and displays them as text.

☐ **Retain Formulas** keeps formulas as formulas.

☑ **Allow Writing to Source File** permits the DataLinkUpdate command to update the external data file when the content of the table is changed in AutoCAD.

☑ **Use Excel Formatting** retains the formatting in the original data file, if any.

 ○ **Keep Table Updated to Excel Formatting** changes the formatting to reflect changes in the data file's formatting, after the DataLinkUpdate command has done its work.

 ⊙ **Start With Excel Formatting, Do Not Update** keeps the original formatting.

OK returns to the original Data Link Manager dialog box.

Click OK to exit the command. The data link has been created; it can now be used by the Table, DataExtraction, and -DataExtraction commands.

- -

DataLinkUpdate Command
Command: datalinkupdate
Select an option [Update data link/Write data link] <Update data link>: *(Enter an option.)*

COMMAND LINE OPTIONS
Update updates tables in AutoCAD, if the external sources have changed.

Write updates the external sources when the tables have changed.

TIPS
■ There's a faster way to link tables in AutoCAD with external data files than using the DataLink and Table commands:

   1. In the spreadsheet program, copy a range of cells to the Clipboard.
   2. In AutoCAD, use the PasteSpec command to paste the cell as "AutoCAD Entities." Ensure that the Paste Link option turned on.

■ There is no need to use Excel; these commands work with plain CSV (comma-separated value) text files output by many programs, as well as with the free spreadsheet software from OpenOffice (www.openoffice.org).

■ When a drawing contains links to external data files, the following icon appears in the tray. (The same icon appears next to the cursor when it hovers over a linked table.)

■ Tables linked to external files are locked by default, and cannot be edited. Use the table editing toolbar to unlock tables.

■ When AutoCAD notices that the data in the tables are different from those in the external data files, it displays this warning balloon in the tray:

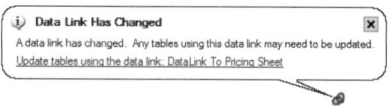

■ "XLSX" refers to XML-format files output by Excel 2007.

■ When linking to a password-protected data file, you may need to repeatedly enter the password.

'DbConnect, DbClose

2000 Opens and closes the dbConnect Manager window to connect objects with rows in external database tables (short for Data Base CONNECTion).

Command	Alias	Ctrl+	F-key	Alt+	Menu Bar	Dashboard
dbconnect	**dbc**	**6**	...	**TPC**	**Tools**	...
					⌦ **Palettes**	
					⌦ **dbConnect**	
dbcclose	...	**6**	...	**TPC**	**Tools**	...
					⌦ **Palettes**	
					⌦ **dbConnect**	

Command: dbconnect

Displays palette. (If a red x appears, the database is disconnected from the drawing.)

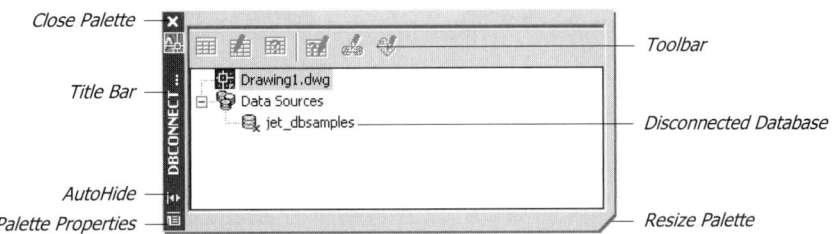

*To connect the drawing with the database, right-click the database icon, and then select **Connect**.*
The icons have the following meaning:

dbConnect Manager Toolbar

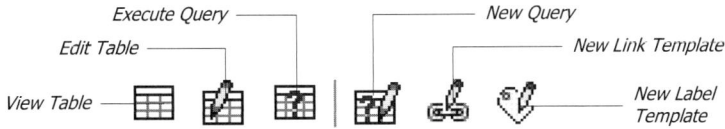

View Table opens an external database table in *read-only* mode; select a table, link template, or label template to make this button available.

Edit Table opens an external database table in *edit* mode; select a table, link template, or label template to make this button available.

Execute Query executes a query; select a previously-defined query to make this button available.

New Query displays the New Query dialog box when a table or link template is selected; displays the Query Editor when a query is selected.

New Link Template displays the New Link Template dialog box when a table is selected; displays the Link Template dialog box when a link template is selected; not available for link templates with links already defined in a drawing.

New Label Template displays the New Label Template dialog box when a table or link template is selected; displays the Label Template dialog box when a label template is selected.

VIEW TABLE AND EDIT TABLE WINDOWS

The View Table and Edit Table windows are identical, with the exception that View Table is read-only; hence all text in columns is grayed-out.

Link! creates a link or a label; click the drop list to select an option:

- **Create Links** turns on link creation mode.
- **Create Attached Labels** turns on the attached label creation mode.
- **Create Freestanding Labels** turns on freestanding label creation mode.

View Link Objects in Drawing highlights objects linked to selected records.

View Linked Records in Data View highlights records that are linked to selected objects in the drawing.

AutoView Linked Objects in Drawing automatically highlights objects that are linked to selected records.

AutoView Linked Records in Data View highlights records linked to selected objects in the drawing.

Query displays the New Query dialog box.

Print Data View prints the data in this window.

Data View and Query Options displays the Data View and Query Options dialog box.

Select a Link Template lists the names of previously-defined link templates.

Select a Label Template lists the names of previously-defined label templates.

New Query dialog box

New query name specifies the name of the query.

Existing query names uses an existing query.

Continue displays the Query Editor dialog box.

QUERY EDITOR DIALOG BOX

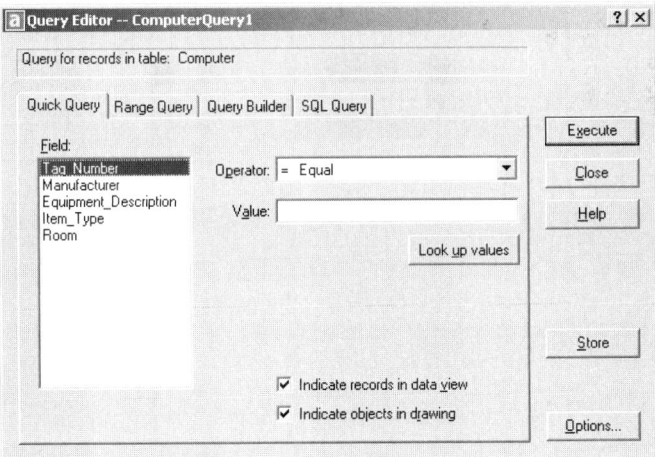

Execute executes the query.

Close closes the dialog box.

Store saves the settings.

Options displays the Data View and Query Options dialog box.

Quick Query tab

Field selects a field name.

Operator specifies a conditional operator:

Operator	Meaning
=	**Equal** — Exactly match the value (default).
<>	**Not equal** — Does not match the value.
>	**Greater than** — Greater than the value.
<	**Less than** — Less than the value.
>=	**Greater than or equal** — Greater than or equal to the value.
<=	**Less than or equal** — Less than or equal to the value.
Like	Contains the value; use the **%** wild-card character (equivalent to * in DOS).
In	Matches two values separated by a comma.
Is null	Does not have a value; used for locating records that are missing data.
Is not null	Has a value; used for excluding records that are missing data.

Value specifies the value for which to search.

Look up values displays a list of existing values:

Range Query Tab

Field lists the names of fields in the current table.

From specifies the first value of the range.

Look Up Values displays the Column Values dialog box.

Through specifies the second value of the range.

☑ **Indicate Records in Data View** highlights records that match the search criteria in the Data View window.

☑ **Indicate Objects in Drawing** highlights objects that match the search criteria in drawings.

Query Builder Tab

(specifies an opening parenthesis, which groups search criteria with parentheses; up to four sets can be nested. **)** specifies a closing parenthesis.

Field specifies a field name; double-click the cell to display the list of fields in the current table.

Operator specifies a logical operator; double-click to display the list of operators.

Value specifies a value for the query; click **...** to display a list of current values.

Logical specifies an And or Or operator; click once to add And; click again to change to Or.

Fields in table displays the fields in the current table; when no fields are selected, the query displays all fields from the table; double-click a field to add it to the list.

Show fields specifies the fields displayed by the Data View window; drag the field out of the list to remove it.

Add adds a field from the Fields in Table list to the Show Fields list.

Sort By specifies the sort order: the first field is the primary sort; to change the sort order, drag the field to another location in the list; press DELETE to remove a field from the list.

Add adds a field from the Fields in Table list to the Sort By list (default = ascending).

▲▼ reverses sort order.

☑ **Indicate Records in Data View** highlights records that match the search criteria in the Data View window.

☑ **Indicate Objects in Drawing** highlights objects that match the search criteria in drawings.

SQL Query Tab

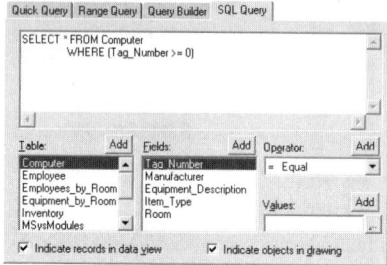

Table lists the names of all database tables available in the current data source.

Add adds the selected table to the SQL text editor.

Fields displays a list of field names in the selected database table.

Add adds the selected field to the SQL text editor.

Operator specifies the logical operator, which is added to the query (default = Equal).

Add adds the selected operator to the SQL text editor.

Values specifies a value for the selected field.

Add adds the value to the SQL text editor.

... lists available values for the field.

☑ **Indicate Records in Data View** highlights records that match the search criteria in the Data View window.

☑ **Indicate Objects in Drawing** highlights objects that match the search criteria in drawings.

DATA VIEW AND QUERY OPTIONS DIALOG BOX

Opened by the Query Editor dialog box:

AutoPan and Zoom

☑ **Automatically Pan Drawing** causes AutoCAD to pan the drawing automatically to display associated objects.

☐ **Automatically Zoom Drawing** causes AutoCAD to zoom the drawing automatically to display associated objects.

Zoom Factor specifies the zoom factor as a percentage of the viewport area; range is 20 - 90; default = 50.

Query Options
Send as Native SQL makes queries to database tables in:

☑ The format of the source table.

☐ SQL 92 format.

☐ **Automatically Store** automatically stores queries when they are executed (default = off).

Record Indication Settings:

◉ **Show Only Indicated Records** displays the records associated with the current AutoCAD selections in the Data View window (default).

○ **Show All Records, Select Indicated Records** displays all records in the current database table.

☑ **Mark Indicated Records** colors linked records to differentiate them from unlinked records.

Marking Color specifies the marking color (default = yellow).

Accumulate Options
Accumulate Selection Set in Drawing

☑ Adds objects to the selection set as data view records are added.

☐ Replaces the selection set each time data view records are selected.

Accumulate Record Set in Data View

☑ Adds records to the selection set as drawing objects are selected.

☐ Replaces the selection set each time drawing objects are selected.

NEW LINK TEMPLATE DIALOG BOX

New link template name specifies the name of the link template.

Start with template reuses an existing template.

Continue displays the Link Template dialog box.

LINK TEMPLATE DIALOG BOX

Key Fields selects one field name; you may select more than one field name, but AutoCAD warns you that too many key fields may slow performance.

NEW LABEL TEMPLATE DIALOG BOX

New label template name specifies the name of the label template.

Start with template reuses an existing template.

Continue displays the Label Template dialog box.

LABEL TEMPLATE DIALOG BOX

Field specifies the names of the fields.

Add adds the field to the label.

Label Offset displays the Label Offset tab:

Start specifies the justification of the label starting point.

Leader offset specifies the x,y distance between the label's starting point and the leader line.

Tip offset specifies the offset distance to the leader tip or label text.

*See the **MText** command for more information.*

RELATED COMMANDS

AttDef creates an attribute definition, akin to an internal database.

EAttExt exports attributes.

TIPS

- Query searches are case sensitive: "Computer" is not the same as "computer."

- OLE DB v2.0 must be installed before you use the dbConnect Manager.

- Leaders must have a length; to get rid of a leader, use a freestanding label.

- SQL is short for "structured query language," a standard method of querying databases.

- The properties of a link template can only be edited if it contains no links, and if the drawing is fully (not partially) loaded.

- The figure below illustrates a database connected with the *db_samp.dwg* sample drawing:

- Before you can edit a record with an SQL Server table, you must define a *primary key*.

- This command cannot be used transparently (within another command) until it has been used at least once non-transparently.

DblClkEdit

Determines whether objects are edited by double-clicking the mouse button (short for DouBLe CLicK EDITing).

Command	Alias	Ctrl+	F-key	Alt+	Menu Bar	Dashboard
dblclkedit	...	...	...	...	...	...

Command: dblclkedit
Enter double-click editing mode [ON/OFF] <ON>: *(Enter ON or OFF.)*

COMMAND LINE OPTIONS
ON enables double-click editing.
OFF disables double-click editing.

RELATED COMMANDS
(Right-click) displays a shortcut menu with editing commands.
Properties displays the Properties palette.

TIPS
- When this command is turned on, double-clicking objects is the equivalent of entering the related editing command:

Object	Command	Dialog Box or Palette
Attribute Definition	**DdEdit**	Edit Attribute
Attribute	**EAttEdit**	Enhanced Attribute Editor
Block	**BEdit**	Block Editor
Dynamic Block	**BEdit**	Block Editor
Hatch	**HatchEdit**	Hatch Edit
Image	**ImageAdjust**	Image Adjust
Mline	**MlEdit**	Multiline Edit Tools
Mtext	**MtEdit**	Multiline Text Editor
Polyline	**PEdit**	*Command-line prompts.*
Section	**LiveSection**	...
Spline	**SplinEdit**	*Command-line prompts.*
Table Cell	**TablEdit**	Multiline Text Editor
Text	**DdEdit**	Edit Text
Viewport	**VpMax**	...
Xref	**RefEdit**	Reference Edit

- For other objects, double-clicking them displays the Properties palette, and highlights them with grips.

- The CUI command's Double-click Edit section allows you to customize the double-click action for any object type.

- In the User Preferences tab of the Options command, you can toggle (turn on or off) double-click editing.

DbList

V. 1.0   Lists information on all objects in the drawing (short for Data Base
LISTing).

Command	Alias	Ctrl+	F-key	Alt+	Menu Bar	Dashboard
dblist	...	...	...	...	...	...

Command: dblist

Sample listing:

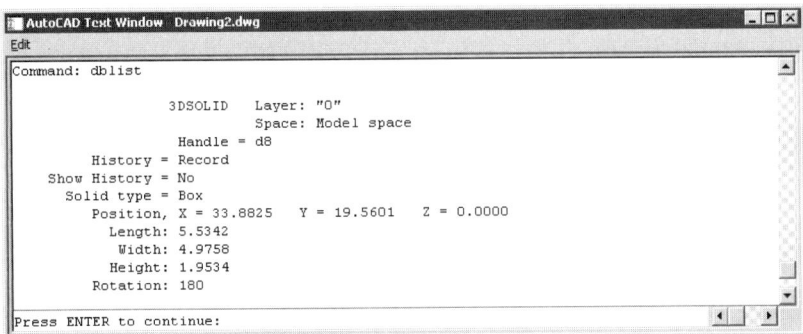

COMMAND LINE OPTIONS

ENTER continues display after pausing.

ESC cancels database listing.

RELATED COMMANDS

Area lists the area and perimeter of objects.

Dist lists the 3D distance and angle between two points.

Id lists the 3D coordinates of a point.

List lists information about selected objects in the drawing.

TIP

- This command is typically used to debug, and has little application for most users.

Renamed Commands

DdAttDef was replaced by AttDef in AutoCAD 2000.

DdAttE was replaced by AttEdit. in AutoCAD 2000.

DdAttExt was replaced by AttExt in AutoCAD 2000.

DdChProp was replaced by Properties in AutoCAD 2000.

DdColor was replaced by Color in AutoCAD 2000.

DdEdit

Rel.11 Edits a single-line text, multiline text, blocks, attribute values, and geometric tolerances (short for Dynamic Dialog EDITor).

Command	Alias	Ctrl+	F-key	Alt+	Menu Bar	Dashboard
ddedit	ed	...	...	MO⬥TE	Modify	...
					⬥Object	
					⬥Text	
					⬥Edit	

Command: ddedit
Select an annotation object or [Undo]: *(Select a text object, or type U.)*

Select single-line text placed by the Text command, and the direct editing field is displayed.

Select paragraph text placed by the MText command, and the in-place text editor is displayed.

Select an attribute definition (not part of a block definition), and the Edit Attribute Definition dialog is displayed.

Select a block with attributes, and the Enhanced Attribute Editor dialog box is displayed.

Select geometric tolerances, and the Geometric Tolerance dialog box is displayed.

Select an annotation object or [Undo]: *(Press Esc to exit command.)*

COMMAND LINE OPTIONS

Undo undoes editing.

Esc ends the command.

SHORTCUT MENU

Right-clicking the text (not mtext) displays the shortcut menu:

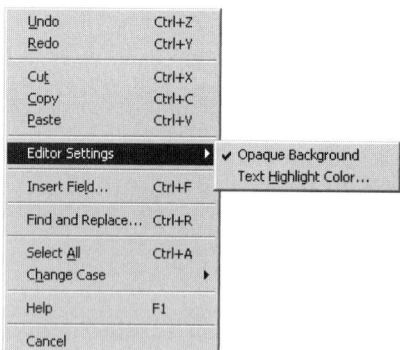

Undo undoes the last action.

Redo redoes the last undo.

Cut copies the selected text to the Clipboard, and then erases it.

Copy copies the selected text to the Clipboard.

Paste pastes text from the Clipboard; unavailable if Clipboard contains no text.

Editor Settings
- **Opaque Background** changes the background color to gray.
- **Text Highlight Color** displays the Select Color dialog box (*new to AutoCAD 2008*).

Insert Field displays the Field dialog box; see the Field command.

Find and Replace displays the Find and Replace dialog box; see the Find command.

Select All selects all text.
Change Case changes the selected text between all UPPERCASE or all lowercase.

Help displays online help for this command.
Cancel ignores editing changes for the current annotation object, and dismisses the dialog box.

RELATED COMMANDS

AttIPedit edits attributes values.

EAttEdit edits all text attributes connected with a block.

Field places automatically-updatable field text.

Find searches for text in drawings.

MtEdit edits paragraph text.

Properties edits all text *properties*, including the text itself.

Spell checks the spelling of words in drawings.

RELATED SYSTEM VARIABLES

DTextEd toggles between direct in-place editing and the old Edit Text dialog box.

FontAlt specifies the name of the font to be used when the required font file cannot be located by AutoCAD.

FontMap maps font names.

MirrText determines how the Mirror command affects text.

TextFill toggles the fill of TrueType fonts during plotting and rendering.

TIPS

- **DdEdit** automatically repeats; press Esc to cancel the command.

- Between AutoCAD 2000 and 2004, this command did edit attribute text.

- As an alternative to this command, double-click text to bring up the appropriate editor.

- AutoCAD 2006 changed this command from a dialog box format to direct editing in drawings.

- Text placed with the Field, Leader, and QLeader commands is mtext; double-click the text to edit.

- Text can also be edited through the Properties window.

. . . .

Renamed Commands

DdGrips was replaced by the Selection tab of the Options command in AutoCAD 2000.

DDim was replaced by DimStyle in AutoCAD 2000.

DdInsert was replaced by Insert in AutoCAD 2000.

DdModify was replaced by Properties in AutoCAD 2000.

. . . .

'DdPtype

Rel.12 Sets the style and size of points (short for Dynamic Dialog Point TYPE).

Command	Alias	Ctrl+	F-key	Alt+	Menu Bar	Dashboard
ddptype	...	...	...	OP	Format	...
					↳Point Style	

Command: ddptype

Displays dialog box:

POINT STYLE DIALOG BOX

Point size sets the size in percentage or pixels.

⊙**Set Size Relative to Screen** sets the size as a percentage of the total viewport height.

○**Set Size in Absolute Units** sets the size in drawing units.

RELATED COMMANDS

Divide draws points along an object at equally-divided lengths.

Point draws points.

Measure draws points a measured distance along an object.

Regen displays the new point format with a regeneration.

RELATED SYSTEM VARIABLES

PdSize specifies the size of points:

0 — Point is 5% of viewport height (default).

positive — Absolute size in drawing units.

negative — Percentage of the viewport size.

PdMode determines the look of points:

TIPS
- Points often cannot be seen in the drawing. To make them visible, change their mode and size.
- The two system variables listed above affect *all* points in the drawing.

Removed Commands

DdEModes was removed from AutoCAD Release 14; it was replaced by the Object Properties toolbar and Object Properties panel on the Dashboard.

DdLModes was removed from AutoCAD 2000; it was replaced by the Layer command.

DdLtype was removed from AutoCAD 2000; it was replaced by the Linetype command.

DdRename was removed from AutoCAD 2000; it was replaced by the Rename command.

DdRModes was removed from AutoCAD 2000; it was replaced by the DSettings cmmand.

DdSelect was removed from AutoCAD 2000; it was replaced by the Selection tab of the Options command.

DdUcs and **DdUcsP** were removed from AutoCAD 2000; they were replaced by UcsMan.

DdUnits was removed from AutoCAD 2000; it was replaced by the Units command.

DdView was removed from AutoCAD 2000; it was replaced by the View command.

DdVPoint

Rel.12 Changes the 3D viewpoint through a dialog box (short for Dynamic Dialog ViewPOINT).

Command	Alias	Ctrl+	F-key	Alt+	Menu Bar	Dashboard
ddvpoint	vp	...	...	V3I	View	...
					⌖3D Views	
					⌖Viewpoint Presets	

Command: ddvpoint

Displays dialog box:

VIEWPOINT PRESETS DIALOG BOX

Set Viewing Angles options

⊙ **Absolute to WCS** sets the view direction relative to the WCS.

○ **Relative to UCS** sets the view direction relative to the current UCS.

From X Axis measures the view angle from the x axis.

From XY Plane measures the view angle from the x,y plane.

Set to Plan View changes the view to plan view in the specified UCS.

RELATED COMMANDS

VPoint adjusts the viewpoint from the command line.

3dOrbit changes the 3D viewpoint interactively.

RELATED SYSTEM VARIABLE

WorldView determines whether viewpoint coordinates are in WCS (1) or UCS (0).

TIPS

- After changing the viewpoint, AutoCAD performs an automatic zoom extents.

- In the image tile shown below, the black arm indicates the new angle.

- In the image tile, a second arm indicates the current angle.

- To select an angle with your mouse:

For fine angle control, click an angle in the inner region of the circle or half-circle.

For coarse angle control, click the outer regions.

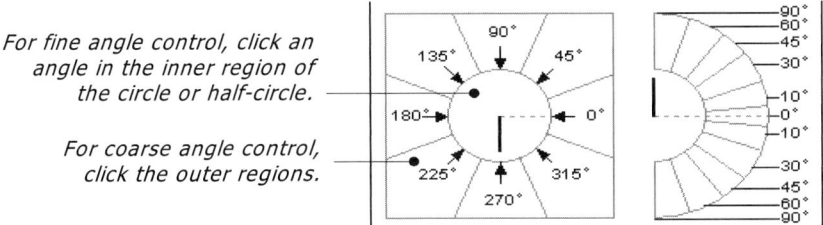

- WCS is short for "world coordinate system."

- UCS is short for "user-defined coordinate system."

- As an alternative, use the 3dOrbit command to set the 3D viewpoint.

- This command is not permitted in paper space (layout mode).

'Delay

<u>**v. 1.4**</u> Delays the next command, in milliseconds.

Command	Alias	Ctrl+	F-key	Alt+	Menu Bar	Dashboard
delay	...	...	...	...	...	...

Command: delay
Delay time in milliseconds: *(Enter the number of milliseconds.)*

COMMAND LINE OPTION

Delay time specifies the number of milliseconds by which to delay the next command.

RELATED COMMAND

Script initiates scripts.

TIPS

- Use Delay to slow down the execution of a script file.

- The maximum delay is 32767, just over 32 seconds.

DetachURL

<u>Rel.14</u> Removes URLs from objects and areas.

Command	Alias	Ctrl+	F-key	Alt+	Menu Bar	Dashboard
detachurl	...	...	...	...	...	...

Command: detachurl
Select objects: *(Select one or more objects.)*
Select objects: *(Press Enter.)*

COMMAND LINE OPTION

Select objects selects the objects from which to remove URL(s).

RELATED COMMANDS

AttachUrl attaches a hyperlink to an object or an area.

Hyperlink attaches and removes hyperlinks via a dialog box.

SelectUrl selects all objects with attached hyperlinks.

TIPS

- When you select a hyperlinked area to detach, AutoCAD reports:

  1. hyperlink ()
  Remove, deleting the Area.
  1 hyperlink deleted...

- When you select an object with no hyperlink attached, AutoCAD reports nothing.

- A URL (short for "universal resource locator") is the universal file naming convention of the Internet; also called a link or hyperlink.

DgnAdjust

2008 Changes the look of MicroStation DGN underlays.

Command	Alias	Ctrl+	F-key	Alt+	Menu Bar	Dashboard
dgnadjust	...	...	...	...	...	...

Command: dgnadjust
Select DGN underlay: *(Select one or more DGN underlays.)*
Enter DGN underlay option [Fade/Contrast/Monochrome] <Fade>: *(Enter an option.)*

COMMAND LINE OPTIONs

Select DGN underlay selects the underlay to adjust.

Enter fade value specifies the amount of fade:

    0 — no fade.
    25 — default value.
    80 — maximum fade.

Enter contrast value specifies the change in contrast, ranging from 0 to 100.

Monochrome toggles monochrome colorization.

RELATED COMMANDS

DgnAttach attaches *.dgn* files as underlays.

DgnClip clipes DGN underlays.

DgnImport imports and translates *.dgn* files into AutoCAD drawing entities.

TIPS

- This command has no effect on the frame around the DGN underlay. Use the DgnFrame system variable to toggle its display.

- Use the Monochrome option to change all of the underlay's colors to black; use the Fade option to change the colors to gray.

DgnAttach

2008 Attaches MicroStation DGN files as underlays.

Command	Alias	Ctrl+	F-key	Alt+	Menu Bar	Dashboard
dgnattach	...	...	...	IG	Insert	...
					⬉DGN Attach	
-dgnattach						

Command: dgnattach

Displays dialog box:

DIALOG BOX OPTIONS

Name specifies the *.dgn* file to attach to the current drawing.

▼ Selects other *.dgn* files.

Browse displays the Select DGN File dialog box to select another *.dgn* file.

Select a design model from the DGN file lists the names of design models in the *.dgn* file; sheets are not listed.

Conversion Units specifies how AutoCAD's units should be matched:
- **Master Units** selects major units such as feet.
- **Sub Units** selects minor units such as inches.

Path Type specifies the type of path to the .dwf file:
- **Full Path** saves the full path name.
- **Relative Path** saves the relative path name.
- **No Path** saves the file name only.

Insertion Point specifies the *.dgn* underlay's insertion point (lower left corner); default =0,0,0.

Specify On-Screen:

☑ prompts you for the insertion point at the command line after closing the dialog box.

☐ enters the x, y, z coordinates in the dialog box.

Scale specifies the DGN underlay's scale factor; default = 1.0

Specify On-Screen:

☑ prompts you for the scale factor at the command line after closing the dialog box.

☐ enters the scale factor in the dialog box.

Angle specifies the DGN underlay's angle of rotation; default = 0 degrees.

Specify On-Screen:

☑ prompts you for the rotation angle at the command line after closing the dialog box.

☐ enters the angle in the dialog box.

Command: -dgnattach
Path to DGN file to attach: *(Enter the path to folder.)*
Enter name of model or [?] <name of design model>: *(Enter the .dgn file name.)*
Specify conversion units [Master/Sub] <Master>: *(Enter an option.)*
Specify insertion point: *(Pick a point.)*
Specify scale factor or [Unit] <1.0000>: *(Enter a value, or type U.)*
Specify rotation <0>: *(Enter a value.)*

COMMAND LINE OPTIONS

Master / Sub specifies MicroStation units, such as feet for master units (MU) and inches for sub units (SU), or user-defined units, such as master = 1 and sub = 1000.

RELATED COMMANDS

DgnAdjust changes the look of attached *.dgn* files as underlays.

DgnClip clips DGN underlays.

DgnImport imports and translates *.dgn* files into AutoCAD drawing entities.

RELATED SYSTEM VARIABLES

DgnFrame toggles display of the rectangular frame around DGN underlays.

DgnOsnap toggles whether osnap recognizes entities in the DGN underlay. The tooltip reporting the osnap mode prefixes "DGN:", as illustrated below:

TIPS

- This command displays 3D *.dgn* files in plan view.

- AutoCAD needs to know MicroStation units to convert them to its own units.

- Double-click the underlay to display its properties; right-click it for options.

- To remove the underlay from the drawing, select it, and then press Del.

- Turn off the display by setting the Properties palette's **Show Underlay** option to off.

DgnClip

2008   Clips DGN underlays to isolate portions of the design file.

Command	Alias	Ctrl+	F-key	Alt+	Menu Bar	Dashboard
dgnclip	...	...	...	...	...	...

Command: dgnclip
Select DGN to clip: *(Select one underlay.)*
Enter DGN clipping option [ON/OFF/Delete/New boundary] <New boundary>: *(Enter an option.)*

COMMAND LINE OPTIONS

ON turns on a previously-created clipping boundary.

OFF turns off the clipping boundary, so that it is hidden.

Delete removes the clipping boundary, restoring the DGN underlay.

New boundary creates a new clipping boundary:

* **Polygonal** creates a multi-side boundary using prompts similar to the PLine command.
* **Rectangular** creates a rectangular boundary based on two pick points.

RELATED COMMANDS

DgnAdjust adjusts the contrast, fade, and color of DGN underlays.

DgnAttach attaches MicroStation *.dgn* files as underlays.

RELATED SYSTEM VARIABLE

DgnFrame toggles the display of the frame around clipped DGN underlays.

TIPS

* After the clipping boundary is in place, you can use grips editing to adjust its size.
* This command can be started by selecting the DGN border, right-clicking, and then selecting DGN Clip from the shortcut menu.

DgnImport, DgnExport

2008 Imports and exports 2D MicroStation *.dgn* files.

Command	Alias	Ctrl+	F-key	Alt+	Menu Bar	Dashboard
dgnimport	...	...	...	...	...	...
dgnexport						

Command: dgnimport

Displays Import DGN File dialog box. Select a .dgn file, and then click Open.

Displays dialog box:

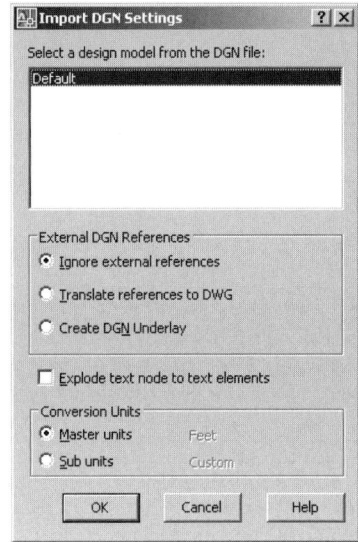

DIALOG BOX OPTIONS

External DGN References

⊙ **Ignore External References** does not import attached *.dgn* and *.dwg* files.

○ **Translate References to DWG** translates xrefs to individual *.dwg* files.

○ **Create DGN Underlay** translates xrefs as DGN files.

☐ **Explode Text Node to Text Elements** retains multi-line text as mtext objects. When on, multi-line text is separated into individual lines of text; required for text placed along curved paths.

Conversion Units

⊙ **Master Units** matches master units to AutoCAD units; sub units are ignored.

○ **Sub Units** matches sub units to AutoCAD units; master units are ignored.

OK imports the design file, translating MicroStation elements into AutoCAD objects.

Command: dgnexport

Displays Export DGN File dialog box. Name the .dgn file, and then click Save.

Displays dialog box:

DIALOG BOX OPTIONS

External DWG References

○ **Translate All DWG References to DGN Files** exports xrefs to *.dgn* files.

○ **Bind All DWG References into one DGN File** combines the parent *.dwg* and all xrefs into one *.dgn* file; xrefs are converted to cells (blocks).

⊙ **Ignore DWG References** leaves out xrefs.

External DGN References

☑ **Export DGN Underlays** includes DGN underlays as *.dgn* references.

Seed File

▧ selects the seed (template) file for the exported *.dgn* file.

⊙ **Master Units** converts AutoCAD units to master units.

○ **Sub Units** converts AutoCAD units to sub units.

RELATED COMMAND

DgnAttach attaches MicroStation *.dgn* files as underlays.

TIPS

■ The DgnImport command fails when the *.dgn* files fails to include at least one design model. It does not import support files, such as *.cel, .s, .h, .cgm, .rdl, .d,* and *.dxf.*

■ AutoCAD converts either MicroStation's master units or sub units to its own units.

■ AutoCAD includes several seed files:

• *Imperial-Seed2D.dgn*
• *Imperial-Seed3D.dgn*
• *Metric-Seed2D.dgn*
• *Metric-Seed3D.dgn*

Dim

<u>V. 1.2</u> Changes the prompt from 'Command' to 'Dim', allowing access to AutoCAD's original dimensioning mode (short for DIMensions).

Command	Alias	Ctrl+	F-key	Alt+	Menu Bar	Dashboard
dim	...	...	...	...	...	...

Command: dim
Dim: *(Enter a dimension command from the list below.)*

COMMAND LINE OPTIONS
Aliases for the dimension commands are shown in UPPERCASE *letters, such as "al" for ALigned.*

ALigned draws linear dimensions aligned with objects *(first introduced with AutoCAD version 2.0)*; replaced by DimAligned.

ANgular draws angular dimensions that measure angles *(ver. 2.0)*; replaced by DimAngular.

Baseline continues dimensions from base points *(ver. 1.2)*; replaced by QDim and DimBaseline.

CEnter draws '+' center marks on circles' and arcs' centers *(ver. 2.0)*; replaced by DimCenter.

COntinue continues dimensions from previous dimensions' extension lines *(ver. 1.2)*; replaced by the QDim and DimContinue commands.

Diameter draws diameter dimensions on circles, arcs, and polyarcs *(ver. 2.0)*; replaced by the QDim and DimDiameter commands.

Exit returns to 'Command' prompt from 'Dim' prompt *(ver. 1.2)*.

HOMetext returns dimension text to its original position *(ver. 2.6)*; replaced by the DimEdit command's Home option.

HORizontal draws horizontal dimensions *(ver. 1.2)*; replaced by DimLinear.

LEAder draws leaders *(ver. 2.0)*; replaced by the Leader and QLeader commands.

Newtext edits text in associative dimensions *(ver. 2.6)*; replaced by the DimEdit command's New option.

OBlique changes the angle of extension lines in associative dimensions *(rel. 11)*; replaced by the DimEdit command's Oblique option.

ORdinate draws x- and y-ordinate dimensions *(Rel. 11)*; replaced by QDim and DimOrdinate.

OVerride overrides current dimension variables *(Rel. 11)*; replaced by DimOverride.

RAdius draws radial dimensions on circles, arcs, and polyline arcs *(ver. 2.0)*; replaced by the QDim and DimRadius commands.

REDraw redraws the current viewport (same as 'Redraw; *ver. 2.0*).

REStore restores dimensions to the current dimension style *(Rel. 11)*; replaced by the -DimStyle command's Restore option.

ROtated draws linear dimensions at any angle *(ver. 2.0)*; replaced by DimLinear.

SAve saves the current settings of dimension styles *(Rel. 11)*; replaced by the -DimStyle command's Save option.

STAtus lists the current settings of dimension variables *(ver. 2.0)*; replaced by the -DimStyle command's Status option.

STYle defines styles for dimensions *(ver. 2.5)*; replaced by DimStyle.

TEdit changes the location and orientation of text in associative dimensions (*Rel. 11*); replaced by DimTEdit.

TRotate changes the rotation angle of text in associative dimensions (*Rel. 11*); replaced by the DimTEdit command's Rotate option.

Undo undoes the last dimension action (*ver. 2.0*); replaced by the Undo command.

UPdate updates selected associative dimensions to the current dimvar setting (*ver. 2.6*); replaced by the -DimStyle command's Apply option.

VAriables lists the values of variables associated with dimension styles, *not* dimvars (*Rel. 11*); replaced by the -DimStyle command's Variables option.

VErtical draws vertical linear dimensions (*ver. 1.2*); replaced by DimLinear.

RELATED DIM VARIABLES

Dimxxx specifies system variables for dimensions; see the DimStyle command.

DimAso determines whether dimensions are drawn associatively.

DimScale determines the dimension scale.

TIPS

- Most dimensions consist of the basic components illustrated below:

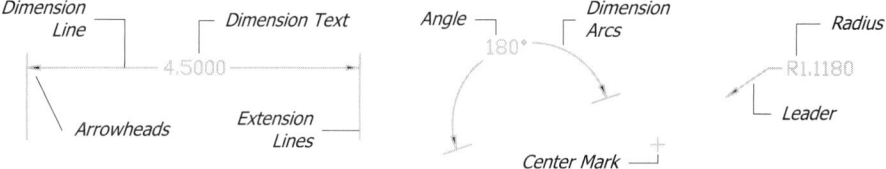

- The 'Dim' prompt dimension commands are included for compatibility with AutoCAD Release 12 and earlier. Only transparent commands and dimension commands work at the 'Dim' prompt. To use other commands, you must exit the 'Dim' prompt with the Exit command or by pressing Esc, which returns you to the 'Command' prompt.

- *Defpoints* (short for "definition points") are used by earlier AutoCAD versions to locate extension lines. Defpoints appear as small dots on the DefPoints layer. When stretching dimensions, ensure defpoints are included; otherwise, the dimensions are not updated automatically.

- The DefPoints layer *never* plots, making it useful for objects and notes that should never plot and for great practical jokes.

- As of AutoCAD 2002, defpoints are not used when DimAssoc = 2 (default); dimensions are instead attached directly to objects.

- Dimension text can have a colored background, which is set globally through the DimStyle command, or overridden locally with the Properties command.

- All the components of an *associative dimension* are treated as a single object; components of a nonassociative dimension are treated as individual objects.

- For additional control over dimensions, the following undocumented commands are described earlier in this book: AiDimFlipArrow, AiDimPrec, AiDimStyle, Ai_Dim_TextAbove, Ai_Dim_TextCenter, Ai_Dim_TextHome, and AiDimTextMove.

Dim1

Displays the 'Dim' prompt for a single dimensioning command, and then returns to the 'Command' prompt (short for DIMension once).

Command	Alias	Ctrl+	F-key	Alt+	Menu Bar	Dashboard
dim1	...	...	...	...	...	...

Command: dim1
Dim: *(Enter a dimension command.)*

COMMAND LINE OPTION

Accepts all "original" dimension commands; see the Dim command for the complete list.

RELATED COMMANDS

DimStyle displays the dialog box for setting dimension variables.

Dim switches to AutoCAD's original dimensioning mode and remains there.

RELATED DIM VARIABLES

DimAso determines whether dimensions are drawn associatively.

DimTxt determines the height of text.

DimScale determines the dimension scale.

TIPS

- Use Dim1 when you require just a single original dimension command.

- This command is rarely used.

DimAligned

V. 2.5 Draws linear dimensions aligned with objects.

Command	Aliases	Ctrl+	F-key	Alt+	Menu Bar	Dashboard
dimaligned	dal	...	...	NG	Dimension	Dimensions
	dimali				⤷Aligned	

Command: dimaligned
Specify first extension line origin or <select object>: *(Pick a point, or press*
ENTER *to enable object selection, as shown by the following prompt.)*
Select object to dimension: *(Select an object.)*
Specify dimension line location or [Mtext/Text/Angle]: *(Pick a point, or enter an option.)*
Dimension text = *nnn*

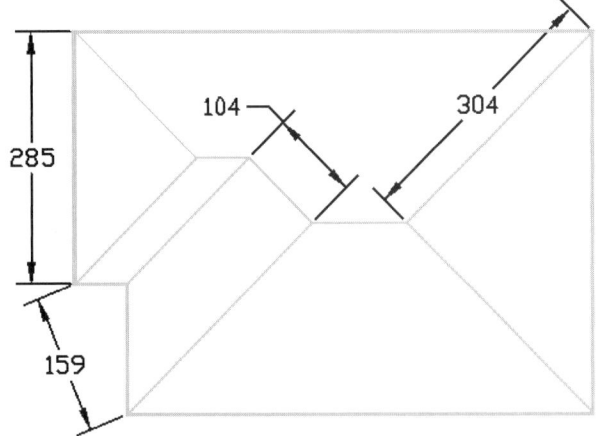

Examples of aligned dimensions.

COMMAND LINE OPTIONS

Specify first extension line origin picks a point for the origin of the first extension line.

Specify second extension line origin picks a point for the origin of the second extension line.

Select object selects an object to dimension, after pressing Enter.

Select object to dimension picks a line, circle, arc, polyline, or explodable object; individual segments of polylines are dimensioned.

Specify dimension line location picks a point from which to locate the dimension line and text.

Mtext changes the wording of the dimension text.

Text changes the position of the dimension text.

Angle changes the angle of the dimension text.

RELATED DIM COMMAND

DimRotated draws rotated dimension lines with perpendicular extension lines.

DimAngular

Rel 13 Draws dimensions that measure angles.

Command	Aliases	Ctrl+	F-key	Alt+	Menu Bar	Dashboard
dimangular	dan	...	...	NA	Dimension	Dimensions
	dimang				⟍Angular	

Command: dimangular
Select arc, circle, line, or <specify vertex>: *(Select an object, or pick a vertex.)*
Specify second angle endpoint: *(Pick a point; prompt varies, depending on object.)*
Specify dimension arc line location or [Mtext/Text/Angle/Quadrant]: *(Pick a point, or enter an option.)*
Dimension text = *nnn*

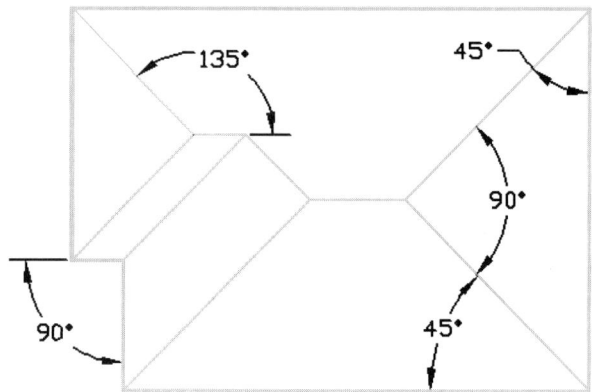

Examples of angular dimensions.

COMMAND LINE OPTIONS

Select arc measures the angle of the arc.

Circle prompts you to pick two points on the circle.

Line prompts you to pick two lines.

Specify vertex prompts you to pick points to make an angle.

Quadrant prompts you to select a point in a quadrant *(new to AutoCAD 2008)*.

Specify dimension arc/line location specifies the location of the angular dimension.

Mtext changes the wording of the dimension text.

Text changes the position of the text.

Angle changes the angle of the dimension text.

RELATED DIM COMMANDS

DimArc dimensions the lengths of arcs.

DimCenter places a center mark at the center of an arc or circle.

DimRadius dimensions the radius of an arc or circle.

DimArc

2006   Dimensions the length along arcs and polyline arcs.

Command	Alias	Ctrl+	F-key	Alt+	Menu Bar	Dashboard
dimarc	dar	...	...	NH	Dimension	Dimensions
					⤷Arc Length	

Command: dimarc
Select arc or polyline arc segment: *(Select an arc.)*
Specify arc length dimension location, or [Mtext/Text/Angle/Partial]: *(Pick a point, or enter an option.)*
Dimension text = 15'-9 3/16"

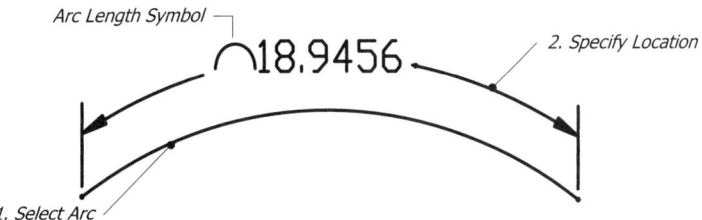

Arc Length Symbol

⌒18.9456

2. Specify Location

1. Select Arc

COMMAND LINE OPTIONS

Select arc measures the length of arcs.

Specify dimension arc length location specifies the location of the arc length dimension.

Mtext changes the wording of the dimension text.

Text changes the position of the text.

Angle changes the angle of the dimension text.

Partial prompts you to pick two points for a partial dimension:
Specify first point for arc length dimension: *(Pick a point.)*
Specify second point for arc length dimension: *(Pick another point.)*

RELATED SYSTEM VARIABLE

DimArcSym specifies the location of the dimension arc symbol.

TIPS

- Specify the placement of the arc length symbol in the Symbols and Arrows tab of the Dimension Style Manager. The arc symbol can be displayed above or in front of the dimension text.

- The extension lines can be orthogonal (angle less than 90 degrees) or radial (more than 90 degrees). Angle can be less than 180 degrees, or between 180 and 360 degrees.

DimBaseline

<u>Rel 13</u> Draws linear dimensions based on previous starting points.

Command	Aliases	Ctrl+	F-key	Alt+	Menu Bar	Dashboard
dimbaseline	dba	...	...	NB	Dimension	Dimensions
	dimbase				⬏ Baseline	

Command: dimbaseline
Specify a second extension line origin or [Undo/Select] <Select>: *(Pick a point, or enter an option.)*
Dimension text = *nnn*
Specify a second extension line origin or [Undo/Select] <Select>: *(Pick a point, enter an option, or press Esc to exit command.)*

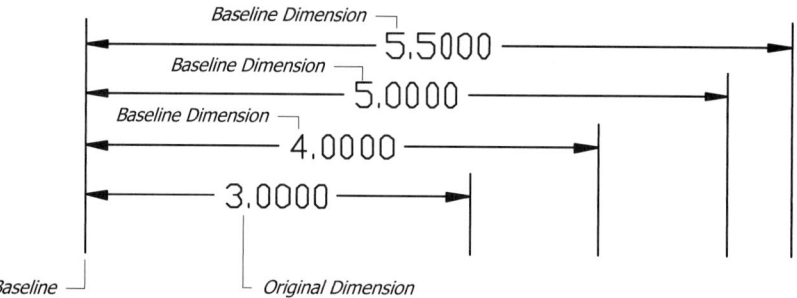

COMMAND LINE OPTIONS

Specify a second extension line origin positions the extension line of the next baseline dimension.

Select prompts you to select the base dimension.

Undo undoes the previous baseline dimension.

Esc exits the command.

RELATED DIM COMMANDS

Continue continues linear dimensioning from the last extension point.

QDim creates continuous or baseline dimensions quickly.

RELATED DIM VARIABLES

DimDli specifies the distance between baseline dimension lines.

DimSe1 suppresses the first extension line.

DimSe2 suppresses the second extension line.

DimBreak

2008 Breaks dimension and extension lines where they cross other objects.

Command	Alias	Ctrl+	F-key	Alt+	Menu Bar	Dashboard
dimbreak	dar	...	...	NK	Dimension	Dimensions
					⮡ Dimension Break	

Command: dimbreak
Select a dimension or [Multiple]: *(Select a dimension, or type M.)*
Select object to break dimension or [Auto/Restore/Manual] <Auto>: *(Select an object, or enter an option.)*
Select object to break dimension: *(Press Enter to exit the command.)*

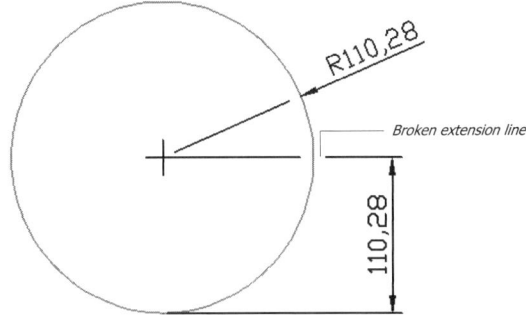

Broken extension line

COMMAND LINE OPTIONS

Select a dimension selects a single dimension for breaking.

Multiple selects multiple dimensions to break.

Select object selects the object that crosses the dimension.

Auto breaks all dimension and extension lines that are crossed.

Restore removes dimension breaks.

Manual specifies two points to indicate the break location.

Multiple Options
    Select dimensions: *(Select one or more, or press Ctrl+A for all.)*
    Select dimensions: *(Press Enter to end object selection.)*
    Enter an option [Break/Restore]: *(Enter an option.)*

Break adds gaps, where necessary.

Restore removes all gaps, healing the dimension and extension lines.

TIPS

- Use the manual option to create breaks even where no objects cross.

- The gap is approximately 0.125 inches (3.75mm), and is affected by DimScale. The gap is associated with the dimension: when the object changes size, so does the gap. There appears to be no system variable for specifying the gap, although it can be changed in DimStyle.

DimCenter

Rel 13 Draws center marks and lines on arcs and circles.

Command	Alias	Ctrl+	F-key	Alt+	Menu Bar	Dashboard
dimcenter	dce	...	...	NM	Dimensions ⮡ Center Mark	Dimensions

Command: dimcenter
Select arc or circle: *(Select an arc, circle, or explodable object.)*

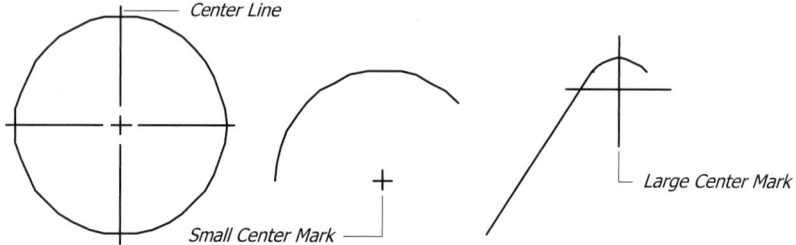

COMMAND LINE OPTION

Select arc or circle places the center mark at the center of the selected arc, circle, or polyarc.

RELATED DIM COMMANDS

DimAngular dimensions arcs and circles.

DimDiameter dimensions arcs and circles by diameter value.

DimRadius dimensions arcs and circles by radius value.

RELATED DIM VARIABLE

DimCen specifies the size and type of the center mark:

negative value — Draws center marks and lines.

0 — Does not draw center marks or center lines.

positive value — Draws center marks.

0.09 — Default value.

TIPS

- Changing the center mark size for a dimension style does not update existing center lines and marks.

- The center mark length defined by DimCen is from the center to one end of the mark; the center line length is the size of gap and extension beyond the circle or arc.

DimContinue

<u>Rel 13</u> Continues dimension from the second extension line of previous dimensions.

Command	Aliases	Ctrl+	F-key	Alt+	Menu Bar	Dashboard
dimcontinue	dco	...	...	NC	Dimension	Dimensions
	dimcont				⤷Continue	

Command: dimcontinue
Specify a second extension line origin or [Undo/Select] <Select>: *(Pick a point, enter an option, or press ENTER to select a base dimension.)*
Dimension text = *nnn*
Specify a second extension line origin or [Undo/Select] <Select>: *(Pick a point, enter an option, or press Esc to exit command.)*

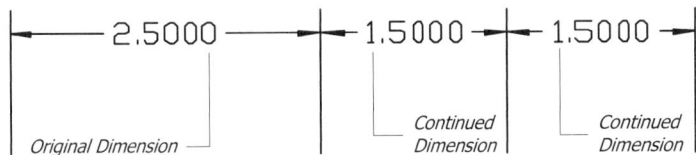

COMMAND LINE OPTIONS

Specify a second extension line origin positions the extension line of the next continued dimension.

Select prompts you to select the originating dimension.

Undo undoes the previous continued dimension.

Esc exits the command.

RELATED DIM COMMANDS

DimBaseline continues dimensioning from the first extension point.

QDim creates continuous or baseline dimensions quickly.

RELATED DIM VARIABLES

DimDli sets the distance between continuous dimension lines.

DimSe1 suppresses the first extension line.

DimSe2 suppresses the second extension line.

DimDiameter

<u>Rel 13</u> Draws diameter dimensions on arcs, circles, and polyline arcs.

Command	Aliases	Ctrl+	F-key	Alt+	Menu Bar	Dashboard
dimdiameter	ddi	...	...	ND	Dimension	Dimensions
	dimdia				⤷Diameter	

Command: diameter
Select arc or circle: (Select *an arc, circle, or explodable object.)*
Dimension text = *nnn*
Specify dimension line location or [Mtext/Text/Angle]: *(Pick point, or enter option.)*

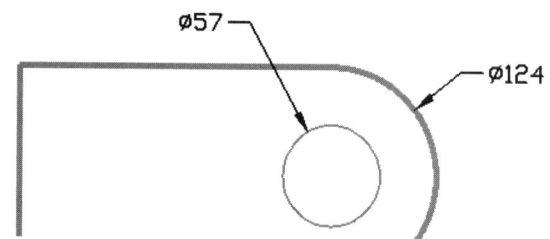

Examples of diameter dimensions.

COMMAND LINE OPTION

Select arc or circle selects an arc, circle, or polyarc.

Specify dimension line location specifies the location of the angular dimension.

Mtext changes the wording of the dimension text.

Text changes the position of the dimension text.

Angle changes the angle of the dimension text.

RELATED DIM COMMANDS

DimCenter marks the center point of arcs and circles.

DimRadius draws the radius dimension of arcs and circles.

TIPS

- The diameter symbol is automatically included. But if you need to add it manually, use the %%d code or the Unicode \U+2205.

- When you drag the dimension line outside an arc, AutoCAD draws an extension to the arc (*new in AutoCAD 2008*):

Extension line

DimDisassociate

<u>2002</u> Converts associative dimensions to non-associative (the command that's most difficult to spell correctly).

Command	Alias	Ctrl+	F-key	Alt+	Menu Bar	Dashboard
dimdisassociate	dda	...	...	...	...	...

Command: dimdisassociate
Select dimensions to disassociate...
Select objects: *(Select one or more dimensions, or enter All to select all dimensions.)*
Select objects: *(Press Enter to end object selection.)*
nn **disassociated.**

COMMAND LINE OPTION

Select objects selects dimensions to convert to non-associative type.

RELATED DIM COMMANDS

DimReassociate converts dimensions from non-associative to associative.

DimRegen updates the location of associative dimensions.

RELATED SYSTEM VARIABLE

DimAssoc determines whether newly-created dimensions are associative:

0 — Dimension is created "exploded," so that all parts (such as dimension lines, arrow heads) are individual, ungrouped objects.

1 — Dimension is created as a single object, but is not associative.

2 — Dimension is created as a single object, and is associative (default).

TIPS

- As you select objects, this command ignores non-dimensions, dimensions on locked layers, and those not in the current space (model or paper).

- The command reports filtered and disassociated dimensions.

- The effects of this command can be reversed with the U and the DimReassociate commands.

DimEdit

Rel 13 Applies editing changes to dimension text.

Command	Aliases Ctrl+	F-key	Alt+	Menu Bar	Dashboard
dimedit	ded ...	...	NQ	Dimension	...
	dimed	⃰Enter		⃰Oblique	

Command: dimedit
Enter type of dimension editing [Home/New/Rotate/Oblique] <Home>: *(Enter an option, or press Enter for Home.)*
Select objects: *(Select one or more objects.)*
Select objects: *(Press Enter.)*

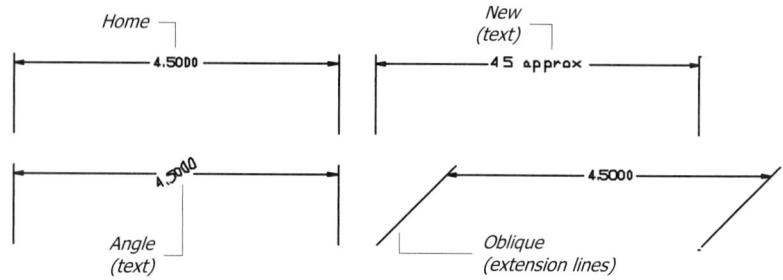

COMMAND LINE OPTIONS

Angle rotates the dimension text.

Home returns the dimension text to its original position.

Oblique rotates the extension lines.

New allows editing of the dimension text.

RELATED DIM COMMANDS

All.

RELATED DIM VARIABLES

Most.

TIPS

- DimEdit operates differently when used from the command line than from the menu bar: Oblique is the default option when selected from the menu bar.

- When you enter dimension text with the DimEdit command's New option, AutoCAD recognizes <> as *metacharacters* representing existing text.

- Use the Oblique option to angle dimension lines by 30 degrees, suitable for isometric drawings; use the Style command to oblique text by the same angle. See the Isometric command.

-DIMINSPECT Command

Command: -diminspect

Add inspection data or [Remove] <Add>: *(Enter an option.)*

Select dimensions: *(Select one or more dimensions.)*

Select dimensions: *(Press Enter to exit dimension selection.)*

Enter shape option [Round/Angular/None] <Round>: *(Enter an option.)*

Enter label data or <None>: *(Enter a value.)*

Enter inspection rate <100%>: *(Enter a value.)*

COMMAND LINE OPTIONS

Add adds inspection notation to the selected dimension.

Remove removes the inspection notations.

Round uses a rounded rectangle for the frame.

Angular uses a rectangle with angled ends.

None leaves out the frame.

Label data specifies the label text.

Inspection Rate specifies the percentage.

DEFINITION

Inspection Rate — specifies how frequently parts should be inspected, as a percentage.

TIPS

- This command works with linear and radial dimensions.

- This command makes AutoCAD drawings more compatible with Inventor mechanical CAD software.

- You can switch dimensions between the three inspection styles by rerunning the DimInspect command, and then selecting another shape.

- This command accepts inspection rates greater than 100%.

DimJogged

<u>**2006**</u> Places jogged radial dimensions; used for very large radii.

Command	Aliases Ctrl+	F-key	Alt+	Menu Bar	Dashboard
dimjogged	jog ...	...	NJ	Dimension	Dimensions
	djo			↳Jogged	

Command: dimjogged
Select arc or circle: *(Pick an arc or a circle.)*
Specify center location override: *(Pick point 2.)*
Dimension text = 28.7205
Specify dimension line location or [Mtext/Text/Angle]: *(Pick point 3.)*
Specify jog location: *(Pick point 4.)*

COMMAND LINE OPTIONS

Select arc or circle measures the radius of arcs and circles.

Specify center location override relocates the center from which to draw the dimension.

Specify dimension arc length location specifies the location of the arc length dimension.

Mtext changes the wording of the dimension text.

Text changes the position of the text.

Angle changes the angle of the dimension text.

Specify job location positions the jog.

RELATED SYSTEM VARIABLE

DimJogAngle specifies the jog angle; default = 45 degrees.

TIPS

- The *transverse angle* of the jog can be set in the Dimension Style Manager.

- Use this command when the center of arcs and circles is located outside the drawing area. It is generally used with very shallow arcs.

- Jogged dimensions can be edited using grips.

- When you drag the dimension line outside an arc, AutoCAD draws an extension to the arc (*new in AutoCAD 2008*):

DimJogLine

2008 Adds jogged dimension lines to linear dimensions; used with very large objects.

Command	Aliases Ctrl+	F-key	Alt+	Menu Bar	Dashboard
dimjogline	djl ...	...	NJJ	Dimension ↳Jogged Linear	Dimensions

Command: dimjogline
Select dimension to add jog or [Remove]: *(Pick a dimension, or type R.)*
Specify jog location (or press ENTER): *(Pick a point for the jog location, or press Enter for automatic placement.)*

COMMAND LINE OPTIONS

Select dimension selects the linear dimensions whose dimension lines should be jogged.

Remove removes jogs from selected dimensions.

Specify job location picks the location of the jog on the dimension lie.

Enter places jogs automatically, midway between the dimension text and the first extension line.

RELATED SYSTEM VARIABLE

DimJogAng specifies the jog angle; default = 45 degrees.

TIP

- You can move the location of the jog using grips editing.

DimLinear

Rel 13   Draws linear dimensions.

Command	Aliases	Ctrl+	F-key	Alt+	Menu Bar	Dashboard
dimlinear	dli	...	...	NL	Dimension	Dimensions
	dimlin				⮡ Linear	

Command: dimlinear
Specify first extension line origin or <select object>: *(Pick a point, or press Enter to select an object.)*
Specify second extension line origin: *(Pick a point.)*
Specify dimension line location or [Mtext/Text/Angle/Horizontal/Vertical/ Rotated]: *(Pick a point, or enter an option.)*
Dimension text = *nnn*

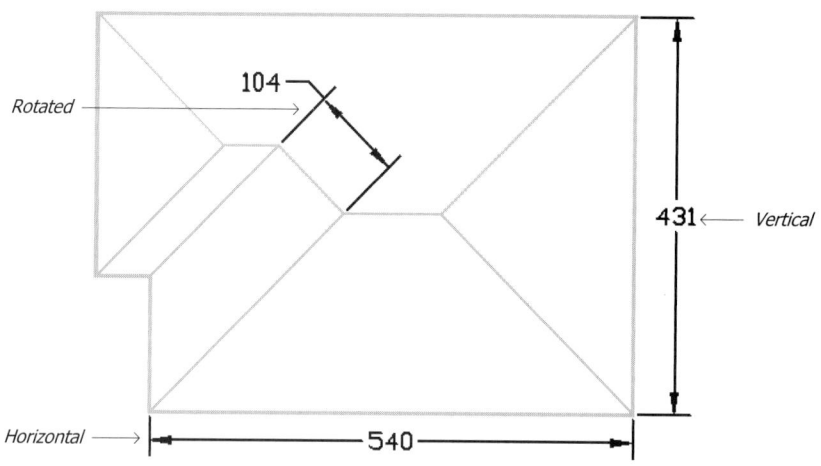

Examples of linear dimensions.

COMMAND LINE OPTIONS

Specify first extension line origin specifies the origin of the first extension line.

Select object dimensions a line, arc, or circle automatically.

Specify second extension line location specifies the location of the second extension line.

Specify dimension line location specifies the location of the dimension line.

Mtext displays the Text Formatting bar, which allows you to modify the dimension text; see the **MText** command.

Text option
Enter dimension text <*nnn*>: *(Enter dimension text, or press Enter to accept default value.)*

Enter dimension text prompts you to replace the dimension text on the command line.

Angle option
Specify angle of dimension text: *(Enter an angle.)*

Specify angle changes the angle of dimension text.

Horizontal option
Specify dimension line location or [Mtext/Text/Angle]: *(Pick a point or enter an option.)*

Horizontal option forces dimension to be horizontal.

Vertical option
Specify dimension line location or [Mtext/Text/Angle]: *(Pick a point or enter an option.)*

Vertical option forces dimension to be vertical.

Rotated option
Specify angle of dimension line <0>: *(Enter an angle.)*

Rotated option rotates the dimension.

Select Object Options

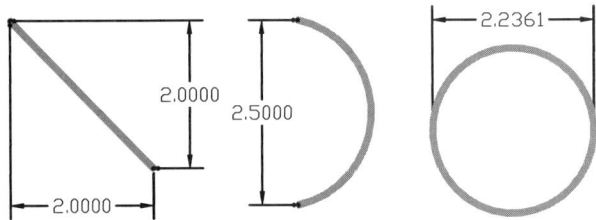

Examples of dimensioned objects: line, arc, and circle.

Specify first extension line origin or <select object>: *(Press ENTER.)*
Select object to dimension: *(Select one object.)*
Specify dimension line location or
[Mtext/Text/Angle/Vertical/Rotated]: *(Pick a point, or enter an option.)*
Dimension text = *nnn*

RELATED DIM COMMANDS

DimAligned draws linear dimensions aligned with objects.

QDim dimensions objects quickly.

DimOrdinate

Rel 13   Draws x and y ordinate dimensions.

Command	Aliases	Ctrl+	F-key	Alt+	Menu Bar	Dashboard
dimordinate	dor	...	...	NO	Dimension	Dimensions
	dimord				⤷Ordinate	

Command: dimordinate
Specify feature location: *(Pick a point.)*
Specify leader endpoint or [Xdatum/Ydatum/Mtext/Text/Angle]: *(Pick a point, or enter an option.)*
**Dimension text = ***nnn*

Examples of ordinate dimensions.

COMMAND LINE OPTIONS

Xdatum forces x ordinate dimension.

Ydatum forces y ordinate dimension.

Mtext displays the Text Formatting toolbar for modifying dimension text; see MText command.

Text prompts you to replace dimension text on the command line: 'Enter dimension text *<nnn>*'.

Angle changes the angle of dimension text: 'Specify angle of dimension text:'.

RELATED DIM COMMANDS

Leader draws leader dimensions.

Tolerance draws geometric tolerances.

TIPS

- Strictly speaking, "ordinate" is the distance from the x-axis only, but AutoCAD and the rest of the drafting world use it to mean from both the x and y axes; distance from y axis is "abcissa."

- The 0,0 point is determined by the UCS origin. Use the UCS Origin command to relocate 0,0.

- Define the x=0 and y=0 ordinate dimensions, and then use DimBaseline for additional ones.

. . . .

DimOverride

<u>Rel 13</u> Overrides the current dimension variables.

Command	Aliases	Ctrl+	F-key	Alt+	Menu Bar	Dashboard
dimoverride	dov	...	...	NV	Dimension	...
	dimover				↳Override	

Command: dimoverride
Enter dimension variable name to override or [Clear overrides]: *(Enter the name of a dimension variable, or type* **C**.*)*
Enter new value for dimension variable <*nnn*>: *(Enter a new value.)*
Select objects: *(Select one or more dimensions.)*
Select objects: *(Press Enter.)*

COMMAND LINE OPTIONS

Dimension variable to override requires you to enter the name of the dimension variable.

Clear removes the override.

New Value specifies the new value of the dimvar.

Select objects selects the dimension objects to change.

RELATED DIM COMMAND

DimStyle creates and modifies dimension styles.

RELATED DIM VARIABLES

All dimension variables.

DimRadius

Rel 13 Draws radial dimensions on circles, arcs, and polyline arcs.

Command	Aliases	Ctrl+	F-key	Alt+	Menu Bar	Dashboard
dimradius	dra	...	...	NR	Dimension	Dimensions
	dimrad				↳Radius	

Command: dimradius
Select arc or circle: *(Select an arc, circle, or explodable object.)*
**Dimension text = ** *nnn*
Specify dimension line location or [Mtext/Text/Angle]: *(Pick a point, or enter an option.)*

Examples of radial dimensions.

COMMAND LINE OPTIONS

Select arc or circle selects the arc, circle, or polyarc to dimension.

Specify dimension line location specifies the location of the angular dimension.

Mtext displays the Text Formatting bar, which allows you to modify the dimension text; see the MText command.

Text prompts you to replace the dimension text at the command line, 'Enter dimension text <*nnn*>'.

Angle changes the angle of dimension text, and prompts, 'Specify angle of dimension text:'.

RELATED DIM COMMANDS

DimCenter draws center marks on arcs and circles.

DimDiameter draws diameter dimensions on arcs and circles.

RELATED DIM VARIABLE

DimCen determines the size of the center mark.

TIP

- When you drag the dimension line outside an arc, AutoCAD draws an extension to the arc (*new in AutoCAD 2008*). See the DimDiameter command.

DimReassociate

2002 Associates dimensions with objects.

Command	Alias	Ctrl+	F-key	Alt+	Menu Bar	Dashboard
dimreassociate	dre	...	...	NN	Dimension	...
					⌞ Reassociate Dimensions	

Command: dimreassociate

Select dimensions to reassociate...

Select objects: *(Select one or more dimensions, or enter All to select all dimensions.)*

nn **found.** *nn* **were on a locked layer.** *nn* **were not in current space.**

Select objects: *(Press Enter to end object selection.)*

The prompts that follow depend on the type of each dimension selected.

Specify first extension line origin or [Select object]: *(Pick a point, or select an object.)*

Select arc or circle <next>: *(Press ENTER to end object selection.)*

Left: Boxed X indicates that the line dimension is associated with the circle.
Right: X indicates the radius dimension is not associated.

COMMAND LINE OPTIONS

Select objects selects the dimensions to reassociate.

Specify first extension line origin picks a point with which to associate the dimension.

Select arc or circle picks a circle or arc with which to associate the dimension.

Next goes to the next circle or arc.

RELATED DIM COMMANDS

DimDisassociate converts dimensions from associative to non-associative.

DimRegen updates the locations of associative dimensions.

TIPS

- As you select objects, this command ignores non-dimensions, dimensions on locked layers, and those not in the current space (model or paper).

- AutoCAD displays a boxed **X** to indicate the object with which the dimension is associated; an unboxed **X** indicates an unassociated dimension. The markers disappear when a wheelmouse performs a zoom or pan.

DimRegen

<u>2002</u>   Updates the locations of associative dimensions.

Command	Alias	Ctrl+	F-key	Alt+	Menu Bar	Dashboard
dimregen	...	...	...	...	...	...

Command: dimregen

COMMAND LINE OPTIONS
None.

RELATED DIM COMMANDS
DimDisassociate converts dimensions from associative to non-associative.

DimReassociate converts dimensions from non-associative to associative.

TIP
- This command is meant for use after three conditions:
  - The drawing has been edited by a version of AutoCAD prior to 2004.
  - The drawing contains dimensions associated with an external reference, and the xref has been edited.
  - The drawing is in layout mode, model space is active, and a wheelmouse has been used to pan or zoom.

DimRotated

2004 Draws rotated dimensions (an undocumented command).

Command	Alias	Ctrl+	F-key	Alt+	Menu Bar	Dashboard
dimrotated	...	...	...	...	...	...

Command: dimrotated
Specify angle of dimension line <0>: *(Enter a rotation angle.)*
Specify first extension line origin or <select object>: *(Pick a point, or press Enter to select one object.)*
Specify second extension line origin: *(Pick a point.)*
Specify dimension line location or [Mtext/Text/Angle]: *(Pick a point, or select an option.)*
Dimension text = *nnn*

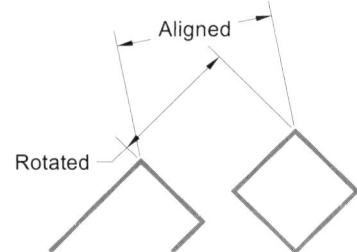

The aligned dimension is aligned with the two corner points.
The rotated dimension is placed at 45 degrees.

COMMAND LINE OPTIONS

Specify angle of dimension line specifies the angle at which the dimension line is rotated.

Specify first extension line origin specifies the origin of the first extension line.

Select object dimensions a line, arc, or circle automatically.

Specify second extension line location specifies the location of the second extension line.

Specify dimension line location specifies the location of the dimension line.

Mtext displays the Text Formatting bar, which allows you to modify the dimension text; see the **MText** command.

Text prompts you to replace the dimension text on the command line: 'Enter dimension text <*nnn*>'.

Angle changes the angle of dimension text: 'Specify angle of dimension text:'.

TIPS

- The DimRotated command draws dimensions at a specified angle, while the DimAligned command draws dimensions that align with the two pick points.

- You can specify the rotation angle by picking two points.

DimSpace

2008 spaces parallel dimensions evenly.

Command	Alias	Ctrl+	F-key	Alt+	Menu Bar	Dashboard
dimspace	...	...	...	NP	Dimension ↳Space	Dimension

Command: dimspace
Select base dimension: *(Select one parallel linear or angular dimension.)*
Select dimensions to space: *(Select other dimensions to space equally.)*
Enter value or [Auto] <Auto>: *(Specify the spacing value, or type A.)*

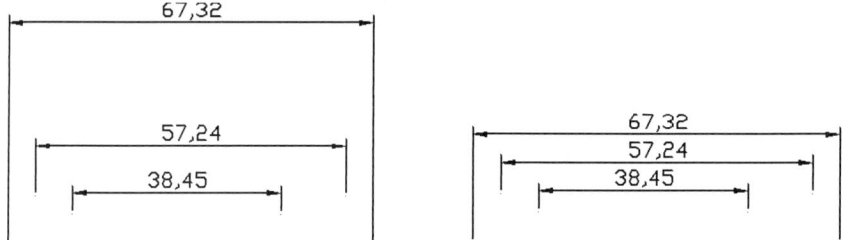

Left: Dimensions before being spaced.
Right: After spacing to the base dimension 38,45.

COMMAND LINE OPTIONS

Select base dimension specifies the dimension by which others are spaced.

Select dimensions to space specifies the dimensions to be relocated.

Enter value spaces the selected dimensions by the amount specified; entering 0 aligns them end to end.

Auto spaces the dimensions using 2x the height of the dimension text of the base dimension.

TIPS

- When dimension lines are overlapping, this command creates the spaced lines, but fails to move the dimensions into the spaces created — possibly a bug.

'DimStyle

Creates and edits dimension styles (short for DIMension STYLE).

Commands	Aliases Ctrl+		F-key	Alt+	Menu Bar	Dashboard
dimstyle	d	...	...	NS	Dimension	Dimension
	dst				⟍Style	
	dimsty			OD	Format	
	ddim				⟍Dimension Style	
-dimstyle	...	...	...	NU	Dimension	Dimension
					⟍Update	

Command: dimstyle

Displays dialog box:

DIMENSION STYLE MANAGER DIALOG BOX

Styles lists the names of dimension styles in the drawing.

List modifies the style names listed under Styles:

- **All styles** lists all dimension style names stored in the current drawing (default).
- **Styles in use** lists only those dimstyles used by dimensions in the drawing.

Don't list styles in Xrefs

☑ Does not list dimension styles found in externally-referenced drawings.

☐ Lists xref dimension styles under Styles (default).

Set Current sets the selected style as the current dimension style.

New creates a new dimension style via Create New Dimension Style dialog box.

Modify modifies an existing dimension style; displays the Modify Dimension Style dialog box.

Override allows temporary changes to dimension styles; displays Override Dimension Style dialog box.

Compare lists the differences between dimension variables of two styles; displays the Compare Dimension Styles dialog box.

· ·

Shortcut Menu

Right-click a dimension style name under Styles:

Set Current sets the selected dimension style as the current style.

Rename renames dimension styles.

Delete erases selected dimension styles from the drawing; you cannot erase the Standard style and styles that are in use.

SELECT ANNOTATION SCALE DIALOG BOX

The first time a dimension is placed in a new drawing, and the style is annotative, AutoCAD displays:

Set the Annotation Scale overrides the annotation scale specified by the dimension style.

☐ **Don't Show Me This Again** prevents the dialog box from appearing in the future.

CREATE NEW DIMENSION STYLE DIALOG BOX

When creating new dimension styles, typically you make a copy of an existing style, and then change it.

New Style Name names the new dimension style.

Start With lists the names of the current dimension style(s), which are used as the template for the new dimension style.

☑ **Annotative** makes the style annotative; see the Fit tab (*new to AutoCAD 2008*).

Use for creates a substyle that applies to a specific type of dimension type: linear, angular, radius, diameter, ordinate, leaders and tolerances.

Continue continues to the next dialog box, New Dimension Style.

Cancel dismisses this dialog box, and returns to the Dimension Style Manager dialog box.

Lines Tab

Sets the format of dimension lines and extension lines.

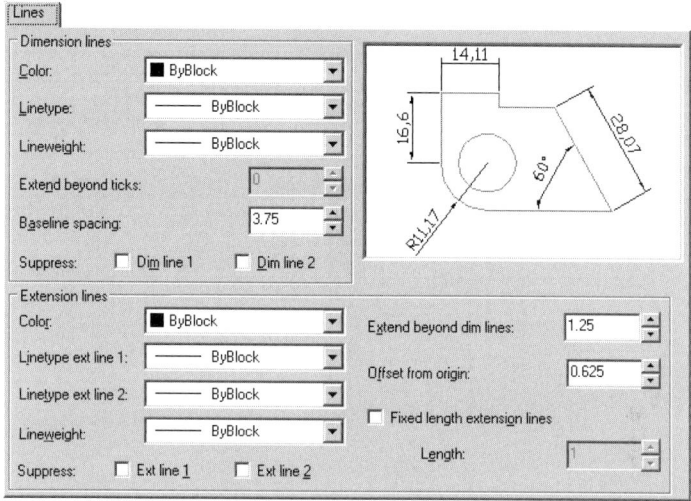

Dimension Lines options

(Related system variables are shown with each option.)

Color specifies the color of the dimension line; select Other to display the Select Color dialog box (stored in dimension variable DimClrD).

Linetype specifies the linetype of the dimension line (DimLtype).

Lineweight specifies the lineweight of the dimension line (DimLwD).

Extend beyond ticks specifies the distance the dimension line extends beyond the extension line; used with oblique, architectural, tick, integral, and no arrowheads (DimDlE).

Baseline spacing specifies the spacing between the dimension lines of a baseline dimension (DimDll); different than DimSpace.

Suppress suppresses dimension lines when outside the extension lines:

☐ **Dim line 1** for portion of dimension line left of text (DimSD1).

☐ **Dim line 2** for portion of dimension line right of text (DimSD2).

Extension Lines options

Color specifies the color of the extension line; select Other to display the Select Color dialog box (stored in dimension variable DimClrE).

Linetype ext 1 specifies the linetype of the first extension line (DimLtEx1).

Linetype ext 2 specifies the linetype of the second extension line (DimLtEx2).

Lineweight specifies the lineweight of the extension line (DimLwE).

Suppress suppresses extension lines:

☐ **Ext line 1** for left extension line (DimSE1).

☐ **Ext line 2** for right extension line (DimSE2).

Extend beyond dim lines specifies the distance the extension line extends beyond the dimension line; used with oblique, architectural, tick, integral, and no arrowheads (DimExe).

Offset from origin specifies the distance from the origin point to the start of the extension lines (DimExO).

☐**Fixed length extension lines** specifies the length of extension lines (DimFxlOn):

   ☐ **Length** is 0.18 by default (DimFXL).

Symbols and Arrows Tab

Sets the format of arrowheads, center marks, and arc lengths; changed in AutoCAD 2008:

Arrowheads options

First specifies name of arrowhead to use for the first end of the dimension line (DimBlk1).

Second specifies arrowhead for the second end of the dimension line (DimBlk2); select User Arrow to display the Select Custom Arrow Block dialog box, illustrated below:

Leader specifies the arrowhead for the leader; select User Arrow to display the Select Custom Arrow Block dialog box (DimLdrBlk).

Arrow Size specifies the size of arrowheads (DimASz).

➤ Closed filled
▷ Closed blank
⇨ Closed
● Dot
✔ Architectural tick
／ Oblique
⇒ Open
-⊖ Origin indicator
-⊚ Origin indicator 2
→ Right angle
⇀ Open 30
◆ Dot small
-○ Dot blank
◇ Dot small blank
⊏ Box
◼ Box filled
◁ Datum triangle
◀ Datum triangle filled
✔ Integral
None

Center Marks options

○ **None** places no center marks or center lines (DimCen=0).

◉ **Mark** places center marks (DimCen > 0).

○ **Line** places center marks and center lines (DimCen < 0).

Size specifies size of center marks or center lines (DimCen; default = 0.09).

Dimension Break option

Break size specifies the gap used in "broken" dimension and extension lines (*new to AutoCAD 2008*). See DimBreak command.

Arc Length Symbol options

⦿ **Preceeding dimension text** places symbol in front of text (DimArcSym).

○ **Above dimension text** places symbol above text.

○ **None** displays no symbol.

Radius Dimension Jog option

Jog Angle specifies the default jog angle for the DimJogged command (DimJogAngle).

Linear Jog Dimension option

Jog Height factor specifies the jog distance as a multiplier of the text height (*new to AutoCAD 2008*).

Text Tab

Text Appearance options

Text style specifies the text style name for dimension text (DimTxSty).

... displays the Text Style dialog box; see the Style command.

Text color specifies the color of the dimension line; select Other to display the Select Color dialog box (DimClrT).

Fill color specifies the color of the background behind the text (DimTFill and DimTFillClr).

Text height specifies the height of the dimension text, when the height defined by the text style is 0 (DimTxt).

Fraction height scale scales fraction text height relative to dimension text; AutoCAD multiples this value by the text height (DimTFac).

☐ **Draw frame around text** draws a rectangle around dimension text; when on, dimension variable **DimGap** is set to a negative value (DimGap).

Text Placement options

◉ **Vertical** specifies the vertical justification of dimension text relative to the dimension line:

Centered — Centers dimension text in the dimension line (DimTad = 0).

Above — Places text above the dimension line (DimTad = 1).

Outside — Places text on the side of the dimension line farthest from the first defining point (DimTad = 2).

JIS — Places text in conformity with JIS (DimTad= 3).

○ **Horizontal** specifies the horizontal justification of dimension text along the dimension and extension lines (DimJust; default = 0):

• **Centered** centers dimension text along the dimension line between the extension lines (DimJust = 0).

• **1st Extension Line** left-justifies the text with the first extension line (DimJust = 1).

• **2nd Extension Line** right-justifies the text with the second extension line (DimJust = 2).

• **Over 1st Extension Line** places the text over the first extension line (DimJust = 3).

• **Over 2nd Extension Line** places the text over the second extension line (DimJust = 4).

○ **Offset from dimension line** specifies the text gap, the distance between the dimension text and the dimension line (DimGap).

Text Alignment options

Horizontal forces dimension text always to be horizontal (DimTih = on; DimToh = off).

Aligned with dimension line forces dimension text to be aligned with the dimension line (DimTih = off; DimToh = off).

ISO Standard forces text to be aligned with the dimension line when inside the extension lines; forces text to be horizontal when outside the extension lines (DimTih = off; DimToh = on).

Fit Tab

Fit options

If there isn't enough room to place both text and arrows inside extension lines, the first thing to move outside the extension lines is:

⊙ **Either the text or the arrows, whichever fits best** places dimension text and arrowheads between the extension lines when space is available; when space is not available for both, the text or the arrowheads are placed outside the extension lines, whichever fits best; if there is room for neither, both are placed outside the extension lines (DimAtFit = 3).

○ **Arrows** places arrowheads between the extension lines when there is not enough room for arrowheads and dimension text (DimAtFit = 2).

○ **Text** places text between extension lines when there is not enough room for arrowheads and dimension text (DimAtFit = 1).

○ **Both text and arrows** places both outside the extension lines when there is not enough room for dimension text and arrowheads (DimAtFit = 0).

○ **Always keep text between ext lines** forces text between the extension lines (DimTix).

□ **Suppress arrows if they don't fit inside the extension lines** suppresses arrowheads when there is not enough room between the extension lines.

Text Placement:

⊙ **Beside the dimension line** places dimension text beside the dimension line (DimTMove =0).

○ **Over the dimension line, with a leader** draws a leader when dimension text is moved away from the dimension line (DimTMove = 1).

○ **Over the dimension line, without a leader** does not draw a leader when dimension text is moved away from the dimension line (DimTMove = 2).

Scale for Dimension Features

□ **Annotative** turns on the annotative property (DimAnno; *new to AutoCAD 2008*).

○ **Scale dimension to layout** determines the scale factor of dimensions in layout mode; based on scale factor between current model space viewport and layout (DimScale).

⊙ **Use overall scale of** specifies the scale factor for all dimensions in the drawing; affects text and arrowhead sizes, distances, and spacing (DimScale).

Fine Tuning options

□ **Place text manually when dimensioning** places text at the position picked at the 'Dimension line location' prompt (DimUpt).

□ **Always draw dim line between ext lines** forces the dimension line between the extension lines (DimTofl).

Primary Units Tab

Linear Dimensions options

Unit Format specifies the linear units format; does not apply to angular dimensions (DimLUnit).

Precision specifies the number of decimal places (or fractional accuracy) for linear dimensions (DimDec).

Fraction Format specifies the stacking format of fractions (DimFrac):

- Horizontal stacked — $\frac{1}{2}$
- Diagonal stacked — ½

Decimal Separator specifies the separator for decimal formats (DimDSep).

Round Off specifies the format for rounding dimension values; does not apply to angular dimensions (DimRnd).

Prefix specifies a prefix for dimension text (DimPost); you can use the following control codes to show special characters:

Control Code	Meaning
%%nnn	Character specified by ASCII number *nnn*.
%%o	Turns on and off overscoring.
%%u	Turns on and off underscoring.
%%d	Degrees symbol (°).
%%p	Plus/minus symbol (±).
%%c	Diameter symbol (Ø).
%%%	Percentage sign (%).

Suffix specifies a suffix for dimension text (DimPost); you can use the control codes listed above to show special characters.

Measurement Scale

Scale factor specifies a scale factor for linear measurements, except for angular dimensions (DimLFac); for example, use this to change dimension values from imperial to metric.

□ **Apply to layout dimensions only** specifies that the scale factor is applied only to dimensions created in layout mode or paper space (stored as a negative value in DimLFac).

Zero Suppression options:

□ **Leading** suppresses leading zeros in all decimal dimensions (DimZin = 4).

□ **Trailing** suppresses trailing zeros in all decimal dimensions (DimZin = 8).

☑ **0 Feet** suppresses zero feet of feet-and-inches dimensions (DimZin = 0).

☑ **0 Inches** suppresses zero inches of feet-and-inches dimensions (DimZin = 2).

Angular Dimensions options

Units Format specifies the format of angular dimensions (DimAUnit).

Precision specifies the precision of angular dimensions (DimADec).

Zero Suppression options

Same as for linear dimensions (DimAZin).

Alternate Units Tab

□ **Display alternate units** adds alternate units to dimension text (DimAlt).

Alternate Units options

Unit Format specifies the alternate units format (DimAltU).

Precision specifies the number of decimal places or fractional accuracy (DimAltD).

Multiplier for alt units specifies the conversion factor between primary and alternate units (DimAltF).

Round distances to specifies the format for rounding dimension values; does not apply to angular dimensions (DimAltRnd).

Prefix specifies a prefix for dimension text (DimAPost); you can use control codes to show special characters.

Suffix specifies a suffix for dimension text (DimAPost).

Zero Suppression options

☐ **Leading** suppresses leading zeros in all decimal dimensions (DimAltZ = 4).

☐ **Trailing** suppresses trailing zeros in all decimal dimensions (DimAltZ = 8).

☑ **0 Feet** suppresses zero feet of feet-and-inches dimensions (DimAltZ = 0).

☑ **0 Inches** suppresses zero inches of feet-and-inches dimensions (DimAltZ = 2).

Placement:

⊙ **After primary units** places alternate units behind the primary units (DimAPost).

○ **Below primary units** places alternate units below the primary units.

Tolerances Tab

Tolerance Format options

Method specifies the tolerance format:

- **None** does not display tolerances (DimTol).
- **Symmetrical** places ± after the dimension (DimTol=0; DimLim=0).
- **Deviation** places + and – symbols (DimTol = 1; DimLim = 1).
- **Limits** places maximum over minimum value (DimTol=0; DimLim=1):
    Maximum value = dimension value + upper value.
    Minimum value = dimension value - lower value.
- **Basic** boxes the dimension text (DimGap=negative value).

Precision specifies the number of decimal places for tolerance values (DimTDec).

Upper value specifies the upper tolerance value (DimTp).

Lower value specifies the lower tolerance value (DimTm).

Scaling for height specifies the scale factor for tolerance text height (DimTFac).

Vertical position specifies the vertical text position for symmetrical and deviation tolerances:

Top — aligns the tolerance text with the top of the dimension text (DimTolJ = 2).

Middle — aligns the tolerance text with the middle of the dimension text (DimTolJ = 1).

Bottom — aligns the tolerance text with bottom of dim text (DimTolJ = 0).

Tolerance Alignment options

Specifies alignment of Deviation and Limits tolerances (new to AutoCAD 2008):

○ **Align Decimal Separators** lines up tolerances by their decimal points.

⊙ **Align Operator Symbols** lines up tolerances by their prefix symbols, + and - .

Zero Suppression options

☐ **Leading** suppresses leading zeros in all decimal dimensions (DimTZin = 4).

☐ **Trailing** suppresses trailing zeros in all decimal dimensions (DimTZin = 8).

☑ **0 Feet** suppresses zero feet of feet-and-inches dimensions (DimTZin = 0).

☑ **0 Inches** suppresses zero inches of feet-and-inches dimensions (DimTZin = 2).

Alternate Unit Tolerance options

Precision specifies the precision (number of decimal places) of tolerance text (DimAltTd).

Zero Suppression options: the same as for tolerance format (DimAltTz).

MODIFY DIMENSION STYLE DIALOG BOX

This dialog box is identical to the New Dimension Style dialog box.

OVERRIDE DIMENSION STYLE DIALOG BOX

This dialog box is identical to the New Dimension Style dialog box.

COMPARE DIMENSION STYLES DIALOG BOX

*The list is blank when AutoCAD finds no differences. When **With** is set to **<none>** or the same style as **Compare**, AutoCAD displays all dimension variables.*

Compare displays the name of one dimension style.

With displays the name of the second dimension style.

Copy to Clipboard copies the style comparison text to the Clipboard, which can be pasted in another Windows application.

Description describes the dimension variable.

Variable names the dimension variable.

Close closes the dialog box.

-DIMSTYLE Command

Command: -dimstyle
Current dimension style: Standard Annotative: No
Enter a dimension style option
[ANnotative/Save/Restore/STatus/Variables/Apply/?] <Restore>: *(Enter an option.)*
Current dimension style: Standard
Enter a dimension style name, [?] or <select dimension>: *(Enter an option.)*
Select dimension: *(Select a dimension in the drawing.)*
Current dimension style: Standard

COMMAND LINE OPTIONS

Annotative toggles the annotative property of dimension styles (*new to AutoCAD 2008*).

Save saves current dimension variable settings as a named dimension style.

Restore retrieves dimvar settings from a named dimension style.

STatus lists current setting of all dimension variables.

Variables lists the current setting for a specific dimension variable.

Apply updates selected dimension objects with current dimstyle settings.

? lists names of dimstyles stored in drawing.

INPUT OPTIONS

~*dimvar* (*tilde prefix*) lists the differences between current and selected dimstyles.

Enter lists the dimension variable settings for the selected dimension object.

RELATED DIMENSION COMMANDS

DDim changes dimension variable settings.

DimScale determines the scale of dimension text.

RELATED DIMENSION VARIABLES

All.

DimStyle contains the name of the current dimension style.

DimAnno toggles annotative scaling in dimensions (*new to AutoCAD 2008*).

TIPS

- At the 'Dim' prompt, the Style command sets the text style for the dimension text, and does *not* select a dimension style.

- Dimension styles cannot be stored to disk, only in drawings. However, you can import dimension styles from other drawings by using the XBind Dimstyle command, or by dragging dimstyle names from Design Center.

- As of AutoCAD 2008, dimensions support annotative scale factors. This means that scale-dependent objects are displayed when their scale factor matches the scale of model space. This includes dimension text and symbols, arrowheads, leaders, and tolerances.

- When the annotative property is turned on, AutoCAD forces associativity in dimensions.

- All new drawings contain two dimension sytles: Annotative and Standard (non-annotative).

DimTEdit

<u>**Rel 13**</u> Changes the location and orientation of text in dimensions dynamically.

Command	Alias	Ctrl+	F-key	Alt+	Menu Bar	Dashboard
dimtedit	dimted ...		...	NX	Dimension ⤷Align Text	...

Command: dimtedit
Select dimension: *(Select a dimension in the drawing.)*
Specify new location for dimension text or [Left/Right/Center/Home/Angle]:
(Pick a point, or enter an option.)

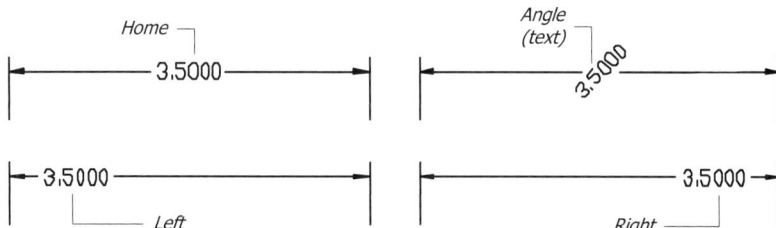

COMMAND LINE OPTIONS

Select dimension selects the dimension to edit.

Angle rotates the dimension text.

Center centers the text on the dimension line.

Home returns the dimension text to the original position.

Left moves the dimension text to the left.

Right moves the dimension text to the right.

RELATED DIM VARIABLES

DimSho specifies whether dimension text is updated dynamically while dragged.

DimTih specifies whether dimension text is drawn horizontally or aligned with the dimension line.

DimToh specifies whether dimension text is forced inside the dimension lines.

TIPS

- This command works only with dimensions created with DimAssoc = 1 or 2; use the DdEdit command to edit text in non-associative dimensions.

- An angle of 0 returns dimension text to its default orientation.

DimVertical

Draw vertical dimensions (an undocumented command).

Command	Alias	Ctrl+	F-key	Alt+	Menu Bar	Dashboard
dimvertical	...	...	...	...	...	...

Command: dimvertical
Specify first extension line origin or <select object>: *(Pick a point, or press Enter to select one object.)*
Specify second extension line origin: *(Pick a point.)*
Specify dimension line location or [Mtext/Text/Angle]: *(Pick a point, or select an option.)*
Dimension text = *nnn*

Examples of vertical dimensions.

COMMAND LINE OPTIONS

Specify first extension line origin specifies the origin of the first extension line.

Select object dimensions a line, arc, circle, or explodable object automatically.

Specify second extension line location specifies the location of the second extension line.

Specify dimension line location specifies the location of the dimension line.

Mtext displays the Multiline Text Editor dialog box, which allows you to modify the dimension text; see the MText command.

Text prompts you to replace the dimension text on the command line: 'Enter dimension text *<nnn>*'.

Angle changes the angle of dimension text: 'Specify angle of dimension text:'.

RELATED DIM COMMANDS
All.

RELATED DIM VARIABLES
Most.

'Dist

V. 1.0 Lists the 3D distances and angles between two points (short for DISTance).

Command	Alias	Ctrl+	F-key	Alt+	Menu Bar	Dashboard
dist	di	...	...	TQD	Tools	...
					↳Inquiry	
					↳Distance	

Command: dist
Specify first point: *(Pick a point.)*
Specify second point: *(Pick another point.)*

Sample result:

Specify first point: Specify second point:
Distance = 19, Angle in XY Plane = 22, Angle from XY Plane = 0
Delta X = 18, Delta Y = 7, Delta Z = 0

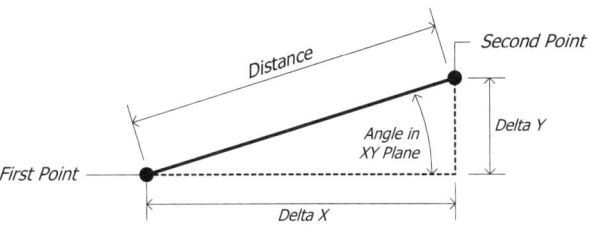

COMMAND LINE OPTIONS

Specify first point determines the start point of distance measurement.

Specify second point determines the endpoint.

RELATED COMMANDS

Area calculates the area and perimeter of objects.

Id lists the 3D coordinates of points.

RELATED SYSTEM VARIABLE

Distance specifies the last calculated distance.

TIPS

- Use object snaps to measure precisely the distance between two geometric features.

- When the z-coordinate is left out, Dist uses the current elevation for z.

- Object snaps give 3D points, and therefore the 3D distance.

Aliased Command

Details about the **DistantLight** command are found under the Light command.

Divide

V. 2.5 Places points or blocks at equal distances along objects.

Command	Alias	Ctrl+	F-key	Alt+	Menu Bar	Dashboard
divide	div	...	...	DOD	Draw	...
					⤷ Point	
					⤷ Divide	

Command: divide
Select object to divide: *(Select one object.)*
Enter the number of segments or [Block]: *(Enter a number, or enter B.)*

COMMAND LINE OPTIONS

Select object to divide selects a single open or closed object.

Enter the number of segments specifies the number of segments; must be a number between 2 and 32767.

Block Options

Enter name of block to insert: *(Enter the name of a block.)*

Align block with object? [Yes/No] <Y>: *(Type Y or N.)*

Enter the number of segments: *(Enter a number.)*

Enter name of block to insert specifies the block to insert. *Caution!* The block must be defined in the current drawing prior to starting this command.

Align block with object aligns the block's x axis with the object.

RELATED COMMANDS

Block creates the block to use with the Divide command.

DdPType controls the size and style of points.

Insert places a single block in the drawing.

MInsert places an array of blocks in the drawing.

Measure places points or blocks at measured distances.

TIPS

- The first dividing point on a closed polyline is its initial vertex; on circles, the first dividing point is in the 0-degree direction from the center.

- The points or blocks are placed in the Previous selection set, so that you can select them with the next 'Select Objects' prompt.

- Objects are unchanged by this command.

- Use the DdPType command to make points visible.

Donut

<u>**V. 2.5**</u>   Draws solid-filled circles as a wide polyline consisting of a pair of arcs.

Commands	Alias	Ctrl+	F-key	Alt+	Menu Bar	Dashboard
donut	do	...	...	DD	Draw	...
	doughnut				↳Donut	

Command: donut
Specify inside diameter of donut <0.5000>: *(Enter a value.)*
Specify outside diameter of donut <1.0000>: *(Enter a value.)*
Specify center of donut or <exit>: *(Pick a point.)*
Specify center of donut or <exit>: *(Press Enter to exit command.)*

Default Donut
Inside Diameter = 0.5
Outside Diameter = 1.0

Polyline Circle
Inside Diameter = 1.0
Outside Diameter = 1.0

Solid Donut
Inside Diameter = 0
Outside Diameter = 1.0

COMMAND LINE OPTIONS

Inside diameter specifies the inner diameter by entering a number or picking two points.

Outside diameter specifies the outer diameter.

Center of donut determines the donut's center point by specifying coordinates, or picking points.

Exit exits the command.

RELATED COMMAND

Circle draws circles.

RELATED SYSTEM VARIABLES

DonutId specifies the default internal diameter.

DonutOd specifies the default outside diameter.

Fill toggles the filling of the donut, as well as wide polylines, hatch, and 2D solids.

TIPS

- This command repeats itself until canceled.

- Donuts are made of two polyline arcs.

'Dragmode

V. 2.0 Controls the display of objects during dragging operations.

Command	Alias	Ctrl+	F-key	Alt+	Menu Bar	Dashboard
dragmode	...	...	...	...	...	...

Command: dragmode
Enter new value [ON/OFF/Auto] <Auto>: *(Enter an option.)*

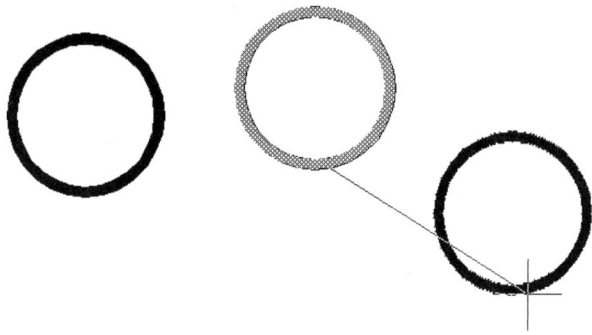

Highlight image (center) and drag image (right).

COMMAND LINE OPTIONS

ON enables dragging display only with the **Drag** option.

OFF turns off all dragging display.

Auto allows AutoCAD to determine when to display drag image.

COMMAND MODIFIER

Drag displays drag images when DragMode = on.

RELATED SYSTEM VARIABLE

DragMode is the current drag setting:

  0 — No drag image.

  1 — On if required.

  2 — Automatic.

TIP

■ Turn off DragMode and Highlight in very large drawings to speed up editing.

DrawingRecovery, DrawingRecoveryHide

Displays and hides the Drawing Recovery Manager palette.

Command	Alias	Ctrl+	F-key	Alt+	Menu Bar	Dashboard
drawingrecovery	drm	...	...	FUD	File	...
					⮡ Drawing Utilities	
					⮡ Drawing Recovery Manager	

drawingrecoveryhide

Command: drawingrecovery

Displays palette (also displayed automatically following a software crash):

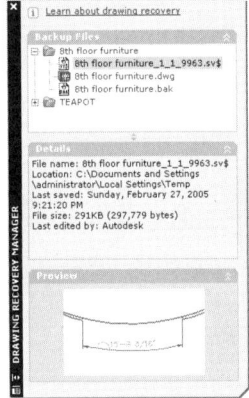

SHORTCUT MENU

Right-click drawing names for shortcut menu:

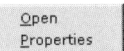

Open opens the drawing in AutoCAD; see the Open command.

Properties displays the Properties dialog box; see the DwgProps command.

Command: drawingrecoveryclose

Closes the palette.

RELATED COMMANDS

Audit determines if drawing files have problems.

Recover recovers drawings at the command line.

RecoverAll recovers drawings and attached xrefs (new to AutoCAD 2008).

RELATED SYSTEM VARIABLES

DrState reports whether the Drawing Recovery palette is open.

RecoveryMode specifies when the Drawing Recovery palette appears, if at all.

DrawOrder

Rel. 14 Controls the display of overlapping objects.

Command	Alias	Ctrl+	F-key	Alt+	Menu Bar	Dashboard
draworder	dr	...	...	TD	Tools	...
					⤷Draw Order	

Command: draworder
Select objects: *(Select one or more objects.)*
Select objects: *(Press Enter.)*
Enter object ordering option
[Above objects/Under objects/Front/Back] <Back>: *(Enter an option.)*

Left: Text **under** *solid.*
Right: Text **above** *solid.*

COMMAND LINE OPTIONS

Select objects selects the objects to be moved.

Above object forces selected objects to appear above the reference object.

Under object forces selected objects to appear below the reference object.

Front forces selected objects to the top of the display order.

Back forces selected objects to the bottom of the display order.

RELATED COMMANDS

BHatch displays hatching in front of or behind other objects.

TextToFront displays text and/or dimensions on top of overlapping objects.

RELATED SYSTEM VARIABLES

DrawOrderCtl controls draw order:

    0 — Turns off draw order.
    1 — Specifies that objects are displayed with draw order.
    2 — Specifies that new objects take on draw order of the object selected first.
    3 — Combines options 1 and 2 (default).

HpDrawOrder specifies display order of hatch patterns and fills; see the Hatch command.

TIPS

- When you pick more than one object for reordering, AutoCAD maintains the relative display order of the selected objects. The order in which you select objects has no effect on display order.

- When DrawOrderCtl is set to 3, editing operations may take longer.

- *Draw order inheritance* means that new objects created from objects with draw order are assigned the display order of the object selected first.

'DSettings

2 0 0 0 Controls the most common settings for drafting operations (short for Drafting SETTINGS).

Commands	Aliases	Ctrl+	F-key	Alt+	Menu Bar	Dashboard
dsettings	ds	...	...	TF	Tools	...
	se, ddrmodes				⤷Drafting Settings	
	os, osnap					

+dsettings

Command: dsettings

Displays dialog box:

DRAFTING SETTINGS DIALOG BOX

Options displays the Options dialog box; does not work during transparent commands.

Snap and Grid tab

☐ **Snap On (F9)** turns on and off snap mode (stored in system variable SnapMode).

☐ **Grid On (F7)** turns on and off the grid display (GridMode).

Snap Spacing options

Snap X Spacing specifies the snap spacing in the x direction (SnapUnit).

Snap Y Spacing specifies the snap spacing in the y direction (SnapUnit).

☑ **Equal X and Y Spacing** sets the y spacing to that of x.

Polar Spacing options

Polar Distance specifies the snap distance, when Snap Type is set to Polar snap; when 0, the polar snap distance is set to the value of Snap X Spacing (PolarDist).

Snap Type options

⊙ **Grid Snap** specifies non-polar snap (SnapType).

    ⊙ **Rectangular Snap** specifies rectangular snap (SnapStyl).

    ○ **Isometric Snap** specifies isometric snap mode (SnapSty).

○ **Polar Snap** specifies polar snap (SnapType).

SnapType	SnapStyl	Meaning
0 (off)	**0** (off)	Rectangular snap.
0 (off)	**1** (on)	Isometric snap.
1 (on)	**0** (off)	Polar snap.

Grid Spacing options

Grid X spacing specifies the spacing of grid lines in the x direction; when 0, the grid spacing is set to the value of Snap X spacing (GridUnit).

Grid Y spacing specifies the spacing of grid lines in the y direction; when 0, the grid spacing is set to the value of Snap Y spacing (GridUnit).

Major Line Every specifies the distance between major grid lines (GridMajor).

Grid Behavior options

☑ **Adaptive Grid** changes the grid spacing, depending on the zoom level (GridDisplay = 2).

☐ **Allow Subdivision Below Grid Spacing** draws sub-grid lines (GridDisplay = 4).

☑ **Display Grid Beyond Limits** draws the grid beyond the extents set by the Limits command (GridDisplay = 1).

☐ **Follow Dynamic UCS** causes the grid to rotate with the dynamic UCS (GridDisplay = 8).

Polar Tracking tab

☑ **Polar Tracking On (F10)** turns on and off polar tracking (AutoSnap; default = off).

Polar Angle Settings options

Increment Angle specifies the increment angle displayed by the polar tracking alignment path; select a preset angle — 90, 60, 45, 30, 22.5, 18, 15, 10, or 5 degrees — or enter a value (PolarAng).

☑ **Additional Angles** allows you to set additional polar tracking angles (PolarMode).

New adds up to ten polar tracking alignment angles (PolarAddAng; default = 0;15;23;45).

Delete deletes added angles.

Object Snap Tracking Settings options

◉ **Track Orthogonally Only** displays orthogonal tracking paths when object snap tracking is on (PolarMode).

○ **Track Using All Polar Angle Settings** tracks cursor along polar angle tracking path when object snap tracking is turned on (PolarMode).

Polar Angle Measurement options

◉ **Absolute** forces polar tracking angles along the current user coordinate system (UCS).

○ **Relative To Last Segment** forces polar tracking angles on the last-created object.

Object Snap tab

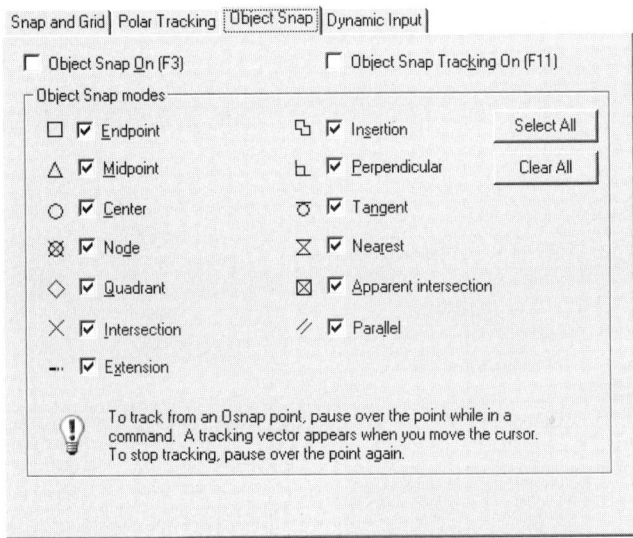

□ **Object Snap On (F3)** turns on and off running object snaps (OsMode).

□ **Object Snap Tracking On (F11)** toggles object snap tracking (AutoSnap).

Object Snap Modes options

☑ **ENDpoint** snaps to the nearest endpoint of a line, multiline, polyline segment, ray, arc, and elliptical arc; and to the nearest corner of a trace, solid, and 3D face.

□ **MIDpoint** snaps to the midpoint of a line, multiline, polyline segment, solid, spline, xline, arc, ellipse, and elliptical arc.

☑ **CENter** snaps to the center of an arc, circle, ellipse, and elliptical arc.

□ **NODe** snaps to a point.

□ **QUAdrant** snaps to a quadrant point (90 degrees) of an arc, circle, ellipse, and elliptical arc.

☑ **INTersection** snaps to the intersection of a line, multiline, polyline, ray, spline, xline, arc, circle, ellipse, and elliptical arc; edges of regions; it does not snap to the edges and corners of 3D solids.

☑ **EXTension** displays an extension line from the endpoint of objects; snaps to the point where two objects would intersect if they were infinitely extended; does not work with the edges and corners of 3D solids; automatically turns on intersection mode. (Do not turn on apparent intersection at the same time as extended intersection.)

☐ **INSertion** snaps to the insertion point of text, block, attribute, or shape.

☐ **PERpendicular** snaps to the perpendicular of a line, multiline, polyline, ray, solid, spline, xline, arc, circle, ellipse, and elliptical arc; snaps from a line, arc, circle, polyline, ray, xline, multiline, and 3D solid edge; in this case *deferred perpendicular mode* is automatically turned on.

☐ **TANgent** snaps to the tangent of an arc, circle, ellipse, or elliptical arc; deferred tangent snap mode is automatically turned on when more than one tangent snap is required.

☐ **NEArest** snaps to the nearest point on a line, multiline, point, polyline, spline, xline, arc, circle, ellipse, and elliptical arc.

☐ **APParent intersection** snaps to the apparent intersection of two objects that do not actually intersect but appear to intersect in 3D space; works with a line, multiline, polyline, ray, spline, xline, arc, circle, ellipse, and elliptical arc; does not work with edges and corners of 3D solids.

☐ **PARallel** snaps to a parallel point when AutoCAD prompts for a second point.

Clear All turns off all object snap modes.

Select All turns on all object snap modes.

Dynamic Input tab

☑ **Enable Pointer Input** toggles coordinate input at the drawing cursor, as illustrated below. Press the Tab key to switch between the x and y coordinates of the cursor position.

☑ **Enable Dimension Input Where Possible** toggles the display of lengths and angles when drawing and editing. Press Tab to switch between distance and angle.

☑ **Show Command Prompting Near Crosshairs** switches the display of commands and prompts between the 'Command:' prompt and the drawing area. Press the Down Arrow key to display additional options.

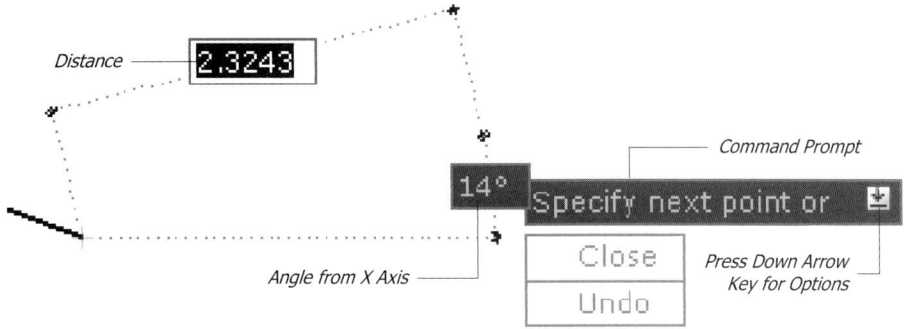

POINT INPUT SETTINGS DIALOG BOX

Format options

For The Second or Next Points, Default To (DynPiFormat and DynPiCoords):

- ◉ **Polar Format** shows distance and angle; press **,** to switch to Cartesian format.

- ○ **Cartesian Format** shows x and y distance; press **<** to change to polar format.

- ◉ **Relative Coordinates** shows coordinates relative to the last point; press **#** to switch to absolute format.

- ○ **Absolute Coordinates** shows coordinates relative to the origin; press **@** to switch to relative format; direct distance entry is disabled when this option is turned on.

Visibility options

Show Coordinate Tooltips (DynPiVis):

- ○ **As Soon As I Type Coordinate Data** displays tooltips when you enter coordinates.

- ◉ **When a Command Asks for a Point** displays tooltips when commands prompt for points.

- ○ **Always - Even When Not in a Command** always displays tooltips.

DIMENSION INPUT SETTINGS DIALOG BOX

Visibility options

When Grip Stretching (DynDiVis):

○ **Show Only 1 Dimension Input Field at a Time** displays only the length dimension.

Length Change

Resulting Dimension

◉ **Show 2 Dimension Input Fields at a Time** displays the length and resulting dimension.

○ **Show the Following Dimension Input Fields Simultaneously** (DynDiGrip):

☑ **Resulting Dimension** displays the length dimension tooltip.

☑ **Length Change** displays the change in length.

☑ **Absolute Angle** displays the angle dimension tooltip.

☑ **Angle Change** displays the change in the angle.

☑ **Arc Radius** displays the radius of arcs.

Angle Change — Length Change

Resulting Dimension — Command Prompt

Absolute Angle

DRAFTING TOOLTIP APPEARANCE DIALOG BOX

Color options
Model Color displays the Select Color dialog box for specifying tooltip colors in model space.

Layout Color displays the same dialog box, but for tooltips in paper space.

Size specifies tooltip size; 0 = default size.

Transparency controls transparency of tooltips; 0% = opaque.

Apply To options
⊙ **Override OS Settings for All Drafting Tooltips** applies settings to all tooltips, overriding settings in the operating system.

○ **Use Settings Only for Dynamic Input Tooltips** applies settings to drafting tooltips only.

· ·

+DSETTINGS Command
Command: +dsettings
Tab Index <0>: *(Enter a digit.)*

COMMAND LINE OPTION
Tab Index displays the Drafting Settings dialog box with the associated tab:

0 — Displays Snap and Grid tab *(default)*.

1 — Displays Polar Tracking tab.

2 — Displays Object Snap tab.

3 — Displays Dynamic Input tab.

RELATED COMMANDS
Grid sets the grid spacing and toggles visibility.

Isoplane selects the working isometric plane.

· ·

Ortho toggles orthographic mode.

Snap sets the snap spacing and isometric mode.

RELATED SYSTEM VARIABLES

AutoSnap controls AutoSnap, polar tracking, and object snap tracking.

CmdInputHistoryMax limits the maximum number of commands stored in history.

CrossingAreaColor specifies the color for crossing area selections.

DynDIGrip determines which dynamic dimensions are displayed during grips editing.

DynDIVis determines which dynamic dimensions are displayed during grips editing.

DynMode toggles dynamic input.

DynPICoords switches pointer input between relative and absolute coordinates.

DynPIFormat switches between polar and Cartesian format for coordinates.

DynPIVis toggles pointer input.

DynPrompt toggles prompts in dynamic input.

DynTooltips determines which tooltips are affected by appearance settings.

GridMode indicates the current grid visibility.

GridUnit indicates the current grid spacing (default = 0.0).

InputHistoryMode specifies where each command history is displayed.

MButtonPan toggles the middle mouse button between panning and *.cui* definition.

PolarAddAng specifies user-defined polar angles, separated by semicolons.

PolarAng specifies the increments of the polar angle.

PolarDist specifies the polar snap distance.

PreviewEffect specifies whether selection preview applies dashes and/or thickness to objects.

PreviewFilter excludes xrefs, tables, mtext, hatches, groups, and objects on locked layers.

SnapAng specifies the current snap rotation angle (default = 0).

SnapBase sets the base point of the snap rotation angle (default = 0,0).

OsMode holds the current object snap modes.

PolarMode holds the settings for polar and object snap tracking.

SelectionArea toggles area selection coloring.

SelectionAreaOpacity specifies the transparency of the area selection color.

SelectionPreview determines when selection preview is active.

SnapIsoPair specifies the current isoplane.

SnapMode sets the current snap mode setting.

SnapStyl specifies the snap style setting.

SnapUnit sets the current snap spacing (default = 1,1).

WindowAreaColor specifies the color for windowed area selections.

TIPS

- Use snap to set the cursor movement increment.
- Use the grid as a visual display to help you better gauge distances.
- Use object snaps to draw precisely to geometric features.
- m2p, a running object snap, snaps to the midpoint of two picked points.

'DsViewer

Rel.13 Displays the bird's-eye view palette; provides real-time pan and zoom (short for DiSplay VIEWer).

Command	Alias	Ctrl+	F-key	Alt+	Menu Bar	Dashboard
dsviewer	av	...	...	VW	View	...
					⮑Aerial View	

Command: dsviewer

Displays the Aerial View palette:

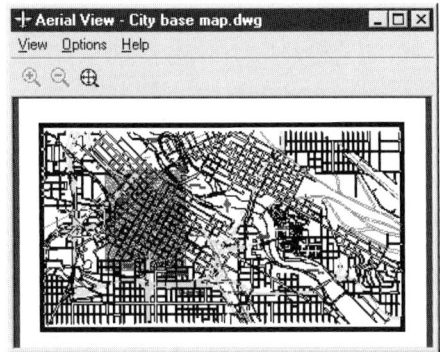

MENU BAR

View Menu

Zoom In increases centered zoom by a factor of 2.

Zoom Out decreases centered zoom by a factor of 2.

Global displays entire drawing in Aerial View palette.

Options Menu

Auto Viewport updates the Aerial View automatically with the current viewport.

Dynamic Update updates the Aerial View automatically with editing changes in the current viewport.

Realtime Zoom updates the drawing in real time as you zoom in the Aerial View palette.

Palette Toolbar

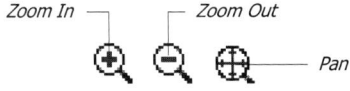

Zoom In — Zoom Out — Pan

RELATED COMMANDS

Pan moves the drawing view.

View creates and displays named views.

Zoom makes the view larger or smaller.

TIPS

- The purpose of the Aerial View is to let you see the entire drawing at all times, and to zoom and pan without entering the Zoom and Pan commands, or selecting items from the menu.

- The parts of the Aerial View palette:

- *Warning!* When in paper space, the Aerial View palette shows only paper space objects.

- To switch quickly between Pan (default) and Zoom modes, click on the Aerial View palette.

- Right-click or press Enter to lock the Aerial View palette and return cursor to editing window.

Renamed Command

DText was reanmed in AutoCAD 2000 as an alias of the Text command.

DView

Rel.10 Zooms and pans 3D drawings dynamically, turns on perspective mode for 3D drawings; superceded by the **3dOrbit** command (short for Dynamic VIEW).

Command	Alias	Ctrl+	F-key	Alt+	Menu Bar	Dashboard
dview	dv	...	...	...	...	...

Command: dview
Select objects or <use DVIEWBLOCK>: *(Select objects, or press Enter.)*
Enter option [CAmera/TArget/Distance/POints/PAn/Zoom/TWist/CLip/Hide/Off/Undo]: *(Enter an option.)*

COMMAND LINE OPTIONS

CAmera indicates the camera angle relative to the target.

Toggle switches between input angles *(available after starting CAmera or TArget)*.

TArget indicates the target angle relative to the camera.

Distance indicates the camera-to-target distance; turns on perspective mode.

POints indicates both the camera and target points.

PAn pans the view dynamically.

Zoom zooms the view dynamically.

TWist rotates the camera (the view).

CLip options
Enter clipping option [Back/Front/Off] <Off>:
Back clip options
Specify distance from target or [ON/OFF] <0.0>:
ON turns on the back clipping plane.
OFF turns off the back clipping plane.

· ·

Distance from target locates the back clipping plane.

Front Clip options
Specify distance from target or [set to Eye(camera)/ON/OFF] <1.0>:

Eye positions the front clipping plane at the camera.

Distance from target locates the front clipping plane.

Off turns off view clipping.

Hide removes hidden lines.

Off turns off the perspective view.

Undo undoes the most recent DView action.

eXit exits DView.

RELATED COMMANDS

Hide removes hidden lines from non-perspective views.

Pan pans non-perspective views.

VPoint selects non-perspective viewpoints of 3D drawings.

Zoom zooms non-perspective views.

3dOrbit creates 3D views interactively, in parallel or perspective mode.

RELATED SYSTEM VARIABLES

BackZ specifies the back clipping plane offset.

FrontZ specifies the front clipping plane offset.

LensLength specifies the perspective view lens length, in millimeters.

Target specifies the UCS 3D coordinates of the target point.

ViewCtr specifies the 2D coordinates of current view center.

ViewDir specifies the WCS 3D coordinates of camera offset from target.

ViewMode specifies the perspective and clipping settings.

ViewSize specifies the height of view.

ViewTwist specifies the rotation angle of current view.

RELATED SYSTEM BLOCK

DViewBlock is the alternate viewing object displayed during DView.

TIPS

- The view direction is from the camera to target.

- Press Enter at the 'Select objects' prompt to display the DViewBlock house. You can replace the house block with your own by redefining the DViewBlock block.

- To view a 3D drawing in one-point perspective, use the Zoom option.

- Menus and transparent zoom and pan are not available during DView. Once the view is in perspective mode, you cannot use any command that requires pointing with the cursor.

Renamed Commands

DwfOut was made part of Plot with AutoCAD 2004.

DwfOutD was removed from AutoCAD 2000; it was combined with DwfOut.

DwfAdjust

2007 Adjusts the contrast, fade, and color of DWF underlays.

Command	Alias	Ctrl+	F-key	Alt+	Menu Bar	Dashboard
dwfadjust	...	...	...	...	...	...

Command: dwfadjust
Select DWF underlay: *(Select objects.)*
Select DWF underlay: *(Press Enter.)*
Enter DWF Underlay option [Fade/Contrast/Monochrome] <Fade>: *(Enter option.)*

DWF underlay with fade set to maximum (80).

COMMAND LINE OPTIONS

Select DWF underlay selects the underlay to adjust.

Enter fade value specifies the amount of fade:

> **0** — no fade.
> **25** — default value.
> **80** — maximum fade.

Enter contrast value specifies the change in contrast, ranging from 0 to 100.

Monochrome toggles monochrome colorization.

RELATED COMMANDS

DwfAttach attaches *.dwf* files as underlays.

DwfClip clips DWF underlays.

DwfLayers toggles the display of layers *(new to AutoCAD 2008)*.

TIPS

- This command has no effect on the frame around the DWF underlay. Use the DwfFrame system variable to toggle its display.

- Use the Monochrome option to change all of the underlay's colors to black; use the Fade option to change the colors to gray.

DwfAttach

<u>2007</u>   Attaches 2D *.dwf* files as xref-like underlays to the current drawing..

Command	Alias	Ctrl+	F-key	Alt+	Menu Bar	Dashboard
dwfattach	...	...	...	IDD	Insert	...
					↳DWF Underlay	

-dwfattach

Command: dwfattach

Displays the Select DWF File dialog box.

Select a file, and then click Open.

Displays the Attach DWF Underlay dialog box:

DIALOG BOX OPTIONS

Name specifies the *.dwf* file to attach to the current drawing; click the droplist for other *.dwf* files. More than one *.dwf* file can be attached to drawings. (This droplist is disabled when the *.dwf* file is stored in Autodesk's Vault software.)

Browse displays the Select DWF File dialog box to select another *.dwf* file.

Select a sheet from the DWF file lists the names of sheets in the *.dwf* file; 3D *.dwf* sheets are not listed. Only one sheet can be attached at a time.

Path Type specifies the type of path to the .dwf file:

• **Full Path** saves the full path name.

• **Relative Path** saves the relative path.

• **No Path** saves the file name only.

Insertion Point specifies the *.dwf* underlay's insertion point (lower left corner); default =0,0,0.

Specify On-Screen:

  ☑ prompts you for the insertion point at the command line after closing the dialog box.

☐ enters the x, y, z coordinates in the dialog box.

Scale specifies the DWF underlay's scale factor; default = 1.0

Specify On-Screen:

☑ prompts you for the scale factor at the command line after closing the dialog box.

☐ enters the scale factor in the dialog box.

Angle specifies the DWF underlay's angle of rotation; default = 0 degrees.

Specify On-Screen:

☑ prompts you for the rotation angle at the command line after closing the dialog box.

☐ enters the angle in the dialog box.

RELATED COMMANDS

DwfAdjust adjusts the contrast, fade, and color of DWF underlays.

DwfClip clips DWF underlays.

DwfLayers toggles the display of layers (*new to AutoCAD 2008*).

Markup places marked-up (redlined) *.dwf* files in the current drawing.

Plot creates single-sheet *.dwf* files of the current drawings.

Publish creates multi-sheet *.dwf* files of multiple drawings.

3dDwfOut creates 3D *.dwf* files.

RELATED SYSTEM VARIABLES

DwfFrame toggles the display of the rectangular frame around DWF underlays.

DWF Underlay

DWF Frame →

DwfOsnap toggles whether osnap recognizes the frame of DWF underlays.

TIPS

- This command cannot import 3D *.dwf* files. It attaches only DWF files with sheets.

- Double-click the underlay to display its properties.

- To remove the underlay from the drawing, select it, and then press Del.

DwfClip

Clips DWF underlays to isolate portions of the drawing.

Command	Alias	Ctrl+	F-key	Alt+	Menu Bar	Dashboard
dwfclip	...	...	...	...	...	...

Command: dwfclip
Select DWF to clip: *(Select one underlay.)*
Enter DWF clipping option [ON/OFF/Delete/New boundary] <New boundary>: *(Enter an option.)*

Clipped DWF Underlay

DWF Clipping Frame

(Hidden portion of DWF underlay)

COMMAND LINE OPTIONS

ON turns on a previously-created clipping boundary.

OFF turns off the clipping boundary, so that it is hidden.

Delete removes the clipping boundary, restoring the DWF underlay.

New boundary creates a new clipping boundary:

- **Polygonal** creates a multi-side boundary using prompts similar to the PLine command.
- **Rectangular** creates a rectangular boundary based on two pick points.

RELATED COMMANDS

DwfAdjust adjusts the contrast, fade, and color of DWF underlays.

DwfAttach attaches *.dwf* files as underlays.

DwfLayers toggles the display of layers *(new to AutoCAD 2008)*.

RELATED SYSTEM VARIABLE

DwfFrame toggles the display of the frame around clipped DWF underlays.

TIPS

- After the clipping boundary is in place, you can use grips editing to adjust its size.
- This command can be started by selecting the DWF border, right-clicking, and then selecting DWF Clip from the shrotcut menu.

DwfLayers

<u>2008</u>   Toggles the display of layers in DWF underlays.

Command	Alias	Ctrl+	F-key	Alt+	Menu Bar	Dashboard
dwflayers	...	...	...	...	...	...

Command: dwflayers
Select DWF underlay: *(Pick one underlay.)*
Displays the DWF Layers dialog box.

DIALOG BOX OPTIONS
Click the lightbulb icon next to a layer name to turn the layer off or on.

Apply changes the visibility of layers without leaving the dialog box.

Shortcut Menu
Layer(s) On creates a new clipping boundary:

Layer(s) Off creates a multi-side boundary using prompts similar to the PLine command.

RELATED COMMAND
DwfAttach attaches *.dwf* files as underlays.

RELATED SYSTEM VARIABLE
DwfFrame toggles the display of the rectangular frame around DWF underlays.

TIP
- Use this command to reduce the amount of information displayed by the DWF. Alternatively, use the DwfAdjust command to fade the underlay.

- You can also access this command by right-clicking a DWF underlay, and then selecting DWF Layers from the shortcut menu.

DwgProps

<u>**2000**</u>   Records and reports information about drawings (short for DraWinG PROPertieS).

Command	Alias	Ctrl+	F-key	Alt+	Menu Bar	Dashboard
dwgprops	...	...	...	FI	File	...
					⮡Drawing Properties	

Command: dwgprops

Displays tabbed dialog box; see below.

DRAWING PROPERTIES DIALOG BOX

Displays information about the drawing obtained from the operating system:

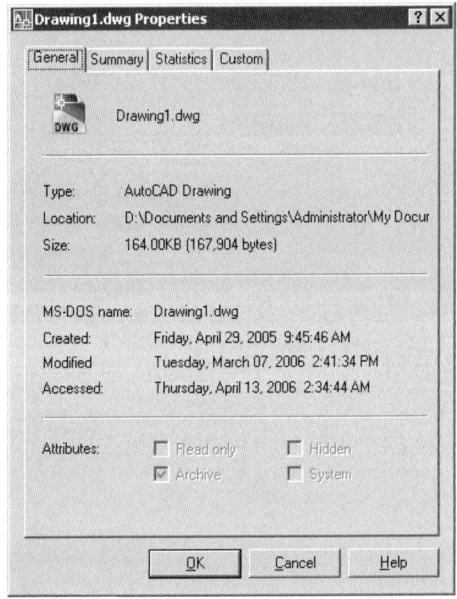

General options

File Type indicates the type of file.

Location indicates the location of the file.

Size indicates the size of the file.

MS-DOS Name provides the MS-DOS file name truncated to eight characters with a three-letter extension.

Created indicates the date and time the file was first saved.

Modified indicates the date and time the file was last saved.

Accessed indicates the date and time the file was last opened.

Attributes options

Read-Only indicates the file cannot be edited or erased.

Archive indicates the file has been changed since it was last backed up.

Hidden indicates the file cannot be seen in file listings.

System indicates the file is a system file; DWG drawing files never have this attribute turned on.

OK records the changes, and exits the dialog box.

Cancel discards the changes, and exits the dialog box.

Summary tab

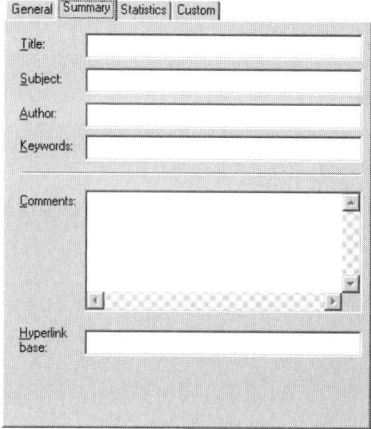

Title specifies a title for the drawing; is usually different from the file name.

Subject specifies a subject for the drawing.

Author specifies the name of the drafter of the drawing.

Keywords specifies keywords used by the operating system's Find or Search commands to locate drawings.

Comments contains comments on the drawing.

Hyperlink Base specifies the base address for relative links in the drawing, such as http://www.upfrontezine.com; may be an operating system path name, such as *c:\autocad 2008*, or a network drive name. (Stored in system variable HyperlinkBase.)

Statistics tab

Displays information about the drawing obtained from the drawing:

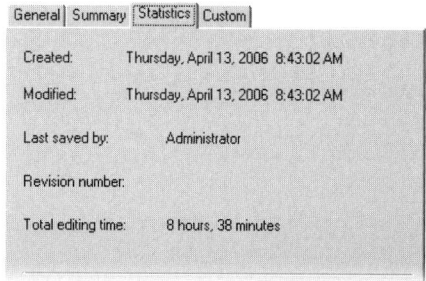

Created indicates the date and time the drawing was first opened (TdCreate).

Modified indicates the date and time the drawing was last opened or modified (TdUpdate).

Last saved by indicates who last accessed the drawing (LoginName).

Revision number indicates the revision number; usually blank.

Total editing time indicates the total amount of time that the drawing has been open (**TdInDwg**).

Custom tab

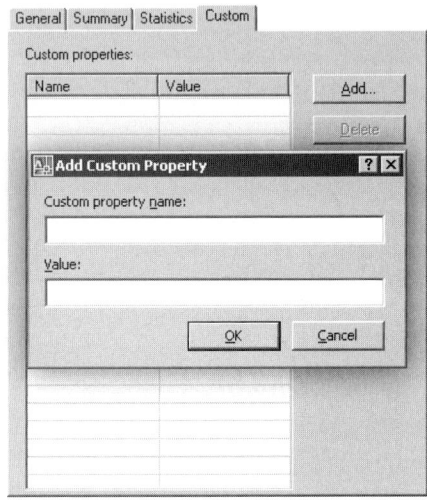

Custom Properties options

Add displays the Add Custom Property dialog box; enter a name and a value. To edit values, click them under Value.

Delete removes the custom property.

RELATED COMMANDS

Properties lists information about objects in the drawing.

Status lists information about the drawing.

TIPS

- Use the Custom Properties tab to include project data in drawings.

- Use the Find button in the Design Center (or Search button in Windows Explorer) to search for drawings containing values stored in the Custom Properties tab, as well as for text in the drawing.

- You can use Google to search for text in *.dwg* files located on the Internet.

- Drawing properties can be exported to spreadsheets with the DataExtraction command (*new to AutoCAD 2008*).

DxbIn

Ver. 2.1 Imports *.dxb* files into drawings (short for Drawing eXchange Binary INput).

Command	Alias	Ctrl+	F-key	Alt+	Menu Bar	Dashboard
dxbin	...	...	...	IE	Insert	...
					⬏ Drawing Exchange Binary	

Command: dxbin

Displays Select DXB File dialog box. Select a .dxb file, and then click Open.

DIALOG BOX OPTION

Open opens the *.dxb* file, and inserts it in the drawing.

RELATED COMMANDS

DxfIn reads DXF-format files.

Plot writes DXB-format files when configured for an ADI plotter.

TIPS

- To produce *.dxb* files, configure AutoCAD with the ADI plotter driver: after starting the PlotterManager command's Add-a-Plotter wizard, select "AutoCAD DXB File" as the manufacturer.

- This command was created for an early software product named CAD\camera, which attempted to convert raster scans into the DXB vector format. CAD\camera had the distinction of being Autodesk's first software release following the success of AutoCAD — and Autodesk's first failure following AutoCAD.

DxfIn, DxfOut

<u>**V. 2.0**</u> Opens and exports *.dxf* files into and from drawings (short for Drawing interchange Format INput / OUTput; undocumented commands).

Command	Alias	Ctrl+	F-key	Alt+	Menu Bar	Dashboard
dxfin	...	...	...	FO	File	...
				⇘DXF	⇘Open	
					⇘DXF	
dxfout	...	...	...	FA	File	...
				⇘DXF	⇘Save As	
					⇘DXF	

Command: dxfin

Displays Select File dialog box. Select a .dxf file, and then click Open.

. . . .

DXFOUT COMMAND
Command: dxfout

Displays Save Drawing As dialog box. Enter a file name, and then click Save.

DIALOG BOX OPTIONS

Save saves the drawing as a *.dxf* file.

Files of type creates a *.dxf* file compatible with these versions of AutoCAD:
* AutoCAD 2007 (compatible with 2008).
* AutoCAD 2004 (compatible with 2005 and 2006).
* AutoCAD 2000 (compatible with 2000, 2000i, and 2002, and AutoCAD LT).
* AutoCAD Release 12 and AutoCAD LT Release 2 (compatible with Release 13 and 14).

SAVEAS OPTIONS DIALOG BOX

From the Tool menu, select Options:

Format
  ⊙ **ASCII** creates files in text format, readable by humans, and imported by most applications.
  ○ **Binary** creates binary files with a smaller file size, but cannot be read by all applications.

. . . .

Additional

☐ **Select objects** selects objects to export, instead of the entire drawing.

☐ **Save thumbnail preview image** includes a preview image in the *.dxf* file.

Decimal places of accuracy (0 to 16) specifies the decimal places of accuracy.

RELATED COMMANDS

DxbIn reads DXB-format files.

AcisOut saves solid model objects in the drawing as ACIS-compatible SAT format.

SaveAs writes drawings in DWG and DXF formats.

TIPS

- Alternatively, you can use the Open command to open *.dxf* files.

- The *.dxf* file comes in two styles: *complete* and *partial*:

  Complete *.dxf* files contain all data required to reproduce a complete drawing.

  Partial *.dxf* files must be imported into existing drawings.

  To load a complete *.dxf* file, AutoCAD requires the current drawing to be empty. Partial *.dxf* files can be imported or inserted into any drawing, empty or not.

- Prior to AutoCAD 2006, DxfIn required a new drawing. (To create an empty drawing, use New with the Start from Scratch option.) As of AutoCAD 2006, *.dxf* files are opened in new drawings automatically.

- If you need to import the complete *.dxf* file into a non-empty drawing, use the Insert command, and insert the *.dxf* file with the Explode option turned on.

- Use the ASCII DXF format to exchange drawings with other CAD and graphics programs. Some applications, such as those for CNC (computer numerically controlled) machines, require 4 decimal places.

- Binary *.dxf* files are much smaller and are created much faster than ASCII binary files; few applications, however, read binary *.dxf* files.

- The AutoCAD Release 12 dialect of DXF is the most compatible with other applications.

- *Warning!* When saving a drawing in DXF format for earlier releases, AutoCAD 2008 erases or converts some objects into simpler objects.

- Autodesk documents the DXF format under Help (Contents > AutoLISP and DXF > DXF Reference) and at at www.autodesk.com/dxf.

EAttEdit

2002 Edits attribute values and properties in a selected block (short for Enhanced ATTribute EDITor).

Command	Alias	Ctrl+	Alt+	Menu Bar	Dashboard
eattedit	...	...	MOAS	Modify	Block Attributes
				⮑Object	
				⮑Attribute	
				⮑Single	

Command: eattedit

If the drawing contains no blocks with attributes, AutoCAD complains, "This drawing contains no attributed blocks," and then the command exits.

When attributed blocks exist, the command continues:

Select a block: *(Select a single block.)*

Displays dialog box:

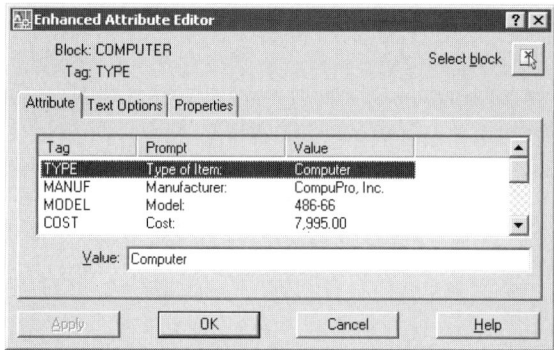

ENHANCED ATTRIBUTE EDITOR DIALOG BOX

Select block selects another block for attribute editing.

Apply applies the changes to the attributes.

Value modifies the value of the selected attribute; neither the tag nor the prompt can be modified by this command.

Text Options tab

Text Style selects a text style name from the list; text styles are defined by the Style command; default is Standard.

Justification selects a justification mode from the list; default is left justification.

Height specifies the text height; can be changed only when height is set to 0.0 in the text style.

Rotation specifies the rotation angle of the attribute text; default = 0 degrees.

□ **Backwards** displays the text backwards.

□ **Upside Down** displays the text upside down.

Width Factor specifies the relative width of characters; default = 1.

Oblique Angle specifies the slant of characters; default = 0 degrees.

Annotative toggles annotative scaling (*new to AutoCAD 2008*).

Boundary Width specifies margins for multi-line attributes (*new to AutoCAD 2008*).

Properties tab

Layer selects a layer name from the list; layers are defined by the Layer command.

Linetype selects a linetype name from the list; linetypes are loaded into the drawing with the **Linetype** command.

Color selects a color from the list; to select from the full 255-color spectrum, select **Other**.

Lineweight selects a lineweight from the list; to display lineweights, click LWT on the status bar.

Plot style selects a plot style name from the list; available only if plot styles are enabled in the drawing.

RELATED COMMANDS

AttDef creates attribute definitions.

Block attaches attributes to objects.

BAttMan manages attributes.

EAttExt extracts attributes to a file.

TIPS

■ This command edits only attribute values and their properties; to edit all aspects of an attribute, use the BAttMan command.

■ If you select a block with no attributes, AutoCAD complains, 'The selected block has no editable attributes.' If you select an object that isn't a block, AutoCAD complains, 'Error selecting entity.'

■ Blocks on locked layers cannot be edited.

· ·

Removed Commands

The **EAttExt** and **-EAttExt** commands are now aliases for the DataExtraction and -DataExtraction commands, as of AutoCAD 2008.

· ·

Edge

Rel.12 Toggles the visibility of 3D faces.

Command	Alias	Ctrl+	F-key	Alt+	Menu Bar	Dashboard
edge	...	...	...	DMME	Draw	...
					⤷Modeling	
					⤷Meshes	
					⤷Edge	

Command: edge
Specify edge of 3dface to toggle visibility or [Display]: *(Pick an edge, or type D.)*
Enter selection method for display of hidden edges [Select/All] <All>: *(Type S or A.)*
Select objects: *(Select one or more objects.)*
Specify edge of 3dface to toggle visibility or [Display]: *(Press Esc to end the command.)*

From left to right: 3D faces; edges selected with Edge; and invisible edges.

COMMAND LINE OPTIONS

Specify edge selects edge to make invisible.

Display Options
Select highlights invisible edges.
All selects all hidden edges, and regenerates them.

RELATED COMMAND

3dFace creates 3D faces.

RELATED SYSTEM VARIABLE

SplFrame toggles visibility of 3D face edges.

TIPS

■ Make edges invisible to improve the appearance of 3D objects.

■ Edge applies only to objects made of 3D faces; it does not work with polyface meshes or solid models.

■ Use the Explode command to convert meshed objects into 3D faces.

■ Re-execute Edge to display an edge that has been made invisible.

■ The command repeats until you press Enter or Esc at the 'Specify edge of 3dface to toggle visibility or [Display]:' prompt.

■ When you hide all edges of a 3D face, it can be difficult to unhide them.

EdgeSurf

Rel.10 Draws 3D meshes as Coons patches between four boundaries (short for EDGE-defined SURFace); does not draw surfaces.

Command	Alias	Ctrl+	F-key	Alt+	Menu Bar	Dashboard
edgesurf	...	...	...	DMME	Draw ⤷Modeling ⤷Meshes ⤷Edge Mesh	...

Command: edgesurf
Current wire frame density: SURFTAB1=6 SURFTAB2=6
Select object 1 for surface edge: *(Pick an object.)*
Select object 2 for surface edge: *(Pick an object.)*
Select object 3 for surface edge: *(Pick an object.)*
Select object 4 for surface edge: *(Pick an object.)*

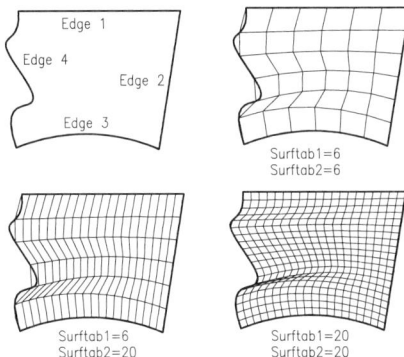

COMMAND LINE OPTION
Select object picks an edge.

RELATED COMMANDS
3dMesh creates 3D meshes by specifying every vertex.
3dFace creates 3D meshes of irregular vertices.
PEdit edits meshes created by EdgeSurf.
TabSurf creates tabulated 3D surfaces.
RuleSurf creates ruled 3D surfaces.
RevSurf creates 3D surfaces of revolution.

RELATED SYSTEM VARIABLES
SurfTab1 stores the current *m*-density of meshing; the maximum mesh density is 32767.
SurfTab2 stores the current *n*-density of meshing; maximum is 32767.

TIP
- The four boundary edges can be made from lines, arcs, and open 2D and 3D polylines; the edges must meet at their end points.

'Elev

V. 2.1 Sets elevation and thickness for creating extruded 3D objects (short for ELEVation).

Command	Alias	Ctrl+	F-key	Alt+	Menu Bar	Dashboard
elev	...	...	...	...	...	...

Command: elev
Specify new default elevation <0.0000>: *(Enter a value for elevation.)*
Specify new default thickness <0.0000>: *(Enter a value for thickness.)*

Elevation = 27"
Elevation = 24"
Thickness = 3"
Thickness = 24"
Elevation = 0"

COMMAND LINE OPTIONS

Elevation changes the base elevation from z = 0.

Thickness extrudes new 2D objects in the z-direction.

RELATED COMMANDS

Change changes the thickness and z coordinate of objects.

Move moves objects, even in the z direction.

Properties changes the thickness of objects.

RELATED SYSTEM VARIABLES

Elevation stores the current elevation setting.

Thickness stores the current thickness setting.

TIPS

- The current value of elevation is used whenever the z coordinate is not supplied.

- Thickness is measured up from the current elevation in the positive z-direction.

Ellipse

V. 2.5 Draws ellipses by four different methods, and draws elliptical arcs and isometric circles.

Command	Alias	Ctrl+	F-key	Alt+	Menu Bar	Dashboard
ellipse	el	...	...	DE	Draw	2D Draw
					⤷ Ellipse	

Command: ellipse
Specify axis endpoint of ellipse or [Arc/Center]: *(Pick a point, or enter an option.)*
Specify other endpoint of axis: *(Pick a point.)*
Specify distance to other axis or [Rotation]: *(Pick a point, or type R.)*

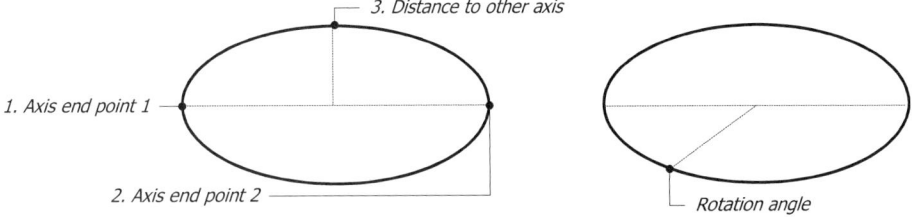

COMMAND LINE OPTIONS

Specify axis endpoint of ellipse indicates the first endpoint of the major axis.

Specify other endpoint of axis indicates the second endpoint of the major axis.

Specify distance to other axis indicates the half-distance of the minor axis.

. . . .

 ## Elliptical Arcs

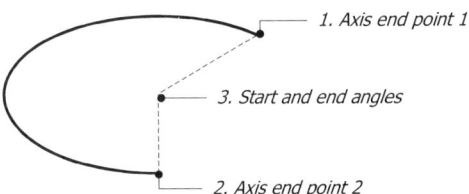

Command: ellipse
Specify axis endpoint of elliptical arc or [Center]: *(Pick a point, or type C.)*
Specify other endpoint of axis: *(Pick a point.)*
Specify distance to other axis or [Rotation]: *(Pick a point, or type R.)*
Specify start angle or [Parameter]: *(Enter an angle, or type P.)*
Specify end angle or [Parameter/Included angle]: *(Enter an angle, or enter an option.)*

COMMAND LINE OPTIONS

Specify start angle indicates the starting angle of the elliptical arc.

Specify end angle indicates the ending angle of the elliptical arc.

. . . .

Parameter indicates the starting angle of the elliptical arc; draws the arc with this formula:
p(u)=**c**+(**a***cos(u))+(**b***sin(u))

Parameter	Meaning
a	Major axis.
b	Minor axis.
c	Center of ellipse.

Included angle indicates an angle measured relative to the start angle, rather than 0 degrees.

Center option
Specify center of ellipse: *(Pick a point.)*

Specify center of ellipse indicates the center point of the ellipse.

Rotation option
Specify rotation around major axis: *(Enter an angle.)*

Specify rotation around major axis indicates a rotation angle around the major axis:

0 degrees — minimum rotation: creates round ellipses, like circles.
89.4 degrees — maximum rotation: creates very thin ellipses.

· ·

Isometric Circles

This option appears only when system variable SnapStyl is set to 1 (isometric snap mode). Use F5 *to switch among the three isometric drawing planes.*

Command: ellipse
Specify axis endpoint of ellipse or [Arc/Center/Isocircle]: *(Type* **I***.)*
Specify center of isocircle: *(Pick a point.)*
Specify radius of isocircle or [Diameter]: *(Pick a point, or type* **D***.)*

COMMAND LINE OPTIONS

Specify center indicates the center point of the isocircle.

Specify radius indicates the radius of the isocircle.

Diameter indicates the diameter of the isocircle.

RELATED COMMANDS

IsoPlane sets the current isometric plane.

PEdit edits ellipses (when drawn with a polyline).

Snap controls the setting of isometric mode.

RELATED SYSTEM VARIABLES

PEllipse determines how the ellipse is drawn.

SnapIsoPair sets the current isometric plane:.

SnapStyl specifies regular or isometric drawing mode.

· ·

TIPS

- Previous to AutoCAD Release 13, Ellipse constructed the ellipse as a series of short polyline arcs. The PEllipse system variable controls how ellipses are drawn. When 0, true ellipses are drawn; when 1, a polyline approximation of an ellipse is drawn.

- When PEllipse = 1, the Arc option is not available.

- Use ellipses to draw circles in isometric mode. When Snap is set to isometric mode, Ellipse's Isocircle option projects a circle into the working isometric drawing plane. Press CTRL+E or F5 to switch between isoplanes.

- The Isocircle option only appears in the option prompt when Snap is set to isometric mode.

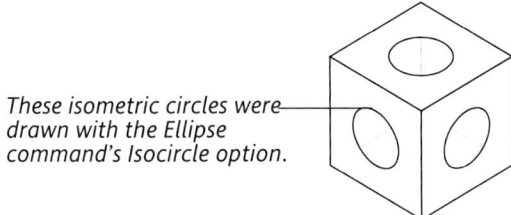

These isometric circles were drawn with the Ellipse command's Isocircle option.

- See the Isoplane command for a tutorial on creating isometric objects.

Removed Commands

End was removed from AutoCAD Release 14; it was replaced by Quit.

EndToday was removed from AutoCAD 2004.

Erase

V. 1.0 Erases objects from drawings.

Command	Alias	Ctrl+	Key	Alt+	Menu Bar	Dashboard
erase	e	...	Del	ME	Modify	2D Draw
					⌖ Erase	
					Edit	
					⌖ Clear	

Command: erase
Select objects: *(Select one or more objects.)*
Select objects: *(Press Enter to end object selection.)*

COMMAND LINE OPTION

Select objects selects the objects to erase.

RELATED COMMANDS

Break erases a portion of a line, circle, arc, or polyline.

Fillet cleans up intersections by erasing selected segments.

Oops returns the most-recently erased objects to the drawing.

Trim cuts off the end of a line, arc, and other objects.

Undo returns the erased objects to the drawing.

TIPS

- The Erase L command erases the last-drawn item visible in the current viewport.

- *Warning!* The Erase All command erases all objects in the current space (model or layout), except on locked, frozen, and/or off layers.

- Oops brings back the most-recently erased objects; use U to bring back other erased objects.

- Objects on locked and frozen layers cannot be erased. To erase them, change the layers to unlocked and thawed.

eTransmit

<u>2000i</u> Collects drawings and related files into packages suitable for transmission by email or CD (short for Electronic TRANSMITtal).

Commands	Alias	Ctrl+	Key	Alt+	Menu Bar	Dashboard
etransmit	...	...	...	FT	File	...
					⇘eTransmit	
-etransmit						

Command: etransmit

If the drawing has not been saved since the last change, displays error dialog box:

OK saves the drawing, and then displays the Create Transmittal dialog box.

Cancel cancels the eTransmit command.

CREATE TRANSMITTAL DIALOG BOX

Files Tree tab

Sheets tab displays sheets included with transmittal.

Files Tree tab displays names of drawing and support files, grouped by category.

Files Table tab displays file names in alphabetical order.

Select a Transmittal Setup lists pre-defined setups.

Transmittal Setups displays the Transmittal Setup dialog box.

Enter Notes to Include With This Transmittal Package provides space for entering notes.

View Report displays the View Transmittal Report dialog box.

Files Tree tab

☑ File included in transmittal.

☐ File excluded from transmittal.

Add File displays the Add File to Transmittal dialog box, to add files to the transmittal.

Files Table tab

TRANSMITTAL SETUP DIALOG BOX

New creates new setups; displays New Transmittal Setup dialog box.

Rename renames setups, except for "Standard."

Modify changes setups; displays Modify Transmittal Setup dialog box.

Delete removes setups, except for "Standard."

NEW TRANSMITTAL SETUP DIALOG BOX

New Transmittal Setup Name specifies the name of the new setup.

Based on selects the setup to copy.

Continue displays the Modify Transmittal Setup dialog box.

MODIFY TRANSMITTAL SETUP DIALOG BOX

New creates new setups; displays New Transmittal Setup dialog box.

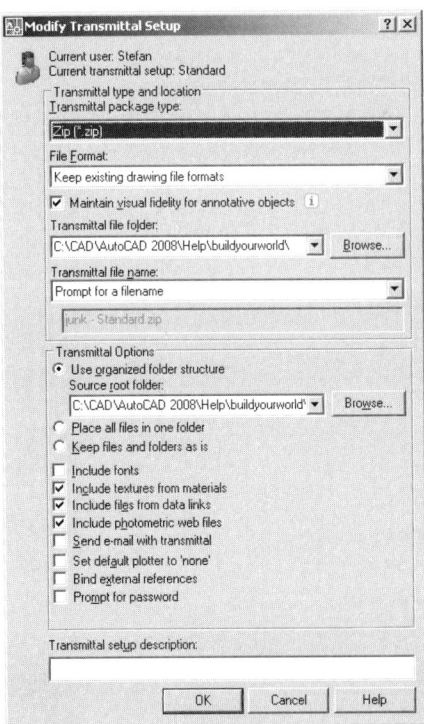

Transmittal Type and Location

Transmittal Package Type:

- **Folder (set of files)** transmits uncompressed files in new and existing folders. Best option for transmitting files on CDs or by FTP.
- **Self-Extracting Executable (*.exe)** transmits files in a compressed, self-extracting file. Uncompress by double-clicking the file.
- **Zip (*.zip)** (*default*) transmits files as a compressed ZIP file. Uncompress the file using the PkZip or WinZip programs. Best option when sending by email.

File Format:

- **Keep existing drawing file formats** transmits files in their version.
- **AutoCAD 2007/LT 2007 Drawing Format** transmits files in AutoCAD 2007/8 format.
- **AutoCAD 2004/LT 2004 Drawing Format** transmits files in AutoCAD 2004/5/6 format.
- **AutoCAD 2000/LT 2000 Drawing Format** transmits files in AutoCAD 2000/i/1 format. *Warning!* AutoCAD 2008 objects and properties not found in AutoCAD 2000-2008 may be changed or lost.

☑ **Maintain Visual Fidelity for Annotative Objects** uses visual fidelity for annotative objects: each scale representation of annotative objects is saved to separate layers (*new in AutoCAD 2008*).

Transmittal File Folder specifies the location in which to collect the files. When no location is specified, the transmittal is created in the same folder as the *.dst* file.

Transmittal File Name:

- **Prompt for a File Name** prompts the user for the file name.
- **Overwrite if Necessary** uses the specified file name, and overwrites the existing file of the same name.
- **Increment File Name if Necessary** uses the specified file name, and appends a digit to avoid overwriting the existing file of the same name.

Transmittal Options

⊙ **Use Organized Folder Structure** preserves the folder structure in the transmittal, but makes the changes listed below; allows you to specify the name of the folder tree.

○ **Place All Files in One Folder** places all files in a single folder.

○ **Keep Files and Folders As Is** preserves the folder structure in the transmittal.

☐ **Include Fonts** includes all *.ttf* and *.shx* font files in the transmittal.

☑ **Include textures from materials** includes material files.

☑ **Include Files from Data Links** includes spreadsheet files linked to the drawing (*new in AutoCAD 2008*)..

☑ **Include Photometric Web Files** includes IEF (photometric web) files associated with web lights placed in the drawing (*new in AutoCAD 2008*).

☐ **Send email with transmittal** opens the computer's default email software.

☐ **Set Default Plotter to 'None'** resets the plotter to None for all drawings in the transmittal.

☐ **Bind External References** merges all xrefs into the drawing.

☐ **Prompt for Password** displays a dialog box for specifying a password; not available when the Folder archive type is selected.

Transmittal Setup Description describes the setup.

-ETRANSMIT command

Command: -etransmit

**Enter an option [Create transmittal package/Report only/CUrrent setup/
CHoose setup/Sheet set] <Report only>:** *(Enter an option.)*

COMMAND LINE OPTIONS

Create Transmittal Package creates transmittal packages from the current drawing and all support files; uses settings in the current transmittal setup.

Report Only displays the Save Report File As dialog box; enter a file name, and then click **Save** saves the report in plain ASCII text format; does not create the transmittal package.

Current Setup lists the name of the current transmittal setup.

Choose Setup selects the transmittal setup.

Sheet Set specifies the sheet set and transmittal setup to use for the transmittal package; available only when a sheet set is open in the drawing.

RELATED COMMANDS

Archive creates archive sets of drawings, support files, and sheet sets.

Publish creates a *.dwf* file with drawing sheets.

PublishToWeb saves the drawing as an HTML file.

RELATED SYSTEM VARIABLE

SaveFidelity toggles visual fidelity for annotative objects.

TIPS

- The Password button is *not* available when you select the Folder option.

- Including a TrueType font (*.ttf*) is touchy, because sending a copy of the font might infringe on its copyright. All *.shx* and *.ttf* files included with AutoCAD may be transmitted. For this reason, the Include Fonts option is turned off by default. In addition, not including fonts saves some file space; recall that smaller files take less time to transmit via the Internet.

- "Self-extracting executable" means that the files are compressed into a single file with the *.exe* extension. The email recipient double-clicks the file to extract (uncompress) the files. The benefit is that recipients do not need to have a copy of PkUnzip or WinZip on their computers; the drawbacks are that a virus could hide in the *.exe* file, and some e-mail programs prevent receipt of attachments with *.exe* extensions..

- If your computer cannot compress the files (or if your recipient cannot uncompress them), you are probably lacking the software needed to uncompress the *.zip* file. (Do not confuse *.zip* files with Iomega's ZIP disk drive; the two having nothing in common, except the name.) You can obtain the PkUnzip and WinZIP utilities as freeware or shareware.

- Visual fidelity attempts to preserve data in drawings translated to earlier releases of AutoCAD, which don't support annotative scaling found in AutoCAD 2008. When visual fidelity is turned on, annotative text appears on separated layers. This ensures the text is there, and can be toggled on and off by thawing and freezing the appropriate layers in each viewport.

- Files occasionally become corrupted when sent by email; the transmittal may need to be resent. If you continue to have problems, you may need to change a setting in your email software. Try changing the attachment encoding method from BinHex or Uuencode to MIME.

Explode

V. 2.5 Explodes polylines, blocks, associative dimensions, hatches, multilines, 3D solids, regions, bodies, and meshes into their constituent objects.

Command	Alias	Ctrl+	F-key	Alt+	Menu Bar	Dashboard
explode	x	...	...	MX	Modify	2D Draw
					⌐Explode	

Command: explode
Select objects: *(Select one or more objects.)*
Select objects: *(Press Enter to end object selection.)*

Left: *Polylines.*
Right: *Exploded into lines and arcs.*

COMMAND LINE OPTION

Select objects selects the objects to explode.

RELATED COMMANDS

Block recreates a block after an explode.

PEdit converts a line into a polyline.

Region converts 2D objects into a region.

Undo reverses the effects of this command.

Xplode controls the explosion process.

RELATED SYSTEM VARIABLE

ExplMode toggles whether non-uniformly scaled blocks can be exploded.

TIPS

- As of Release 13, AutoCAD can explode blocks inserted with unequal scale factors, mirrored blocks, and blocks created by the MInsert command.

- You cannot explode xrefs and dependent blocks (blocks from xref drawings), or single-line text.

- Parts making up exploded blocks and associative dimensions of BYBLOCK color and linetype are displayed in color White (or Black when the background color is white) and Continuous linetype.

- This command is not a transparent command, but Xplode is.

- The Explode command alters objects, as follows:

Objects	Exploded Into
Arcs in non-uniformly scaled blocks	Elliptical arc.
Associative dimensions	Lines, solids, and text.
Blocks	Constituent parts.
Circles in non-uniformly scaled blocks	Ellipse.
Mtext and field text	Text
Multilines	Lines.
Polygon meshes	3D faces.
Polyface meshes	3D faces, lines, and points.
Tables	Lines.
2D polylines	Lines and arcs; width and tangency information lost.
3D polylines	Lines.
3D solids	Regions and surfaces.
Regions	Lines, arcs, ellipses, and splines.
Bodies	Single bodies, regions, and curves.

- Resulting objects become the previous selection set.
- Flat faces of 3D solids become regions; curved faces become surfaces.

Export

Rel.13 Saves drawings in formats other than DWG and DXF.

Command	Alias	Ctrl+	F-key	Alt+	Menu Bar	Dashboard
export	exp	...	...	FEE⬐	File	...
					⬐Export	

Command: export

Displays dialog box:

EXPORT DATA DIALOG BOX

Save in selects the folder (subdirectory) and drive into which to export the file.

Back returns to the previous folder (alternatively, press the Alt+1 shortcut keystroke).

Up One Level moves up one level in the folder structure (Alt+2).

Search the Web displays a simple Web browser that accesses the Autodesk Web site (Alt+4).

Delete erases the selected file(s) or folder (Del).

Create New Folder creates a new folder (Alt+5).

Views displays files and folders in a list or with details.

Tools lists several additional commands, including the Options dialog box — available only with encapsulated PostScript; see the PsOut command.

File name specifies the name of the file, or accepts the default.

Save as type selects the file format in which to save the drawing.

Save saves the drawing.

Cancel dismisses the dialog box, and returns to AutoCAD.

RELATED COMMANDS

AttExt exports attribute data in the drawing in CDF, SDF, or DXF formats.

CopyClip exports the drawing to the Clipboard.

CopyHist exports text from the text screen to the Clipboard.

DgnExport exports drawings as MicroStation design files (*new to AutoCAD 2008*).

Import imports several vector and raster formats.

LogFileOn saves the command line text as ASCII text in the *acad.log* file.

MassProp exports the mass property data as ASCII text in an *.mpr* file.

MSlide exports the current viewport as an *.sld* slide file.

Plot exports the drawing in many vector and raster formats.

SaveAs saves the drawing in AutoCAD's DWG format.

SaveImg exports the rendering in TIFF, Targa, or BMP formats.

TIPS

- The Export command exports the current drawing in the following formats:

Extension	File Type	Related Command
3D DWF	3D Studio file	**3dDwf**
BMP	Device-independent bitmap	**BmpOut**
DGN	MicroStation V8 design file	**DgnExport**
DWG	AutoCAD drawing file	**WBlock**
DXX	Attribute extract DXF file	**AttExt**
EPS	Encapsulated PostScript file	**PsOut**
SAT	ACIS solid object file	**AcisOut**
STL	Stereolithography	**StlOut**
WMF	Windows metafile	**WmfOut**

- This command acts as a "shell": it launches other AutoCAD commands that perform the actual export function, as noted above.

- When drawings are exported in formats other than *.dwg* or *.dxf*, information is lost, such as layers and attributes.

- AutoCAD 2007 removes 3DS (3D Studio) and WMF formats, but adds 3D DWF; AutoCAD 2008 adds the DGN format.

Removed (and Restored) Command

ExpressTools (also known as "bonus CAD tools" in earlier versions of AutoCAD) were removed from AutoCAD 2002, but returned to AutoCAD 2004.

--/ Extend

Extends the length of lines, rays, open polylines, arcs, and elliptical arcs to boundary objects.

Command	Alias	Ctrl+	F-key	Alt+	Menu Bar	Dashboard
extend	ex	...	...	MD	Modify ⮑Extend	2D Draw

Command: extend
Current settings: Projection=UCS Edge=None
Select boundary edges ...
Select objects: *(Select one or more objects.)*
Select objects: *(Press ENTER.)*
Select object to extend or shift-select to trim or [Project/Edge/Undo]: *(Select an object, or enter an option.)*
Select object to extend or shift-select to trim or [Project/Edge/Undo]: *(Press ENTER to end the command.)*

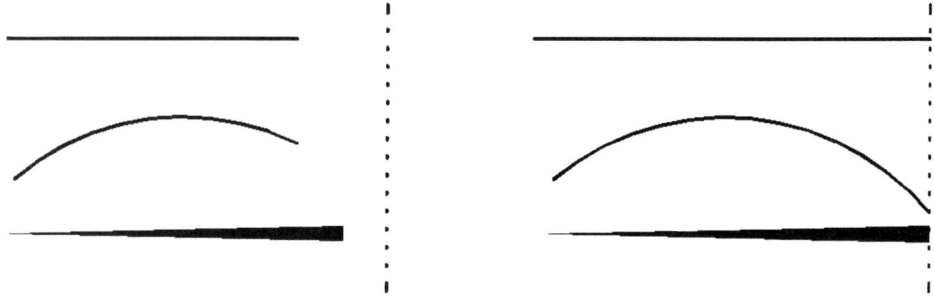

Left: Line, arc, and variable-width polyline.
Right: Extended to dotted line.

COMMAND LINE OPTIONS

Select objects selects the objects to use for the extension boundary.

Select objects to extend selects the objects to be extended.

Shift-select to trim trims objects when you hold down the SHIFT key.

Undo undoes the most recent extend operation.

Project options
Enter a projection option [None/Ucs/View] <Ucs>: *(Enter an option.)*

None extends objects to boundary (Release 12-compatible).

Ucs extends objects in the x,y-plane of the current UCS.

View extends objects in the current view plane.

Edge options
Enter an implied edge extension mode [Extend/No extend] <No extend>:
(Enter an option.)

Extend extends to implied boundary.

No extend extends only to actual boundary (Release 12-compatible).

RELATED COMMANDS

Change changes the length of lines.

Lengthen changes the length of open objects.

SolidEdit extends the face of a solid object.

Stretch stretches objects wider and narrower.

Trim reduces the length of lines, polylines and arcs.

RELATED SYSTEM VARIABLES

EdgeMode toggles boundary mode for the Extend and Trim commands.

ProjMode toggles projection mode for the Extend and Trim commands:

0 — None; Release 12 compatible.

1 — Current UCS (default).

2 — Current view plane.

TIPS

- The following objects (even when inside blocks) can be used as boundaries:

2D polylines	Lines
3D polylines	Rays
Arcs	Regions
Circles	Splines
Ellipses	Text
Floating viewports	Xlines
Hatches	

- When a wide polyline is the edge, Extend extends to the polyline's centerline.

- Pick the object a second time to extend it to a second boundary line.

- Circles and other closed objects are valid edges: the object is extended in the direction nearest to the pick point.

- Extending a variable-width polyline widens it proportionately; extending a splined polyline adds a vertex.

ExternalReferences, ExternalReferencesClose

2007 Displays the External References palette for centrally managing all xref'ed drawings, images, DGN, and DWF underlays.

Command	Aliases Ctrl+	F-key	Alt+	Menu Bar	Dashboard
externalreferences	er ...	...	TPN	Modify	2D Draw
	image, im			⬐Palettes	
	xref, xr			⬐External References	
externalreferencesclose					

Command: externalreferences

Opens the External References palette:

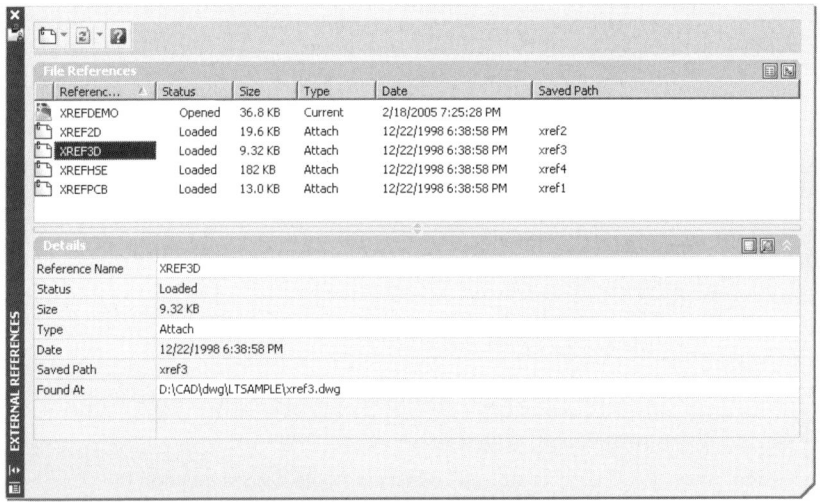

Command: externalreferencesclose

Closes the External References palette.

PALETTE TOOLBARS

Attach DWG attaches *.dwg* files as external references; displays the Select Reference File dialog box, followed by the External Reference dialog box. See XAttach command.

Attach Image attaches raster files as underlays; displays the Select Image File dialog box, followed by the Image dialog box. See ImageAttach command.

Attach DWF attaches *.dwf* files as underlays; displays the Select DWF File dialog box, followed by the Attach DWF Underlay dialog box. See DwfAttach command.

Attach DGN attaches MicroStation.*dgn* files as underlays; displays the Select DGN File dialog box, followed by the Attach DGN Underlay dialog box (*new to AutoCAD 2008*). See DgnAttach command. (To import the design files, use the DgnImport command.)

File References

 List View displays each external reference with information about its status, and so on.

 Tree View shows the connections between external references.

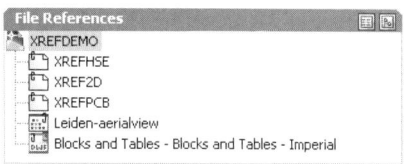

Right-click an externally-referenced file to display this shortcut menu:

Open opens the selected external reference so that it can be edited:
- *.dwg* files are opened in another AutoCAD window; see the XOpen command.
- Image files are opened in the default external paint program, such as PaintShop Pro.
- *.dwf* files are opened in the external DWF Viewer program.
- Option unavailable for *.dgn* files.

Attach displays the related dialog box so that changes can be made to the attachment status:
- *.dwg* files display the External Reference dialog box.
- Image files display the Image dialog box.
- *.dwf* files display the Attach DWF Underlay dialog box.
- *.dgn* files display the Attach DGN Underlay dialog box.

Unload prevents the external reference from being displayed in the drawing, but preserves attachment information; use the Reload option to bring it back.

Reload updates the external reference, if it has been unloaded, changed its content or moved to another folder.

Detach removes the external reference from the drawing.

Bind merges the external reference with the drawing (for *.dwg* files only); see the XBind command.

Details

Details lists information about the selected external reference.

Preview shows a preview image of the selected external reference; no preview is available for old versions of *.dwg* files.

RELATED COMMANDS

DgnAttach attaches *.dgn* files (*new to AutoCAD 2008*).

DwfAttach attaches *.dwf* files.

ImageAttach attaches raster files.

XAttach attaches *.dwg* files.

RELATED SYSTEM VARIABLES

ErState reports whether this palette is open or closed.

TIPS

- To reload all external references at once, right-click a blank area in the File References pane, and then select Reload All References from the shortcut menu.

- If you have access to Autodesk's Vault software, the Attach from Vault option allows you to attach files from the Vault.

- The Status column displays brief reports that have the following meaning:

Status	Meaning
Loaded	File is attached to the drawing.
Unloaded	File will be unloaded.
Not Found	File cannot be found on search paths.
Unresolved	File cannot be read.
Orphaned	File is attached to an unresolved file.

- *.Dwg* xrefs can be attached or overlaid; see the XRef command. *.Dgn* files can be attached or imported; see the DgnImport command.

Extrude

Rel.11 Creates 3D solids and surfaces by extruding 2D objects.

Command	Alias	Ctrl+	F-key	Alt+	Menu Bar	Dashboard
extrude	ext	...	...	DMX	Draw	3D Make
					⮡ Modeling	
					⮡ Extrude	

Command: extrude
Select objects: *(Select one or more objects.)*
Select objects: *(Press Enter.)*
Specify height of extrusion or [Direction/Path/Taper angle] <1>: *(Enter a height or an option.)*

COMMAND LINE OPTIONS

Select objects selects the 2D objects to extrude, called the "profile."

Specify height of extrusion specifies the extrusion height. As of AutoCAD 2007, the extrusion dynamically changes height as you move the cursor.

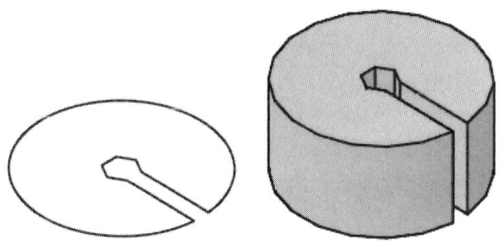

Left: *Original object (the profile).*
Right: *Extrusion by height.*

Direction specifies the extrusion height, direction, and angle by picking two points.

Extrusion by direction (picking two points).

Path extrudes the profile along another object, called the "path."

Left: *Original object (the profile).*
Center: *Extrusion by path.*
Right: *The path object (a helix).*

Taper extrudes the profile at an angle; ranges from -90 to +90 degrees.

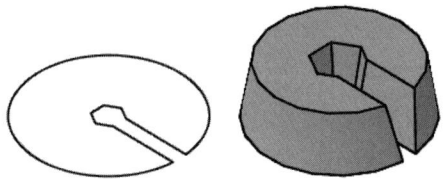

Extrusion by 10-degree taper (specifying an angle for the walls).
Notice that the outside tapers in, while the inside tapers out.

RELATED COMMANDS

Revolve creates a 3D solid by revolving a 2D object.

Elev gives thickness to non-solid objects to extrude them.

Thicken creates 3D solids by thickening surfaces.

TIPS

■ Prior to AutoCAD 2007, this command worked only with closed objects, creating 3D solids. It now works with open and closed objects: open ones, such as arcs, create 3D surfaces; while closed ones, such as circles, create 3D solids.

■ When extruding an object along a path, AutoCAD relocates the path to the object's centroid.

■ This command extrudes the following objects: lines, traces, arcs, elliptical arcs, ellipses, circles, 2D polylines, 2D splines, 2D solids, regions, planar 3D faces, planar surfaces, and planar faces on solids. To select the face of a solid object, hold down the Ctrl key when picking.

■ This command can use the following objects as extrusion paths: lines, circles, arcs, ellipses, elliptical arcs, 2D and polylines, 2D and 3D splines, helixes, solid edges, and edges of surfaces.

■ You *cannot* extrude polylines with less than 3, or more than 500, vertices; similarly, you cannot extrude crossing or self-intersecting polylines.

■ Objects within a block cannot be extruded; use the Explode command first.

■ The taper angle must be between 0 (default) and +/-90 degrees; positive angles taper in from the base; negative angles taper out.

■ This command also does not work if the combination of angle and height makes the object's extrusion walls intersect.

Field

2005 Places automatically updatable field text in drawings.

Command	Alias	Ctrl+	F-key	Alt+	Menu Bar	Dashboard
field	...	...	...	IF	Insert	...
					⮡ Field	

Command: field

Displays dialog box, the content of which varies depending on the field name selected:

Select a field name and its options.

Click OK. AutoCAD places the field as mtext:

MTEXT Current text style: "Standard" Text height: 0.2000
Specify start point or [Height/Justify]: *(Pick a point, or enter an option.)*

DRAWING1.DWG

If the field has no value, AutoCAD inserts dashes as placeholders

FIELD DIALOG BOX

Field category lists groups of fields.

Field names lists the names of fields that can be placed in drawings.

Format lists optional formats available.

Field Expression illustrates field codes inserted in the drawing; cannot be edited directly.

Additional options appear, depending on the field name selected.

RELATED COMMANDS

UpdateField forces selected fields to update their values.

Find finds field text in drawings.

MText places text in drawings.

AttDef defines attribute text.

RELATED SYSTEM VARIABLES

FieldDisplay toggles the gray background to fields (default = 1, on).

FieldEval determines when fields are updated (default = 31, all options on).

TIPS

- When FieldEval=31 (all on), all fields are supposed to update whenever you use the Plot, Save, and other *trigger* commands. In practice, this occurs when the field is also a system variable. For example, if the field accesses the Date system variable, it updates correctly; when it is the Date option of the Date & Time category, UpdateField must be used to force the update.

- To edit: (1) double-click a field; (2) in the mtext editor, right-click the field text; (3) from the shortcut menu, select Edit Field. To convert fields to mtext, select Convert Field to Text.

- The gray background helps identify field text in drawings.

- AutoCAD supports the following field names:

Field Category	Field Names	Comments
Date & Time	Create Date	Date drawing created.
	Date	Current date and time.
	Plot Date	Date last plotted.
	Save Date	Date last saved.
Document	Author	Data stored in the Drawing Properties Comments dialog box; see DwgProps command.
	Filename	
	Filesize	
	HyperlinkBase	
	Keywords	
	LastSavedBy	
	Subject	
	Title	
Linked	Hyperlink	Links to other drawings and files.
Objects	Block	Placeholder; can be used in Block Editor only.
	NamedObject	Blocks, layers, linetypes, etc.
	Object	Object selected from drawing.
	Formula	Average, Sum, Count, Cell, and Formula; for use in tables only.
Other	Diesel Expression	Allows use of Diesel macros.
	LISP Variable	All AutoLISP and VLISP variable names.
	System Variable	Access to all system variables.
Plot	DeviceName	Name of the plotter.
	Login	Login name.
	PageSetup	Page setup name.
	PaperSize	Size of paper.
	PlotDate	Same as under Date & Time.
	PlotOrientation	Orientation of the plot.
	PlotScale	Scale of the plot.
	PlotStyleTable	Name of the plot style table.
Sheetset	SheetSet	Information about sheet sets.
	SheetSetPlaceholder	Placeholder for future sheet sets.
	SheetView	CurrentSheetSetSubset
	CurrentSheetNumber	CurrentSheetSet
	CurrentSheetNumberandTitle	CurrentSheetSubset
	CurrentSheetTitle	CurrentSheetCategory
	CurrentSheetIssuePurpose	CurrentSheetRevisionDate
	CurrentSheetRevisionNumber	CurrentSheetCategory
	CurrentSheetCustom	CurrentSheetDescription
	Current SheetSetCustom	CurrentSheetSetDescription
	CurrentSheetProjectMilestone	CurrentSheetSetProjectNumber
	CurrentSheetProjectName	CurrentSheetSetProjectPhase

- Field text can have an annotative property when created with an annotative style or changed with the Properties or MText commands.

FileOpen

Rel.12 Opens drawing files without dialog boxes (an undocumented command).

Command	Alias	Ctrl+	F-key	Alt+	Menu Bar	Dashboard
fileopen	...	...	...	...	...	...

Command: fileopen
Enter name of drawing to open <filename.dwg>: *(Enter a file name.)*

COMMAND LINE OPTIONS

Enter name of drawing specifies the name of the *.dwg* file to open.

RELATED COMMANDS

Close closes the current drawing.

Open opens multiple drawing files.

RELATED SYSTEM VARIABLES

DbMod detects whether the drawing was changed since being opened.

SDI allows only one drawing at a time to be opened in AutoCAD .

TIPS

- Use this command in menu and toolbar macros to open a drawing file when you don't want to display a dialog box.

- This command can only be used in SDI (single drawing interface) mode; if you use this command when two or more drawings are open, AutoCAD complains, "The SDI variable cannot be reset unless there is only one drawing open. Cannot run FILEOPEN if SDI mode cannot be established."

- Use QSave before using FileOpen; otherwise, FileOpen displays the following dialog box:

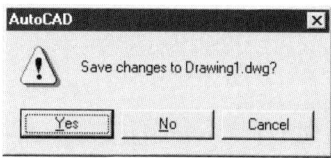

When you click Yes, AutoCAD closes the current drawing, and displays the Select Drawing File dialog box.

Removed Command

The **Files** command was removed from Release 14. In its place, use Windows Explorer.

'Fill

<u>V. 1.4</u>   Toggles whether hatches and wide objects — traces, multilines, solids, and polylines — are displayed and plotted with fills or as outlines.

Command	Alias	Ctrl+	F-key	Alt+	Menu Bar	Dashboard
fill	...	...	...	...	...	...

Command: fill
Enter mode [ON/OFF] <ON>: *(Type* **ON** *or* **OFF**.*)*

Left: *Fill on.*
Right: *Fill turned off with a wide polyline, donut, and 2D solid.*

COMMAND LINE OPTIONS

ON turns on fill after the next regeneration.

OFF turns off fill after the next regeneration.

RELATED SYSTEM VARIABLE

FillMode holds the current setting of fill status..

RELATED COMMAND

Regen changes the display to reflect the current fill or no-fill status.

TIPS

- The state of fill (or no fill) does not come into effect until the next regeneration:

  Command: regen
  Regenerating model.

- Traces, solids, and polylines are only filled in plan view, regardless of the setting of Fill. When viewed in 3D, these objects lose their fill.

- Fill affects objects derived from polylines, including donuts, polygons, rectangles, and ellipses, when created with PEllipse = 1.

- Fill does *not* affect TrueType fonts, which have their own system variable — TextFill — which toggles their fill/no-fill status for plots only.

- Fill does not toggle in rendered mode.

Fillet

V. 1.4 Joins intersecting lines, polylines, arcs, circles, and faces of 3D solids with a radius.

Command	Alias	Ctrl+	F-key	Alt+	Menu Bar	Dashboard
fillet	f	...	...	MF	Modify ⌐ Fillet	2D Draw

Command: fillet
Current settings: Mode = TRIM, Radius = 0.0
Select first object or [Polyline/Radius/Trim/mUltiple]: *(Select an object, or enter an option.)*
Select second object: *(Select another object.)*

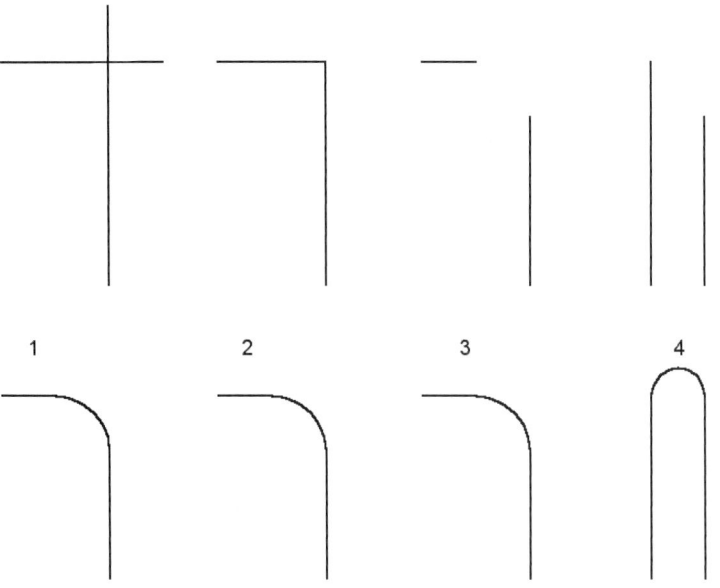

*Lines before (top) and after (bottom) applying the **Fillet** command:*
1. Crossing lines.   2. Touching lines.   3. Non-intersecting lines.   4. Parallel lines.

COMMAND LINE OPTIONS

Select first object selects the first object to be filleted.

Select second object selects the second object to be filleted.

mUltiple prompts you to select additional pairs of objects to fillet.

Polyline option
Select 2D polyline: *(Pick a polyline.)*

Select 2D polyline fillets all vertices of a 2D polyline; 3D polylines cannot be filleted.

Radius option
Specify fillet radius <0.0>: *(Enter a value.)*

Specify fillet radius specifies the filleting radius.

Trim option
Enter Trim mode option [Trim/No trim] <Trim>: *(Type T or N.)*

Trim trims objects when filleted.

No trim does not trim objects.

Filleting 3D Solids
Select an edge or [Chain/Radius]: *(Pick an edge, or enter an option.)*
Select an edge or [Chain/Radius]: *(Press Enter.)*
n edge(s) selected for fillet.

Left: *Box before filleting.*
Right: *After applying* ***Fillet*** *to the 3D solid.*

Edge option
Select an edge or [Chain/Radius]: *(Pick an edge.)*

Select edge selects a single edge.

Chain option
Select an edge or [Chain/Radius]: *(Type C.)*
Select an edge chain or [Edge/Radius]: *(Pick an edge.)*

Select an edge chain selects all tangential edges.

Radius option
Select an edge or [Chain/Radius]: *(Type R.)*
Enter fillet radius <1.0000>: *(Enter a value.)*

Enter fillet radius specifies the fillet radius.

RELATED COMMANDS

Chamfer bevels intersecting lines or polyline vertices.

SolidEdit edits 3D solid models.

RELATED SYSTEM VARIABLES

FilletRad specifies the current filleting radius.

TrimMode toggles whether objects are trimmed.

TIPS

- Pick the end of the object you want filleted; the other end will remain untouched.

- The lines, arcs, or circles need not touch.

- As a faster alternative to the Extend and Trim commands, use the Fillet command with a radius of zero.

- If the lines to be filleted are on two different layers, the fillet is drawn on the current layer.

- The fillet radius must be smaller than the length of the lines. For example, if the lines to be filleted are 1.0m long, the fillet radius must be less than 1.0m.

- Use the **Close** option of the PLine command to ensure a polyline is filleted at all vertices.

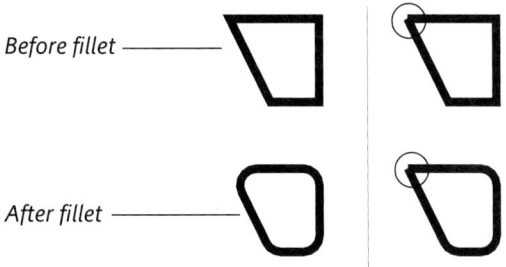

Before fillet —————

After fillet —————

Left: *Polyline closed with Close option...*
Right: *...and without.*

- Filleting polyline segments from different polylines joins them into a single polyline.

- Filleting a line or an arc with a polyline joins it to the polyline.

- You cannot apply internal and external fillets in the same operation, as illustrated below:

- Filleting a pair of circles does not trim them.

- As of AutoCAD Release 13, the Fillet command fillets a pair of parallel lines; the radius of the fillet is automatically determined as half the distance between the lines.

'Filter

Creates filter lists that can be applied when creating selection sets.

Command	Alias	Ctrl+	F-key	Alt+	Menu Bar	Dashboard
filter	fi	...	...	...	...	...

Command: filter

Displays dialog box:

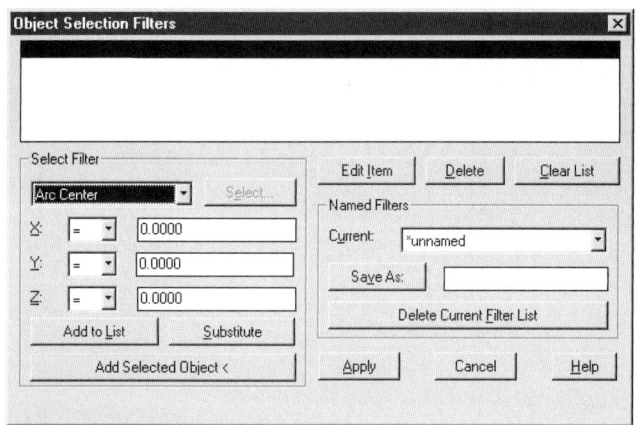

After you click Apply, AutoCAD continues in the command area:

Applying filter to selection.
Select objects: *(Select one or more objects.)*
Select objects: *(Press Enter to end object selection.)*
Exiting filtered selection.

AutoCAD highlights filtered objects with grips.

OBJECT SELECTION FILTERS DIALOG BOX

Select Filter options

Select displays all items of the specified type in the drawing.

X, Y, Z specifies the object's coordinates.

Add to List adds the current select-filter option to the filter list.

Substitute replaces a highlighted filter with the selected filter.

Add Selected Object selects the object to be added from the drawing.

Edit Item edits the highlighted filter item.

Delete deletes the highlighted filter item.

Clear list clears the entire filter list.

Named Filter options

Current selects the named filter from the list.

Save As saves the filter list with a name and the *.nfl* extension.

Delete Current Filter List deletes the named filter.

Apply closes the dialog box, and applies the filter operation.

COMMAND LINE OPTION

Select options selects the objects to be filtered; use the **All** option to select all non-frozen objects in the drawing.

RELATED COMMANDS

Any AutoCAD command with a 'Select objects' prompt.

QSelect creates a selection set quickly via a dialog box.

Select creates a selection set via the command line.

RELATED FILE

*.nfl is a named filter list.

TIPS

- The selection set created by Filter is accessed via the P (previous) selection option.

- Alternatively, 'Filter is used transparently at the 'Select objects:' prompt.

- Filter uses the following grouping operators:

**Begin OR	*with*	**End OR
**Begin AND	*with*	**End AND
**Begin XOR	*with*	**End XOR
**Begin NOT	*with*	**End NOT

- Filter uses the following relational operators:

Operator	Meaning
<	Less than.
<=	Less than or equal to.
=	Equal to.
!=	Not equal to.
>	Greater than.
>=	Greater than or equal to.
*	All values.

- Save selection sets by name to an *.nfl* (short for *named filter*) file on disk for use in other drawings or editing sessions.

Find

2000 Finds, and optionally replaces, text in drawings.

Command	Alias	Ctrl+	F-key	Alt+	Menu Bar	Dashboard
find	...	...	...	EF	Edit	Text
					⤷Find	

Command: find

Displays dialog box:

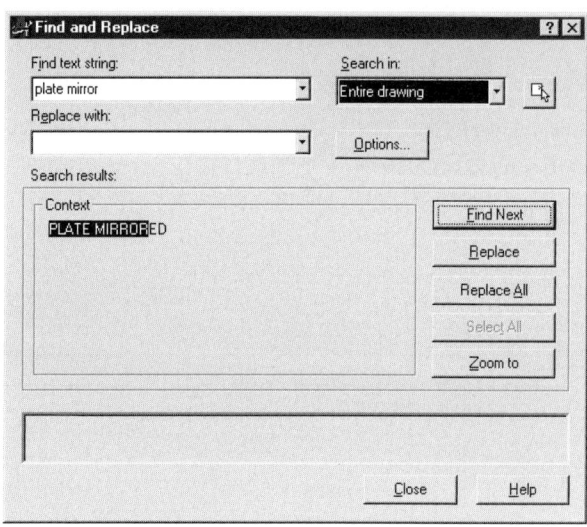

FIND AND REPLACE DIALOG BOX

Find text string specifies the text to find; enter text or click the down arrow to select one of the most recent lines of text searched for.

Replace with specifies the text to be replaced (only if replacing found text); enter text or click the down arrow to select one of the most recent lines of text searched for.

Search in:
  • **Current Selection** searches the current selection set for text; click the Select objects button to create a selection set, if necessary.
  • **Entire Drawing** searches the entire drawing.

Select objects allows you to select objects in the drawing; press Enter to return to dialog box.

Options displays the Find and Replace Options dialog box.

Search Results options

Context displays the found text in its context.

Find Next finds the text entered in the Find text string field.

Replace replaces a single found text with the text entered in the Replace with field.

Replace All replaces all instances of the found text.

Select All selects all objects containing the text entered in the Find text string field; AutoCAD displays the message "AutoCAD found and selected *n* objects that contain "...".”

Zoom to zooms in to the area of the current drawing containing the found text.

FIND AND REPLACE OPTIONS DIALOG BOX

Include options

☑ **Block Attribute Value** finds text in attributes.

☑ **Dimension Annotation Text** finds text in dimensions.

☑ **Text (Mtext, DText, Text)** finds text in paragraph text placed by the **MText** command, single-line text placed by the **Text** command, and field text placed by the **Field** command.

☑ **Table Text** finds text in tables.

☑ **Hyperlink Description** finds text in the description of a hyperlink.

☑ **Hyperlink** finds text in a hyperlink.

☐ **Match case** finds text that exactly matches the uppercase and lowercase pattern; for example, when searching for "Quick Reference," AutoCAD would find "Quick Reference" but not "quick reference."

☐ **Find whole words only** finds text that exactly matches whole words; for example, when searching for "Quick Reference," AutoCAD would find "Quick Reference" but not "Quickly Reference."

RELATED COMMANDS

AttDef creates attribute text.

DdEdit edits text.

Dim*xxx* commands create dimension text.

Field places updatable text.

Hyperlink creates hyperlinks and hyperlink descriptions.

MText creates paragraph text.

Properties edits selected text.

QSelect finds text objects.

Text creates single-line text.

TIPS

- To find database links, use the dbConnect command.

- The QSelect command places text in a selection set.

. . . .

FlatShot

2004 Creates 2D blocks flattened from 3D models in the current viewport; short for "FLATten snapSHOT"; formerly an Express Tool.

Command	Alias	Ctrl+	F-key	Alt+	Menu Bar	Dashboard
flatshot	fshot	...	...	...	...	3D Make

Command: flatshot

Displays dialog box:

*Click Create. The command continues with **-Insert** command-like prompts:*

Units: Unitless   Conversion: 1.0000

Specify insertion point or [Basepoint/Scale/X/Y/Z/Rotate]: *(Pick a point, or enter an option.)*

Enter X scale factor, specify opposite corner, or [Corner/XYZ] <1>: *(Press Enter, or enter an option.)*

Enter Y scale factor <use X scale factor>: *(Press Enter, or enter a scale factor.)*

Specify rotation angle <0>: *(Press Enter, or enter a rotation angle.)*

DIALOG BOX OPTIONS

Destination options

Insert As New Block inserts the flattened model as a block.

Replace Existing Block replaces the existing block, if it already exists.

Select Block selects the block to replace; press Enter to return to the dialog box.

Export to a File saves the flattened model to a *.dwg* file on disk.

Foreground Lines options

Color specifies the color of non-hidden (non-obscured) lines; click Select Color to select other colors.

Linetype specifies the linetype of non-hidden lines; click Select Linetype to load additional linetypes into the drawing.

Obscured Lines options

Show toggles the display of hidden-lines.

Color specifies the color of hidden (obscured) lines.

Linetype specifies the linetype of hidden lines.

Create closes the dialog box, and then creates the flattened view; the original model is retained.

TIPS

- This command works with solids and surfaces only; it does not work with other kinds of 3D models, such as those made of faces and thickened 2D objects.

Left: Original 3D model.
Right: 2D block created by the FlatShot command.

- The command creates a block from the 3D geometry; the block can be exploded with the Explode command for further editing.

- This command flattens the current screen view onto the current UCS. For the best result, set the UCS to View before using this command. This ensures the flattened view looks like the screen view. Finally, insert the new block into the WCS.

. . . .

Relocated and Removed Commands

FreePoint, **FreeSpot**, and **FreeWeb** commands (new to AutoCAD 2008) are found with the Light command.

Fog was removed from AutoCAD 2007; now an alias for the RenderEnvironment command.

GiffIn was removed from Release 14. In its place, use Image.

. . . .

GeographicLocation

<u>**2007**</u> Positions the sun (light) according to the selected location.

Command	Alias	Ctrl+	F-key	Alt+	Menu Bar	Dashboard
geographiclocation	geo	...	...	VELG	View	Lights
	north				⮡ Render	
	northdir				⮡ Light	
					⮡ Geographic Location	

Command: geographiclocation

Displays dialog box:

DIALOG BOX OPTIONS

Latitude sets the latitude from 0 to 90 degrees; when a city is selected from the map, its latitude is shown here.

Direction selects North or South of the equator.

Longitude sets the longitude from 0 to 180.

Direction selects East or West from the Prime Meridian.

Angle specifies the direction of North; default is the positive y axis.

Map options

You can click the map to locate the latitude and longitude; the red cross shows the current location.

☑ **Nearest Big City** displays the name of the nearest city to the latitude and longitude values entered.

Region selects a region or the entire world.

Nearest City selects a city.

Time Zone specifies the time zone.

RELATED COMMAND

SunProperties opens the Sun Properties palette.

RELATED SYSTEM VARIABLES

Latitude stores the current latitude.

Longitude stores the current longitude.

NorthDirection stores the angle from the positive y axis.

Timezone stores the current time zone.

TIPS

- You can access the sunlight through the Light control panel on the Dashboard:

- In addition to GeographicLocation, you can also control the sunlight by date and time.

- This command uses names and coordinates of cities found in the *.map* files found in *AutoCAD 2008\Support*, which are no longer user-editable.

'GotoUrl

<u>2000</u> Goes to hyperlinks contained by objects.

Command	Alias	Ctrl+	F-key	Alt+	Menu Bar	Dashboard
gotourl	...	...	...	...	...	...

Command: gotourl
Select objects: *(Select one or more objects.)*
Select objects: *(Press Enter to end object selection.)*
**browser Enter Web location (URL) <http://www.autodesk.com>: http://
www.upfrontezine.com**
AutoCAD launches your computer's default Web browser, and attempts to access the URL.

COMMAND LINE OPTION

Select objects selects one or more objects containing a hyperlink.

RELATED COMMANDS.

Browser launches the Web browser.

Hyperlink attaches, edits, and removes hyperlinks from objects.

TIPS

- This command is meant for use by macros and menus.

- AutoCAD uses this command for the shortcut menu's Hyperlink | Open option.

Gradient

2006 Floods areas with gradient fills.

Command	Alias	Ctrl+	F-key	Alt+	Menu Bar	Dashboard
gradient	gd	...	...	...	Draw	2D Draw
					↳Gradient	

Command: gradient

Displays the Gradient tab of the dialog box:

HATCH AND GRADIENT DIALOG BOX

Color options

⊙**One color** produces color-shade gradients.

Shade - Tint slider varies the second color between white (shade) and black (tint).

○**Two color** produces two-color gradients.

Color 1 selects the first color.

Color 2 selects the second color.

.... displays the Select Color dialog box.

Orientation options

☑ **Centered** centers the gradient in the hatch area.

☐ **Centered** moves the gradient up and to the left.

Angle rotates the gradient.

See the BHatch command for other options.

RELATED SYSTEM VARIABLES

GfAng specifies the angle of the gradient fill; ranges from 0 to 360 degrees.

GfClr1 specifies the first gradient fill color in RGB format, such as "RGB 000, 128, 255."

GfClr2 specifies the second gradient fill color in RGB format.

GfClrLum specifies the luminescence of one-color gradient fills, from 0.0 (black) to 1.0 (white).

GfClrState specifies whether the gradient fill is one-color or two-color.

GfName specifies the gradient fill pattern:

1 — Linear.
2 — Cylindrical.
3 — Inverted cylindrical.
4 — Spherical.
5 — Inverted spherical.
6 — Hemispherical.
7 — Inverted hemispherical.
8 — Curved.
9 — Inverted curved.

GfShift specifies whether the gradient fill is centered or shifted to the upper-left.

TIP

- The Gradient command simply displays the Gradient tab of the Hatch and Fill dialog box.

'GraphScr

<u>V. 2.1</u> Places the graphics window in front of the text window.

Command	Alias	Ctrl+	F-key	Alt+	Menu Bar	Dashboard
graphscr	...	...	F2	...	...	...

Command: graphscr

Before AutoCAD displays the drawing window:

RELATED COMMANDS

CopyHist copies text from the Text window to the Clipboard.

TextScr switches from the graphics window to the Text window.

RELATED SYSTEM VARIABLE

ScreenMode indicates whether the current screen is in text or graphics mode. Since the advent of Windows, only holds the value of 3: dual screen, displaying both text and graphics.

TIP

- The Text window appears frozen when a dialog box is active. Click the dialog box's OK or Cancel button to regain access to the Text window.

'Grid

V. 1.0 Displays a grid of reference dots within the current drawing limits.

Command	Alias	Ctrl+	F-key	Alt+	Status Bar	Dashboard
grid	...	G	F7	...	GRID	...

Command: grid
Specify grid spacing(X) or [ON/OFF/Snap/Major/aDaptive/Limits/Follow/ Aspect] <5'-0">: *(Enter a value, or enter an option.)*

When the visual style is set to 2D wireframe, the grid is displayed as dots.
In all other visual styles, the grid is displayed as lines.

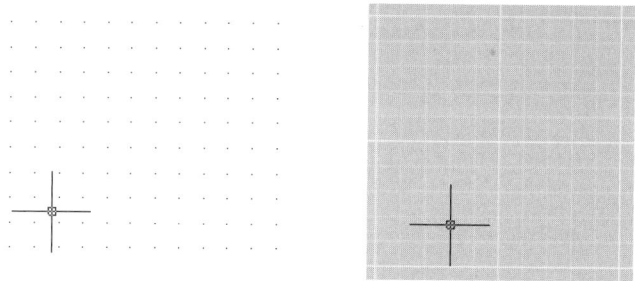

Left: *Grid of dots displayed by 2D wireframe visual style.*
Right: *Grid of lines displayed by other visual styles.*

COMMAND LINE OPTIONS

Specify grid spacing(X) sets the x and y direction spacing; an **x** following the value sets the grid spacing to a multiple of the current snap setting, such as 2x.

ON turns on grid markings.

OFF turns off grid markings.

Snap makes the grid spacing the same as the snap spacing.

Major specifies the number of minor grid lines per major line; default = 5.

aDaptive displays fewer grid lines when zoomed out

Limits determines whether the grid is limited to the rectangle defined by the Limits command.

Follow matches the grid plane to the dynamic UCS.

Aspect specifies different spacing for grid lines/dots in the x and y directions.

RELATED COMMANDS

DSsettings sets the grid via a dialog box.

Options sets the colors for grid lines.

Limits sets the limits of the grid in WCS.

Snap sets the snap spacing.

VsCurrent changes grid from dots (2D wireframe) to lines (other modes).

RELATED SYSTEM VARIABLES

GridDisplay determines how the grid is displayed

0 — grid restricted to the area specified by the Limits command.

1 — grid goes beyond the Limits command.

2 — adaptive display limits grid density when drawing is zoomed out.

4 — grid lines are more closely spaced when zoomed in; requires 2 turned on.

8 — grid plane follows the x,y-plane of the dynamic UCS.

GridMajor specifies the number of minor grid lines per major lines.

GridMode toggles the grid in the current viewport.

GridUnit specifies the current grid x,y-spacing.

LimMin specifies the x,y coordinates of the lower-left corner of the grid display.

LimMax specifies the x,y coordinates of the upper-right corner of the grid display.

SnapStyl displays a normal or isometric grid.

TIPS

- The grid is most useful when set to the snap spacing, or to a multiple of the snap spacing.

- When the grid spacing is set to 0, it matches the snap spacing.

- The Options command sets the color of grid lines. The grid line along the x axis is colored red, while the y axis line is green.

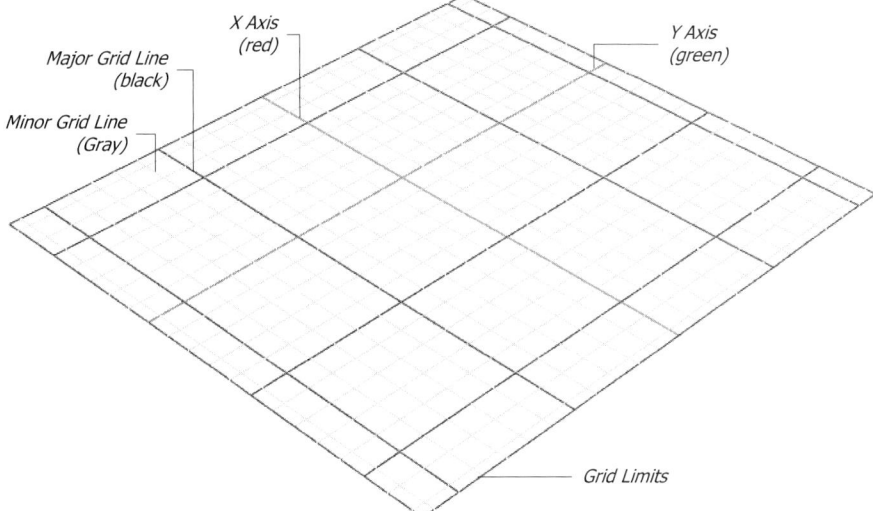

- The Snap command's Isometric option displays an isometric grid; when perspective mode is turned on, the grid is also displayed in perspective.

- You can set a different grid spacing in each viewport, and a different grid spacing in the x and y directions.

- Rotate the grid with the Snap command's Rotate option.

- Grid markings are not plotted; to create a plotted grid, use the Array command to place an array of points or lines.

Group

Rel.13 Creates named selection sets of objects.

Commands	Aliases	Ctrl	F-key	Alt+	Menu Bar	Dashboard
group	g	...	...	...	...	...
-group	-g					

Command: group

Displays dialog box:

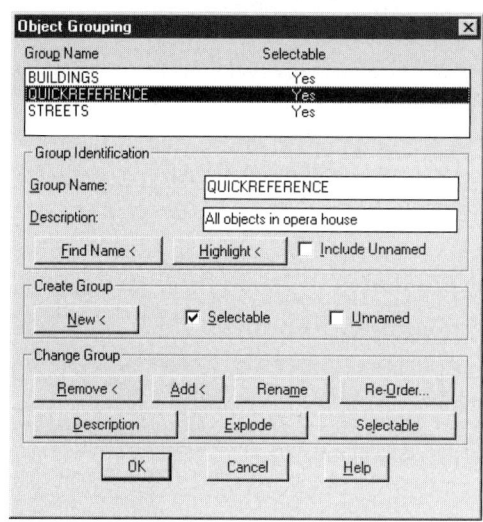

OBJECT GROUPING DIALOG BOX

Group Name lists the names of groups in the drawing.

Group Identification options

Group Name displays the name of the current group.

Description describes the group; may be up to 64 characters long.

Find Name lists the name(s) of group(s) that a selected object belongs to.

Highlight highlights the objects included in the current group.

☐ **Include Unnamed** lists unnamed groups in the dialog box.

Create Group options

New selects objects for the new group.

☑ **Selectable** toggles selectability: picking one object picks the entire group.

☐ **Unnamed** creates an unnamed group; AutoCAD gives the name ***A***n*, where *n* is a number that increases with each group.

Change Group options

Remove removes objects from the current group.

Add adds objects to the current group.

Rename renames the group.

Re-order changes the order of objects in the group; displays Order Group dialog box.

Description changes the description of the group.

Explode removes the group description; does not erase group members.

Selectable toggles selectability.

ORDER GROUP DIALOG BOX

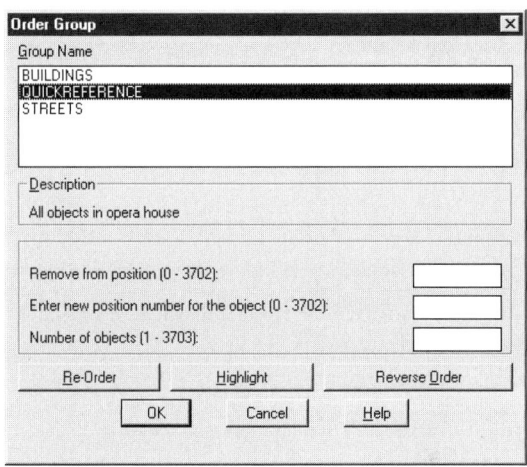

Group Name lists the names of groups in the current drawing.

Description describes the selected group.

Remove from position (0 - *n*) selects the object to move.

Replace at position (0 - *n*) moves the group name to a new position.

Number of objects (1 - *n*) lists the number of objects to reorder.

Re-Order applies the order changes.

Highlight highlights the objects in the current group.

Reverse Order reverses the order of the groups.

. . . .

-GROUP Command
Command: -group
Enter a group option
[?/Order/Add/Remove/Explode/REName/Selectable/Create] <Create>:
(Enter an option.)

COMMAND LINE OPTIONS

? lists the names and descriptions of currently-defined groups.

Order changes the order of objects within the group.

Add adds objects to the group.

. . . .

Remove removes objects from the group.

Explode removes the group definition from the drawing.

REName renames the group.

Selectable toggles whether the group is selectable.

Create creates a newly-named group from the objects selected.

RELATED COMMANDS

Block creates named symbols from a group of objects.

Select creates selection sets.

RELATED SYSTEM VARIABLE

PickStyle toggles whether groups are selected by the usual selection process:

0 — Groups and associative hatches are not selected.
1 — Groups are included in selection sets (default).
2 — Associative hatches are included in selection sets.
3 — Both are selected.

TIPS

- You can toggle groups on and off with the Ctrl+Shift+A shortcut keystroke.

- Use Ctrl+H to toggle between selecting the entire group, and objects within the group.

- Consider a group as a named selection set; unlike a regular selection set, a group is not "lost" when the next group is created.

- Group descriptions can be up to 64 characters long.

- Anonymous groups are unnamed; AutoCAD refers to them as *A*n*.

. . . .

Changed Command

As of AutoCAD 2006, the **Hatch** command acts like the BHatch command, while the **-Hatch** command is like the -BHatch command.

. . . .

Hatch

<u>Rel.12</u>   Applies associative hatch pattern to objects within boundaries.

Commands	Aliases Ctrl+	F-key	Alt+	Menu Bar	Dashboard
hatch	bh ...	...	DH	Draw	2D Draw
	bhatch			⬦Hatch	
	h				
-hatch	-bhatch				

Command: hatch

Displays dialog box:

HATCH AND GRADIENT DIALOG BOX

Type and Pattern options

▼ **Type** selects the pattern type:

- **Predefined** uses hatches stored in *acad.pat* and *acadiso.pat* files.
- **User Defined** creates parallel-line hatches with spacing defined by you.
- **Custom** uses hatches defined by *.pat* files added to AutoCAD's search path.

▼ **Pattern** selects hatch pattern.

... displays Hatch Pattern Palette dialog box, showing sample pattern types.

Swatch displays a non-scaled preview of the hatch pattern; click to display the Hatch Pattern Palette dialog box.

▼ **Custom Pattern** lists the custom patterns, if available.

Angle and Scale options

▼ **Angle** specifies the hatch pattern rotation; default = 0 degrees.

▼ **Scale** specifies the hatch pattern scale; default = 1.0.

Double (available when User-defined type is selected):

☑ User-defined hatch is applied a second time at 90 degrees to the first pattern.

☐ User-defined hatch is applied once.

Relative to Paper Space (available when hatch is applied through layout viewport):

☑ Hatch patterns scale relatively to layout scale.

☐ Hatch patterns scale independently of layout scale.

Spacing specifies the spacing between the lines of a user-defined hatch pattern.

▼ **ISO Pen Width** patterns scale according to pen width; available for ISO patterns only.

Hatch Origin options

⊙ **Use Current Origin** uses x,y coordinates stored in the HpOrigin system variable.

○ **Specified Origin** specifies new hatch origins:

Click to Set New Origin prompts "Specify origin point:" to locate new origin in drawing.

☑ **Default to Boundary Extents** places the new origin at one of the four corners of the extents or at the center of the rectangular *extents* of the hatch.

☐ **Store as Default Origin** stores the coordinates in the HpOrigin system variable.

Boundaries options

Add: Pick Points prompts, "Pick internal point or [Select objects/remove Boundaries]:". Pick a point inside an area; AutoCAD automatically detects the boundary, which will be filled with the hatch pattern.

Add: Select Objects prompts, "Select objects or [picK internal point/remove Boundaries]:". Pick one or more objects, which will be filled with the hatch pattern.

Remove Boundaries prompts "Select objects or [Add boundaries]:". Pick one or more boundaries to be removed from the hatch pattern selection set.

Recreate Boundaries places polylines or regions around selected hatches.

View Selections views hatch pattern selection set.

Options options

☐ **Annotative** toggles annotative scaling for hatch patterns (*new to AutoCAD 2008*).

☑ **Associative** creates hatches that update automatically when their boundaries or properties are modified; pattern is created as a hatch object.

Create separate hatches:

☑ Each hatch pattern is independent.

☐ Patterns are a single pattern, when multiple hatches are created at once.

▼ Draw Order:

- **Do not assign** places hatch normally.
- **Send to back** places hatch behind all other overlapping objects in the drawings.
- **Bring to front** places hatch in front of all other overlapping objects.
- **Send behind boundary** places hatch behind its boundary.
- **Bring in front of boundary** places hatch in front of its boundary.

Inherit Properties sets the hatch pattern parameters from an existing hatch pattern. **Preview** previews the hatch pattern.

▶ More Options

Islands options

Island detection:

☑ Island detection is turned on.

☐ Island detection is turned off.

Island display style:

⊙ **Normal** turns hatching off and on each time it crosses a boundary; text is not hatched.

○ **Outer** hatches only the outermost areas; text is not hatched.

○ **Ignore** hatches everything within the boundary; text is hatched.

Boundary Retention options
Retain Boundaries:

☑ Boundary polylines or regions (created during the boundary hatching process) are kept after Hatch finishes.

☐ Boundaries are discarded after hatching.

▼ **Object Type** constructs the boundary from:

- **Polylines** forms the boundary from a polyline.
- **Region objects** forms the boundary from a 2D region.

Boundary Set options

▼ **Boundary Set** defines how objects are analyzed for defining boundaries; not available when Select Objects is used to define the boundary (default = current viewport).

⬚ **New** creates new boundary sets; prompts you to select objects at the command line.

Gap Tolerance options

Tolerance specifies the maximum gap between boundary objects. (Objects don't need to touch to form a valid hatching boundary.) Range is 0 to 5000 units.

Gradient tab
See Gradient command.

. . . .

-HATCH Command
Command: -hatch
Current hatch pattern: ANSI31
Specify internal point or [Properties/Select/draW boundaries/remove Boundaries/Advanced/DRaw order/Origin/Annotative]: *(Pick a point, or specify an option.)*

COMMAND LINE OPTIONS

Internal Point option
Pick a point creates a boundary of the area surrounding the pick point.

Property options
Enter a pattern name or [?/Solid/User defined] <ANSI31>: *(Enter a name, or select an option.)*
Specify a scale for the pattern <1.0000>: *(Enter a scale factor.)*
Specify an angle for the pattern <0>: *(Enter an angle.)*

Enter a pattern name allows you to name the hatch pattern.

? lists the names of available hatch patterns.

Solid floods the area with a solid fill in the current color.

User defined creates a simple, user-defined hatch pattern.

Specify a scale specifies the hatch pattern angle (default = 0 degrees).

Specify an angle specifies the hatch pattern scale (default: = 1.0).

Select options
Select objects: *(Select one or more objects.)*
Select objects: *(Press Enter.)*

Select objects selects one or more objects to fill with hatch pattern.

. . . .

Draw Boundary options
Retain polyline boundary? [Yes/No] <N>: *(Type* **Y** *or* **N**.)
Specify start point: *(Pick point.)*
Specify next point or [Arc/Length/Undo]: *(Pick point, or enter an option.)*
Specify next point or [Arc/Close/Length/Undo]: *(Pick point or enter an option.)*
Specify start point for new boundary or <Accept>: *(Press Enter to end option.)*

Retain polyline boundary:

- **On** keeps the boundary created during the hatching process after -Hatch finishes.
- **Off** discards the boundary.

Specify start point / next point prompts to draw a closed polyline.

Remove Boundaries options
Select objects or [Add boundaries]: *(Select one or more boundary objects.)*
Select objects or [Add boundaries/Undo]: *(Press Enter, or type U.)*

Select objects selects hatch boundaries to remove.

Undo adds the removed island.

Advanced options
**Enter an option [Boundary set/Retain boundary/Island detection/Style/
Associativity/Gap tolerance]:** *(Specify an option.)*

Boundary set defines the objects analyzed when a boundary is defined by a specified pick point.

Retain boundary:

- **On** the boundary created during the hatching process is kept after -Hatch finishes.
- **Off** the boundary is discarded.

Island detection:

- **On** objects within the outermost boundary are used as boundary objects.
- **Off** all objects within the outermost boundary are filled.

Style selects the hatching style: ignore, outer, or normal.

Associativity:

- **On** hatch pattern is associative.
- **Off** hatch pattern is not associative.

Gap tolerance specifies the largest gap permitted in the boundary:
Specify a boundary gap tolerance value <0>: *(Enter a value between 0 and 5000.)*

Draw Order options
**Enter draw order [do Not assign/send to Back/bring to Front/send beHind
boundary/bring in front of bounDary] <send beHind boundary>:** *(Specify an
option.)*

do **N**ot assign places hatches normally.

send to **B**ack places hatches behind all other overlapping objects in the drawing.

bring to **F**ront places hatches in front of all other overlapping objects.

send be**H**ind boundary places hatches behind its boundary.

bring in front of boun**D**ary places hatch in front of the boundary.

Hatch patterns included with AutoCAD

ANSI Patterns:

ANSI31 ANSI32 ANSI33

ANSI34 ANSI35 ANSI36

ANSI37 ANSI38

ISO Patterns:

ISO02W100 ISO03W100 ISO04W100

ISO05W100 ISO06W100 ISO07W100

ISO08W100 ISO09W100 ISO10W100

ISO11W100 ISO12W100 ISO13W100

ISO14W100 ISO15W100

Other Predefined Patterns:

SOLID ANGLE AR-B816 AR-B816C AR-B88 AR-BRELM AR-BRSTD

AR-CONC AR-HBONE AR-PARQ1 AR-RROOF AR-RSHKE AR-SAND BOX

BRASS BRICK BRSTONE CLAY CORK CROSS DASH

DOLMIT DOTS EARTH ESCHER FLEX GRASS GRATE

GRAVEL HEX HONEY HOUND INSUL LINE MUDST

NET NET3 PLAST PLASTI SACNCR SQUARE STARS

STEEL SWAMP TRANS TRIANG ZIGZAG

Origin options
[Use current origin/Set new origin/Default to boundary extents] <Use current origin>: *(Enter an option.)*

Use current origin uses the x,y coordinates stored in the HpOrigin system variable.

Set new origin prompts to pick a point in the drawing as the new origin.

Default to boundary extents places the new origin at one of the four corners of the extents or at the center of the rectangular *extents* of the hatch.

Annotative options
Create annotative hatch [Yes/No] <No>: *(Enter an option.)*

Yes / No does (or doesn't) apply the annotative property to the hatch pattern *(new to AutoCAD 2008)*.

RELATED COMMANDS

AdCenter places hatch patterns from other drawings.

Boundary traces polylines automatically around closed boundaries.

Convert converts Release 13 (and earlier) hatch patterns into Release 14-2008 format.

Hatch places nonassociative hatch patterns.

HatchEdit edits hatch patterns.

Properties changes the properties of hatch patterns.

PsFill floods closed polylines with PostScript fill patterns.

ToolPalette stores and places selected hatch patterns and fill colors.

RELATED SYSTEM VARIABLES

DelObj toggles whether boundary is erased after hatch is placed.

FillMode determines whether hatch patterns are displayed.

HpAng specifies the current hatch pattern angle; default = 0.

HpAssoc determines whether or not hatches are associative.

HpBound specifies the hatch boundary object: polyline or region.

HpDouble specifies single or double hatching.

HpDrawOrder controls the display order of the hatch pattern relative to other overlapping objects.

HpGalTol reports the current gap tolerance.

HpInherit determines whether the MatchProp command copies hatch origins from the source objects.

HpMaxlines specifies the maximum number of hatch lines to draw. Range is 10 to 10,000,000; default = 1,000,000 (*new to AutoCAD 2008*).

HpName names the current hatch pattern.

HpObjWarning specifies the maximum number of hatch boundary objects that can be selected before AutoCAD sounds the alert; default = 10000.

HpOrigin specifies the x,y coordinates for the origin of new hatches.

HpOriginMode determines the default hatch origin point.

    0 — Uses HpOrigin.
    1 — Bottom left corner of boundary's rectangular extents.
    2 — Bottom right corner of boundary's rectangular extents.
    3 — Top right corner of boundary's rectangular extents.
    4 — Top left corner of boundary's rectangular extents.
    5 — Center of boundary's rectangular extents.

HpScale specifies the current hatch pattern scale factor; default = 1.

HpSeparate determines whether a single hatch object or separate hatch objects are created when operating on several closed boundaries.

HpSpace specifies the current hatch pattern spacing factor; default = 1.

OsnapHatch determines whether object snap snaps to hatch patterns.

PickStyle controls the selection of hatch patterns:

0 — Neither groups nor hatches selected.
1 — Groups selected (default).
2 — Associative hatches selected.
3 — Both selected.

SnapBase specifies the starting coordinates of hatch pattern (default = 0,0).

RELATED FILES

acad.pat contains the ANSI and other hatch pattern definitions.

acadiso.pat contains the ISO hatch pattern definitions.

TIPS

- This command first generates a boundary, and then hatches the inside area. Use the Boundary command to create just the boundary.

- Hatch stores hatching parameters in the pattern's extended object data.

- Bring the pattern to the front (through the Draw Order option) for easier editing of the hatch.

- ANSI patterns are all defined as 45° lines. To keep at 45°, the angle should be specified as 0.

- The Trim command trims hatches.

- When you pick a non-hatchable area, AutoCAD displays a dialog box with helpful tips:

- Set the OsnapHatch system variable to 0 (default) to avoid object snapping to hatch and fill patterns.

HatchEdit

Rel.13   Edits associative hatch objects.

Commands	Alias	Ctrl+	F-key	Alt+	Menu Bar	Dashboard
hatchedit	he	...	...	MO⍉H	Modify	...
					⍉Object	
					⍉Hatch	

-hatchedit

Command: hatchedit
Select associative hatch object: *(Select one hatch object.)*
   Displays dialog box:

HATCH EDIT DIALOG BOX
*See the **BHatch** and **Gradient** commands for options.*

. . . .

-HATCHEDIT Command
Command: -hatchedit
Select associative hatch object: *(Select one object.)*
**Enter hatch option [DIsassociate/Style/Properties/DRaw order/ADd
boundaries/Remove boundaries/recreate Boundary/ASsociate/separate
Hatches/Origin/ANnotative] <Properties>:** *(Enter an option.)*

. . . .

COMMAND LINE OPTIONS

Select selects a single associative hatch pattern.

DIsassociate removes associativity from hatch pattern.

Style options
Enter hatching style [Ignore/Outer/Normal] <Normal>: *(Enter an option.)*

Normal hatches alternate boundaries (default).

Outer hatches only outermost boundary.

Ignore hatches everything within outermost boundary.

Properties options
Enter a pattern name or [?/Solid/User defined] <ANSI31>: *(Enter name or type an option.)*
Specify a scale for the pattern <1.0000>: *(Enter scale factor.)*
Specify an angle for the pattern <0>: *(Enter pattern angle.)*

Pattern name specifies the name of a valid hatch pattern.

? lists the available hatch pattern names.

Solid replaces the hatch pattern with solid fill.

User defined creates a new on-the-fly hatch pattern.

Scale changes the hatch pattern scale.

Angle changes the hatch pattern angle.

Draw Order options
Enter draw order [do Not assign/send to Back/bring to Front/send beHind boundary/bring in front of bounDary] <send beHind boundary>: *(Specify an option.)*

do Not assign places hatch normally.

send to Back places hatch behind all other overlapping objects in the drawings.

bring to Front places hatch in front of all other overlapping objects.

send beHind boundary places hatch behind its boundary.

bring in front of bounDary places hatch in front of its boundary.

Add Boundaries options
Specify internal point or [Select objects]: *(Pick a point or select objects.)*

Specify internal point selects points within closed objects.

Select objects selects one or more objects to add to the boundary collection.

Remove Boundaries options
Select objects or [Add boundaries]: *(Pick a point or select objects.)*

Select objects selects one or more objects to remove from the boundary collection.

Add boundaries switches to add boundary mode.

Recreate Boundaries options
Enter type of boundary object [Region/Polyline] <Polyline>: *(Type R or P.)*
Reassociate hatch with new boundary? [Yes/No] <N>: *(Type Y or N.)*

Enter type of boundary object:

• **Region** switches boundary to region object.

• **Polyline** switches boundary to polyline object.

Reassociate hatch with new boundary?

- **Yes** associates hatch with changed boundary type.

- **No** does not associate hatch with changed boundary type.

Associate options

Associates hatch with the selected boundary objects.

Specify internal point or [Select objects]: *(Pick a point or select objects.)*

Specify internal point selects points within closed objects.

Select objects selects one or more boundary objects to associate with hatch.

Origin options

Associates hatch with the selected boundary objects.

[Use current origin/Set new origin/Default to boundary extents] <Use current origin>: *(Enter an option.)*

Use current origin retains the current origin.

Set new origin prompts for a new origin point, and then asks whether to store it as the new default origin.

Default to boundary extents specifies one of bottom Left (default), bottom Right, top rIght, top lEft, or Center.

Annotative options

Create annotative hatch [Yes/No] <No>: *(Enter an option.)*

Yes / No does (or doesn't) apply the annotative property to the hatch pattern (*new to AutoCAD 2008*).

Separate Hatches separates a single object consisting of multiple hatch areas into independent hatch objects.

RELATED COMMANDS

BHatch applies associative hatch patterns.

Explode explodes a hatch pattern block into lines.

Gradient applies associative fills to areas.

MatchProp matches the properties of one hatch to other hatches.

Properties changes properties of hatch patterns.

TIPS

- HatchEdit works with associative and non-associative hatch objects.

- AutoCAD cannot change a non-associative hatch to associative; uncheck the **Associative** option to convert a non-associative hatch to associative. Also, you can use the Explode command to reduce hatches to lines.

- As an alternative to entering this command, double-click hatch patterns to display the Hatch Edit dialog box.

- To select a solid fill, click the outer edge of the hatch pattern, or use a crossing window selection on top of the solid fill.

- Even though the Hatch Edit dialog box looks identical to the BHatch dialog box, many options are not available in the Hatch Edit dialog box.

Helix

2007 Draws helixes.

Commands	Alias	Ctrl+	F-key	Alt+	Menu Bar	Dashboard
helix	...	...	...	DX	Draw	3D Make
					⬉ Helix	

Command: helix
Number of turns = 3.0000   Twist=CCW
Specify center point of base: *(Pick point 1.)*
Specify base radius or [Diameter] <1.0>: *(Enter radius 2, or type D.)*
Specify top radius or [Diameter] <1.5>: *(Enter radius 3, or type D.)*
Specify helix height or [Axis endpoint/Turns/turn Height/tWist] <1.0>: *(Enter height 4, or enter an option.)*

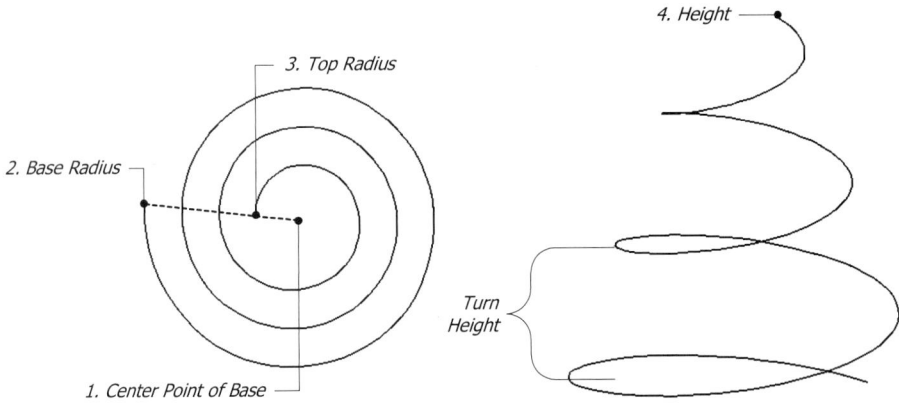

COMMAND OPTIONS

Center point of base specifies the coordinates of the helix's base; when z is not specified, the helix is drawn at the current elevation.

Base radius specifies the radius of the base.

Diameter specifies the diameter of the base.

Top radius specifies the radius of the top; value must be positive.

Helix height specifies the height of the helix; if negative, the helix is drawn downwards from its base.

Axis endpoint specifies a second point (from the base center point) to draw the helix at an angle.

Turns specifies the number of turns the helix makes; default = 3.

turn Height specifies the height of one turn.

tWist switches between drawing the helix clockwise and counterclockwise (default).

TIP

- To draw a cylindrical spring, make the top and bottom radii the same.

'Help

V. 1.0 Lists information for using AutoCAD's commands.

Command	Alias	Ctrl+	F-key	Alt+	Menu Bar	Dashboard
help	'?	...	F1	HH	Help	...
					⬏Help	

Command: help

Displays window:

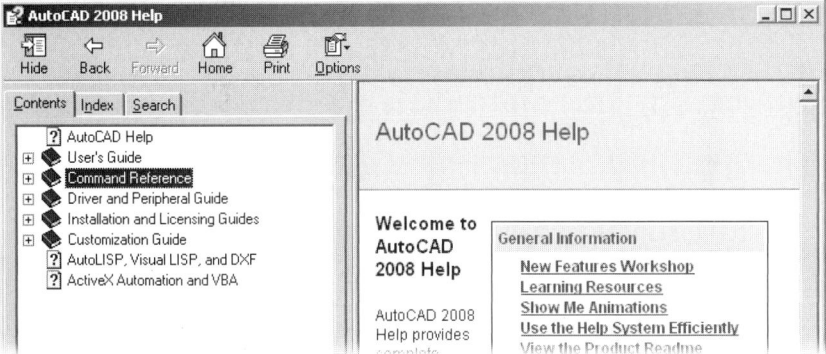

RELATED COMMANDS

All.

RELATED FILE

acad171.chm is AutoCAD 2008's primary help file.

TIPS

- Because 'Help, '?, and F1 are transparent commands, you can use them during another command to get help with the command's options.

- AutoCAD 2008 adds the Infocenter to the menu bar:

- Pressing F1 during a command displays context-sensitive help.

. . . .

Removed Commands

Hide now operates only in 2D wireframe mode; as of AutoCAD 2007 it is an alias for the **VsCurrent** command in all other modes, such as 3D wireframe and Realistic.

HlSettings was removed from AutoCAD 2007; it was replaced by the VisualStyles command.

HpConfig was removed from AutoCAD 2000; it was replaced by the Plot command.

. . . .

Hyperlink

2000 Attaches hyperlinks (URLs) to objects in drawings.

Commands	Alias	Ctrl+	F-key	Alt+	Menu Bar	Dashboard
hyperlink	...	k	...	IH	Insert	...
					↳ Hyperlink	

-hyperlink

Command: hyperlink
Select objects: *(Select one or more objects.)*
Select objects: *(Press Enter to end object selection.)*
Displays dialog box:

COMMAND LINE OPTION

Select object selects the objects to which to attach the hyperlink.

INSERT HYPERLINK DIALOG BOX

Existing File or Web Page page

Text to display describes the hyperlink; the description is displayed by the tooltip; when blank, the tooltip displays the URL.

Type the file or Web page name specifies the hyperlink (URL) to associate with the selected objects; the hyperlink may be any file on your computer, on any computer you can access on your local network, or on the Internet.

Or select from list:

* **Recent files** lists drawings recently opened in AutoCAD.
* **Browsed pages** lists pages recently viewed with Web browsers, and other software.
* **Inserted links** lists URLs recently entered in Web browsers.

File opens the Browse the Internet - Select Hyperlink dialog box.
Web Page starts up a simple Web browser.

Target specifies a location in the file, such as a target in an HTML file, a named view in AutoCAD, or a page in a spreadsheet document. Displays Select Place in Document dialog box.

Path displays the full path and filename to the hyperlink; only the file name appears when **Use relative path for hyperlink** is checked.

☑ **Use relative path for hyperlink** toggles use of the path for relative hyperlinks in the drawing; when "" (null), the drawing paths stored in AcadPrefix are used.

☑ **Convert DWG hyperlinks to DWF** preserves hyperlinks when drawings are exported in DWF format.

Remove link removes the hyperlink from the object; this button appears only if you select an object that already has a hyperlink.

View of This Drawing page

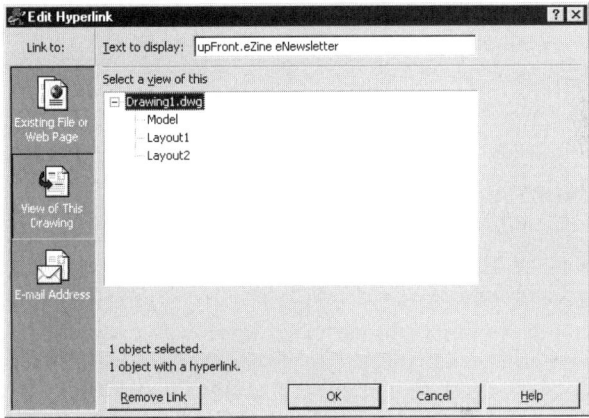

Select a view of this selects a layout, named view, or named plot.

Email Address page

Email address specifies the email address; the *mailto:* prefix is added automatically.

Subject specifies the text that will be added to the Subject line.

Recently used e-mail addresses lists the email addresses recently entered.

-HYPERLINK Command

Command: -hyperlink

Enter an option [Remove/Insert] <Insert>: *(Enter an option.)*
Enter hyperlink insert option [Area/Object] <Object>: *(Enter an option.)*
Select objects: *(Select one or more objects.)*
Select objects: *(Press Enter to end object selection.)*
Enter hyperlink <current drawing>: *(Enter hyperlink address.)*
Enter named location <none>: *(Optional: Enter a bookmark.)*
Enter description <none>: *(Optional: Enter a description.)*

COMMAND LINE OPTIONS

Remove removes a hyperlink from selected objects or areas.

Insert adds a hyperlink to selected objects or areas.

Select objects selects the object to which the hyperlink will be added.

Enter hyperlink specifies the filename or hyperlink address.

Enter named location (*optional*) specifies a location within the file or hyperlink.

Enter description (*optional*) describes the hyperlink.

RELATED COMMANDS

HyperlinkOptions toggles the display of the hyperlink cursor, shortcut menu, and tooltip.

HyperlinkOpen opens hyperlinks (URLs) via the command line.

HyperlinkBack returns to the previous URL.

HyperlinkFwd moves forward to the next URL; works only when the HyperlinkBack command was used, otherwise AutoCAD complains, "** No hyperlink to navigate to **".

HyperlinkStop stops the display of the current hyperlink.

GoToUrl displays specific Web pages.

PasteAsHyperlink pastes hyperlinks to selected objects.

RELATED SYSTEM VARIABLE

HyperlinkBase specifies the path for relative hyperlinks in the drawing; when "" (null), the drawing paths stored in AcadPrefix are used.

TIPS

- If the drawing has never been saved, AutoCAD is unable to determine the default *relative folder*. For this reason, the Hyperlink command prompts you to save the drawing.

- By using hyperlinks, you can create a *project document* consisting of drawings, contacts (word processing documents), project timelines, cost estimates (spreadsheet pages), and architectural renderings. To do so, create a "title page" of an AutoCAD drawing with hyperlinks to the other documents.

- To edit a hyperlink with Hyperlink, select object(s), make editing changes, and click OK.

- To remove a hyperlink with Hyperlink, select the object, and then click Remove Link.

- An object can have just one hyperlink attached to it; more than one object, however, can share the same hyperlink.

- The alternate term for "hyperlink" is *URL*, which is short for "universal resource locator," the universal file naming system used by the Internet.

'HyperlinkOpen/Back/Fwd/Stop

__2000__ Controls the display of hyperlinked pages (undocumented commands).

Commands	Alias	Ctrl+	F-key	Alt+	Menu Bar	Dashboard
hyperlinkopen	...	...	...	...	...	...
hyperlinkback						
hyperlinkfwd						
hyperlinkstop						

Command: hyperlinkopen
Enter hyperlink <current drawing>: *(Enter a hyperlink option.)*
Enter named location <none>: *(Optional: Enter a bookmark.)*
Displays the specified Web page or file, if possible.

Command: hyperlinkback
Returns to the previous hyperlinked page.

Command: hyperlinkstop
Stops displaying the Web page.

Command: hyperlinkfwd
Hyperlinks to the next page; can be used only after the HyperLinkBack command.

COMMAND LINE OPTIONS

Enter hyperlink enters a URL (uniform resource locator) or a filename.

Enter named location enters a named view or other valid target.

RELATED COMMANDS

Hyperlink attaches a hyperlink to objects.

HyperlinkOptions specifies the options for hyperlinks.

GoToUrl displays a specific Web page.

RELATED TOOLBAR ICONS

Go Back goes to the previous hyperlink; executes the HyperlinkBack command.

Go Forward goes to the next hyperlink; executes the HyperlinkFwd command.

Stop Navigation stops loading the current hyperlink file; executes the HyperlinkStop command.

Browse the Web displays the Web browser; executes the Browser command.

'HyperlinkOptions

<u>2000</u>   Toggles the display of the hyperlink cursor, shortcut menu, and tooltip.

Command	Alias	Ctrl+	F-key	Alt+	Menu Bar	Dashboard
hyperlinkoptions	...	...	...	...	...	

Command: hyperlinkoptions
Display hyperlink cursor tooltip and shortcut menu? [Yes/No] <Yes>: *(Type Y or N.)*

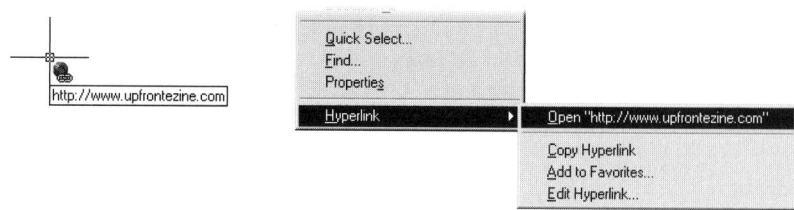

Left: Hyperlink cursor and tooltip.
Right: Hyperlink shortcut menu.

COMMAND LINE OPTIONS

Display hyperlink cursor and shortcut menu toggles the display of the hyperlink cursor and shortcut menu.

Display hyperlink tooltip toggles the display of the hyperlink tooltip.

SHORTCUT MENU

Select an object containing a hyperlink; then right-click to display the cursor menu.

Hyperlink options:

Open "url" launches the appropriate applications, and loads the file referenced by the URL.

Copy Hyperlink copies hyperlink data to the Clipboard; use the PasteAsHyperlink command to paste the hyperlink to selected objects.

Add to Favorites adds the hyperlink to a favorites list.

Edit Hyperlink displays the Edit Hyperlink dialog box; see the Hyperlink command.

RELATED COMMANDS

Hyperlink attaches a hyperlink to objects.

Options determines options for most other aspects of AutoCAD.

RELATED SYSTEM VARIABLE

HyperlinkBase specifies the path for relative hyperlinks in the drawing; when "" (null), the drawing paths stored in AcadPrefix are used.

TIP

■ Answering "N" to this command makes hyperlinks unavailable.

'Id

V. 1.0 Identifies the 3D coordinates of specified points (short for IDentify).

Command	Alias	Ctrl+	F-key	Alt+	Menu Bar	Dashboard
id	...	...	...	TQI	Tools	...
					⮤ Inquiry	
					⮤ Id Point	

Command: id
Specify point: *(Pick a point.)*

Sample output:
X = 1278.0018   Y = 1541.5993   Z = 0.0000

COMMAND LINE OPTION
Specify point picks a point.

RELATED COMMANDS
List lists information about picked objects.

Point draws points.

RELATED SYSTEM VARIABLE
LastPoint contains the 3D coordinates of the last picked point.

TIPS
- The Id command stores the picked point in the LastPoint system variable. Access that value by entering @ at the next prompt for a point value.

- Invoke the Id command to set the value of the LastPoint system variable, which can be used as relative coordinates in another command.

- If a 2D point is specified, the z-coordinate displayed by Id is the current elevation setting; otherwise, the z-coordinate is that of the specified point.

- When you use Id with an object snap, the z-coordinate is the object-snapped value.

- It is faster to enter "id" at the command line than to access it through menu picks.

Replaced Command
Image was removed from AutoCAD 2007; it was replaced by the ExternalReferences command.

-Image

Rel.14 Controls the attachment of raster images at the command line.

Commands	Alias	Ctrl+	F-key	Alt+	Menu Bar	Dashboard
-image	-im	...	...	...	...	...

Command: -image
Enter image option [?/Detach/Path/Reload/Unload/Attach] <Attach>: *(Enter an option.)*

COMMAND LINE OPTIONS

? lists currently-attached image files.

Detach erases images from drawings.

Path lists the names of images in the drawing.

Reload reloads image files into drawings.

Unload removes images from memory without erasing them.

Attach displays the Attach Image File dialog box; see the ImageAttach command.

RELATED COMMANDS

ExternalReferences replaces the dialog box version of the Image command.

ImageAdjust controls the brightness, contrast, and fading of images.

ImageAttach attaches images to the current drawing.

ImageClip creates clipping boundaries around images.

ImageFrame toggles the display of image frames.

ImageQuality toggles between draft and high-quality mode.

RELATED SYSTEM VARIABLE

ImageHlt toggles whether the entire image is highlighted.

TIPS

- The -Image command loads files with the following extensions: *.bil, .bmp, .cal, .cg4, .dib, .flc, .fli, .gif, .gp4, .igs, .jpg, .mil, .pct, .pcx, .png, .rlc, .rle, .rst, .tga,* and *.tif.* Note that *.igs* is not IGES, but a raster format used for mapping. Autodesk's RasterDesign software permits viewing of additional raster formats, such as *.sid* and *.ecw.* This command handles raster images of these color depths:
  Bitonal — black and white (monochrome).
  8-bit gray — 256 shades of gray.
  8-bit color — 256 colors.
  24-bit color — 16.7 million colors.

- AutoCAD can display one or more images in any viewport. There is no theoretical limit to the number and size of images.

- Images do not display in shaded visual modes.

- The DrawOrder command can place images "behind" other objects.

- Images are always xrefs; the source image file must be present when the drawing is opened.

- The Image command was replaced by the ExternalReferences command AutoCAD 2007.

ImageAdjust

Rel.14 Controls brightness, contrast, and fade of attached raster images.

Commands	Alias	Ctrl+	F-key	Alt+	Menu Bar	Dashboard
imageadjust	iad	...	...	MOIA	Modify	...
					⌐Object	
					⌐Image	
					⌐Adjust	

-imageadjust

Command: imageadjust
Select image(s): *(Select one or more image objects.)*
Select image(s): *(Press Enter to end object selection.)*
Displays dialog box:

IMAGE ADJUST DIALOG BOX

Brightness options
  Dark reduces the brightness of the image as values approach 0.
  Light increases the brightness of the image as values approach 100.

Contrast options
  Low reduces the image contrast as values approach 0.
  High increases the image contrast as values approach 100.

Fade options
  Min reduces the image fade as values approach 0.
  Max increases the image fade as values approach 100.

  Reset resets the image to its original parameters; default values are:
    Brightness — 50
    Contrast — 50
    Fade — 0

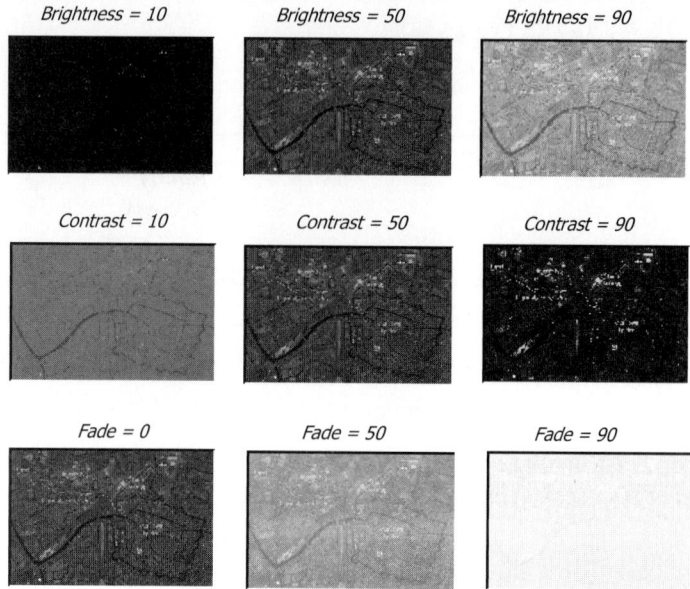

Brightness = 10   Brightness = 50   Brightness = 90

Contrast = 10   Contrast = 50   Contrast = 90

Fade = 0   Fade = 50   Fade = 90

-IMAGEADJUST Command

Command: -imageadjust

Select image(s): *(Select one or more images.)*

Select image(s): *(Press Enter to end object selection.)*

Enter image option [Contrast/Fade/Brightness] <Brightness>: *(Enter an option.)*

COMMAND LINE OPTIONS

Contrast option

Enter contrast value (0-100) <50>: *(Enter a value.)*

 Enter contract value adjusts the contrast between 0% contrast and 100%; default = 50.

Fade Ooption

Enter fade value (0-100) <0>: *(Enter a value.)*

 Enter fade option adjusts the fading between 0% faded and 100%; default = 0.

Brightness option

Enter brightness value (0-100) <50>: *(Enter a value.)*

 Enter brightness value adjusts the brightness between 0% brightness and 100%; default = 50.

RELATED COMMANDS

 ExternalReferences controls the loading of raster image files in drawings.

 ImageAttach attaches images to the current drawing.

 ImageClip creates clipping boundaries on images.

 ImageFrame toggles display of image frames.

 ImageQuality toggles display between draft and high-quality mode.

ImageAttach

__Rel.14__ Selects raster files to attach to drawings.

Command	Alias	Ctrl+	F-key	Alt+	Menu Bar	Dashboard
imageattach	iat	...	...	II	Insert	...
					↳Raster Image	

Command: imageattach

Displays file dialog box:

*Select an image file, and then click **Open**.*

Displays dialog box:

DIALOG BOX OPTIONS

Name selects names from a list of previously-attached images.

Browse selects files; displays Select Image File dialog box.

Path Type options

Full Path saves the path to the image file.

Relative Path saves the path relative to the current drawing file.

No Path does not save the path.

Insertion Point options

☑ **Specify On-Screen** specifies the insertion point of the image in the drawing, after the dialog box is dismissed.

X, Y, Z specifies the x, y, z coordinates of the lower-left corner of the image.

Scale options

☑ **Specify on-screen** specifies the scale of the image (relative to the lower-left corner) in the drawing after the dialog box is dismissed.

Scale specifies the scale of the image; a positive value enlarges the image, while a negative value reduces it.

Rotation options

☐ **Specify on-screen** specifies the rotation angle of the image about the lower-left corner in the drawing, after the dialog box is dismissed.

Angle specifies the angle to rotate the image; positive angles rotate the image counterclockwise.

Details expands the dialog box to display information about the image:

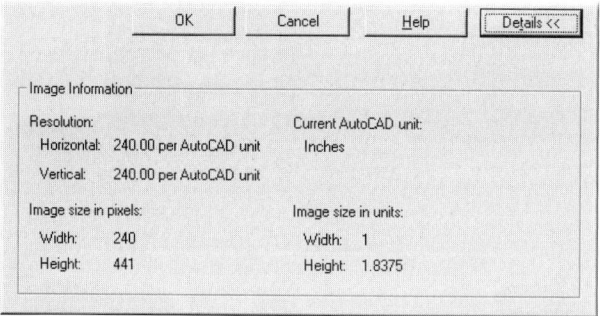

RELATED COMMAND

ExternalReferences controls the loading of raster image files in drawings.

RELATED SYSTEM VARIABLE

InsUnits specifies the drawing units for the inserted image.

TIPS

■ For a command-line version of the ImageAttach command, use the -Image command's Attach option.

■ This dialog box no longer selects units from the Current AutoCAD Unit list box; as of AutoCAD 2000, use the InsUnits system variable.

■ You can specify different x and y scales if you cheat using these instructions provided by the technical editor:
a. Insert the image.
b. Turn the image into a block.
c. Insert the block using different x and y scale factors.

ImageClip

<u>Rel.14</u>   Clips raster images.

Command	Alias	Ctrl+	F-key	Alt+	Menu Bar	Dashboard
imageclip	icl	...	...	MC⦚I	Modify ⦚Clip ⦚Image	...

Command: imageclip
Select image to clip: *(Select one image object.)*
Enter image clipping option [ON/OFF/Delete/New boundary] <New>: *(Enter an option.)*

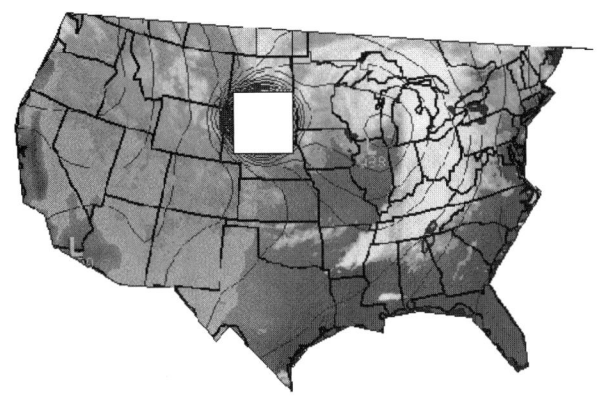

COMMAND LINE OPTIONS

Select image selects one image to clip.

ON turns on a previous clipping boundary.

OFF turns off the clipping boundary.

Delete erases the clipping boundary.

New Boundary options
Enter clipping type [Polygonal/Rectangular] <Rectangular>: *(Enter an option.)*

Polygonal creates a polygonal clipping path.

Rectangular creates a rectangular clipping boundary.

Polygonal options
Specify first point: *(Pick a point.)*
Specify next point or [Undo]: *(Pick a point, or type U.)*
Specify next point or [Undo]: *(Pick a point, or type U.)*
Specify next point or [Close/Undo]: *(Pick a point, or enter an option.)*
Specify next point or [Close/Undo]: *(Enter C to close the polygon.)*

Specify first point specifies the start of the first segment of the polygonal clipping path.

Specify next point specifies the next vertex.

Undo undoes the last vertex.

Close closes the polygon clipping path.

· ·

Rectangular options
Specify first corner point: *(Pick a point.)*
Specify opposite corner point: *(Pick a point.)*

Specify first corner point specifies one corner of the rectangular clip.

Specify opposite corner point specifies the second corner.

When you select an image with a clipped boundary, AutoCAD prompts:

Delete old boundary? [No/Yes] <Yes>: *(Type N or Y.)*

Delete old boundary?

Yes removes the previously-applied clipping path.

No exits the command.

RELATED COMMANDS

ExternalReferences controls the loading of raster image files in the drawing.

ImageAdjust controls the brightness, contrast, and fading of the image.

ImageAttach attaches an image in the current drawing.

ImageFrame toggles the display of the image's frame.

ImageQuality toggles the display between draft and high-quality mode.

XrefClip clips *.dwg* drawing files attached as externally-referenced files.

TIPS

- You can use object snap modes on the image's frame, but not on the image itself.

- To clip a hole in the image, create the hole, and then double back on the same path:

ImageFrame

Rel.14 Toggles the frame around raster images for display and plotting.

Command	Alias	Ctrl+	F-key	Alt+	Menu Bar	Dashboard
imageframe	...	...	...	MO⁃IF	Modify	...
					⁃Object	
					⁃Image	
					⁃Frame	

Command: imageframe
Enter image frame setting [0, 1, 2] <1>: *(Enter a number.)*

Left: Image frame turned on.
Right: Turned off.

COMMAND LINE OPTIONS

0 hides image frames.

1 displays and plots image frames.

2 displays, but does not plot, image frames.

RELATED COMMANDS

ExternalReferences controls the loading of raster image files in the drawing.

ImageAdjust controls the brightness, contrast, and fading of the image.

ImageAttach attaches an image in the current drawing.

ImageClip creates a clipping boundary of the image.

ImageQuality toggles the display between draft and high-quality mode.

TIPS

- *Warning!* When ImageFrame is turned off, you cannot select the image.

- Frames are turned on (or off) in all viewports.

- This command applies to all image frames; they cannot be selectively toggled.

ImageQuality

Rel.14 Toggles the quality of raster images.

Command	Alias	Ctrl+	F-key	Alt+	Menu Bar	Dashboard
imagequality	...	...	...	MO⟱IQ	Modify	...
					⟱Object	
					⟱Image	
					⟱Quality	

Command: imagequality
Enter image quality setting [High/Draft] <High>: *(Type H or D.)*

COMMAND LINE OPTIONS

High displays images at a higher quality.

Draft displays images at a lower quality.

RELATED COMMANDS

ExternalReferences controls the loading of raster image files in the drawing.

ImageAdjust controls the brightness, contrast, and fading of the image.

ImageAttach attaches an image in the current drawing.

ImageClip creates a clipping boundary on the image.

ImageFrame toggles the display of the image's frame.

TIPS

- High quality displays the image more slowly; draft quality displays the image more quickly.

- I find that *draft* quality looks better (crisper) than high quality (blurred), but technical editor Bill Fane finds the quality varies with the type of image — lines versus shades, and insertion scale versus image file resolution.

- This command affects the display only; AutoCAD always plots images in high quality.

Import

Rel.13 Imports vector files into drawings.

Command	Alias	Ctrl+	F-key	Alt+	Menu Bar	Dashboard
import	imp	...	...	...	...	2D Draw

Command: import

Displays dialog box:

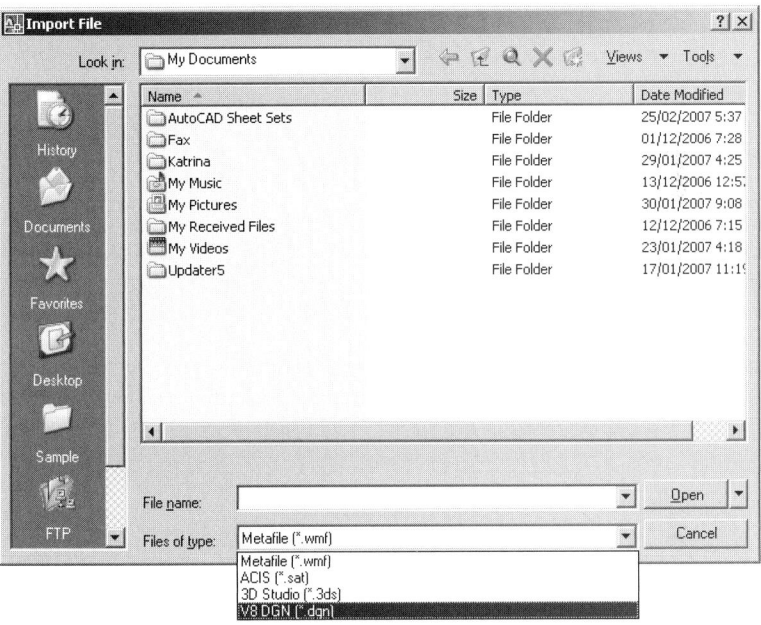

IMPORT FILE DIALOG BOX

Look in selects the folder (subdirectory) and drive from which to import the file.

File name specifies the name of the file, or accepts the default.

File of type selects the file format in which to import the file:

- Metafile (*.wmf).
- ACIS (*.sat).
- 3D Studio (*.3ds).
- V8 DGN (*.dgn).

Open imports the file.

Cancel dismisses the dialog box, and returns to AutoCAD.

RELATED COMMANDS

AppLoad loads AutoLISP, VBA, and ObjectARX routines.

DgnImport imports MicroStation V8 DGN design files (*new to AutoCAD 2008*).

DxbIn imports a DXB file.

ExternalReferences displays *.dwg*, image, and *.dwf* files in the current drawing.

Export exports the drawing in several vector and raster formats.

Load imports SHX shape objects.

Insert places another drawing in the current drawing as a block.

InsertObj places an OLE object in the drawing via the Clipboard.

LsLib imports landscape objects.

MatLib imports rendering material definitions.

MenuLoad loads customization and menu files into AutoCAD.

Open opens AutoCAD (any version) *.dwg* and *.dxf* files.

PasteClip pastes objects from the Clipboard.

PasteSpec pastes or links objects from the Clipboard.

Replay displays renderings in TIFF, Targa, or GIF formats.

VSlide displays *.sld* slide files.

XBind imports named objects from another *.dwg* file.

TIPS

- The Import command acts as a "shell" command; it launches other AutoCAD commands that perform the actual import function. Other options may be available with the actual command, such as insertion point and scale.

Format	Meaning	Related Command
Metafile	Windows metafile WMF	**WmfIn**
ACIS	ASCII SAT	**AcisIn**
3D Studio	3D Studio	**3dsIn**
V8 DGN	MicroStation V8	**DgnImport**

- To import *.dxf* files, use the Open command.

- AutoCAD no longer imports PostScript and EPS files.

Removed Commands

INetCfg was removed from AutoCAD 2000; it was replaced by Windows' Internet configuration.

INetHelp was removed from AutoCAD 2000; it was replaced by AutoCAD's standard online help.

Imprint

<u>**2007**</u>   Imprints objects on the faces of 3D solids; updated from the SolidEdit command's Body/Imprint option.

Command	Alias	Ctrl+	F-key	Alt+	Menu Bar	Dashboard
imprint	...	...	...	MNI	Modify	3D Make
					↳Solid Editing	
					↳Imprint Edges	

Command: imprint
Select a 3D solid: *(Select one 3D solid.)*
Select an object to imprint: *(Select another 2D, 3D, surface, or solid object.)*
Delete the source object [Yes/No] <N>: *(Type Y or N.)*
Select an object to imprint: *(Press Enter to exit the command.)*

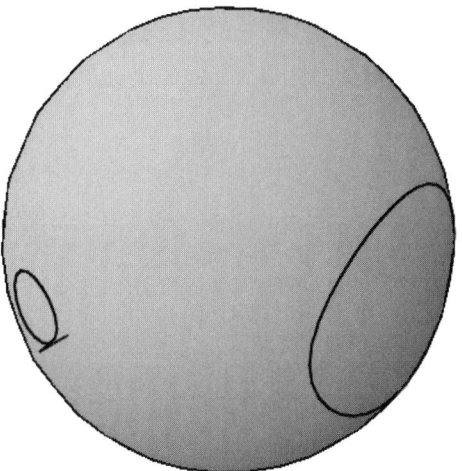

The two circles are the result of imprinting a cone on the sphere.

COMMAND OPTIONS

Select a 3D solid selects the 3D solid to be imprinted.

Select an object selects the object that will do the imprinting; can be an arc, circle, line, 2D polyline, ellipse, spline, region, and 3D polyline, or 3D solid. Also called the "source" object.

Delete the source object toggles deletion of the source object.

RELATED COMMAND

SolidEdit performs the same function as its Body/Imprint option.

Ink...

2006 Group of commands that enable ink and gestures on TabletPC-supported versions of AutoCAD (undocumented commands).

Commands	Aliases	Ctrl+	F-key	Alt+	Menu Bar	Dashboard
Ink...	...	...	...	...	...	...

Warning! These commands do not operate unless AutoCAD is running on a TabletPC or on a regular computer loaded with the TabletPC SDK, and the asdkTPCtest.arx application is loaded.

- -

INKBORDERDISPLAY Command

Toggles the display of a border around the ink objects.

Command: inkborderdisplay

Enter Border Display (0 = None, 1 = Bounding Box, 2 = Spine): *(Enter a number.)*

0 suppresses the display of the bounding box.

1 surrounds the ink objects with a bounding box.

2 indicates ink objects with a vertical line.

- -

INKCOLOR Command

Specifies the color of the ink.

Command: inkcolor

Enter Red component: *(Enter a number between 0 and 255.)*

Enter Green component: *(Enter a number between 0 and 255.)*

Enter Blue component: *(Enter a number between 0 and 255.)*

Red specifies the amount of red for the ink color; 0 = no red, 255 = full red.

Green specifies the amount of green for the ink color.

Blue specifies the amount of blue for the ink color.

- -

INKGESTURE Command

Enables ink gesture mode in AutoCAD.

Command: inkgesture

Enable/Disable gestures <Enable>: *(Type E or D.)*

Enable enables gestures. *Warning!* This option causes AutoCAD to crash on computers without TabletPC support.

Disable disables gesture mode.

- -

INKHILITE Command

Toggles highlight mode.

Command: inkhilite

- -

INKOLECREATE Command

Creates OLE (object linking and embedding) objects of ink.

Command: inkolecreate

Enter CreateOle (0 = No 1 = yes): *(Type 0 or 1.)*

0 turns off Create OLE mode.

1 turns on Create OLE mode.

INKPENWIDTH Command

Specifies the width of ink.

Command: inkpenwidth

Enter pen width (HIMETRIC): *(Enter a value.)*

Enter pen width specifies the width of the ink lines.

INKRECO Command

Converts ink to text by attempting to recognize the handwriting.

Command: inkreco

INKRLINE Command

Toggles redline mode.

Command: inkrline

INKTRANSPARENCY Command

Toggles redline mode.

Command: inktransparency

Enter transparency ([0, 255], 0 = opaque): *(Enter a value.)*

Enter transparency specifies the transparency of the ink lines.

RELATED COMMANDS

Sketch draws freehand lines and polylines.

Tablet provides support for tablet input devices.

TIPS

- Before using these command, you must load the *asdkTPCtest.arx* application into AutoCAD with the AppLoad command.

- These commands are for demonstration purposes, to show what is possible using ink and gestures with AutoCAD. These commands work only in 2D; they do not work in USCs and views other than the WCS.

- Some of these commands will crash AutoCAD if it is not running on a TabletPC or on a regular computer with the TabletPC SDK installed.

- More information on ink is available at msdn.microsoft.com/library/en-us/dntab101/html/Tab101C03.asp. Vista includes support for TabletPCs.

- Autodesk recommends that drawings with ink objects not be saved.

Insert

V. 1.0 Inserts previously-defined blocks into drawings.

Commands	Aliases	Ctrl+	F-key	Alt+	Menu Bar	Dashboard
insert	i	...	...	IB	Insert	2D Draw
	inserturl				↳Block	
-insert	-i					

Command: insert

Displays dialog box:

INSERT DIALOG BOX

Name selects the name from a list of previously-inserted blocks.

Browse displays the Select Drawing File dialog box; select a block in either file format:

- **Drawing (*.dwg)** AutoCAD drawing file.
- **DXF (*.dxf)** drawing interchange file.

Insertion Point options

☑ **Specify On-Screen** specifies the insertion point in the drawing, after closing dialog box.

X, Y, Z specifies the x, y, z coordinates of the lower-left corner of the block.

Scale options

☑ **Specify on-screen** specifies the scale of the block (relative to the lower-left corner) in the drawing, after the dialog box is dismissed.

X, Y, Z specifies the x,y,z scale of the block; positive values enlarge the block, while negative values reduce it.

☐ **Uniform Scale** forces the y and z scale factors to be the same as the x scale factor.

Rotation options

☐ **Specify on-screen** specifies the rotation angle of the block (about the lower-left corner) in the drawing after the dialog box is dismissed.

Angle specifies the angle to rotate the block; positive angles rotate the block counterclockwise.

Unit Block options

Unit specifies the units for the block based on the value stored in the InsUnit variable.

Factor reports the scale factor based on the InsUnit variable.

☐ **Explode** explodes the block upon insertion.

. . . .

-INSERT Command

Enter block name or [?]: *(Enter name, or **?**.)*

Specify insertion point or [Basepoint/Scale/X/Y/Z/Rotate]: *(Pick a point, or enter an option.)*

Enter X scale factor, specify opposite corner, or [Corner/XYZ] <1>: *(Enter a scale factor, or enter an option.)*

Enter Y scale factor <use X scale factor>: *(Enter a scale factor, or press Enter.)*

Specify rotation angle <0>: *(Enter a rotation angle.)*

COMMAND LINE OPTIONS

Block name specifies the name of the block to be inserted.

? lists the names of blocks stored in the drawing.

Specify insertion point specifies the lower-left corner of the block's insertion point.

Basepoint places the block temporarily to allow you to specify a new base point. AutoCAD prompts:

Specify base point: *(Pick a point.)*

Scale specifies the scale for the x, y, and z factors equally. AutoCAD prompts:

Specify scale factor for XYZ axes: *(Enter a scale factor.)*

X indicates the x scale factor.

Y indicates the y scale factor.

Z indicates the z scale factor.

Rotate specifies the rotation angle.

The following options are hidden:

PScale supplies a predefined x, y, and z scale factor.

PX supplies a predefined x scale factor.

PY supplies a predefined y scale factor.

PZ supplies a predefined z scale factor.

PRotate supplies a predefined rotation angle.

Corner indicates the x and y scale factors by pointing on the screen.

XYZ displays the x, y, and z scale submenu.

INPUT OPTIONS

In response to the 'Block Name' prompt, you can enter:

~ — Display a dialog box of drawings stored on disk: **Block name:** ~

* — Insert block exploded: **Block name:** *filename

= — Redefine existing block with a new block: **Block name:** oldname=newname

In response to the 'Insertion point' prompt, you can enter:

Scale — Specify x, y, and z scale factors.

PScale — Preset x, y, and z scale factors.

XScale — Specify x scale factor.

. . . .

PxScale — Preset x scale factor.
YScale — Specify y scale factor.
PyScale — Preset y scale factor.
ZScale — Specify z scale factor.
PzScale — Preset z scale factor.
Rotate — Specify rotation angle.
PRotate — Preset rotation angle.

RELATED COMMANDS

Block creates a block out of a group of objects.

Explode reduces inserted blocks to their constituent objects.

ExternalReferences displays drawings stored on disk in the drawing.

MInsert inserts a block as a blocked rectangular array.

Rename renames blocks.

WBlock writes blocks to disk.

RELATED SYSTEM VARIABLES

ExplMode toggles whether non-uniformly scaled blocks can be exploded.

InsBase specifies the name of the most-recently inserted block.

InsUnits specifies the drawing units for the inserted block.

TIPS

- You can insert any other AutoCAD drawing into the current drawing.

- A *preset* scale factor or rotation means the dragged image is shown at that scale, but you can enter a new scale when inserting.

- Drawings are normally inserted as a block; prefix the filename with an * (*asterisk*) to insert the drawing as separate objects.

- Redefine all blocks of the same name in the current drawing by adding the = (*equal*) suffix after its name at the 'Block name' prompt.

- Insert a mirrored block by supplying a negative x or y scale factor, such as:

  X scale factor: -1

- AutoCAD converts a negative z scale factor into its absolute value, which always makes it positive.

- As of AutoCAD Release 13, you can explode a mirrored block and a block inserted with different scale factors, when the system variable ExplMode is turned on.

- As of AutoCAD 2004, the Insert command no longer imports drawings in XML format (designXML).

- As of AutoCAD 2008, the preview image shows the annotation 🔺 icon for blocks with annotative scaling. Blocks with annotative scaling cannot have non-uniform scaling.

- When inserting blocks with attached xrefs, the xrefs are retained.

InsertObj

Rel.13 Places OLE objects as linked or embedded objects in drawings (short for INSERT OBJect).

Command	Alias	Ctrl+	F-key	Alt+	Menu Bar	Dashboard
insertobj	io	...	...	IO	Insert	2D Draw
					↳OLE Object	

Command: insertobj

Displays dialog box:

INSERT OBJECT DIALOG BOX

⊙**Create New** creates new objects in other applications, and then embeds them in the drawing.

○**Create from File** selects a file to embed in or link to the current drawing.

Object Type selects an object type from the list; the related application automatically launches if you select the **Create New** option.

☑ **Display As Icon** displays the object as an icon, rather than as itself.

Change Icon selects another icon.

RELATED COMMANDS

OleLinks controls the OLE links.

PasteSpec places an object from the Clipboard in the drawing as a linked object.

RELATED SYSTEM VARIABLES

MsOleScale determines the scale of OLE objects placed in model space.

OleHide toggles the display of OLE objects.

OleQuality determines the plot quality of OLE objects.

OleStartup specifies whether the source apps of embedded OLE objects load for plotting.

RELATED WINDOWS COMMANDS

Edit | Copy copies an object to the Clipboard for use in other Windows applications.

File | Update updates an OLE object from another application.

· ·

Renamed Command

InsertUrl was removed from AutoCAD 2000; it was replaced by the Insert command's Browse | Search the Web option.

· ·

Interfere

Rel.11 Determines the interference of two or more 3D solid objects; optionally creates a 3D solid body of the common volumes.

Command	Alias	Ctrl+	F-key	Alt+	Menu Bar	Dashboard
interfere	inf	...	...	M3I	Modify	3D Make
					↳3D Operations	
					↳Interference Checking	

-interfere

Command: interfere
Select first set of objects or [Nested selection/Settings]: *(Select one or more solid objects, or enter an option.)*
Select second set of objects or [Nested selection/checK first set] <checK>: *(Press Enter to check, or enter an option.)*

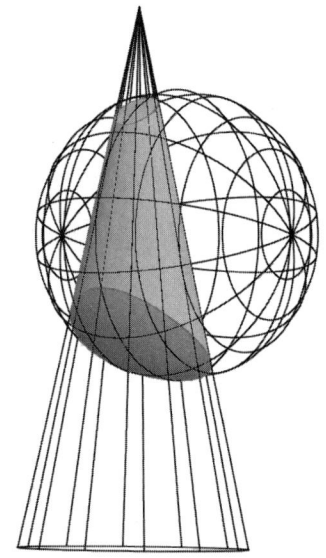

An interfering solid (shaded at center).

COMMAND LINE OPTIONS

Select first set of objects checks solids in a single selection set for interference with one another.

Nested selection selects solids nested in blocks and xrefs.

Settings displays the Interference Settings dialog box.

checK displays the Interference Checking dialog box.

DIALOG BOX OPTIONS

Interference Settings dialog box

Visual Style specifies the visual style of interference objects.

Color specifies the color of interference objects.

⊙ **Highlight Interfering Pair** highlights the pair of solids that interfere with each other.

○ **Highlight Interference** highlights the interference object(s) created from the interfering pair.

Visual Style specifies the visual style while checking for interference.

Interference Checking dialog box

Previous highlights the previous interference object.

Next highlights the next interference object.

☑ **Zoom to Pair** zooms to interference object(s) when clicking Previous and Next.

☑ **Delete Interference Objects Created on Close** deletes the interference object(s) after closing the dialog box.

Zoom temporarily dismisses the dialog box, and then starts the Zoom command.

Pan temporarily dismisses the dialog box, and then starts the Pan command.

3D Orbit temporarily dismisses the dialog box, and then starts the 3dOrbit command.

-Interfere Command

Command: -interfere

Select first set of objects or [Nested selection]: *(Select one or more solid objects, or enter an option.)*

Select second set of objects or [Nested selection/checK first set] <checK>: *(Press Enter to check, or enter an option.)*

Comparing 2 objects with each other.

Interfering objects (first set): 2

               (second set): 0

Interfering pairs: 1

Create interference objects? [Yes/No] <No>: *(Type **Y** or **N**.)*

RELATED COMMANDS

Intersect creates a new volume from the intersection of two volumes.

Section creates a 2D region from a 3D solid.

Slice slices a 3D solid with a plane.

RELATED SYSTEM VARIABLES

InterfereColor specifies the color of interference objects.

InterfereObjVs specifies the visual style of interference objects.

InterferenceVpSs specifies the visual style for the current viewport during interference checking.

⊚ Intersect

Rel.11 Creates 3D solids of 2D regions through the Boolean intersection of two or more solids or regions.

Command	Alias	Ctrl+	F-key	Alt+	Menu Bar	Dashboard
intersect	in	...	...	MNI	Modify ⤷Solids Editing ⤷Intersect	3D Make

Command: intersect
Select objects: *(Select one or more solid objects.)*
Select objects: *(Press Enter to end object selection.)*

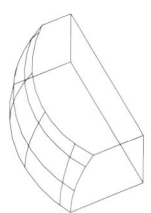

Left: Two intersecting solids.
Right: The result of applying the Intersect command.

COMMAND LINE OPTION

Select objects selects two or more objects to intersect.

RELATED COMMANDS

Interfere creates a new volume from the interference of two or more volumes.

Subtract subtracts one 3D solid from another.

Union joins 3D solids into a single body.

RELATED SYSTEM VARIABLES

ShowHist toggles display of history in solids.

SolidHist toggles the retention of history in solids.

TIPS

- You can use this command on 2D regions and 3D solids.

- The Interference and Intersect commands may seem similar. Here is the difference between the two:

  Intersect *erases* all of the 3D solid parts that do not intersect.

  Interfere *creates a new object* from the intersection; it does not erase the original objects.

'Isoplane

V. 2.0 Changes the crosshair orientation and grid pattern among the three isometric drawing planes.

Command	Alias	Ctrl+	F-key	Alt+	Menu Bar	Dashboard
isoplane	...	E	F5	...	...	...

Command: isoplane
Enter isometric plane setting [Left/Top/Right] <Top>: *(Enter an option, or press Enter.)*

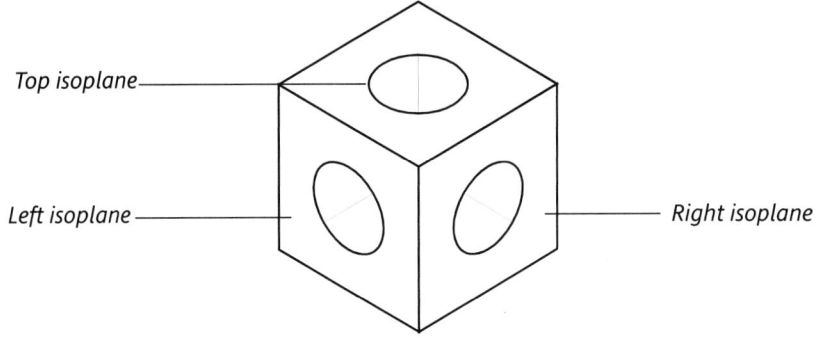

Top isoplane

Left isoplane

Right isoplane

COMMAND LINE OPTIONS

Left switches to the left isometric plane.

Top switches to the top isometric plane.

Right switches to the right isometric plane.

Enter switches to the next isometric plane in the following order: left, top, right.

RELATED COMMANDS

Options displays a dialog box for setting isometric mode and planes.

Ellipse draws isocircles.

Snap turns on isometric drawing mode.

RELATED SYSTEM VARIABLES

SnapIsoPair contains the current isometric plane.

GridMode toggles grid visibility.

GridUnit specifies the current grid x,y spacing.

LimMin specifies the x,y coordinates of the lower-left corner of the grid display.

LimMax holds the x,y coordinates of the upper-right corner of the grid display.

SnapStyl specifies a normal or isometric grid:
   0 — Normal (default).
   1 — Isometric grid.

JogSection

Adds 90-degree jogs to section planes.

Command	Alias	Ctrl+	F-key	Alt+	Menu Bar	Dashboard
jogsection	...	...	...	...	...	...

Command: jogsection

This command requires that at last one section be placed in the drawing with the SectionPlane command.
Select section object: *(Select one section objects.)*
Specify a point on the section line to add jog: *(Pick a point along the section line.)*
Specify a point on the section line to add jog: *(Press Enter to exit the command.)*

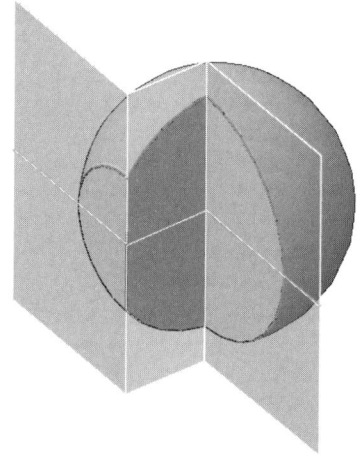

COMMAND LINE OPTIONS

Select section object selects the section to be jogged.

Specify a point on the section line to add jog picks the jog location; the jog angle is always 90 degrees.

RELATED COMMANDS

LiveSection activates section planes.

SectionPlane places section planes in drawings.

TIPS

- Use osnap modes to help position the jog more accurately.

- To create the jog section illustrated above, use the commands in this order:
    1. **SectionPlane** command places the section plane.
    2. **LiveSection** command edits the section plane.
    3. **JogSection** command adds jogs to the section plane.

- After the jog is placed, you can use grips to modify it.

- Jogs can be added while placing the section plane during the SectionPlane command; use this command to place jogs after the section plane is in place.

Join

2006   Joins open objects to make one object or closed objects.

Command	Alias	Ctrl+	F-key	Alt+	Menu Bar	Dashboard
join	j	...	...	MJ	Modify Join	2D Draw

Command: join
Select source object: *(Select an object.)*
Select *objects* **to join to source or [cLose]:** *(Select other objects, or enter L to close open object.)*

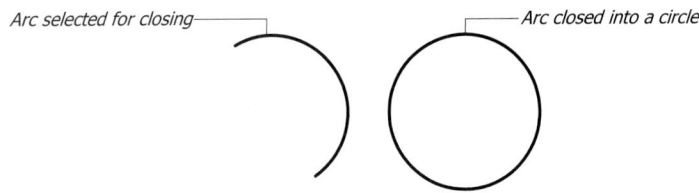

Arc selected for closing — Arc closed into a circle

COMMAND LINE OPTIONS

Select source object selects object to be joined or closed.

Select objects to join joins similar objects. The following objects can be joined, even when there is a gap between them:
- Colinear lines — joined into a single line.
- Arcs in the same imaginary circle — joined into a single arc.
- Elliptical arcs in the same imaginary ellipse — joined into a single elliptical arc.

These objects can only be joined when no gap exists between them:
- Polylines, lines, and arcs — joined into a single polyline.
- Splines — joined into a single spline.

cLose closes arcs and elliptical arcs:
- Arcs become circles.
- Elliptical arcs become ellipses.

RELATED COMMANDS

Break turns circles and ellipses into arcs.

PEdit joins lines, arcs, and polylines with its **Join** option.

TIP

- If arcs have been dimensioned with the DimArc command, the closed circles are disassociated from the dimension.

JpgOut

<u>2004</u> Exports drawings in JPEG format.

Command	Alias	Ctrl+	F-key	Alt+	Menu Bar	Dashboard
jpgout	...	...	...	...	...	...

Command: jpgout

*Displays Create Raster File dialog box. Specify a filename, and then click **Save**.*

Select objects or <all objects and viewports>: *(Select objects, or press Enter to select all objects and viewports.)*

COMMAND LINE OPTIONS

Select objects selects specific objects.

All objects and viewports selects all objects and all viewports, whether in model space or in layout mode.

RELATED COMMANDS

BmpOut exports drawings in BMP (bitmap) format.

Image places raster images in the drawing.

Plot outputs drawings in JPEG and other raster formats.

PngOut exports drawings in PNG (portable network graphics) format.

TifOut exports drawings in TIFF (tagged image file format) format.

TIPS

- JPEG files are often used by digital cameras and Web pages.

- For more control over the resulting *.jpg* file, use the Plot command's PublishToWebJpg driver. For example, it can specify a high resolution such as 2550x3300, whereas the JpgOut command is limited to the resolution of your computer's monitor (1280x1024, typically).

- When using the JpgOut command to save a rendering, the rendering effects of the VisualStyles command are preserved, but not those of the Render command.

- The drawback to saving drawings in JPEG format is that the image is less clear (due to artifacts) than in other formats; the advantage is that JPEG files are highly compressed.

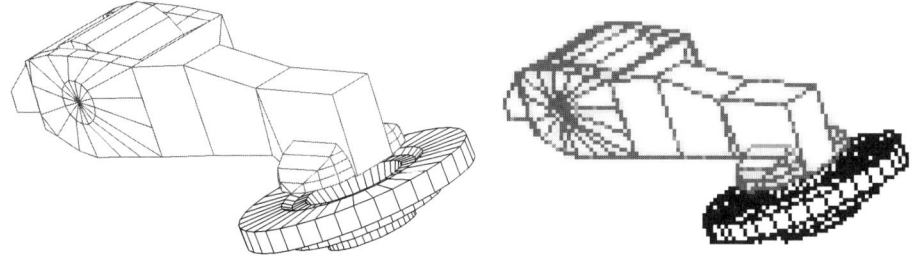

Left: *Original AutoCAD image zoomed-in.*
Right: *Enlarged JPEG image with artifacts.*

- This command provides no options for specifying the level of compression.

- JPEG is short for "joint photographic expert group."

JustifyText

2001 Changes the justification of text.

Command	Alias	Ctrl+	F-key	Alt+	Menu Bar	Dashboard
justifytext	...	...	...	MOTJ	Modify	...
					↳Object	
					↳Text	
					↳Justify	

Command: justifytext
Select objects: *(Select one or more text objects.)*
Select objects: *(Press Enter to end object selection.)*
Enter a justification option
[Left/Align/Fit/Center/Middle/Right/TL/TC/TR/ML/MC/MR/BL/BC/BR]
<Left>: *(Enter an option, or press Enter.)*

COMMAND LINE OPTIONS

Select objects selects one or more text objects in the drawing.

Align aligns the text between two points with adjusted text height.

Fit fits the text between two points with fixed text height.

Center centers the text along the baseline.

Middle centers the text horizontally and vertically.

Right right-justifies the text.

TL justifies to top-left.

TC justifies to top-center.

TR justifies to top-right.

ML justifies to middle-left.

MC justifies to middle-center.

MR justifies to middle-right.

BL justifies to bottom-left.

BC justifies to bottom-center.

BR justifies to bottom-right.

RELATED COMMANDS

Text places text in the drawing.

DdEdit edits text.

ScaleText changes the size of text.

Style defines text styles.

Properties changes the justification, but moves text.

TIPS

- This command works with text, mtext, leader text, and attribute text.

- When the justification is changed, the text does not move to reflect the changed insertion point.

Lay...

2007 Manipulates layers; a group of commands formerly in Express Tools.

Command	Alias	Ctrl+	F-key	Alt+	Menu Bar	Dashboard
...	...	...	...	))	Format	...
					⮩ **Layer Tools**	

This group of commands performs very focused operations on layers and objects. Many commands are paired, one undoing the action of another.

LayOff turns off the layer of the selected object.
LayOn turns on all layers, except frozen layers.

LayIso turns off all layers except the ones holding selected objects; see LayIso command.
LayVpi isolates the selected object's layer in the current viewport
LayWalk displays objects on selected layers, turning off all other layers.
LayUnIso turns on layers that were turned off with the LayIso command

LayFrz freezes the layers of the selected objects.
LayThw thaws all layers.

LayLck locks the layer of the selected object.
LayULk unlocks the layer of a selected object.

LayCur changes the layer of selected objects to the current layer.
LayMch changes the layers of selected objects to that of another selected object.
LayMCur makes the selected object's layer current (similar to Ai_Molc command).

CopyToLayer copies objects to another layer; see CopyToLayer command.
LayMrg moves objects to another layer.

LayDel erases all objects from the specified layer, and then purges the layer from the drawing.

. . . .

LayCur

LayCur changes the layer of selected objects to the current layer (short for "LAYer CURrent").
Command: laycur
Select objects to be changed to the current layer: *(Select one or more objects.)*
Select objects to be changed to the current layer: *(Press Enter.)*
One object changed to layer *layername* **(the current layer).**

. . . .

LayDel and -LayDel

LayDel and -LayDel erase all objects from the specified layer, and then purge the layer from drawing (short for "LAYer DELete"). The current layer cannot be deleted. If you make a mistake, use the U command to restore the purged layer name and its erased objects.

Command: laydel

Select object on layer to delete or [Name]: *(Select an object, or type N.)*

Selected layers: *layername*.

Select object on layer to delete or [Name/Undo]:

******** WARNING ********

You are about to delete layer "*layername***" from this drawing.**

Do you wish to continue? [Yes/No] <No>: *(Type Y or N.)*

Deleting layer "*layername***".**

1 layer deleted.

COMMAND OPTIONS

Name displays the Delete Layers dialog box; select one or more layers to delete and purge.

Undo undoes the last layer delete.

LayFrz

LayFrz freezes the layers of the selected objects (short for "LAYer FReeZe").

Command: layfrz

Current settings: Viewports=Vpfreeze, Block nesting level=Block

Select an object on the layer to be frozen or [Settings/Undo]: *Select an object, or enter an option.)*

Layer "*layername***" has been frozen.**

Select an object on the layer to be frozen or [Settings/Undo]: *(Press Enter, or enter an option.)*

COMMAND OPTIONS

Settings offers these options:

Enter setting type for [Viewports/Block selection] selects options for viewports or blocks.

In paper space viewport use determines how to freeze layers in viewports:

- **Freeze** freezes the layers in all viewports.
- **Vpfreeze** freezes just the current viewport.

Enter Block Selection nesting lever: determines how to freeze layers in blocks and xrefs:

- **Block** freezes the entire block's layer, but freezes the object's layer in xrefs.
- **Entity** freezes the object's layer
- **None** freezes neither the block nor xref.

Undo undoes the last freeze action.

. . . .

LayLck

LayLck locks the layer of the selected object (short for "LAYer LoCK").
Command: laylck
Select an object on the layer to be locked: *(Select an object.)*
Layer "*layername***" has been locked.**

. . . .

LayMch and -LayMch

LayMch and -LayMch change the layers of selected objects to that of a selected object (short for "LAYer MatCH").
Command: laymch
Select objects to be changed: *(Pick one or more objects.)*
Select objects: *(Press Enter.)*
Select object on destination layer or [Name]: *(Pick an object on another layer.)*
One object changed to layer "*layername***"**

COMMAND OPTION

Name displays the Change to Layer dialog box; see LayDel command.

In the -LayMch command, the Name option displays the following prompt:

Enter layer name: *(Type the valid name of a layer.)*

. . . .

LayMCur

LayMCur makes the selected object's layer current, just like the Ai_Molc command (short for "LAYer Make CURrent").
Command: laycur
Select object whose layer will become current: *(Pick an object.)*
layername **is now the current layer.**

. . . .

LayMrg and -LayMrg

LayMrg and -LayMrg move objects to another layer (short for "LAYer MeRGe").
Command: laymrg
Select object on layer to merge or [Name]: *(Pick an object.)*
Selected layers: *layername.*
Select object on layer to merge or [Name/Undo]: *(Pick another object.)*
Selected layers: *layername, anothername.*
Select object on layer to merge or [Name/Undo]: *(Press Enter.)*
Select object on target layer or [Name]: *(Pick an object.)*

. . . .

COMMAND OPTION

Name displays the Merge to Layer dialog box; see LayDel command.

In the -LayMrg command, the Name option displays the following prompt:

Enter layer name: *(Type the name of a layer.)*

. . . .

LayOff

LayOff turns off the layer of the selected object.

Command: layoff

Current settings: Viewports=Vpfreeze, Block nesting level=Block

Select an object on the layer to be turned off or [Settings/Undo]: *(Select an object, or enter an option.)*

Layer "0" has been turned off.

See LayFrz for the meaning of the Setting and Undo options.

. . . .

LayOn

LayOn turns on all layers, except frozen ones.

Command: layon

Warning: layer "*layername***" is frozen. Will not display until thawed.**

All layers have been turned on.

. . . .

LayThw

LayThw thaws all layers (short for "LAYer THaW").

Command: laythw

All layers have been thawed.

. . . .

LayULk

LayULk unlocks the layer of a selected object (short for "LAYer UnLocK").

Command: layulk

Select an object on the layer to be unlocked: *(Pick an object on a locked layer.)*

Layer "*layername***" has been unlocked.**

. . . .

LayUnIso

LayUnIso turns on layers that were turned off with the last LayIso command (short for "LAYer UNISOlate").

Command: layuniso

Layers isolated by LAYISO command have been restored.

. . . .

LayVpi

LayVpi isolates the selected object's layer in the current viewport (short for "LAYer ViewPort Isolate"). This command works only in paper space with two or more viewports.

Command: layvpi

Current settings: Viewports=Vpfreeze, Block nesting level=Block

Select objects on the layer to be isolated in viewport or [Settings/Undo]: *(Select one or more objects, or enter an option.)*

Layer *layername* has been frozen in all viewports but the current one.

. . . .

Select an object on the layer to be isolated in viewport or [Settings/Undo]:
(Press Enter to exit the command.)

See LayFrz for the meaning of the Setting and Undo options.

. . . .

LayWalk

LayWalk displays objects on selected layers.

Command: laywalk

Displays dialog box:

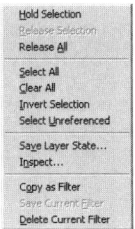

Left: *Doors are displayed because layer Doors is selected.*
Right: *LayerWalk dialog box.*

DIALOG BOX OPTIONS

Select objects selects one or more objects in the drawing, and reports their name(s).

☐ **Filter** filters the layer list according to the rules of the selected filter.

Purge removes the selected layer, if it contains no objects.

☑ **Restore on Exit** restores layer display upon exiting the dialog box.

Close closes the dialog box and exits the command.

SHORTCUT MENU OPTIONS

Right-click the LayerWalk dialog box to access the following shortcut menu:

Hold Selection adds an asterisk (*) to the layer name, and always displays it.

Release Selection removes the asterisk from the selected layers.

Release All removes all asterisks, and then turns off the Always Show option.

Select All selects all layers, and displays them.

Clear All unselects all layers, hiding them.

. . . .

Invert Selection selects unselected layers, and vice versa.

Select Unreferenced selects all layers with no objects; the Purge button removes them from the drawing.

Save Layer State saves the selected layers for use by the Layer States Manager.

Inspect displays the Inspect dialog box, which reports the total number of layers, the number of selected layers, and the total number of objects on the selected layers.

Copy as Filter adds the selected layer's name to the Filter text box; edit the name with the ? and * wildcard characters.

Save Current Filter saves the filter for reuse.

Delete Current Filter erases the filter from the filter list.

RELATED COMMANDS

Layer controls layer creation and properties.

CopyToLayer copies objects from one layer to another.

RELATED SYSTEM VARIABLE

CLayer is the name of the current layer.

TIPS

- In the LayerWalk dialog box, double-click a layer name to keep it on always; an asterisk appears beside its name.

- Also in the LayerWalk dialog box, hold down the **Ctrl** key to select more than one layer at a time.

'Layer

V. 1.0 Controls the creation, status, and visibility of layers.

Commands	Aliases	Ctrl+	F-key	Alt+	Menu Bar	Dashboard
layer	la	...	...	OL	Format	Layers
	ddlmodes				⬫ Layer	
-layer	-la					
+layer						

Command: layer

Displays dialog box:

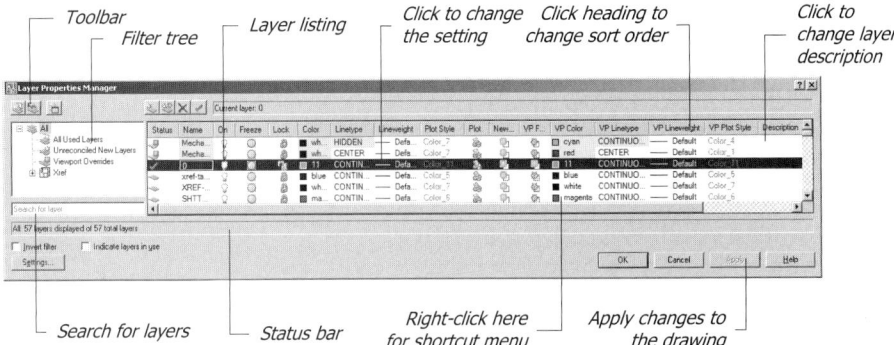

LAYER PROPERTIES MANAGER DIALOG BOX

Click a header name to sort alphabetically (A-Z); click a second time for reverse-alphabetical sort (Z-A).

Status reports the layer status: current layer, empty layer, layer in use, or layer filter.

Name lists the names of layers in the current drawing.

On toggles layers between on and off.

Freeze toggles layers between thawed and frozen in all viewports.

Lock toggles layers between unlocked and locked.

Color specifies the color of objects on layers.

Linetype specifies the linetype for objects on layers.

Lineweight specifies the lineweight for objects on layers.

Plot Style specifies the plot style for objects on layers; available only when plot styles are enabled in the drawing.

Plot toggles layers between plot and no-plot.

Description provides space to describe the layer.

The following options appear only in layout mode (paper space):

Current VP Freeze specifies whether the layer is frozen in the current viewport.

NewVP Freeze specifies whether the layer is frozen in new viewports.

VP Color overrides the color property for the current viewport (*new in AutoCAD 2008*).

VP Linetype overrides the linetype property for the current viewport (*new in AutoCAD 2008*).

VP Lineweight overrides the lineweight property for the current viewport (*new in AutoCAD 2008*).

VP Plot Style overrides the plot style for the current viewport (*new in AutoCAD 2008*).

☐ **Invert Filter** inverts the display of layer names; for example, when Show all used layers is selected, the Invert filter option causes all layers with no content to be displayed (ALT+I).

☑ **Indicate Layers in Use** displays icons next to layers that contain objects.

Apply applies the changes to the layers without exiting the dialog box.

OK exits the dialog box, and applies changes to the layers.

Cancel exits the dialog box, and leaves layers unchanged.

Dialog Box toolbar

New Group Filter (Alt+G) — *New Layer (Alt+N)* — *New Layer with VP Frozen in all viewports.*

New Property Filter (Alt+P) — *Make Layer Current (Alt+C)*

Layer States Manager (Alt+S) — *Delete Layer (Alt+D)*

New Property Filter displays New Filter Properties dialog box.

New Group Filter adds an item to the group filter list, initially named "Group Filter 1."

Layer States Manager displays Layer States Manager dialog box; see LayerStates command.

New Layer adds layers to the drawing, initially named "Layer1."

New Layer VP Frozen in All Viewports freezes the new layer in all viewports (*new to AutoCAD 2008*).

Delete Layer purges the selected layer; some layers cannot be deleted, as described by the warning dialog box:

Make Current Layer sets the selected layer as current.

Shortcut Menus

Most layer options are found in right-click menus.

Column Headers shortcut menu

Maximize Column changes the selected column width to display all column content.

Maximize All Columns changes the width of all columns to display content of all columns. To display complete headers, you must resize the column by dragging the header separators. You can resize the dialog box.

Filter Tree shortcut menu

Depending on where you right-click in the filter tree area, some menu options may be grayed-out, indicating they are unavailable.

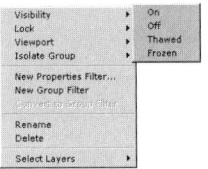

Visibility toggles the visibility of all layers in the selected filter or group:

- **On** displays, plots, and regenerates objects; includes objects during hidden-line removal.
- **Off** does not display or plot objects; does not include objects during hidden-line removal, and the drawing is not regenerated when the layer is turned on.
- **Thawed** reverses the Frozen option.
- **Frozen** does not display or plot objects; includes objects during hidden-line removal, and the drawing is regenerated when the layer is thawed.

Lock determines whether objects can be edited:

- **Lock** prevents objects from being edited; all other operations that don't involve editing are permitted, such as object snaps.
- **Unlock** allows objects to be edited.

Viewport specifies whether layers are frozen in layout mode (unavailable in model space):

- **Freeze** applies Current VP Freeze to all layers (in the filter or group) of the current viewport.
- **Thaw** turns off Current VP Freeze for layers in the filter or group.

Isolate Group turns off all layers not part of the filter or group; only layers that are part of the filter or group are visible in the drawing. In model space, this option applies to all layers; in layout mode, this option applies selectively, depending on the suboption selected:

- **All Viewports** freezes all layers not in the filter or group.
- **Active Viewport Only** freezes all layers (not in the filter or group) in the current viewport only.

New Properties Filter displays the Layer Filter Properties dialog box.

New Group Filter creates a new layer group filter; to add layers to groups:

1. Hold down the Ctrl key, and then select layer names in the layer list (right-hand pane).
2. Drag the layers onto the group filter name (left-hand pane).

Convert to Group Filter converts the selected filter to a group; the name does not change, but the icon changes to indicate a group filter.

Rename changes the name of the filter or group. As an alternative, click the name twice, slowly, and then enter a new name.

Delete erases the filter or group; does not erase the layers in the filter or group. The All, All Used Layers, and Xref filters cannot be erased.

Properties displays the Layer Filter Properties dialog box (available only when a filter is selected).

Select Layers allows you to add and remove layers by selecting objects in the drawing (available only when a group is selected):

- **Add** allows you to select objects in the drawing; press Enter to return to the layer dialog box.
- **Replace** removes existing layers from the group, and adds the newly-selected layers.

Layer List shortcut menu

The content of this shortcut menu varies, depending on where you right-click:

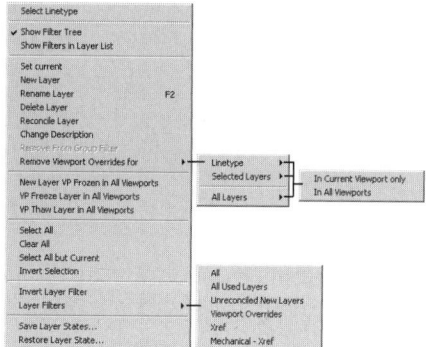

Select Linetype/Color/Lineweight/Plotstyle opens the related dialog box.

Show Filter Tree toggles filter tree view, left-hand pane of this dialog box.

Show Filters in Layer List toggles the display of filters in the layer list view (right-hand pane); when off (no check mark), only layers are shown.

Set Current sets the selected layer as current. Alternatively, press Alt+C or click the Set Current button on the dialog box's toolbar.

New Layer creates a new layer, names it "Layer1," and gives it the properties of the currently-selected layer. As alternatives, press Alt+N or click the New Layer button.

Rename Layer renames the selected layer; alternatively, press F2 (*new to AutoCAD 2008*).

Delete Layer erases the selected layers from the drawing; layers 0, Defpoints, the current layer, xref-dependent layers, and those containing objects cannot be erased. As alternatives, press Alt+D or click the Delete Layer button on the dialog box's toolbar.

Reconcile Layer adds the layer to the list of acceptable layers (*new to AutoCAD 2008*).

Change Description adds and changes layer description text. As an alternative, click the description twice, and then edit the text.

Remove from Group Filter removes the selected layers from the select group.

Remove Viewport Overrides for Linetype/Color/Lineweight/Plotstyle/Selected Layer/All Layers returns overriden properties to ByLayer (*new to AutoCAD 2008*):

- **In Current Viewport Only** changes overriden layer properties to ByLayer in the current viewport only.
- **In All Viewports** changes overriden layer properties to ByLayer in all viewports.

New Layer VP Frozen in All Viewports creates a new layer, with its status frozen in all viewports (*new to AutoCAD 2008*).

VP Freeze Layer in All Viewports freezes selected layer(s) in all viewports (*new to AutoCAD 2008*).

VP Thaw Layer in All Viewports thaws selected layer(s) in all viewports (*new to AutoCAD 2008*).

Select All selects all layers; alternatively, press Ctrl+A.

Clear All unselects all layers; alternatively, click a single layer.

Select All but Current selects all layers, except the current layer.

Invert Selection unselects selected layers, and selects all other layers.

Invert Layer Filter displays all layers not in the selected filter; alternatively, press ALT+I.

Layer Filters displays a submenu listing the names of filters; alternatively, turn on the filter tree view (left-hand pane). Select a filter name to apply it to the layer list.

Save Layer States displays the New Layer State to Save dialog box; see LayerSates command.

Restore Layer State displays the Layer States Manager dialog box; alternatively, press Alt+S.

LAYER FILTERS PROPERTIES DIALOG BOX

Filter name names the filter.

Show Example displays only help.

Filter definition defines the parameters of the filter.

LAYER SETTINGS DIALOG BOX

New Layer Notification Settings options
(New to AutoCAD 2008.)

☑ **Evaluate New Layers Added to Drawing** checks for new layers being added to drawings (aka "unreconciled layers"), when:

⊙ **Evaluate New Xref Layers Only** checks for new layers added through attached xrefs.

○ **Evaluate All New Layers** checks for new layers added to drawings, including from xrefs.

☑ **Notify when New Layers Are Present** toggles new layer notification, when:

☑ **Open** when opening drawings.

☑ **Attach/Reload Xrefs** when attaching and reloading xrefs.

☑ **Restore Layer State** when restoring layer states.

☑ **Save** when saving drawings.

☐ **Insert** when inserting blocks.

☑ **Display Alert for Plot When New Layers are Present** displays icon and balloon in the tray.

Dialog Settings options

☑ **Apply Layer Filter to Layer Toolbar** toggles display of layers affected by the current layer filter.

Viewport Override Background Color chooses the background color for viewport overrides.

. . . .

+LAYER Command

Command: +layer

Specify filter [All/Properties Filter1/All Used Layers/Unreconciled New Layers/Viewport Overrides] <All>: *(Enter an option.)*

Specify Filter reports the names of layers associated with filters; an undocumented command.

. . . .

-LAYER Command

Command: -layer

Current layer: "0"

Enter an option [?/Make/Set/New/ON/OFF/Color/Ltype/LWeight/Plot/ PStyle/Freeze/Thaw/LOck/Unlock/stAte]: *(Enter an option.)*

COMMAND LINE OPTIONS

Groups and filters cannot be created by this command.

Color indicates the color for all objects drawn on the layer.

Freeze disables the display of the layer.

LOck locks the layer.

Ltype indicates the linetype for all objects drawn on the layer.

LWeight specifies the lineweight.

Make creates a new layer, and makes it current.

. . . .

New creates a new layer.

OFF turns off the layer.

ON turns on the layer.

Plot determines whether the layer is plotted.

PStyle specifies the plot style (available only when plot styles are attached to the drawing).

Set makes the layer current.

stAte sets and saves layer states.

Thaw unfreezes the layer.

Unlock unlocks the layer.

? lists the names of layers in the drawing.

stAte options
Enter an option [?/Save/Restore/Edit/Name/Delete/Import/EXport]: *(Enter an option).*

? lists the names of layer states in the drawing.

Save saves the layer state and properties by name; properties include on, frozen, lock, plot, newvpfreeze, color, linetype, lineweight, and plot style.

Restore restores a named state.

Edit changes the settings of named states.

Name renames named states.

Delete erases named states from the drawing.

Import opens layer states *.las* files.

Export saves a selected named state to a *.las* file.

RELATED COMMANDS

Lay... manipulates layers; a group of commands.

LayerP returns to the previous layer.

LayerState creates and restores layer states.

Change moves objects to different layers via the command line.

Properties moves objects to different layers via a dialog box.

LayTrans translates layer names.

Purge removes unused layers from drawings.

Rename renames layers.

SetByLayers forces VP override properties to ByLayer *(new to AutoCAD 2008).*

View controls visibility of layers with named views.

VpLayer controls the visibility of layers in paper space viewports.

RELATED SYSTEM VARIABLES

CLayer contains the name of the current layer.

LayerFilterAlert removes extraneous filters.

LayerEval controls when AutoCAD checks the Unreconciled New Layer filter list for new layers *(new to AutoCAD 2008).*

LayerNotify specifies when alerts are displayed for unreconciled layers*(new to AutoCAD 2008).*

SetByLayerMode specifies which properties are changed by the SetByLayer command(*new to AutoCAD 2008*).

ShowLayerUsage toggles the layer usage icons.

VpLayerOverrides reports whether any layers in the current viewport have VP (viewport) property overrides (*new to AutoCAD 2008*).

VpLayerOverridesMode toggles whether viewport property overrides are displayed or plotted (*new to AutoCAD 2008*).

RELATED FILE

*.las are layer state files, which use the DXF format.

TIPS

- A *frozen* layer cannot be seen or edited; a *locked* layer can be seen but not edited.
- Layer "Defpoints" never plots, no matter its setting.
- To create more than one new layer at a time in -Layer, use commas to separate layer names.
- For new layers to take on properties of an existing layer, select the layer before clicking New.
- If layer names appear to be missing, they may have been filtered from the list.
- When layers are added automatically, such as through xrefs, AutoCAD now reports these as "unreconciled layers." A warning icon and balloon appear in the tray:

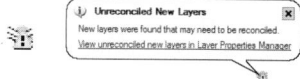

- Click the icon to see the layer dialog box with all unreconciled layers listed.
- Right-click the icon to control when you are warned about unreconciled layers:

- Turn off Plot on several random layers, and then drag the header until the Plot column is no longer shown. According to Lynn Allen, this is guaranteed to drive coworkers nuts!
- The layer manager in AutoCAD is incompatible with the layer manager found earlier in Express Tools.

LayerP, LayerPMode

Undoes changes made to layer settings (short for LAYER Previous).

<u>2002</u> Toggles layer-previous mode (short for LAYER Previous MODE).

Command	Alias	Ctrl+	F-key	Alt+	Menu Bar	Dashboard
layerp	...	...	...	...	...	Layers
layerpmode						

Command: layerp
Restored previous layer status

. . . .

LAYERPMODE Command
Command: layerpmode
Enter LAYERP mode [ON/OFF] <ON>: *(Type ON or OFF.)*

COMMAND LINE OPTIONS
On turns on layer previous mode (default).

Off turns off layer previous mode.

RELATED COMMANDS
Layer creates and sets layers and modes.

LayerState creates and edits layer states.

Lay... manipulates layers; a group of commands.

View stores layer settings with named views.

RELATED SYSTEM VARIABLE
CLayer specifies the name of the current layer.

TIPS
- This command acts like an "undo" command for changes made to layers only, for example, the layer's color or lineweight.

- LayerPMode command must be turned on for the LayerP command to work.

- The LayerP command does not undo the renaming, deleting, purging, or creating of layers.

- When layer previous mode is on, AutoCAD tracks changes to layers.

LayerState

Saves, restores, and manages named layer states.

Command	Alias	Ctrl+	F-key	Alt+	Menu Bar	Dashboard
layerstate	las	...	...	OA	Format	Layers
	lman				↳ Layer States Manager	

Command: layerstate

Displays dialog box:

LAYER STATES MANAGER DIALOG BOX

To access this dialog box, right-click in the Layer List, and select Restore Layer States from the shortcut menu (alternatively, click the Layer States Manager button, or press Alt+S).

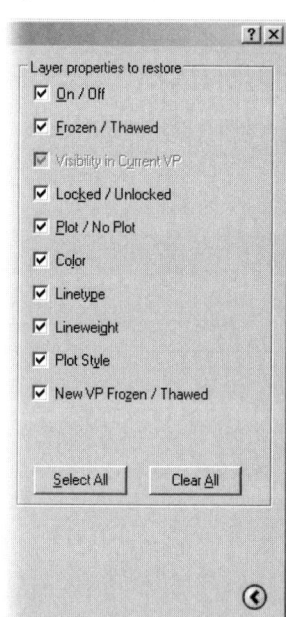

Layer states lists the names of layer states that can be restored:

To rename a layer state, click its name twice, and then edit the name.

To edit the description, click the description twice, and then edit the text.

To sort layer states, click the headers – Name, Space, and Description.

New displays the New Layer State to Save dialog box (detailed later).

Save saves the layer state settings.

Edit edits the selected layer state (detailed later).

Delete deletes the selected layer state.

Rename renames the selected layer state.

Import imports layer states from DWG drawings, DWS CAD standards, DWT templates, and LAS layer state files.

Export exports the selected layer state to LAS files.

☐ **Don't List Layer States in Xrefs** hides layer states stored in externally-referenced drawings.

Restore Options
☑ **Turn off layers not found in layer state** turns off layers with unsaved settings.

☑ **Apply Properties as Viewport Overrides** applies property overrides to the current viewport (available only when this command is started in a viewport.)

Restore closes the dialog box, and restores the selected layer state.

⊙ MORE RESTORE OPTIONS

Layer Properties to Restore toggles the properties to restore with the layers.

Select All turns on all properties.

Close All turns off all properties.

NEW LAYER STATE TO SAVE DIALOG BOX

New Layer State Name specifies the name of the layer state.

Description provides an optional description of the layer state.

OK displays the Layer States Manager dialog box (described above).

EDIT LAYER STATE DIALOG BOX

Click any icon to change the property. For example, click an On icon to turn off the layer.
Click headings to change the sort order. For example, click Color to sort by colors.

☐ **Add Layer to Layer State** adds layers to the layer state; displays the Select Layers to Add to Layer State dialog box (detailed later). This option does not work when all layers are included in the state.

✗ **Remove Layer From Layer State** removes selected layers from the layer state.

EDIT LAYER STATE DIALOG BOX

Select one or more layer names, and then click OK. Hold down the Ctrl key to select more than one layer.
OK returns to the Edit Layer State dialog box.

RELATED COMMANDS

Layer creates and sets layers and modes.

LayerP returns to the previous layer.

View stores layer settings with named views.

TIPS

■ This is not a new feature in AutoCAD 2008; instead, this command makes layer states more accessible by giving them their own command. You can still access this dialog box from the layers dialog box.

■ Use LAS layer state files to share layer states among drawings.

■ To control layer states at the command line, such as in macros, use the stAte option of the -Layer command.

LayIso

2007 LayIso freezes or fades all layers, except those holding selected objects (short for "LAYer ISOlate").

Command	Alias	Ctrl+	F-key	Alt+	Menu Bar	Dashboard
layiso	...	...	...	OOI	Format	Layers
					⮡ Layer Tools	
					⮡ Layer Isolate	

Command: layiso
Current setting: Lock layers, Fade = 90
Select objects on the layer(s) to be isolated or [Settings]: *(Select an object, or type S).*
Select objects on the layer(s) to be isolated or [Settings]: *(Press Enter.)*
Layer *layername* **has been isolated.**
Isolated layers appear normal; all other layers appear faded, or disappear from view:

COMMAND OPTIONS

Select objects selects the objects whose layers will be isolated from all other layers.

Settings offers these options:

Enter setting for layers not isolated [Off/Lock and fade] <Lock>: *(Enter an option.)*

Off turns off all unselected layers; prompts you:

In paperspace viewport use [Vpfreeze/Off]: *(Enter V to freeze objects in just the selected viewport, or O to turn off layers in all viewport.)*

Lock and fade locks all unselected layers, and makes them look faded; prompts you:

Enter fade value (0-90): *(Enter a value between 0 for no fading and 90 for maximum fading.)*

-Layout

2000 Creates and deletes paper space layouts on the command line.

Commands	Aliases	Ctrl+	F-key	Alt+	Menu Bar	Dashboard
-layout	layout	...	...	IL	Insert	...
	lo				⬚Layout	

Command: -layout
**Enter layout option [Copy/Delete/New/Template/Rename/SAveas/Set/?]
<set>:** *(Enter an option.)*

COMMAND LINE OPTIONS

Copy copies a layout to create a new layout.

Delete deletes a layout; the Model tab cannot be deleted.

New creates a new layout tab, automatically generating the name for the layout (default = first unused tab in the form Layout*n*), which you may override.

Template displays the Select File dialog box, which allows you to select a *.dwg* drawing or *.dwt* template file to use as a template for a new layout. If the file has layouts, it displays the Insert Layout(s) dialog box.

Rename renames a layout.

SAveas saves the layouts in a drawing template (*.dwt*) file. The last current layout is used as the default for the layout to save.

Set makes a layout current.

? lists the layouts in the drawing in a format similar to the following:

```
Active Layouts:
Layout: Layout1      Block name: *PAPER_SPACE.
Layout: Layout2      Block name: *Paper_Space5.
```

Shortcut Menu

Right-click any layout tab:

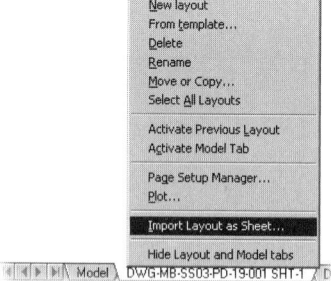

New layout creates a new layout with the default name of "Layout*n*."

From template displays the Select File and Insert Layout dialog boxes.

Delete deletes the selected layout; displays a warning dialog box:

Rename displays the Rename Layout dialog box:

Move or Copy displays the Move or Copy dialog box. You can move tabs by dragging them, and copy tabs by holding down the Ctrl key while dragging them.

Select All Layouts selects all layouts.

Activate Previous Layout returns to the previously-accessed layout; useful when a drawing has many layouts.

Activate Model Tab returns to the Model tab.

Page Setup displays the Page Setup dialog box; see the PageSetup command.

Plot displays the Plot dialog box; see the Plot command.

Import Layout as Sheet turns the layout into a sheet; available only when the Sheet Set Manager is operating (*new to AutoCAD 2008*). Displays the Import Layouts as Sheets dialog box. A layout can be assigned only to one sheet; if needed, make copies of layouts for other sheets.

Hide Layout and Model tabs moves the tabs to the status bar.

INSERT LAYOUT(S) DIALOG BOX

Layout names(s) lists the names of layouts found in the selected template or drawing; you may select more than one layout at a time by holding down the **Ctrl** key.

OK adds the selected layouts to the current drawing.

Cancel dismisses the dialog box and cancels the command.

MOVE OR COPY DIALOG BOX

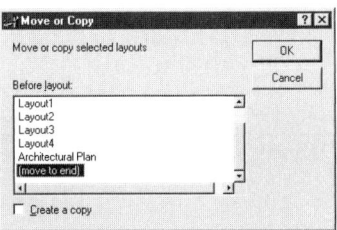

Before layout lists the names of layouts; select one to appear before the current layout.

(move to end) moves the current layout to the end of layouts.

☐ **Create a copy** copies selected layouts.

RELATED COMMAND

LayoutWizard creates and deletes paper space layouts via a wizard.

STARTUP SWITCH

/layout specifies the layout tab to show when AutoCAD starts up.

TIPS

- "Layout" is the name for paper space as of AutoCAD 2000. A layout name can be up to 255 characters long; the first 31 characters are displayed in the tab. Drawings can have up to 255 layouts each. To switch between layouts, click tabs located below the drawing.

- Copy tabs by holding down the Ctrl and dragging them; move them by dragging them. To copy or move more than one tab, hold down the Shift key to select them.

- Layout tabs can be moved to the status bar: right-click a tab, and then select **Hide Layout and Model tabs**:

- The Model tab cannot be deleted, renamed, or copied.

LayoutWizard

2000 Creates and deletes paper space layouts via a wizard.

Command	Alias	Ctrl+	F-key	Alt+	Menu Bar	Dashboard
layoutwizard	...	...	...	ILW	Insert	...
					↳Layout	
					↳Create Layout Wizard	
				TZC	Tool	
					↳Wizards	
					↳Create Layout	

Command: layoutwizard

Displays dialog box:

CREATE LAYOUT DIALOG BOX

Enter a name specifies the name for the layout.

Back displays the previous dialog box.

Next displays the next dialog box.

Cancel cancels the command.

Printer page

Select a configured plotter selects a printer or plotter to output the layout.

Paper Size page

Select a paper size... selects a size of paper supported by the output device.

Enter the paper units

⊙ **Millimeters** measures paper size in metric units.

○ **Inches** measures paper size in Imperial units.

○ **Pixels** measures paper size in dots per inch.

Orientation page

Select the orientation

• **Portrait** plots the drawing vertically.

• **Landscape** plots the drawing horizontally.

Title Block page

Select a title block... specifies a title border for the drawing as a block or an xref:

⊙**Block** inserts the title border drawing.

○**Xref** references the title border drawing.

Define Viewports page

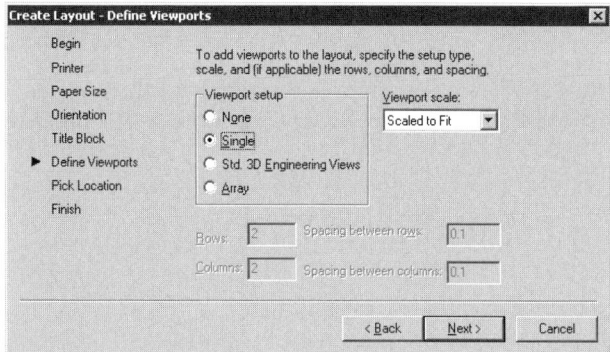

Viewport Setup

○**None** creates no viewport.

⊙**Single** creates a single paper space viewport.

○**Std. 3D Engineering Views** creates top, front, side, and isometric views.

○**Array** creates a rectangular array of viewports.

Viewport scale

• **Scaled to Fit** fits the model to the viewport.

• *mm*:*nn* specifies a scale factor, ranging from 100:1 to 1/128":1'0".

Rows specifies the number of rows for arrayed viewports.

Columns specifies the number of columns for arrayed viewports.

Spacing between rows specifies the vertical distance between viewports.

Spacing between columns specifies the horizontal distance between viewports.

Pick Location page

Select location specifies the corners of a rectangle holding the viewports. AutoCAD prompts:

Specify first corner: *(Pick a point.)*
Specify opposite corner: *(Pick a point.)*

AutoCAD returns to the Create Layout dialog box.

Finish page

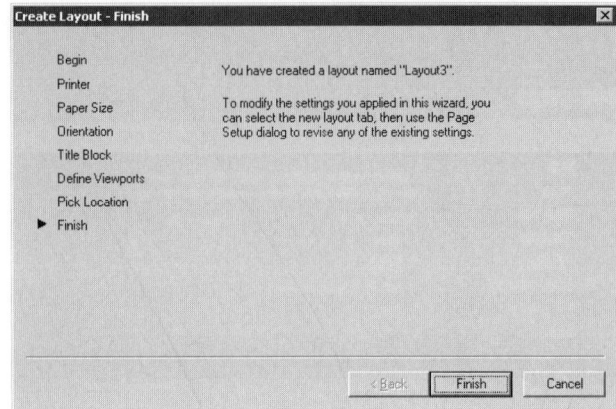

Finish exits the dialog box and creates the layout.

RELATED COMMANDS

Layout creates a layout on the command line.

RELATED SYSTEM VARIABLES

CTab contains the name of the current tab.

LayTrans

2002 Translates layer names (short for LAYer TRANSlation).

Command	Alias	Ctrl+	F-key	Alt+	Menu Bar	Dashboard
laytrans	...	...	...	TSL	Tools	...
					⸦CAD Standards	
					⸦Layer Translator	

Command: laytrans

Displays dialog box:

LAYER TRANSLATOR DIALOG BOX

Translate From

Translate From lists the names of layers in the current drawing; icons indicate whether the layer is being used (referenced):

🗇 Layer contains at least one object.

🗇 Layer contains no objects, and can be purged.

Selection Filter specifies a subset of layer names through wildcards.

Select highlights the layer names that match the selection filter.

Map maps the selected layer(s) in the Translate From column to the selected layer in the Translate To column.

Map same maps layers automatically with the same name.

Translate To

Translate To lists layer names in the drawing opened with the Load button.

Load accesses the layer names in another drawing via the Select Drawing File dialog box.

New creates a new layer via the New Layer dialog box.

Layer Translation Mappings options

Edit edits the linetype, color, lineweight, and plot style settings via the Edit Layer dialog box; identical to the New Layer dialog box.

Remove removes the selected layer from the list.

Save saves the matching table to a *.dws* (drawing standard) file.

Settings specifies translation options via the Settings dialog box.

Translate changes the names of layers, as specified by the Layer Translation Mappings list.

NEW LAYER DIALOG BOX

Identical to the Edit Layer dialog box.

Name specifies the name of the layer, up to 255 characters long.

Linetype selects a linetype from those available in the drawing.

Color selects a color; select Other for the Select Color dialog box.

Lineweight selects a lineweight.

Plot style selects a plot style from those available in the drawing; this option is not available if plot styles have not been enabled in the drawing.

SETTINGS DIALOG BOX

☑ **Force object color to Bylayer** forces every translated layer to take on color Bylayer.

☑ **Force object linetype to Bylayer** forces every translated layer to take on linetype Bylayer.

☑ **Translate objects in blocks** forces objects in blocks to take on new layer assignments.

☑ **Write transaction log** writes the results of the translation to a *.log* file, using the same filename as the drawing. When command is complete, AutoCAD reports:

Writing transaction log to *filename*.log.

☐ **Show layer contents when selected** lists the names of selected layers only in the Translate From list.

RELATED COMMANDS

Standards creates the standards for checking drawings.

CheckStandards checks the current drawing against a list of standards.

Layer creates and sets layers and modes.

RELATED FILES

__.dws__* drawing standard file; saved in DWG format.

__.log__* log file recording layer translation; saved in ASCII format.

TIPS

- You can purge unused layers (those prefixed by a white icon) within the Layer Translator dialog box:

    1. Right click any layer name in the Translate From list.

    2. Select Purge Layers. The layers are removed from the drawing.

- You can load layers from more than one drawing file; duplicate layer names are ignored.

Leader

Rel.13 Draws leader lines with one or more lines of text.

Command	Alias	Ctrl+	F-key	Alt+	Menu Bar	Dashboard
leader	lead	...	...	...	...	...

Command: leader
Specify leader start point: *(Pick a starting point, such as at 1 illustrated below.)*
Specify next point: *(Pick the shoulder point — point 2.)*
Specify next point or [Annotation/Format/Undo] <Annotation>: *(Press Enter to start the annotation, or pick another point — point 3 — or enter an option.)*
Specify next point or [Annotation/Format/Undo] <Annotation>: *(Press Enter to specify the text — 4.)*
Enter first line of annotation text or <options>: *(Enter text — 5 — or press Enter for options.)*
Enter next line of annotation text: *(Press Enter.)*

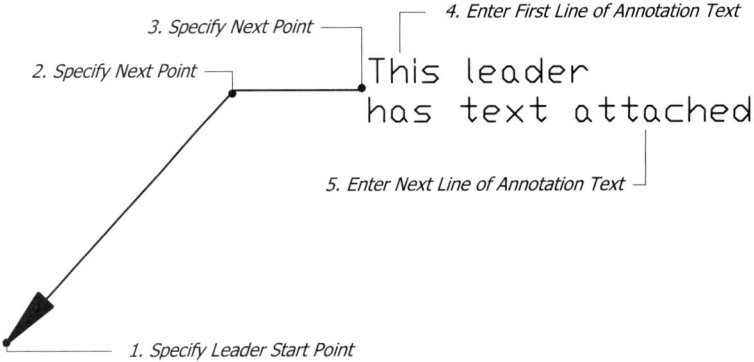

3. Specify Next Point

2. Specify Next Point

4. Enter First Line of Annotation Text

This leader has text attached.

5. Enter Next Line of Annotation Text

1. Specify Leader Start Point

COMMAND LINE OPTIONS

Specify start point specifies the location of the arrowhead.

Specify next point positions the leader line's vertex.

Undo undoes the leader line to the previous vertex.

Format options
Enter leader format option [Spline/STraight/Arrow/None] <Exit>: *(Enter an option.)*

Spline draws the leader line as a NURBS (short for non uniform rational Bezier spline) curve; see Spline command.

STraight draws a straight leader line (default).

Arrow draws the leader with an arrowhead (default).

None draws the leader with no arrowhead.

Annotation options
Enter first line of annotation text or <options>: *(Press* ENTER.*)*
Enter an annotation option [Tolerance/Copy/Block/None/Mtext] <Mtext>:
(Enter an option.)

Enter first line of annotation text specifies the leader text.

Tolerance places one or more tolerance symbols; see the Tolerance command.

Copy copies text from another part of the drawing.

Block places a block; see the -Insert command.

None specifies no annotation.

MText displays the Text Formatting toolbar; see the MText command.

RELATED DIMENSION VARIABLES

DimAsz specifies the size of the arrowhead and the hookline.

DimBlk specifies the type of arrowhead.

DimClrd specifies the color of the leader line and the arrowhead.

DimGap specifies the gap between hookline and annotation (gap between box and text).

DimScale specifies the overall scale of the leader.

RELATED COMMANDS

DimStyle defines styles for leaders.

MLeader draws multi-line leaders.

QLeader provides more options than Leader.

TIPS

- This command draws several types of leader:

- The text in a leader is an mtext (multiline text) object.

- Use the \P metacharacter to create line breaks in leader text.

- Autodesk recommends using the QLeader command, which has a dialog box for settings and more options.

- As of AutoCAD 2008, leaders can take on the annotative property.

- For leaders with multiple arrowheads, use the MLeader command.

Lengthen

Rel.13 Lengthens and shortens open objects by several methods.

Command	Alias	Ctrl+	F-key	Alt+	Menu Bar	Dashboard
lengthen	len	...	...	MG	Modify	...
					⤷Lengthen	

Command: lengthen
Select an object or [DElta/Percent/Total/DYnamic]: *(Select an open object.)*
Current length: n.nnnn

COMMAND LINE OPTIONS

Select an object displays length; does not change the object.

DElta options
Enter delta length or [Angle] <0.0000>: *(Enter a value, or type A.)*
Specify second point: *(Pick a point.)*
Select an object to change or [Undo]: *(Select an open object, or type U.)*

Enter delta length changes the length by the specified amount.

Angle changes the angle by the specified value.

Undo undoes the most recent lengthening operation.

Percent options
Enter percentage length <100.0000>: *(Enter a value.)*
Select an object to change or [Undo]: *(Select an open object, or type U.)*

Enter percent length changes the length to a percentage of the current length.

Undo undoes the most recent lengthening operation.

Total options
Specify total length or [Angle] <1.0000)>: *(Enter a value, or type A.)*
Select an object to change or [Undo]: *(Select an open object, or type U.)*

Specify total length changes the length by an absolute value.

Angle changes the angle by the specified value.

Undo undoes the most recent lengthening operation.

DYnamic options
Select an object to change or [Undo]: *(Select an open object, or type U.)*
Specify new end point: *(Pick a point.)*

Specify new end point dynamically changes the length by dragging.

Undo undoes the most-recent lengthening operation.

TIPS
- Lengthen command only works with open objects, such as lines, arcs, and polylines; it does not work with closed objects, such as circles, polygons, and regions.
- DElta changes lengths using these measurements: (1) distance from endpoint of selected object to pick point; or (2) incremental length measured from endpoint of angle.
- Angle option applies only to arcs.

Light

<u>Rel.12</u>   Places point, spot, distant, target, and web lights in drawings for use by the Render command.

Command	Aliases	F-key	Alt+	Menu Bar	Dashboard
light	pointlight	...	VEL	View	Lights
	distantlight			⃗Render	
	spotlight			⃗Light	
	weblight				
	targetpoint				
	freepoint				
	freespot				
	freeweb				

Command: light
Enter light type [Point/Spot/Web/Targetpoint/Freespot/freeweB/Distant] <Freespot>: *(Enter an option.)*

COMMAND LINE OPTIONS

Point places point lights, which radiate light in all directions; you can also use the PointLight alias.

Spot places spot lights, which direct light at a target point; you can also use the SpotLight alias.

Web places web lights, which use IEF data to produce a precise illumintaion distribution; you can also use the WebLight alias *(new to AutoCAD 2008)*.

Targetpoint places target point lights; you can also use the TargetPoint alias *(new to AutoCAD 2008)*.

Freespot places spot lights without target points; you can also use the FreeSpot alias *(new to AutoCAD 2008)*.

freeweB places web lights without target points; you can also use the FreeWeb alias *(new to AutoCAD 2008)*.

Distant places distant lights, which do not attenuate, like the sun; you can also use the DistantLight alias.

 Point light options

When system variable LightingUnits = 0 (default), displays the following prompts:
Specify source location <0,0,0>: *(Pick a point, or enter x,y,z coordinates.)*
Enter an option to change [Name/Intensity/Status/shadoW/Attenuation/ Color/eXit] <eXit>: *(Press Enter, or enter an option.)*

Glyph placed by Pointline command.
Point light glyphs can be relocated with their Position grip.

Name names the light; if you do not give it a name, Autodesk assigns generic names, such as "Pointlight1."

Intensity specifies the light's brightness, ranging from 0 to the largest real number.

Status toggles the light on and off.

Shadow determines how the light casts shadows:

- **Off** — shadows are turned off.
- **Sharp** — shadows have sharp edges.
- **Soft** — shadows have soft edges; most realistic.

Attenuation specifies how the light diminishes or decays with distance:

- **None** — light does not diminish.
- **Inverse Linear** — light intensity diminishes by the inverse of the linear distance.
- **Inverse Squared** — light intensity diminishes by the inverse of the square of the distance.
- **Attenuation Start Limit** — specifies where light starts (offset from the light's center); *Warning!* Attenuation limit options are not supported by OpenGL display drivers.
- **Attenuation End Limit** — specifies where light ends; no light is cast beyond this point.

Color specifies the light's color.

eXit exits the command and places the light glyph in the drawing.

The prompt changes when system variable LightingUnits = 1 or 2, as follows (new to AutoCAD 2008):

Enter an option to change [Name/Intensity factor/Status/Photometry/ shadoW/Attenuation/filterColor/eXit] <eXit>: *(Enter an option.)*

Prompts that differ are listed here:

Intensity factor is the same as Intensity.

Photometry is a more accurate method of specifying the brightness and color of lights.

filterColor is the same as Color.

Photometric options

Enter a photometric option to change [Intensity/Color/eXit] <I>:

Enter intensity (Cd) or enter an option [Flux/Illuminance] <1500>:

Enter color name or enter an option [?/Kelvin] <D65>:

Intensity factor specifies brightness in terms of candelas (Cd), flux, or illuminance.

Color specifies color by name or by degrees Kelvin (absolute temperature).

Freepoint options

Options for Freepoint lights are identical to those of point lights, except that the target prompt is missing (new to AutoCAD 2008).

Spotlight and Targetpoint options

When system variable LightingUnits = 0, displays the following prompts:

Specify source location <0,0,0>: *(Pick a point.)*
Specify target location <0,0,-10>: *(Pick another point.)*
**Enter an option to change [Name/Intensity/Status/Hotspot/Falloff/shadoW/
Attenuation/Color/eXit] <eXit>:** *(Press Enter, or enter an option.)*

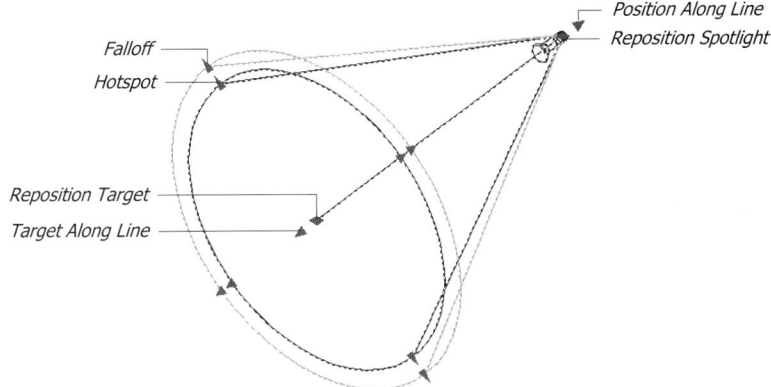

Glyph placed by Spotlight and Targetlight commands. These lights can be extensively edited using grips:

Name names the light; if you don't give it a name, generic names are assigned, such as "Spotlight1."

Intensity specifies the light's brightness, ranging from 0 to the largest real number.

Status toggles the light on and off.

Hotspot defines the angle of the brightest (inner) cone of light (a.k.a. beam angle); ranges from 0 to 160 degrees.

Falloff defines the angle of the full (outer) cone of light (field angle); range is 0 to 160 degrees.

Shadow determines how the light casts shadows:
- **Off** — shadows are turned off.
- **Sharp** — shadows have sharp edges.
- **Soft** — shadows have soft edges; most realistic.

Attenuation specifies how the light diminishes or decays with distance:
- **None** — light does not diminish.
- **Inverse Linear** — light intensity diminishes by the inverse of the linear distance.
- **Inverse Squared** — light intensity diminishes by the inverse of the square of the distance.
- **Attenuation Start Limit** — specifies where light starts (offset from the light's center).
- **Attenuation End Limit** — specifies where light ends; no light is cast beyond this point.

Color specifies the light's color.

eXit exits the command and places the light glyph in the drawing.

The prompts changes when system variable LightingUnits = 1 or 2 (new to AutoCAD 2008). See Pointlight options for details.

Freespot option

Options for Freespot lights are identical to those of spot lights, except that the target prompt is missing (new to AutoCAD 2008).

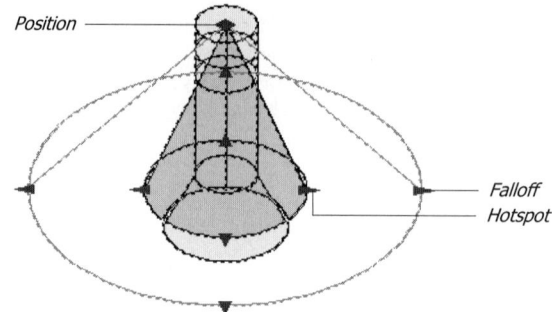

Position ——

—— Falloff
—— Hotspot

Glyph placed by Freespot commands. The light can be edited using grips:

 Weblight options

(New to AutoCAD 2008) This option is not available when system variable LightingUnits = 0. When LightingUnits = 1 or 2:

Specify source location <0,0,0>: *(Pick a point, or enter x,y,z coordinates.)*
Specify target location <0,0,0>: *(Pick a point, or enter x,y,z coordinates.)*
Enter an option to change [Name/Intensity factor/Status/Photometry/weB/ filterColor/eXit] <eXit>: *(Press Enter, or enter an option.)*

Options are the same as point lights, with the following exception:

weB specifies the IES file to use.

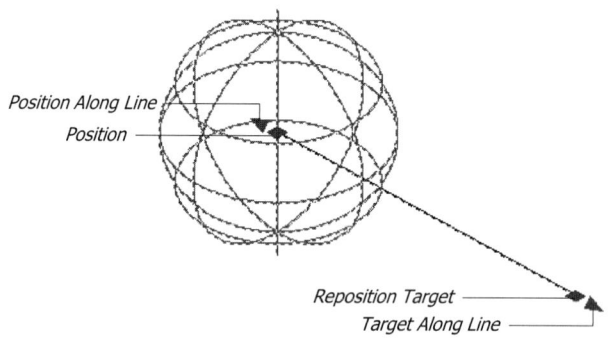

Position Along Line
Position

Reposition Target ——
Target Along Line ——

Above: *Glyph placed by Weblight command without IES data.*
Below: *And with IES data; the shape of the glyph mimics the shape of IES data.*

Freeweb options

Options for Freeweb lights are identical to those of web lights, except that the target prompt is missing (new to AutoCAD 2008).

Distant light options

When system variable LightingUnits = 0, displays the following prompts:

Specify light direction FROM <0,0,0> or [Vector]: *(Pick a point, or type V.)*
Specify light direction TO <1,1,1>: *(Pick another point.)*
**Enter an option to change [Name/Intensity/Status/shadoW/Color/eXit]
<eXit>:** *(Press Enter, or enter an option.)*

The prompt changes when system variable LightingUnits = 1 or 2 (new to AutoCAD 2008). See Pointlight options, above.

From specifies the location of the distant light.

Vector prompts you to enter a vector direction, such as 0,-1,1.

To specifies the direction of the light.

Name names the light; Autodesk assigns generic names, such as "Pointlight1."

Intensity specifies the light's brightness, ranging from 0 to the largest real number.

Status toggles the light on and off.

shadoW determines how the light casts shadows:
- **Off** — shadows are turned off.
- **Sharp** — shadows have sharp edges.
- **Soft** — shadows have soft edges; most realistic.

Color specifies the light's color.

eXit exits the command and places the light in the drawing; distant lights show no glyphs.

RELATED COMMANDS

LightList displays a list of lights defined in the current drawing.

GeographicLocation determines the position of the sun based on longitude and latitude.

SunProperties specifies the properties of sunlight.

Dashboard provides additional controls for lights, sun position, and shadows.

Options changes the colors of light glyphs.

RELATED SYSTEM VARIABLES

LightGlyphDisplay toggles the display of light glyphs.

LightingUnits toggles photometric lighting.

ShadowPlaneLocation locates a ground plane onto which shadows can be cast.

3dConversionMode controls how material and light definitions are converted when a pre-AutoCAD 2008 drawing is opened.

TIPS

- This command works in model space only.
- Lights can be edited through the Properties palette, through grips editing, or through the Light List palette; see the LightList command.
- As of AutoCAD 2007, light glyphs replace blocks to represented lights.
- When a drawing has no lights defined, AutoCAD assumes ambient light, which ensures every object in the scene has illumination; ambient light is an omnipresent light source.
- Use the Dashboard's Light control panel to turn on the sun light.

- Use point lights to simulate light bulbs.

- New drawings have a bright light over your left shoulder and a dimmer one just over your right one.

- Turning on shadows slows down AutoCAD's display speed.

Shadows cast onto the shadow plane by the VsShadows system variable.

- Light can be named using upper- and lowercase letters, numbers, spaces, hyphens, and underscores to a maximum of 256 characters.

- Attenuation start limits and end limits are not supported by graphics boards using the OpenGL driver (*wopengl9.hdi*). Use the 3dConfig command's View Tune Log to identify the driver your computer's graphics board uses.

DEFINITIONS

Candela (cd) — international unit of luminous intensity.

Constant light — attenuation is 0; default intensity is 1.

Extents distance — distance from minimum lower-left coordinate to the maximum upper-right.

Falloff — angle of the full light cone; field angle ranges from 0 to 160 degrees.

Foot-candle (fc) — American unit of illuminance.

HLS color — changes colors by hue (color), lightness, and saturation (amount of gray).

Hotspot — brightest cone of light; beam angle ranges from 0 to 160 degrees.

IES — short for Illuminating Engineering Society.

Inverse linear light — light strength decreases to ½-strength two units of distance away, and ¼-strength four units away; default intensity is ½ extents distance.

Inverse square light — light strength decreases to ¼-strength two units away, and $1/_8$-strength four units away; default intensity is ½ the square of the extents distance.

Kelvin (K) — absolute temperature, where 0K = absolute zero.

Lux (lx) — international unit of illuminance.

RGB color — three primary colors (red, green, blue) shaded from black to white.

Web lights — use IES files that define the real-world light distribution of commercial lamps.

LightList, LightListClose

2007 Edits the properties of lights placed in drawings.

Command	Alias	Ctrl+	F-key	Alt+	Menu Bar	Dashboard
lightlist	...	...	...	VELL	View	Lights
					⇘Render	
					⇘Light	
					⇘Light List	

lightlistclose

This command is for editing the properties of lights after they have been placed in drawings.

Command: lightlist

Displays palette (the sun light does not appear here; see the SunProperties command):

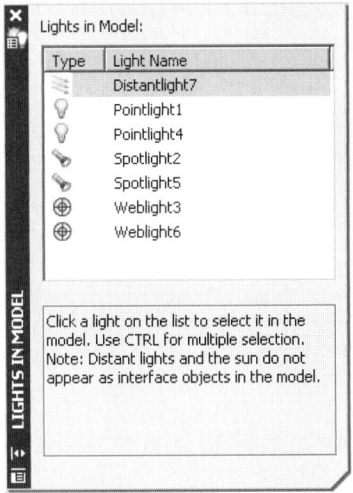

Command: lightlistclose

Closes the palette.

RIGHT-CLICK MENU

Right-click a light name to display the shortcut menu:

Delete Light
Properties

Delete Light erases the light from the drawing.

Properties displays the Properties palette.

DISTANT LIGHT PROPERTIES

General properties

Name names the light; if you did not name it, a generic name is assigned, such as "Distantlight1."

On/Off Status toggles the light on and off.

Shadows toggles the display of shadows.

Intensity Factor specifies the light's brightness, ranging from 0 to the largest real number.

Filer Color specifies the light's filter color.

Photometric properties

These properties are displayed only when LightingUnits = 1 or 2 (new to AutoCAD 2008):

Lamp Intensity specifies brightness in terms of candelas (Cd), flux, or illuminance.

Lamp Color specifies color by name or by degrees Kelvin (absolute temperature).

Geometry properties

From Vector X, Y, Z specifies the x, y, z coordinates of the distant light.

To Vector X, Y, Z specifies the x, y, z coordinates of the light's target.

Source Vector X, Y, Z specifies x, y, z coordinates of light source based on altitude and azimuth.

Rendered Shadow Details properties

Type determines how the light casts shadows:
- **Off** — shadows are turned off.
- **Sharp** — shadows have sharp edges.
- **Soft** — shadows have soft edges; most realistic.

Map size specifies the size of the shadow map; range is 64x64 to 4096x4096.

Softness determines the softness of the shadow's edges.

PROPERTIES OF POINT AND TARGETPOINT LIGHTS

General Properties

See distant light properties.

Plot Glyph determines whether the light glyph is plotted.

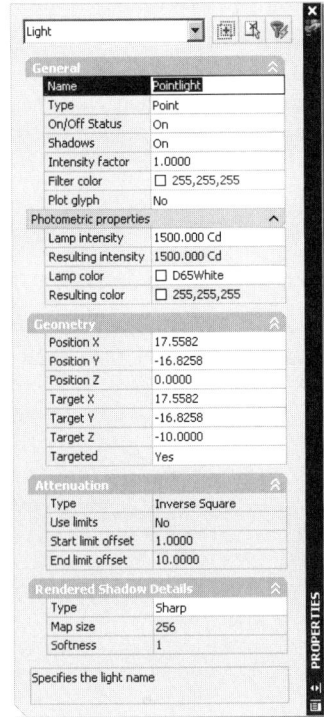

Left: Properties for Point lights.
Right: Properties for Targetpoint lights.

Photometric properties

See distant light properties.

Geometry properties

Position X, Y, Z specifies the x, y, z coordinates of the light.

New to AutoCAD 2008:

Target X, Y, Z specifies the x, y, z coordinates of the light's target (targetpoint light only).

Targeted toggles this light type between point and targetpoint.

Attenuation

Type specifies how the light diminishes or decays with distance:

- **None** — light does not diminish.
- **Inverse Linear** — light intensity diminishes by the inverse of the linear distance.
- **Inverse Squared** — light intensity diminishes by the inverse of the square of the distance.

Use Limits toggles whether attenuation limits are employed:

Start Limit Offset — specifies where the light beam starts (offset from the light's center).

End Limit Offset — specifies where the light beam ends; no light is cast beyond this point.

Rendered Shadow Details properties
See distant light properties.

PROPERTIES OF SPOT AND FREESPOT LIGHTS

General Properties
See distant light properties.

Hotspot Angle defines the angle of the brighter, inner cone of light; from 0 to 160 degrees.
Falloff Angle defines the angle of the full, outer cone of light; ranges from 0 to 160 degrees.
Plot Glyph determines whether the light glyph is plotted.

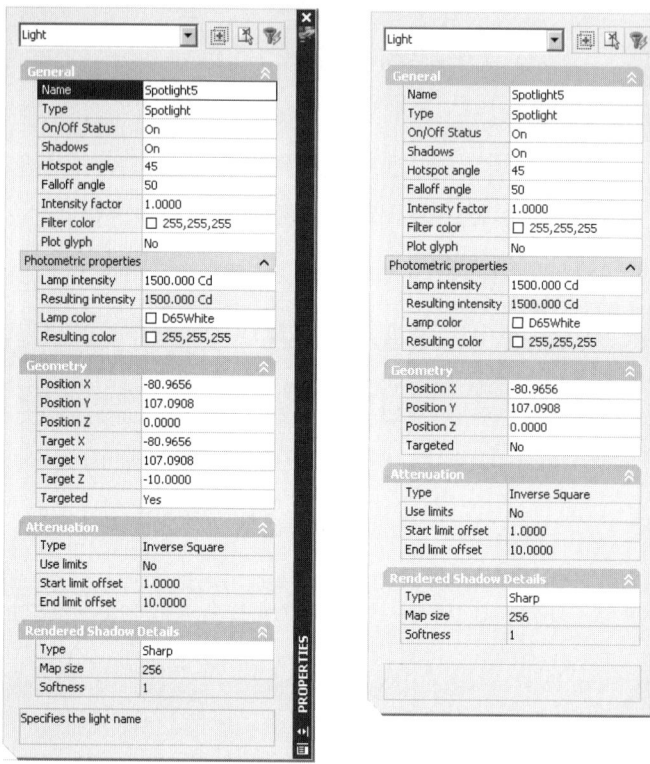

Left: Properties for Spotlights.
Right: Properties for Freespot lights.

Geometry & Photometric properties
See distant light properties.

Attenuation
Type specifies how the light diminishes or decays with distance:
- **None** — light does not diminish.
- **Inverse Linear** — light intensity diminishes by the inverse of the linear distance.
- **Inverse Squared** — light intensity diminishes by the inverse of the square of the distance.

Use Limits toggles whether attenuation limits are employed:

Start Limit Offset — specifies where the light beam starts (offset from the light's center).

End Limit Offset — specifies where the light beam ends; no light is cast beyond this point.

Rendered Shadow Details properties
See distant light properties.

PROPERTIES OF WEB AND FREEWEB LIGHTS

General Properties
See distant light properties.

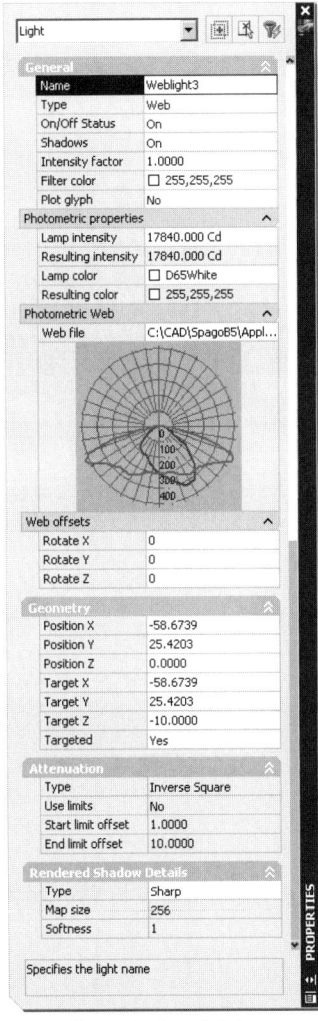

Properties for web lights.

Photometric properties
See point light properties.

Photometric Web properties
Web File selects the *.ies* file for specifying the light's 3D distribution of illumination.

Web Offsets properties
Rotate X, Y, Z rotates the web distribution orientation.

Geometry, Attenuation, and Rendered Shadow Details properties
See distant and spot light properties.

PHOTOMETRIC LAMP DIALOG BOXES

Clicking the ▤ *button displays additional dialog boxes:*

Lamp Color dialog box

⊙ **Standard Colors** lists a catalog of standard colors (spectra):

○ **Kelvin Colors** specifies the color temperature of the light; range is 1000 to 20,000.

Filter Color specifies the color of the light's filter. Resulting Color shows the result of mixing the colors of the light and its filters.

Lamp Intensity dialog box

⊙ **Intensity (Candela)** specifies candelas of luminous intensity.

○ **Flux (Lumen)** specifies the rate of total energy leaving the lamp.

○ **Illuminance (Foot-candles)** specifies energy per area at the surface

  Distance specifies the distance; controlled by the InsUnits system variable.

Intensity Factor specifies a factor for the intensity of the lamp; can be used to darken or brighten lights. Resulting Intensity reports the intensity x the factor.

RELATED COMMANDS

Light places regular and photometric lights in the current drawing.

Dashboard provides additional controls for lights.

RenderExposure changes the overall brightness of rendered scenes (*new to AutoCAD 2008*).

RELATED SYSTEM VARIABLES

LightGlyphDisplay toggles the display of light glyphs.

New in AutoCAD 2008:

LightingUnits toggles between generic and photometric lighting.

LightsInBlocks toggles lights in blocks for rendering.

LinearBrightness determines the global brightness of generic lights.

LinearContrast determines the global contrast of generic lights.

LogExpBrightness determines the global brightness of photometric lighting.

LogExpContrast determines the global contrast of photometric lighting.

LogExpDaylight toggles global daylight of photometric lighting.

LogExpMidtones determines the global midtones of photometric lighting.

RenderUserLights toggles lights for renderings.

3dConversionMode controls how material and light definitions are converted when a pre-AutoCAD 2008 drawing is opened.

'Limits

V. 1.0 Defines the 2D limits in the WCS for the grid markings and the Zoom All command; optionally prevents specifying points outside of limits.

Command	Alias	Ctrl+	F-key	Alt+	Menu Bar	Dashboard
limits	...	...	...	OII⬚	Format	...
					⬚Drawing Limits	

Command: limits
Reset Model *(or* **Paper***)* **space limits:**
Specify lower left corner or [ON/OFF] <0.0,0.0>: *(Pick a point, or type ON or OFF.)*
Specify upper right corner <12.0,9.0>: *(Pick another point.)*

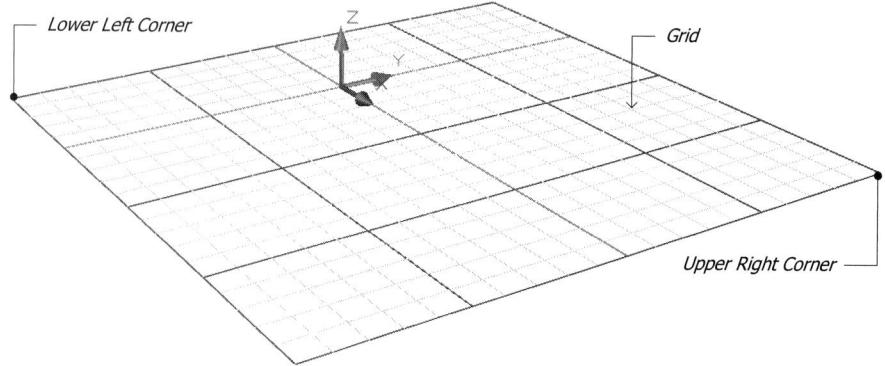

COMMAND LINE OPTIONS

OFF turns off limits checking.
ON turns on limits checking.
Enter retains limits values.

RELATED COMMANDS

Grid displays grid dots, which are bounded by limits.
Zoom displays the drawing's extents or limits (whichever is larger) with the All option.

RELATED SYSTEM VARIABLES

GridDisplay determines if the grid is constrained by the limits.
LimCheck toggles the limit's drawing check.
LimMin specifies the lower-right 2D coordinates of current limits.
LimMax specifies the upper-left 2D coordinates of current limits.

TIP

- As of AutoCAD 2007, the grid can ignore the limits set by this command; see the DSettings and Grid commands.

Line

V. 1.0 Draws straight 2D and 3D lines.

Command	Alias	Ctrl+	F-key	Alt+	Menu Bar	Dashboard
line	l	...	...	DL	Draw ↳Line	2D Draw

Command: line
Specify first point: *(Pick a starting point.)*
Specify next point or [Undo]: *(Pick another point, or type U.)*
Specify next point or [Undo]: *(Pick another point, or type U.)*
Specify next point or [Close/Undo]: *(Pick another point, or enter an option.)*
Specify next point or [Close/Undo]: *(Press Enter to end the command.)*

COMMAND LINE OPTIONS

Close closes the line from the current point to the starting point.

Undo undoes the last line segment drawn.

Enter continues the line from the last endpoint at the 'From point' prompt; terminates the Line command at the 'To point' prompt.

RELATED COMMANDS

MLine draws up to 16 parallel lines.

PLine draws polylines and polyline arcs.

Trace draws lines with width.

Ray creates semi-infinite construction lines.

Sketch draws continuous line segments, with up-down control.

XLine creates infinite construction lines.

RELATED SYSTEM VARIABLES

Lastpoint specifies the last-entered coordinate triple (x,y,z-coordinate).

Thickness determines the thickness of the line.

TIPS

- To draw 2D lines, enter x,y coordinate pairs; the z coordinate takes on the value of the Elevation system variable.

- To draw 3D lines, enter x,y,z coordinate triples.

- When system variable Thickness is not zero, the line has thickness, which makes it a plane perpendicular to the current UCS.

'Linetype

V. 2.0 Loads linetype definitions into the drawing, creates new linetypes, and sets the working linetype.

Commands	Aliases Ctrl+	F-key	Alt+	Menu Bar	Dashboard
linetype	lt ...	...	ON	Format	Object Properties
	ltype			⅏Linetype	
	ddltype				
-linetype	-lt				
	-ltype				

Command: linetype

Displays dialog box:

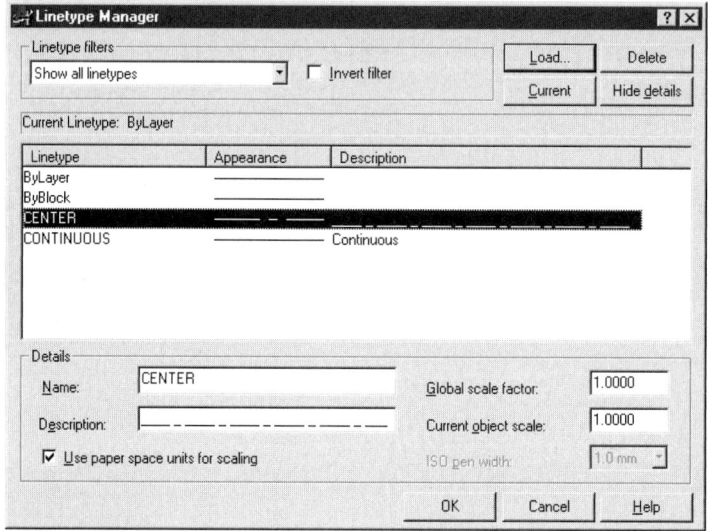

LINETYPE MANAGER DIALOG BOX

Linetype Filters options
- **Show all linetypes** displays all linetypes defined in the current drawing.
- **Show all used linetypes** displays all linetypes being used.
- **All xref dependent linetypes** displays linetypes in externally-referenced drawings.

☐ **Invert filter** inverts the display of layer names; for example, when Show all used linetypes is selected, the Invert filter option displays all linetypes not used in the drawing.

Loads displays the Load or Reload Linetypes dialog box.

Current sets the selected layer as current.

Delete purges the selected linetypes; some linetypes cannot be deleted, as described by the warning dialog box:

Show/Hide Details toggles the display of the Details portion of the Linetype Properties Manager dialog box.

Details options

Name names the selected linetype.

Description describes the linetype.

☑ **Use paper space units for scaling** specifies that paper space linetype scaling is used, even in model space.

Global scale factor specifies the scale factor for all linetypes in the drawing.

Current object scale specifies the individual object scale factor for all subsequently-drawn linetypes, multiplied by the global scale factor.

ISO pen width applies standard scale factors to ISO (international standards) linetypes.

LOAD OR RELOAD LINETYPES DIALOG BOX

File names the *.lin* linetype definition file.

SHORTCUT MENU

Right-click any linetype name in the Linetype Manager dialog box:

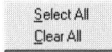

Select All selects all linetypes.

Clear All selects no linetypes.

-LINETYPE Command

Command: -linetype

Enter an option [?/Create/Load/Set]: *(Enter an option.)*

COMMAND LINE OPTIONS

Create creates new user-defined linetypes.

Load loads linetypes from linetype definition *(.lin)* files.

Set sets the working linetype.

? lists the linetypes loaded into the drawing.

RELATED COMMANDS

Change changes objects to a new linetype; changes linetype scale.

ChProp changes objects to a new linetype.

LtScale sets the scale of the linetype.

Rename changes the name of the linetype.

RELATED SYSTEM VARIABLES

CeLtype specifies the current linetype setting.

LtScale specifies the current linetype scale.

MsLtScale matches the linetype scale to the model view scale *(new to AutoCAD 2008)*.

PsLtScale specifies the linetype scale relative to paper scale.

PlineGen controls how linetypes are generated for polylines.

TIPS

- The only linetypes defined initially in a new AutoCAD drawing are:
    Continuous draws unbroken lines.
    Bylayer specifies linetype by the layer setting.
    Byblock specifies linetypes by the block definition.

- Linetypes must be loaded from *.lin* definition files before being used in a drawing.

- When loading one or more linetypes, it is faster to load all linetypes, and then use the Purge command to remove the linetype definitions that have not been used in the drawing.

- As of AutoCAD Release 13, objects can have independent linetype scales.

- To use MsLtScale correctly, set the value of other linetype scale system variables to 1:

    LtScale = 1 (or another size correct for plotting)
    CeLtScale = 1
    PsLtScale = 1
    MsLtScale = 1

- Changes in annotation scale appear with the next regeneration (Regen command).

RELATED FILE

The following standard linetypes are in *autocad 2008**support**acad.lin*:

ACAD_ISO02w100	ISO dash __ __ __ __ __ __ __ __ __ __ __
ACAD_ISO03w100	ISO dash space __ __ __ __ __ __
ACAD_ISO04w100	ISO long-dash dot ____ . ____ . ____ . ____ . _
ACAD_ISO05w100	ISO long-dash double-dot ____ .. ____ .. ____ .
ACAD_ISO06w100	ISO long-dash triple-dot ____ ... ____ ... ____
ACAD_ISO07w100	ISO dot . . . .
ACAD_ISO08w100	ISO long-dash short-dash ____ __ ____ __ ____ _
ACAD_ISO09w100	ISO long-dash double-short-dash ____ __ __ ____
ACAD_ISO10w100	ISO dash dot __ . __ . __ . __ . __ . __ .
ACAD_ISO11w100	ISO double-dash dot __ __ . __ __ . __ __ . __ __
ACAD_ISO12w100	ISO dash double-dot __ . . __ . . __ . . __
ACAD_ISO13w100	ISO double-dash double-dot __ __ . . __ __ . .
ACAD_ISO14w100	ISO dash triple-dot __ . . . __ . . . __ .
ACAD_ISO15w100	ISO double-dash triple-dot __ __ . . . __ __ .
BATTING	Batting SSSSSSSSSSSSSSSSSSSSSSSSSSSSSSSSSSSSSSSSS
BORDER	Border __ __ . __ __ . __ __ . __ __ .
BORDER2	Border (.5x) __.__.__.__.__.__.__.__.
BORDERX2	Border (2x) ____ ____ . ____ ____ . ____
CENTER	Center ____ _ ____ _ ____ _ ____ _ ____
CENTER2	Center (.5x) ___ _ ___ _ ___ _ ___ _ ___ _ ___
CENTERX2	Center (2x) _____ __ _____ __ ____
DASHDOT	Dash dot __ . __ . __ . __ . __ . __ .
DASHDOT2	Dash dot (.5x) _._._._._._._._._._._._._.
DASHDOTX2	Dash dot (2x) ____ . ____ . ____ . ____
DASHED	Dashed __ __ __ __ __ __ __ __ __ __ _
DASHED2	Dashed (.5x) _ _ _ _ _ _ _ _ _ _ _ _ _ _ _ _ _
DASHEDX2	Dashed (2x) ____ ____ ____ ____ ____ __
DIVIDE	Divide ____ . . ____ . . ____ . . ____
DIVIDE2	Divide (.5x) __._.__._.__._.__._.__._._
DIVIDEX2	Divide (2x) _____ . . _____ . . _
DOT	Dot . . . .
DOT2	Dot (.5x) . . . .
DOTX2	Dot (2x) . . . .
FENCELINE1	Fenceline circle ----O-----O----O----O----O---
FENCELINE2	Fenceline square ----[]-----[]----[]-----[]----
GAS_LINE	Gas line ----GAS----GAS----GAS----GAS----GAS---
HIDDEN	Hidden __ __ __ __ __ __ __ __ __ __ __ __ __
HIDDEN2	Hidden (.5x) _ _ _ _ _ _ _ _ _ _ _ _ _ _ _ _ _
HIDDENX2	Hidden (2x) ____ ____ ____ ____ ____ ____
HOT_WATER_SUPPLY	Hot water supply ---- HW ---- HW ---- HW ----
PHANTOM	Phantom _____ __ __ ____ __ __ ____
PHANTOM2	Phantom (.5x) ___ _ _ ___ _ _ ___ _ _
PHANTOMX2	Phantom (2x) _____ ____ ____ _
TRACKS	Tracks -I-I-I-I-I-I-I-I-I-I-I-I-I-I-I-I-I
ZIGZAG	Zig zag /\\/\\/\\/\\/\\/\\/\\/\\/\\/\\/\\/\\/\\/\\/

List

V. 1.0 Lists information about selected objects in the drawing.

Command	Alias	Ctrl+	F-key	Alt+	Menu Bar	Dashboard
list	li	...	...	TQL	Tools	...
	ls				� Inquiry	
	showmat				� List	

Command: list
Select objects: *(Select one or more objects.)*
Select objects: *(Press Enter to end object selection.)*

Sample output for a linear dimension:

```
                    DIMENSION  Layer: "0"
                        Space: Model space
                Handle = 521
        associative: no
        type: horizontal
        1st extension  defining point: X=2476.0511  Y=1012.8004  Z=  0.0000
        2nd extension  defining point: X=2533.2927  Y=1012.8004  Z=  0.0000
        dimension line defining point: X=2533.2927  Y=1021.6790  Z=  0.0000
        default text position: X=2504.6719  Y=1023.7623  Z=  0.0000
        default text
        dimension style: "Annotative"
        dimension style overrides:
        Annotative: Yes
        Annotative scale: 1:1
```

COMMAND LINE OPTIONS

Enter continues the display.

Esc cancels the display.

F2 returns to graphics screen.

RELATED COMMANDS

Area calculates the area and perimeter of selected objects.

DbList lists information about *all* objects in the drawing.

MassProp calculates the properties of 2D regions and 3D solids.

TIPS

- This command lists the following information only under certain conditions:

    Color — when not set BYLAYER.
    Linetype — when not set BYLAYER.
    Thickness — when not 0.
    Elevation — when z coordinate is not 0.
    Extrusion direction — when z axis differs from current UCS.

- Object handles are described by hexadecimal numbers.

- Alternatives to this command include the Properties, Dist, and Id commands.

. . . .

Removed Command

ListURL was removed from AutoCAD 2000; it was replaced by -Hyperlink.

. . . .

LiveSection

<u>2007</u>   Activates section planes to show the interior of 3D models.

Command	Alias	Ctrl+	F-key	Alt+	Menu Bar	Dashboard
livesection	...	...	...	...	...	...

As an alternative, you can double-click a section plane to make it live.
Command: livesection
Select section object: *(Select one selection plane.)*
Selecting the live section activates its dynamic grips illustrated below.

Section Plane

Change Type of
Section Plane

Mirror Plane

Move Plane

Stretch Plane

Live section planes can be edited.

COMMAND LINE OPTION

Select section object selects the section plane to make live.

SHORTCUT MENU OPTIONS

Right-click the live section plane for these additional options:

Activate live sectioning toggles live sectioning on and off.
Show cut-away geometry displays the hidden portions in another color.
Live section settings displays the Section Settings dialog box.

Generate 2D/3D section displays the Generate Section/Elevation dialog box, and then creates a second model of the 2D section or 3D elevation.

Add jog to section adds a 90-degree jog to the section plane; see the JogSection command.

DYNAMIC GRIP MENU
Click the lookup grip:

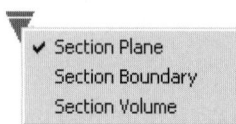

Section Plane displays the section line and cutting plane; the plane extends to "infinity" on all four sides.

Section Boundary displays a 2D box limiting the x,y extents of the cutting plane.

Section Volume displays a 3D box limiting the extent of the cutting plane in all directions.

Left: Section boundary.
Right: Volume boundary.

SECTION SETTINGS DIALOG BOX

☑ **Activate Live Section** turns on live sectioning for the selected section.

Intersection Boundary options

Color specifies the color of the intersection boundary.

Linetype specifies the linetype.

Linetype Scale specifies the linetype scale.

Lineweight specifies the lineweight.

Intersection Fill options

Show toggles the display of the intersection fill.

Face Hatch determines if the fill is solid, a hatch pattern, or none.

Angle specifies the angle of the hatch pattern.

Hatch Scale specifies the hatch pattern scale factor.

Hatch Spacing specifies the spacing between hatch lines.

Color specifies the color of the intersection fill.

Linetype specifies the linetype.

Linetype Scale specifies the linetype scale.

Lineweight specifies the lineweight.

Surface Transparency changes the translucency of the intersection fill; ranges from 0 to 100.

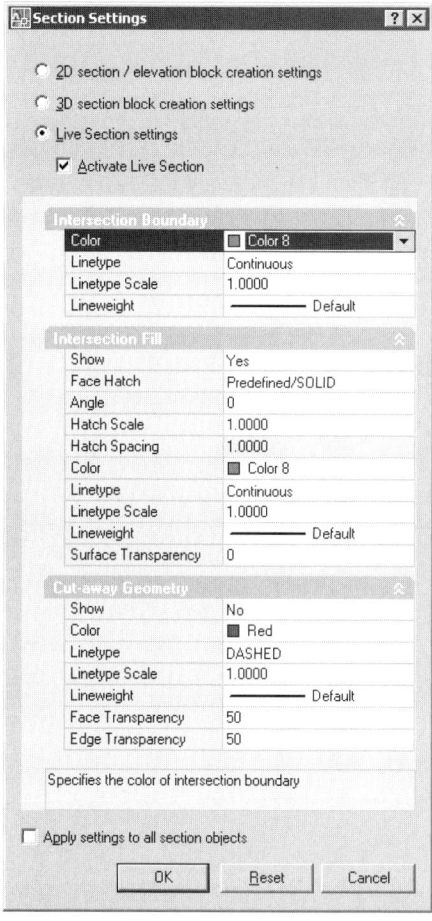

Cut-away Geometry options

Show toggles the display of cut-away geometry (removed by the live section plane).

Color specifies the color of the intersection fill.

Linetype specifies the linetype.

Linetype Scale specifies the linetype scale.

Lineweight specifies the lineweight.

Face Transparency changes the translucency of the cutaway faces.

Edge Transparency changes the translucency of the cutaway edges.

☑ **Apply Settings to All Section Objects** applies settings to future sectionings.

Reset reverts settings to default values.

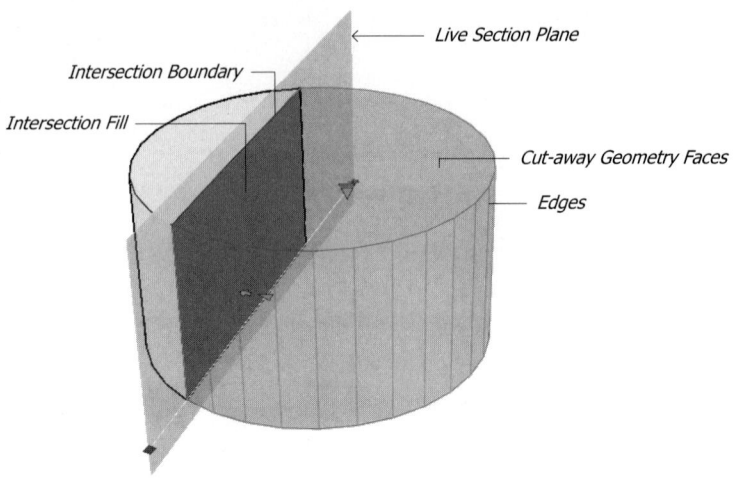

Live Section Plane

Intersection Boundary

Intersection Fill

Cut-away Geometry Faces

Edges

GENERATE SECTION/ELEVATION DIALOG BOX

2D/3D options

⊙ **2D Section/Elevation** generates 2D sections (a.k.a. elevations).

○ **3D Section** generates 3D sections.

The following options are visible only when the More button is clicked:

Source Geometry options

⊙ **Include All Objects** sections all 3D solids, surfaces, and regions in the drawing, blocks, and xrefs.

○ **Select Objects to Include** selects manually the 3D solids, surfaces, and regions to be sectioned.

🔲 **Select Objects** allows you to pick the objects to section with a cursor.

Destination options

⊙ **Insert as New Block** inserts the section as a block in the drawing.

○ **Replace Existing Block** replaces an existing section block.

🔲 **Select Block** selects the block to replace.

○ **Export to a File** saves the section as an AutoCAD drawing on disc.

Filename and Path specifies the *.dwg* file name and path by which the section is saved.

Section Settings displays the Section Settings dialog box.

Create creates the section.

RELATED AUTOCAD COMMANDS

SectionPlane places section planes in 3D objects.

JogSection adds 90-degree jogs to section planes.

TIPS

- Once live, section planes can be edited with dynamic grips, as illustrated above.

- Live sectioning works on 3D objects and regions in model space only.

- Only one section plane can be live at a time.

Load

V. 1.0 Loads SHX-format shape files into drawings.

Command	Alias	Ctrl+	F-key	Alt+	Menu Bar	Dashboard
load	...	...	...	...	...	...

Command: load

Displays Load Shape File dialog box. Select an .shx file, and the click **Open**.

COMMAND LINE Options

None.

RELATED AUTOCAD COMMAND

Shape inserts shapes into the current drawing.

RELATED FILES

.shp are source code files for shapes.

.shx are compiled shape files.

In \autocad 2008\support folder:

gdt.shx and **gdt.shp** are geometric tolerance shapes used by the Tolerance command.

ltypeshp.shx and **ltypeshp.shp** are linetype shapes used by the Linetype command.

TIPS

- Shapes are more efficient than blocks, but are harder to create.

- The Load command cannot load *.shx* files meant for fonts. For example, AutoCAD complains, "gdt.shx is a normal text font file, not a shape file."

- Do not confuse this command with the AutoLISP **(load)** function, which loads *.lsp* files.

Loft

<u>**2007**</u>   Creates 3D surface or solid lofts between two or more objects.

Command	Alias	Ctrl+	F-key	Alt+	Menu Bar	Dashboard
loft	...	...	...	DML	Draw Modeling Loft	3D Make

Command: loft
Select cross-sections in lofting order: *(Pick two or more cross-section objects.)*
Select cross-sections in lofting order: *(Press Enter.)*
Enter an option [Guides/Path/Cross-sections only] <Cross-sections only>:
(Press Enter, or enter an option.)

   AutoCAD creates an initial lofting, and then displays the Loft Settings dialog box:

*Click **OK**. AutoCAD creates the final lofting:*

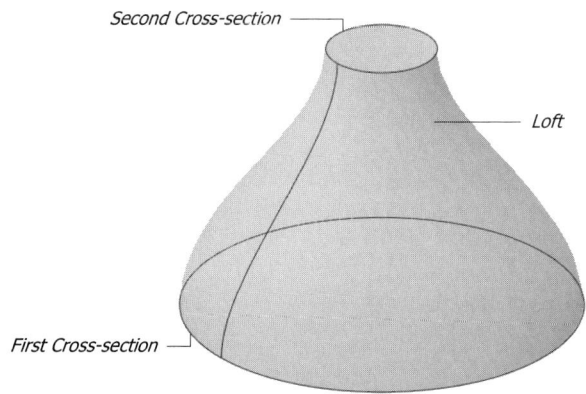

COMMAND LINE OPTIONS

Select cross-sections selects the objects that define the lofting surface: lines, arcs, elliptical arcs, circles, ellipses, polylines, and points (for the first and last cross-sections only).

Cross-sections only uses the cross-section objects selected earlier.

Path selects a single path object that guides the loft: a line, arc, elliptical arc, circle, ellipse, spline, helix, polyline, or 3D polyline.

Guides selects two or more guide-curve objects that define the loft; guides must start on the first cross-section, intersect each cross-section, and end at the last cross-section. Guide objects can be lines, arcs, elliptical arcs, 2D and 3D polylines, and splines.

DIALOG BOX OPTIONS

⊙ **Ruled** creates smoothly-ruled lofts between cross sections, but sharp-edged rules at cross sections.

○ **Smooth Fit** creates smoothly-ruled lofts between cross sections, but sharp edges at the starting and ending cross sections.

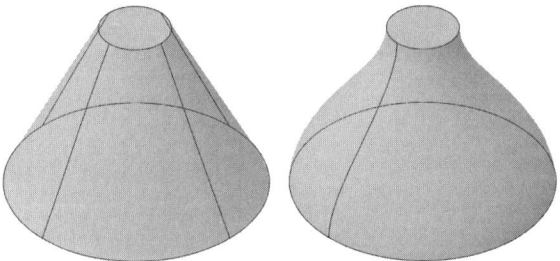

Left: *Ruled loft.*
Right: *Normal to start and end cross-sections.*

○ **Normal to** controls the surface normal where the loft passes through the cross-sections:

- **All Cross Sections** — surface normal is normal to all cross-sections.
- **Start Cross Section** — surface normal is normal to the starting cross-section.
- **End Cross Section** — surface normal is normal to the ending cross-section.
- **Start and End Cross Sections** — surface normal is normal to both starting and ending cross-sections.

○ **Draft Angles** controls the draft angle and magnitude of the first and last cross-sections of the lofted solid or surface.

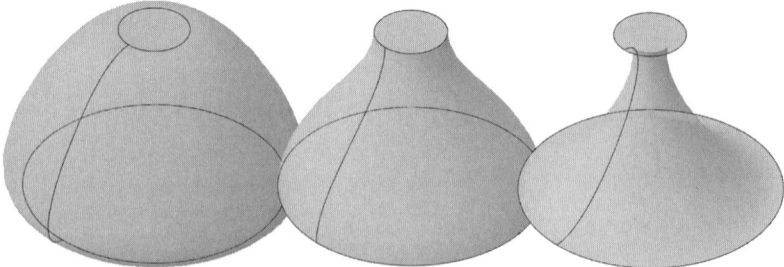

Left to right: *Loft angles of 0, 80 and 180 degrees.*

Start Angle specifies the draft angle of the starting cross-section.

Start Magnitude controls the distance between the starting cross-section in the direction of the draft angle but before the loft bends toward the next cross-section.

End Angle specifies the draft angle of the starting cross-section.

End Magnitude controls the distance between the ending cross-section in the direction of the draft angle but before the loft bends toward the previous cross-section.

☐ **Close Surface or Solid** (when off) creates open lofts; when on, closes the loft.

☑ **Preview Changes** shows changes to the loft as options are changed in this dialog box.

RELATED AUTOCAD COMMAND

Shape inserts shapes into the current drawing.

RELATED SYSTEM VARIABLES

DelObj toggles whether the cross-section geometry is erased after the loft is created.

LoftAng1 stores the default for the first lofting angle.

LoftAng2 stores the default for the last lofting angle.

LoftMag1 specifies the magnitude of the first lofting angle.

LoftMag2 specifies the magnitude of the last lofting angle.

LoftNormals specifies where lofting normals are placed.

LoftParam specifies the loft shape.

SurfU specifies the number of m-direction isolines on lofts.

SurfV specifies the number of n-direction isolines on lofts.

TIPS

- The cross-section curves must be all open or all closed; open ones create surfaces, while closed ones create solids.

- After the loft is created, its shape can be edited with grips, like spline curves.

- Use this command to create surfaces from contour maps and boat hulls.

LogFileOn, LogFileOff

<u>Rel.13</u> Turns on and off the command logging to *.log* files.

Command	Alias	Ctrl+	F-key	Alt+	Menu Bar	Dashboard
logfileon	...	...	...	...	...	...
logfileoff	...	...	...	...	...	...

Command: logfileon

AutoCAD begins recording command-line text to the log file.

Command: logfileoff

AutoCAD stops recording command-line text, and closes the log file.

COMMAND LINE OPTIONS
None.

RELATED AUTOCAD COMMAND
CopyHist copies all command text from the Text window to the Clipboard.

RELATED SYSTEM VARIABLES
LogFileName specifies the name of the log file.

LogFilePath specifies the path for the log files for all drawings in a session.

LogFileMode toggles whether text window is written to log file:

  0 — text not written to file (default).
  1 — text written to file.

TIPS
- AutoCAD places a dashed line at the end of each log file session.

- If log file recording is left on, it resumes when AutoCAD is next loaded, which can result in very large log files.

- The default log file name is the same as the drawing name, and is stored in folder *C:\Documents and Settings\username\Local Settings\Application Data\Autodesk\AutoCAD 2008\R17.1\enu\.* You can give the file a different folder and name with the Options command's Files tab, or with system variables LogFileName and LogFilePath.

- *Historical note:* In some early versions of AutoCAD, Ctrl+Q meant "quick screen print," which output the current screen display to the printer. Ctrl+Q reappeared in AutoCAD Release 14 to record command text to a file. As of AutoCAD 2004, the Ctrl+Q shortcut quits AutoCAD, instead of toggling the log file — curious, given that there is already a keyboard shortcut, Alt+F4, that quits AutoCAD.

Removed Commands
The **LsEdit**, **LsLib**, and **LsNew** commands were removed from AutoCAD 2007.

'LtScale

Sets the global scale factor of linetypes (short for Line Type SCALE).

Command	Alias	Ctrl+	F-key	Alt+	Menu Bar	Dashboard
ltscale	lts	...	...	...	...	...

Command: ltscale
Enter new linetype scale factor <1.0000>: *(Enter a scale factor.)*
Regenerating drawing.

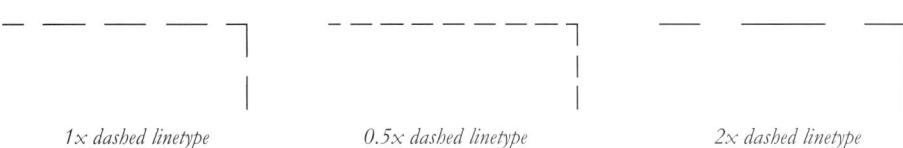

1x dashed linetype   *0.5x dashed linetype*   *2x dashed linetype*

COMMAND LINE OPTION

Enter new linetype scale factor changes the global scale factor of all linetypes in drawings.

RELATED COMMANDS

ChProp changes the linetype scale of one or more objects.

Properties changes the linetype scale of objects.

Linetype loads, creates, and sets the working linetype.

RELATED SYSTEM VARIABLES

LtScale contains the global linetype scale factor.

MsLtScale matches the linetype scale to the model space scale factor *(new to AutoCAD 2008)*.

CeLtScale specifies the current object linetype scale factor relative to the global scale.

PlineGen controls how linetypes are generated for polylines.

PsLtScale specifies that the linetype scale is relative to paper space.

TIPS

- If the linetype scale is too large, the linetype appears solid.

- If the linetype scale is too small, the linetype appears as a solid line that redraws very slowly.

- In addition to setting the scale with the LtScale command, the *acad.lin* file contains each linetype in three scales: normal, half-size, and double-size.

- To use MsLtScale correctly, set the value of other linetype scale system variables to 1:

    LtScale = 1 (or another size correct for plotting)
    CeLtScale = 1
    PsLtScale = 1
    MsLtScale = 1

    The annotative scaling does not come into effect until the next regeneration,.

- You can change the linetype scaling of individual objects, which is then multiplied by the global scale factor specified by the LtScale command.

'LWeight

2000 Sets the current lineweight (display width) of objects.

Commands	Aliases	Status Bar	Alt+	Menu Bar	Dashboard
lweight	lw	LWT	OW	Format	Object properites
	lineweight			⍦Lineweight	
-lweight					

Command: lweight

Displays dialog box::

LINEWEIGHT SETTINGS DIALOG BOX

Lineweights lists lineweight values.

Units for Listing options

⊙**Millimeters (mm)** specifies lineweight values in millimeters.

○**Inches (in)** specifies lineweight values in inches.

☐ **Display Lineweight** toggles the display of lineweights; when checked, lineweights are displayed.

Default specifies the default lineweight for layers (default = 0.01" or 0.25 mm).

Adjust Display Scale controls the scale of lineweights in the Model tab, which displays lineweights in pixels.

SHORTCUT MENU

Right-click LWT on the status bar to display shortcut menu:

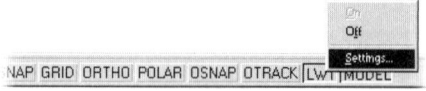

On turns on lineweight display.

Off turns off lineweight display.

Settings displays Lineweight Settings dialog box.

-LWEIGHT Command

Command: -lweight

Enter default lineweight for new objects or [?]: *(Enter a value, or type ?.)*

COMMAND LINE OPTIONS

Enter default lineweight specifies the current lineweight; valid values include Bylayer, Byblock, and Default.

? lists the valid values for lineweights:

```
ByLayer ByBlock Default
0.000" 0.002" 0.004"  0.005" 0.006" 0.007"
0.008" 0.010" 0.012"  0.014" 0.016" 0.020"
0.021" 0.024" 0.028"  0.031" 0.035" 0.039"
0.042" 0.047" 0.055"  0.062" 0.079" 0.083"
```

RELATED COMMAND

Properties changes lineweights of selected objects.

RELATED SYSTEM VARIABLES

LwDefault specifies the default linewidth; default = 0.01" or 0.25 mm.

LwDisplay toggles the display of lineweights in the drawing.

LwUnits determines whether the lineweight is measured in inches or millimeters.

TIPS

■ To create custom lineweights for plotting, use the Plot Style Table Editor.

■ A lineweight of 0 plots the lines at the thinnest width of which the plotter is capable, usually one pixel or one dot wide.

■ Linewieghts are "hardwired" into AutoCAD, and cannot be customized.

■ Lineweights can be set through layers and the Properties toolbar.

Replaced Command

MakePreview was removed from AutoCAD Release 14; it was replaced by the RasterPreview system variable, which controls the creation of previews when drawings are saved.

Markup, MarkupClose

<u>2005</u> Opens and closes the Markup Set Manager palette.

Command	Alias	Ctrl+	F-key	Alt+	Menu Bar	Dashboard
markup	msm	7	...	TPK	Tools	...
					↳Palettes	
					↳Markup Set Manager	
markupclose		7				

Command: markup

Displays palette:

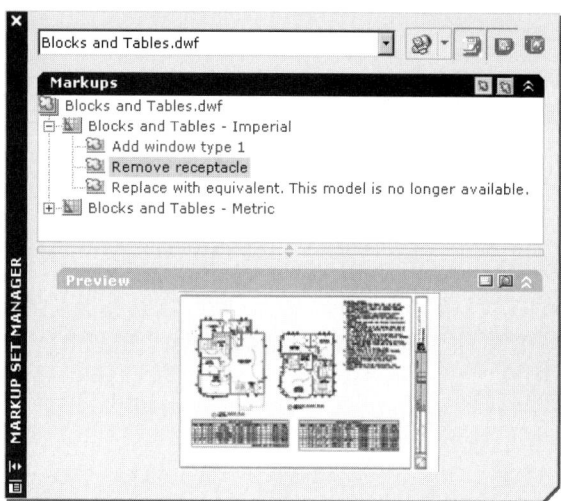

To open a markup set, select Open. AutoCAD displays the Open Markup DWF dialog box. Select a .dwf file, and then click Open.

If the file contains no markup data, AutoCAD complains, 'Filename.dwf does not contain any markup data. Would you like to open this DWF file in the viewer?' Click Yes to open the file in the DWF Viewer; click No to cancel.

Command: markupclose

Closes the palette.

MARKUP SET MANAGER ICONS

Close Window closes the Markup Set Manager palette; reopen with Ctrl+7.

Show All Sheets shows all sheets with markups.

Hide Non-Markup Sheets hides all sheets without markups.

Collapse collapses the preview/details area.

Details displays details about the markup sheet.

Preview displays a bitmap image of the sheet.

Palette toolbar

View Redline Geometry (Alt+2)

View DWG Geometry (Alt+3)

Republish Markup DWF (Alt+1)

View DWF Geometry (Alt+3)

Republish Markup DWF displays menu (Alt+1):
- **Republish All Sheets** republishes all sheets as a background job using Publish.
- **Republish Markup Sheets** republishes only sheets with markups.

View Redline Geometry toggles the display of redline markup objects (Alt+2).

View DWG Geometry toggles the display of *.dwg* drawing files (Alt+3).

View DWF Geometry toggles the display of *.dwf* files (Alt+4).

SHORTCUT MENU
Right-click a markup file:

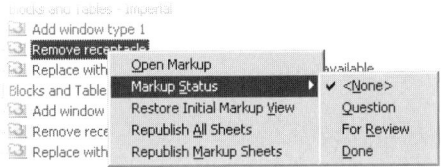

Open Markup opens the markup data in AutoCAD; alternatively, double-click the markup.

Markup Status displays a submenu for changing the status of a markup sheet:
- **None** indicates no change in status.
- **Question** indicates markup has questions to be answered.
- **For Review** indicates markup changes need to be reviewed.
- **Done** indicates markup is done.

Restore Initial Markup View resets all markup views to their original state.

Republish All Sheets republishes all sheets as a background job using Publish.

Republish Markup Sheets republishes only sheets with markups.

RELATED COMMANDS
OpenDwfMarkup opens *.dwf* files containing markup data.

DwfAttach attaches *.dwf* files as underlays.

3dDwfOut exports 3D drawings in DWF format.

RELATED SYSTEM VARIABLE
MsmState reports whether the Markup Set Manager palette is open.

TIPS
- MSM is short for "markup set manager."
- This command reads only *.dwf* files marked-up with Design Review (nee DWF Composer); it does not read other *.dwf* files.

MassProp

Rel.11 Reports the mass properties of 3D solid models, bodies, and 2D regions (short for MASS PROPerties).

Command	Alias	Ctrl+	F-key	Alt+	Menu Bar	Dashboard
masprop	...	...	...	TQM	Tools	...
					⤷Inquiry	
					⤷Region/Mass Properties	

Command: masprop
Select objects: *(Select one or more regions and/or solid model objects.)*
Select objects: *(Press Enter.)*
Write analysis to a file? [Yes/No] <N>: *(Type Y or N.)*

Example output of a solid sphere:

```
---------------     SOLIDS     ---------------

Mass:                  9.9378
Volume:                9.9378
Bounding box:        X: 4.7662  --  7.1926
                     Y: 4.9973  --  8.1426
                     Z: -3.8483 --  1.6878
Centroid:            X: 5.8455
                     Y: 6.6391
                     Z: -0.2218
Moments of inertia:  X: 455.3659
                     Y: 354.4693
                     Z: 784.7290
Products of inertia: XY: 384.7960
                     YZ: -10.0200
                     ZX: -14.3752
Radii of gyration:   X: 6.7692
                     Y: 5.9723
                     Z: 8.8862
Principal moments and X-Y-Z directions about centroid:
                     I: 17.1241 along [0.9515 0.3078 0.0000]
                     J: 16.6049 along [-0.2741 0.8472 -0.4552]
                     K: 4.6438 along [-0.1401 0.4331 0.8904]

Write analysis to a file? [Yes/No] <N>:
```

COMMAND LINE OPTIONS

Select objects selects the solid model objects (2D regions, 3D solids, and bodies) to analyze.

Write analysis to a file:

Yes writes mass property reports to *.mpr* files.

No does not write reports to file.

RELATED COMMANDS

Area calculates the area and perimeter of non-solid objects.

Properties displays information about all objects.

RELATED FILE

.mpr is the file to which MassProp writes its results (mass properties report).

TIPS

- This command can be used with 2D regions as well as 3D solids; it cannot be used with 3D surface models or 2D non-region objects.

- Distances are relative to the current UCS. For them to be meaningful, you should orient the UCS to a significant feature on the object.

- Mass properties are computed as if the selected regions were unioned, and as if the selected solids were unioned.

- As of Release 13, AutoCAD's solid modeling no longer allows you to apply a material density to a solid model. All solids and bodies have a density of 1.

- AutoCAD only analyzes regions coplanar (lying in the same plane) with the first region selected.

DEFINITIONS

Bounding Box

– the lower-right and upper-left coordinates of a rectangle enclosing 2D regions.

– the x,y,z coordinate triple of a 3D box enclosing 3D solids or bodies.

Centroid

– the x,y,z coordinates of the center of 2D regions.

– the center of mass of 3D solids and bodies.

Mass – equal to the volume, because density = 1; not calculated for regions.

Moment of Inertia

– for 2D regions = Area * $Radius^2$

– for 3D bodies = Mass * $Radius^2$

Perimeter – total length of inside and outside loops of 2D regions (not calculated for 3D solids).

Product of Inertia

– for 2D regions = Mass * Distance (of centroid to y,z axis) * Distance (of centroid to x,z axis).

– for 3D bodies = Mass * Distance (of centroid to y,z axis) * Distance (of centroid to x,z axis)

Radius of Gyration – for 2D regions and 3D solids = $(MomentOfInertia / Mass)^{1/2}$

Volume – 3D space occupied by a 3D solid or body (not calculated for regions).

MatchCell

2005 Matches the properties of table cells.

Command	Alias	Ctrl+	F-key	Alt+	Menu Bar	Dashboard
matchcell	...	...	...	...	...	...

Command: matchcell
Select source cell: *(Select a cell in a table.)*
Select destination cell: *(Select one or more cells.)*

COMMAND LINE OPTIONS

Select source cell gets property settings from the source cell.

Select destination cell passes property settings to the destination cells.

RELATED COMMANDS

Table creates new tables in drawings.

TableStyle defines table styles.

MatchProp copies properties between objects other than cells.

RELATED SYSTEM VARIABLE

CTable Style specifies the name of the current table style.

TIPS

- Use this command to copy formatting from one cell to another.

- This command can also be launched by right-clicking a selected cell, and then choosing Match Cell from the shortcut menu.

- Use the MatchProp command to copy properties from one table to another.

'MatchProp

Rel.14 Matches the properties between selected objects (short for MATCH PROPerties).

Command	Alias	Ctrl+	F-key	Alt+	Menu Bar	Dashboard
matchprop	ma	...	...	MM	Modify	...
	painter				⌖Match Properties	

Command: matchprop
Select source object: *(Select a single object.)*
Current active settings: Color Layer Ltype Ltscale Lineweight Thickness
PlotStyle Dim Text Hatch Polyline Viewport Table Material Shadow display
Multileader
Select destination object(s) or [Settings]: *(Pick one or more objects, or type S.)*
Select destination object(s) or [Settings]: *(Press Enter to exit command.)*

COMMAND LINE OPTIONS

Select source object gets property settings from the source object.

Select destination object(s) passes property settings to the destination objects.

Settings displays dialog box:

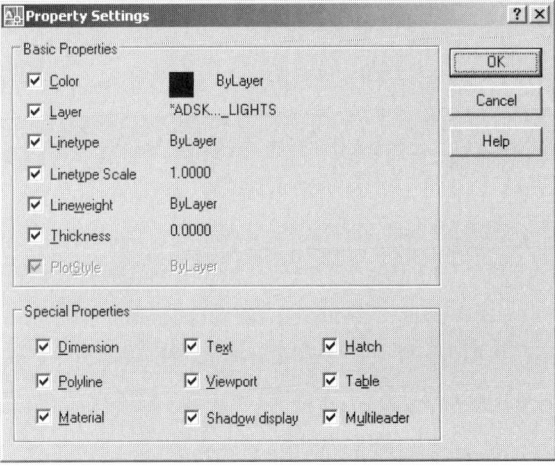

PROPERTY SETTINGS DIALOG BOX

Basic Properties

☑ **Color** specifies the color for destination objects; not available when OLE objects are selected.

☑ **Layer** specifies the layer name for destination objects; not available when OLE objects are selected.

☑ **Linetype** specifies the linetype for destination objects; not available when attributes, hatch patterns, mtext, OLE objects, points, or viewports are selected.

☑ **Linetype Scale** specifies the linetype scale for destination objects; not available when attributes, hatch patterns, mtext, OLE objects, points, or viewports are selected.

☑ **Lineweight** specifies the lineweight for destination objects.

☑ **Thickness** specifies the thickness for destination objects; available only for objects that can have thickness: arcs, attributes, circles, lines, mtext, points, 2D polylines, regions, text, and traces.

☑ **Plot Style** specifies the plot style; not available when PStylePolicy = 1 (color-dependent plot style mode), or when OLE objects are selected.

Special Properties

☑ **Dimension** copies the dimension styles of dimensions, leaders, and tolerance objects.

☑ **Text** copies the text style of text and mtext objects.

☑ **Hatch** copies the hatch pattern of hatched objects.

☑ **Polyline** copies the width and linetype generation of polylines; curve fit, elevation, and variable width properties are not copied.

☑ **Viewport** copies all properties of viewport objects, except clipping, UCS-per-viewport, freeze-thaw settings, and viewprot scale.

☑ **Table** copies style of table objects.

☑ **Material** copies the material name.

☑ **Shadow display** copies type of shadow casting.

☑ **Multileader** copies the multileader styles (*new to AutoCAD 2008*).

RELATED COMMAND

Properties changes most properties of selected objects.

RELATED SYSTEM VARIABLE

PStylePolicy determines whether the PlotStyle option is available.

TIPS

- In other Windows applications, this command is known as Format Painter.

- Use this command to make pattern properties the same among multiple hatches.

- Use the MatchCell command to copy properties from one table cell to another.

Removed Command

MatLib command was removed from AutoCAD 2007; it is replaced by the Materials group of the ToolPalette command.

MaterialAttach, MaterialAssign

2007 Attaches rendering materials to layers, objects.

Command	Alias	Ctrl+	F-key	Alt+	Menu Bar	Dashboard
materialattach	...	...	...	...	...	Materials
materialassign						

Command: materialattach

Before using this command, add materials to the drawing through the Materials group of the Tool palette.

Displays dialog box:

To attach a material to a layer, drag the material name onto the layer name.

DIALOG BOX OPTIONS

Material Name lists the names of materials defined in the drawing.

Layer lists the names of layers in the current drawing.

Material lists the material attached to the layer.

✗ *(remove)* removes the material from the layer; the layer reverts to the Global material.

MATERIALASSIGN Command
Command: materialassign
Select objects: *(Select one or more objects.)*

Select objects assigns the current material to the selected objects *(undocumented command)*.

RELATED COMMANDS

Materials creates and edits materials.

Render displays the materials applied to layers.

RELATED SYSTEM VARIABLE

CMaterial specifies the default material for newly-created objects.

TIPS

- Materials can be matched between objects using the MatchProp command.

- Layers without a user-attached material are automatically assigned the Global material.

- Materials can be removed from drawings with the Purge command.

MaterialMap

2007 Maps materials to the faces of objects, and adjusts the mapping.

Command	Alias	Ctrl+	F-key	Alt+	Menu Bar	Dashboard
materialmap	setuv	...	...	VEA	View	Materials
					⮡Render	
					⮡Mapping	

*To see the effect of this command, change the visual style to Realistic with the **vs** alias.*
Command: materialmap
Select an option [Box/Planar/Spherical/Cylindrical/copY mapping to/Reset mapping]<Box>: *(Enter an option.)*
Select faces or objects: *(Pick one or more objects.)*
Select faces or objects: *(Press Enter.)*
Accept the mapping or [Move/Rotate/reseT/sWitch mapping mode]: *(Press Enter, or enter an option.)*

COMMAND LINE OPTIONS

Planar maps the material once over the entire object, scaling it vertically and horizontally; meant for use on faces.

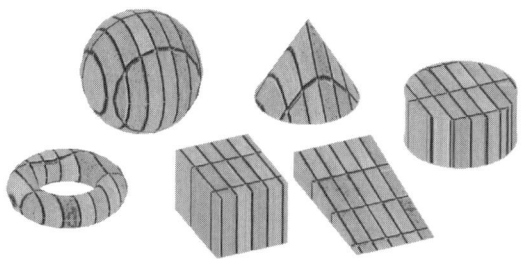

Planar mapping.

Box repeats the mapping on each face of objects; meant for rectangular objects; default mapping for new objects.

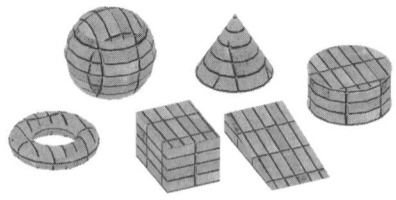

Box mapping.

Cylindrical wraps the material horizontally, and scales the material vertically; meant for use on cylindrical objects.

Cylindrical mapping.

Spherical wraps the material horizontally and vertically; meant for use on round objects.

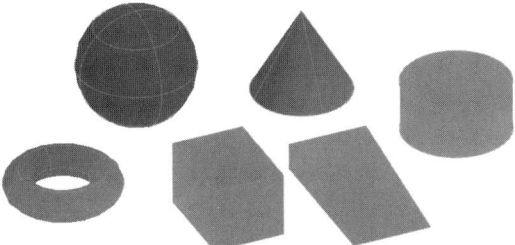

Spherical mapping.

copY mapping to copies the mapping style to other objects.

Reset mapping (and **reseT**) resets the u,v-mapping coordinates to their default values.

Select faces or objects selects the faces or objects to which the mapping style should be applied; hold down the **Ctrl** key to select faces.

Accept the mapping exits the command.

Move moves the material mapping interactively; see the Move3D command. Use the triangular grips to change the size of the map in the x, y, and z directions.

Rotating the material map interactively.

Rotate rotates the material mapping interactively; see the Rotate3D command.

Rotating the material map interactively.

sWitch mapping mode switches to one of the other mapping modes: box, planar, spherical, or cylindrical.

RELATED COMMANDS

Materials creates and edits materials.

Render displays the materials applied to layers.

RELATED SYSTEM VARIABLE

CMaterial specifies the default material for newly-created objects; default = bylayer.

TIP

- Materials can be mapped only to 3D solids, 3D surfaces, faces, and 2D objects with thickness.

Materials, MaterialsClose

2007

Creates and edits materials for use by the Render command and the Realistic visual style.

Command	Aliases Ctrl+	F-key	Alt+	Menu Bar	Dashboard
materials	mat ...	...	VEM	View	Materials
	rmat			⌐Render	
	finish			⌐Materials	

materialsclose

Command: materials

Displays palette, displayed on the next page (substantially changed in AutoCAD 2008).

Command: materialsclose

Closes the palette.

MATERIALS PALETTE

Materials Editor options

Type specifies the generic type of material:

- **Realistic** [*noted by* R *below*] provides a basic set of parameters.
- **Realistic Metal** [*noted by* RM *below*] provides realistic parameters suitable for metal; leaves out Refraction Index, for example.
- **Advanced** [*noted by* A *below*] provides more options than Realistic; includes Reflection, for example.
- **Advanced Metal** [*noted by* AM *below*] provides advanced parameters suitable for metal.

Template reuses existing material definitions.

The following options vary, depending on the Type selected:

[A | AM] **Ambient** specifies color for faces lighted only by ambient light.

[R | RM | A | AM] **Diffuse** specifies the main color of the material (button appears when By Object is off).

[A] **Specular** specifies the color of highlights on shiny materials.

[R | A | AM] **By Object** toggles whether the material color is based on the object's color.

[A | AM] **Ambient-Diffuse Lock** sets the ambient color the same as the diffuse color.

[A] **Diffuse-Specular Lock** sets the specular color the same as the diffuse color (padlock icon).

[R | RM | A | AM] **Shininess** adjusts the amount of shininess by changing the size of the highlight; very shiny objects have smaller and brighter shine spots. Range is 0 to 100.

[R] **Refraction Index** adjusts the refraction index (how light bends going through translucent objects):

- 1.0 — no distortion.
- 3.0 — increased distortion.

[R | A] **Translucency** adjusts translucency (see-through-ness) and light scattering within the material:

- 0 — opaque.
- 100 — fully translucent.

[R | RM | A | AM] **Self-illumination** causes objects to give off light. Range is 0 to 100; when 0, there is no no self-illumination.

Maps options

[R | RM | A | AM] ☑ **Diffuse Map** toggles diffuse mapping:

- **Texture map** — applies an image file: *.tga*, *.bmp*, *.rle*, *.dib*, *.png*, *.jpg*, *.jpeg*, *.tif*, *.gif*, and *.pcx*. AutoCAD stores 340 JPEG texture files in *c:\documents and settings\all users\application data\autodesk\autocad 2008\R17.1\enu\textures*.

- **Checker, Marble, Noise, Speckle, Tiles, Weaves, Wood** — displays parameters suitable for each material. For example, marble has parameters suitable for simulating marble veins.

From left to right: None; texture map applied; wood grain; and marble vein.

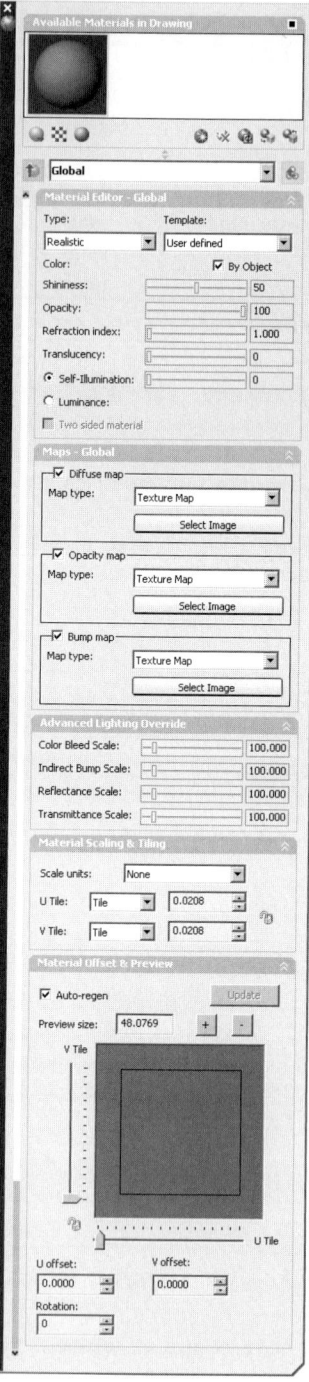

After a texture map is opened, these buttons appear:

Adjust Bitmap Button changes the scale, tiling, offset, and rotation of the image; displays the Adjust Bitmap dialog box.

Delete Map Information removes diffuse map from the material.

After other diffuse maps are selected, these buttons appear:

Edit Map displays a window for editing the material's look, such as wood grain.

Slider adjusts the intensity of the material's effect.

Opacity, Reflection, and Bump Map options

From left to right: *Opacity (transparency); reflection; and bump mapping.*

[R | RM | A | AM] **Opacity** specifies areas of transparency. Select a file that defines transparency:

- **Black** — opaque areas.
- **Gray** — translucent areas.
- **White** — transparent areas.

Slider adjust the amount of overall opacity.

Select Image selects the image that defines the areas of transparency; displays the Select Image File dialog box. AutoCAD stores a single JPEG opacity (cutout) file in the hidden *C:\Documents and Settings\All Users\Application Data\Autodesk\AutoCAD 2008\R17.1 \enu\Textures\Cutout* folder.

Opacity (cutout) image for trellis.

[RM | A | AM] **Reflection** adjusts the amount of scene reflection on metal surfaces. Range is 0 to 100; the higher the value, the more reflective the material.

Slider adjusts the amount of reflectivity.

Select Image selects the image to be reflected off the material's surface; displays the Select Image File dialog box.

[R | RM | A | AM] **Bump Map** adds surface bumpiness.

Slider adjusts the amount of bump texture. Range is -1000 (inverts the bump map) to 1000.

Select Image selects the image to be applied to the material's surface; displays the Select Image File dialog box. AutoCAD stores 115 JPEG bump files in the hidden *C:\Documents and Settings\All Users\ Application Data\Autodesk\AutoCAD 2008\R17.1\enu\Textures\Bump* folder.

Bump map image for woven basket.

Advanced Lighting Override options
New to AutoCAD 2008:

Color Bleed Scale adjusts the saturation of color reflected from materials.

Indirect Bump Scale changes the amount of bump mapping in indirectly lit areas of base materials.

Reflection Scale adjusts the energy reflected by materials.

Transmittance Scale adjusts the energy transmitted by transparent materials into scenes.

Material Scaling and Tiling options
(Formerly part of the Adjust Bitmap dialog box).

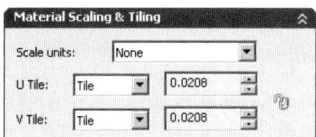

Scale Units sizes the image (bitmap) to the object, ranging from Fit to Gismo (formerly Fit to Object) to Kilometers.

U Tile and **V Tile** repeat the image in the x (and y) direction; range is from 0.02 (magnified) to 50 (reduced):

- **None** displays a single copy of the image.
- **Tiled** arrays the image to fit the object.
- **Mirrored** reverses the image *(new to AutoCAD 2008)*.

Lock Aspect Ratio makes the height and width (U and V) the same.

Material Offset and Preview options
(Formerly part of the Adjust Bitmap dialog box).

☑ **Auto-regen** automatically regenerates the image after changes are made (*new to AutoCAD 2008*).

Update updates the image when Auto-regen is turned off.

Preview Size zooms the preview image; range is from 0.0001 (zoomed in, large image) to 495176003 (zoom out, small image). You can drag the preview image to change its offsets.

+ and **-** change the size of the image.

U Tile and **V Tile** sliders change the amount of tiling along the U and V axes.

Lock preserves the image's aspect ratio.

U Offset and **V Offset** relocate the image's starting point (lower left corner) along the U and V axes.

Rotation rotates the image around the W axis (equivalent of the z axis). This does not work for spherical and cylindrical mapping; use the MaterialMap command for them instead.

RELATED COMMANDS

MaterialAttach attaches materials by layer.

MaterialMap adjusts mapping on surfaces.

RELATED SYSTEM VARIABLES

MatState reports whether the palette is open.

3dConversionMode controls how material and light definitions are converted when a pre-AutoCAD 2008 drawing is opened.

TIPS

- *Materials* define the look of rendered objects through many parameters, such as coloring, reflection, shininess, roughness, and ambient reflection. Examples of material definitions included with AutoCAD are:

   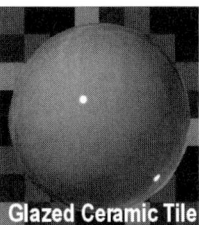

Clear Glass Glossy Paint Translucent Paper Glazed Ceramic Tile

- Materials appear when visual style is set to Realistic and when the Render command is used.

- By default, a drawing contains a single material definition, called *GLOBAL*, with the default parameters for color, reflection, roughness, and ambience.

- Materials do not define the density of 3D solids and bodies.

Measure

V. 2.5 Places points or blocks at constant intervals along lines, arcs, circles, and polylines.

Command	Alias	Ctrl+	F-key	Alt+	Menu Bar	Dashboard
measure	me	...	...	DOM	Draw ⮡ Point ⮡ Measure	...

Command: measure
Select object to measure: *(Pick a single object.)*
Specify length of segment or [Block]: *(Enter a value, or type **B**.)*

COMMAND LINE OPTIONS

Select object selects a single object for measurement.

Specify length of segment indicates the distance between markers.

Block Options
Enter name of block to insert: *(Enter name.)*
Align block with object? [Yes/No] <Y>: *(Type Y or N.)*
Specify length of segment: *(Enter a value.)*

Enter name of block indicates the name of the block to use as a marker; the block must already exist in the drawing.

Align block with object? aligns the block's x axis with the object.

RELATED COMMANDS

Block creates blocks that can be used with the Measure command.

Divide divides objects into a specific number of segments.

RELATED SYSTEM VARIABLES

PdMode controls the shape of a point.

PdSize controls the size of a point.

TIPS

- You must define the block before it can be used with this command.

- The **Measure** command does not place points (or blocks) at the beginning and end of measured objects.

Removed Command

MeetNow was removed from AutoCAD 2004.

Menu

<u>**V. 1.0**</u> Loads *.cui*, *.mns*, and *.mnu* menu files.

Command	Alias	Ctrl+	F-key	Alt+	Menu Bar	Dashboard
menu	...	...	...	...	...	...

Command: menu

Displays the Select Menu File dialog box.

Select a .cui, .mns, or .mnu file, and then click Open.

COMMAND LINE OPTIONS

None.

RELATED COMMANDS

Cui controls most aspects of customization.

CuiLoad loads a partial menu file.

Tablet configures digitizing tablet for use with overlay menus.

RELATED SYSTEM VARIABLES

MenuName specifies the name of the currently-loaded menu file.

MenuCtl determines whether sidescreen menu pages switch in parallel with commands entered at the keyboard.

MenuEcho suppresses menu echoing.

ScreenBoxes specifies the number of menu lines displayed on the side menu.

RELATED FILES

.cui source file for menus, toobars, and so on; stored in XML format.

.mns source for obsolete menu files; stored in ASCII format.

.mnu source for obsolete menu template files; stored in ASCII format.

TIPS

- AutoCAD automatically converts *.mns* and *.mns* files into *.cui* files.

- The *.cui* file defines the function of the screen menu, menu bar, cursor menu, icon menus, digitizing tablet menus, pointing device buttons, toolbars, help strings, and the AUX: device.

- AutoCAD includes the *custom.cui* file for customizing menus independently of *acad.cui*.

. . . .

Replaced Commands

MenuLoad and **MenuUnload** commands became aliases for the CuiLoad and CuiUnload commands in AutoCAD 2006.

. . . .

MInsert

V. 2.5   Inserts an array of blocks as a single block (short for Multiple INSERT).

Command	Alias	Ctrl+	F-key	Alt+	Menu Bar	Dashboard
minsert	...	...	...	...	...	...

Command: minsert
Enter block name or [?]: *(Enter a name, type ?, or enter ~ to select a .dwg file.)*
Specify insertion point or [Scale/X/Y/Z/Rotate]: *(Pick a point, or enter an option.)*
Enter X scale factor, specify opposite corner, or [Corner/XYZ] <1>: *(Enter a value, pick a point, or enter an option.)*
Enter Y scale factor <use X scale factor>: *(Enter a value, or press Enter.)*
Specify rotation angle <0>: *(Enter a value, or press Enter.)*
Enter number of rows (---) <1>: *(Enter a value.)*
Enter number of columns (|||) <1>: *(Enter a value.)*
Enter distance between rows or specify unit cell (---): *(Enter a value.)*
Specify distance between columns (|||): *(Enter a value.)*

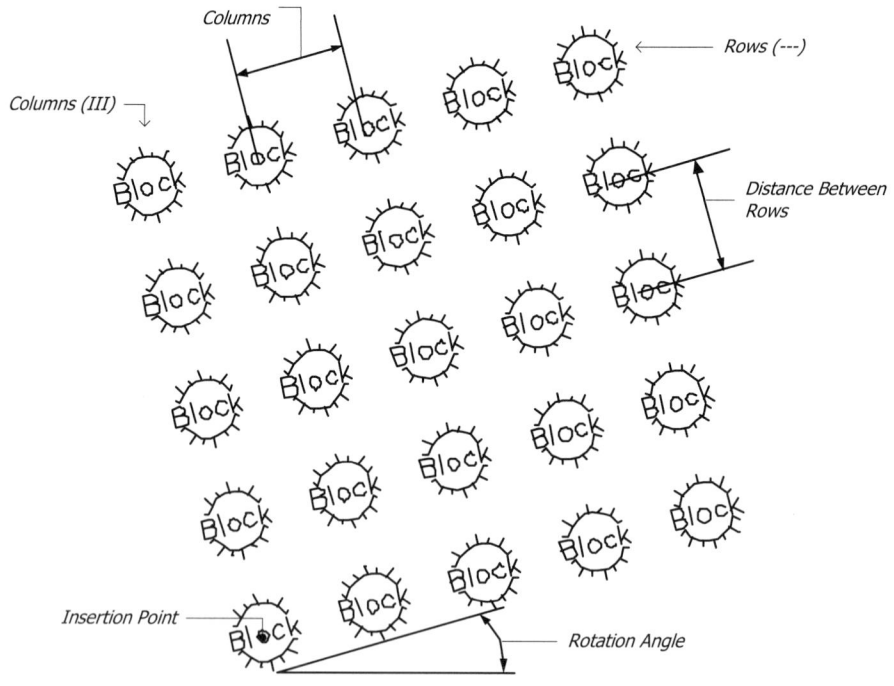

COMMAND LINE OPTIONS

Enter block name names the block to be inserted; the block must already exist in the drawing.

? lists the names of blocks stored in the drawing.

Specify insertion point specifies the x,y coordinates of the first block.

X scale factor indicates the x scale factor.

Specify opposite corner specifies a second point that indicates the x,y-scale factor.

Corner indicates the x and y scale factors by picking two points on the screen.

XYZ specifies x, y, and z scaling.

Specify rotation angle specifies the angle of the array.

Number of rows specifies the number of horizontal rows.

Number of columns specifies the number of vertical columns.

Distance between rows specifies the distance between rows.

Specify unit cell shows the cell distance by picking two points on the screen.

Distance between columns specifies the distance between columns.

RELATED COMMANDS

3dArray creates 3D rectangular and polar arrays.

Array creates 2D rectangular and polar arrays.

Block creates a block.

Insert inserts single blocks.

TIPS

- The array placed by this command is a single block.

- You *cannot* explode blocks created by this command.

- You may redefine blocks created by this command.

Mirror

V. 2.0   Creates a mirror copy of a group of objects in 2D space.

Command	Alias	Ctrl+	F-key	Alt+	Menu Bar	Dashboard
mirror	mi	...	...	MI	Modify ⌖Mirror	2D Draw

Command: mirror
Select objects: *(Pick one or more objects.)*
Select objects: *(Press Enter to end object selection.)*
Specify first point of mirror line: *(Pick a point.)*
Specify second point of mirror line: *(Pick another point.)*
Delete source objects? [Yes/No] <N>: *(Type Y or N.)*

Mirror Line ——— *Original Object* — *Mirrored Object*

COMMAND LINE OPTIONS

Select objects selects the objects to mirror.

First point specifies the starting point of the mirror line.

Second point specifies the endpoint of the mirror line.

Delete source objects deletes selected objects.

RELATED COMMANDS

Copy creates non-mirrored copies of objects.

Mirror3d mirrors objects in 3D-space.

RELATED SYSTEM VARIABLE

MirrText determines whether text is mirrored by this command. Since AutoCAD 2005, this variable defaults to 0 (text is not mirrored).

TIPS

■ This command is excellent for cutting drafting work by half for symmetrical objects. For doubly-symmetrical objects, use Mirror twice.

■ Although you can mirror a viewport in paper space, this does not mirror the model space objects inside the viewport.

■ Turn on Ortho or Polar mode to ensure that the mirror is perfectly horizontal or vertical.

■ The mirror line becomes a mirror plane in 3D; it is perpendicular to the x,y plane of the UCS containing the mirror line.

Mirror3d

<u>Rel.11</u>   Mirrors objects about a plane in 3D space.

Command	Alias	F-key	Alt+	Menu Bar	Dashboard
mirror3d	3dmirror	...	M3D	Modify	...
				⤷3D Operation	
				⤷3D Mirror	

Command: mirror3d
Select objects: *(Pick one or more objects.)*
Select objects: *(Press Enter to end object selection.)*
Specify first point of mirror plane (3 points) or
[Object/Last/Zaxis/View/XY/YZ/ZX/3points] <3points>: *(Pick a point, or enter an option.)*
Delete old objects? <N>: *(Type Y or N.)*

COMMAND LINE OPTIONS

Select objects selects the objects to be mirrored in space.

Specify first point specifies the first point of the mirror plane.

Object selects a circle, arc or 2D polyline segment as the mirror plane.

Last selects the last-picked mirror plane.

View specifies that the current view plane is the mirror plane.

XY specifies that the x,y plane is the mirror plane.

YZ specifies that the y,z plane is the mirror plane.

ZX specifies that the z,x plane is the mirror plane.

Zaxis defines the mirror plane by a point on the plane and the normal to the plane, i.e., the z axis.

3points defines three points on the mirror plane.

RELATED COMMANDS

Align translates and rotates objects in 2D planes and 3D space.

Mirror mirrors objects in 2D space.

Rotate3d rotates objects in 3D space.

RELATED SYSTEM VARIABLE

MirrText determines whether text is mirrored by the Mirror and Mirror3D commands:

0 — Text is not mirrored about the horizontal axis *(default)*.

1 — Text is mirrored.

MLeader

<u>**2008**</u>   Draws multiline leaders.

Command	Alias	Ctrl+	F-key	Alt+	Menu Bar	Dashboard
mleader	mld	...	...	NE	Dimension ⮩Multileader	Multileaders

Command: mleader
Specify leader arrowhead location or [leader Landing first/Content first/
Options]: *(Pick a point, or enter an option.)*
Specify leader landing location: *(Pick a point.)*
  Displays the MText toolbar; enter text, and then click OK.

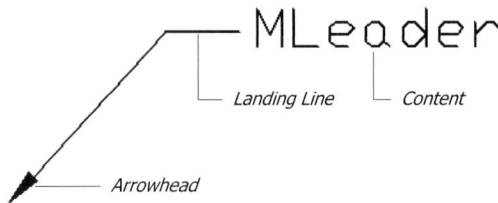

COMMAND LINE OPTIONS

  The option you pick here affects all subsequent uses of this command, until changed:

  Specify leader arrowhead location locates the start of the mleader.

  leader Landing first draws the mleader starting with its landing line.

  Content first draws the mleader's content first, such as its text or block

Options options

Enter an option [Leader type/leader lAnding/Content type/Maxpoints/First
angle/Second angle/eXit options]:

  Leader type specifies the type of leader line to use: straight, spline, or no leader line.

  leader lAnding toggles use of the horizontal landing line.

  Content type specifies the type of content: block or none.

  Maxpoints specifies the maximum number of vertices.

  First angle specifies the angle of the first point.

  Second angle specifies the second angle.

  eXit options returns to the earlier prompt line.

RELATED COMMANDS

  MLeaderAlign aligns selected multileaders to a specified justification.

  MLeaderCollect organizes selected multileaders (with blocks) as a group of blocks attached to a single leader.

  MLeaderEdit edits mleaders, adding and removing leader lines.

  MLeaderStyle displays the Multileader Style Manager dialog box for creating new styles and modifying existing ones.

Leader draws single-line leaders.

QLeader draws single-line leaders with additional options.

RELATED SYSTEM VARIABLE

CMleaderStyle reports the current multileader style.

TIPS

- Use the MLeaderEdit command to add leader lines to mleaders.

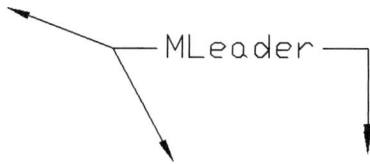

- Mleaders can be edited using grips:

MLeaderAlign

2008 Aligns two or more multiline leaders by their text.

Command	Alias	Ctrl+	F-key	Alt+	Menu Bar	Dashboard
mleaderalign	mla	...	...	...	...	Multileaders

Command: mleaderalign
Select multileaders: *(Pick one or more multileaders.)*
Select multileader to align to or [Options]: *(Pick the reference multileader or type O.)*
Specify direction: *(Move cursor to locate alignment.)*

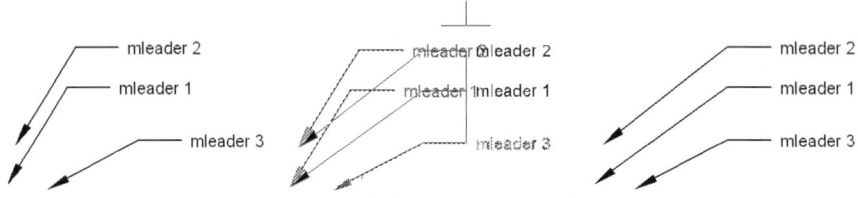

Left: *Three leaders before being aligned.*
Center: *The cursor alignubg the mleaders to the reference, mleader 3.*
Right: *The text of all mleaders aligned.*

COMMAND LINE OPTIONS

Select multileaders selects one or more multileaders to be aligned.

Select multileader to align to picks the reference mleader; the others will line up with this one.

Options displays further prompts.

Options options
Enter an option [Distribute/make leader segments Parallel/specify Spacing/ Use current]: *(Enter an option.)*

Distribute spaces the mleaders evenly between two selected points.

Make Leader Segments Parallel places content so that each of the last leader segments in the selected multileaders are parallel.

Specify Spacing specifies spacing between the landing lines of the selected multileaders.

Use Current uses the current spacing between multileader content.

RELATED COMMANDS

MLeader draws multiline leaders.

MLeaderEdit edits mleaders, adding and removing leader lines.

MLeaderStyle displays the Multileader Style Manager dialog box for creating new styles and modifying existing ones.

TIP

- This command aligns leaders by the text or block; the arrowheads remain in place.

MLeaderCollect

<u>2008</u> Collects the blocks of several multiline leaders into a single mleader with a row of blocks.

Command	Alias	Ctrl+	F-key	Alt+	Menu Bar	Dashboard
mleadercollect	mlc	...	...		Modify	Multileaders

Command: mleadercollect
Select multileaders: *(Pick one or more multiline leaders.)*
Specify collected multileader location or [Vertical/Horizontal/Wrap]: *(Pick a point or enter an option.)*

COMMAND LINE OPTIONS

Specify collected multileader location picks the upper-left corner of the collected blocks.

Vertical collects the blocks into a vertical stack.

Horizontal collects the blocks into a vertical stack.

Wrap specifies a width for wrapped collections; prompt, 'Specify wrap width or [Number]:'.

Number specifies the maximum number of blocks per row.

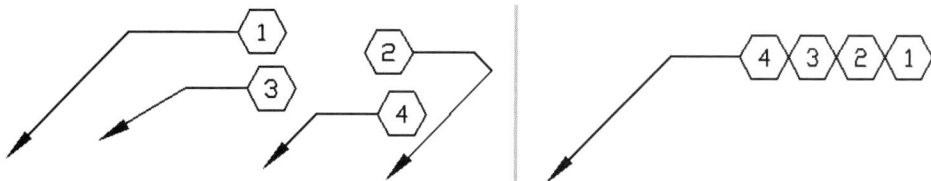

Left: Several multiline leaders with block content.
Right: Collected into a single mleader.

RELATED COMMANDS

MLeader draws multiline leaders.

MLeaderEdit edits mleaders, adding and removing leader lines.

MLeaderStyle displays the Multileader Style Manager dialog box for creating new styles and modifying existing ones.

TIPS

■ This command works only with mleaders that have blocks for their content. If you select mleaders with text, AutoCAD complains, '1 was filtered out, 0 total'.

■ This command is useful for combining multiple, drawing-crowding leaders into a single object.

MLeaderEdit

<u>**2008**</u> Add and removes leaders from multileader objects.

Command	Alias	Ctrl+	F-key	Alt+	Menu Bar	Dashboard
mleaderedit	mle	...	...	...	...	Multileaders

Command: mleaderedit
Select a multileader: *(Select one mleader.)*
Select an option [Add leader/Remove leader]: *(Enter an option.)*
Specify leader arrowhead location: *(Pick a point.)*
Specify leader arrowhead location: *(Press Enter to exit command.)*

COMMAND LINE OPTIONS

Add Leader adds one leader line to the selected multileader; the new leader line is added on the left or right side, depending on where the cursor is:

Move cursor left or right of text to place new leader line.

Remove Leader removes one or more leader lines; prompts you, 'Specify leaders to remove:'.

RELATED COMMANDS

MLeader draws multiline leaders.

MLeaderAlign aligns selected multileaders to a specified justification.

MLeaderCollect organizes selected multileaders (with blocks) as a group of blocks attached to a single leader.

MLeaderStyle displays the Multileader Style Manager dialog box for creating new styles and modifying existing ones.

TIPS

- The Remove option can removes all leader lines from an mleader; you can use the Add option to add leader lines to the remaining mleader text.

- Mleaders can be edited with grips editing and with the Properties palette.

- You can apply annotative scaling, colors, linetypes, lineweights, plot styles, and other properties to mleaders.

MLeaderStyle

2008 Creates and edits multiline leader styles.

Command	Alias	Ctrl+	F-key	Alt+	Menu Bar	Dashboard
mleaderstyle	mls	...	...	OL	Format	Multileaders
					↳Multileader Style	

Command: mleaderstyle

Displays dialog box:

DIALOG BOX OPTIONS

Styles lists the names of multiline styles in the drawing.

List modifies the style names listed under Styles:

- **All styles** lists all multiline style names stored in the current drawing.
- **Styles in use** lists only those multiline styles used by the drawing.

Set Current sets the selected style as the current multiline style.

New creates a new multiline style via Create New Multiline Style dialog box.

Modify modifies an existing multiline style; displays the Modify Multiline Style dialog box.

Delete erases the selected style, if it is not in use or is not named Standard.

NEW DIALOG BOX

New Style Name names the new style.

Start With specifies an existing style as the template.

☐ **Annotative** toggles annotative scaling for multiline leaders.

Continue displays the Modify Multileader Style dialog box (detailed below).

Leader Format tab

Type specifies the shape of the leader line: straight, spline, or none.

Color specifies the color of the leader line.

Linetype specifies the linetype of the leader line.

Lineweight specifies the lineweight of the leader line.

Symbol selects the arrowhead symbol.

Size specifies the size of arrowhead.

Break Size specifies the gap size when leader lines are broken by the DimBreak command.

Leader Structure tab

☑ **Maximum Leader Points** specifies a maximum number of vertices on the leader line.

☐ **First Segment Angle** specifies the increment angle of the first vertex (angle between the first and second leader lines): 0, 15, 30, 45, 60 or 90 degrees.

☐ **Second Segment Angle** specifies the increment angle of a second vertex.

☑ **Automatically Include Landing** includes the horizontal landing line.

☑ **Set Landing Distance** specifies a fixed length for the landing line.

☐ **Annotative** toggles annotative scaling.

The following options are available only when Annotative is turned off:

○ **Scale Multileaders to Layout** scales the mleader based on the scale factor between model space and the current viewport.

⊙ **Specify Scale** specifies the scale factor.

Content tab

Left: Options for mleader styles with text.
Center : Options for mleader styles with blocks.
Right: No options for mleader styles with no content.

Multileader Type chooses between text, blocks, or no content.

Text options

Default Text specifies default text for the mleader. Click ... to access the mtext editor.

Text Style selects the text style.

Text Angle specifies the angle of the multileader text: horizontal, as inserted, or right-reading.

Text Color specifies the color of the multileader text.

Text Height specifies the height of the text.

☐ **Always Left Justify** forces text always to be left justified.

☐ **Frame Text** places a box around the text.

Left Attachment specifies where the landing line is attached to the multileader text:

Right Attachment specifies where the landing line is attached to the multileader text; see above.

Landing Gap specifies the distance between the landing line and the text.

Block options

Source Block selects the predefined block used in places of multileader text:

Attachment specifies where the blocks are attached: extents, insertion point, or center point of the block.

Color specifies the color of the block content.

RELATED COMMANDS

MLeader draws multiline leaders.

MLeaderAlign aligns selected multileaders to a specified justification.

MLeaderCollect organizes selected multileaders (with blocks) as a group of blocks attached to a single leader.

MLeaderEdit edits mleaders, adding and removing leader lines.

RELATED SYSTEM VARIABLE

CMleaderStyle reports the current multileader style.

TIPS

- Set a multileader style before drawing mleaders.

- You can apply mleader styles from the Multileader and Styles toolbars: select a mleader, and then choose a style name from one of the droplists.

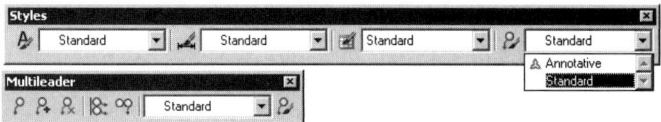

- Multileader styles can be overridden by the Properties palette.

- Changes to the multileader style retroactively updates existing mleaders.

MlEdit

Rel.13 Edits multilines (*short for MultiLine EDITor*).

Command	Alias	Ctrl+	F-key	Alt+	Menu Bar	Dashboard
mledit	...	...	...	MO⤷M	Modify	...
					⤷Object	
					⤷Multiline	

-mledit

Command: mledit

Displays dialog box:

MULTILINE EDIT TOOLS DIALOG BOX

Closed Cross closes the intersection of two multilines.

Open Cross opens the intersection of two multilines.

Merged Cross merges a pair of multilines: opens exterior lines; closes interior lines.

Closed Tee closes T-intersections.

Open Tee opens T-intersections.

Merged Tee merges T-intersection by opening exterior lines and closing interior lines.

Corner Joint creates corner joints with pairs of intersecting multilines.

Add Vertex adds vertcies (*joints*) to multiline segments.

Delete Vertex removes vertices from multiline segments.

Cut Single places gaps in a single line of multilines.

Cut All places gaps in all lines of multilines.

Weld All removes gaps from multilines.

-MLEDIT Command

Command: -mledit

Enter mline editing option [CC/OC/MC/CT/OT/MT/CJ/AV/DV/CS/CA/WA]:
(Enter an option.)

COMMAND LINE OPTIONS

AV adds vertices.

DV deletes vertices.

CC closes crossings.

OC opens crossings.

MC merges crossings.

CT closes tees.

OT opens tees.

MT merges tees.

CJ creates corner joints.

CS cuts a single line.

CA cuts all lines.

WA welds all lines.

U undoes the most-recent multiline edit.

RELATED COMMANDS

MLine draws up to 16 parallel lines.

MlStyle defines the properties of a multiline.

RELATED SYSTEM VARIABLES

CMlJust specifies the current multiline justification mode.

CMlScale specifies the current multiline scale factor (default = 1.0).

CMlStyle specifies the current multiline style name (default = " ").

RELATED FILE

***.mln** is the multiline style definition file.

TIPS

- Use the Cut All option to open up a gap before placing door and window symbols in a multiline wall.

- Use the Weld All option to close up a gap after removing the door or window symbol in a multiline.

- Use the Stretch command to move a door or window symbol in a multiline wall.

- When you open a gap in a multiline, AutoCAD does not cap the sides of the gap. You may need to add the end caps with the Line command.

MLine

<u>Rel.13</u>   Draws up to 16 parallel lines (short for Multiple LINE).

Command	Alias	Ctrl+	F-key	Alt+	Menu Bar	Dashboard
mline	ml	...	...	DU	Draw ⮡ Multiline	...

Command: mline
Current settings: Justification = Top, Scale = 1.00, Style = STANDARD
Specify start point or [Justification/Scale/STyle]: *(Pick a point, or enter an option.)*
Specify next point: *(Pick a point.)*
Specify next point or [Undo]: *(Pick a point, or type U.)*
Specify next point or [Close/Undo]: *(Pick a point, or else enter an option.)*

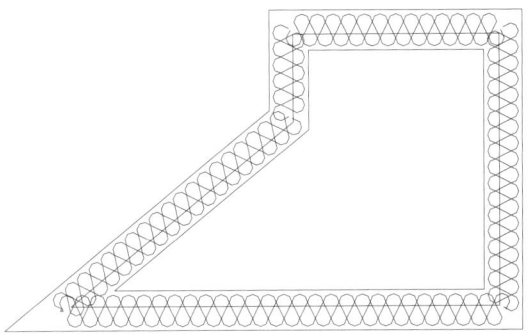

COMMAND LINE OPTIONS

Specify start point indicates the start of the multiline.

Specify next point indicates the next vertex.

Undo removes the most recently-added segment.

Close closes the multiline to its start point.

Justification options
Enter justification type [Top/Zero/Bottom] <top>: *(Enter an option.)*

Top draws the top line of the multiline at the cursor; remainder of multiline is "below" the cursor.

Zero draws the center *(zero offset point)* of the multiline at the cursor.

Bottom draws the bottom of the multiline at the cursor; the remainder of the multiline is "above" the cursor.

Scale options
Enter mline scale <1.00>: *(Enter a value.)*

Enter mline scale specifies the scale of the width of the multiline; see Tips for examples.

STyle options
Enter mline style name or [?]: *(Enter style name, or type ?.)*

Enter mline style name specifies the name of the multiline style.

? lists the names of the multiline styles defined in drawing.

RELATED COMMANDS

MlEdit edits multilines.

MlProp defines the properties of a multiline.

RELATED SYSTEM VARIABLES

CMlJust specifies the current multiline justification:

0 — Top (default).

1 — Middle.

2 — Bottom.

CMlScale specifies the current multiline scale factor (default = 1.0).

CMlStyle specifies the current multiline style name (default = " ").

RELATED FILE

*.mln is the multiline style definition file.

TIPS

- Examples of scale factors:

Scale	Meaning
1.0	Default scale factor.
2.0	Draws multiline twice as wide.
0.5	Draws multiline half as wide.
-1.0	Flips multiline.
0	Collapses multiline to a single line.

- Multiline styles are stored in *.mln* files in DXF-like format.

MlStyle

<u>Rel.13</u> Defines the characteristics of multilines (short for MultiLine STYLE).

Command	Alias	Ctrl+	F-key	Alt+	Menu Bar	Dashboard
mlstyle	...	...	...	OM	Format ↳Multiline Style	...

Command: mlstyle

Displays dialog box:

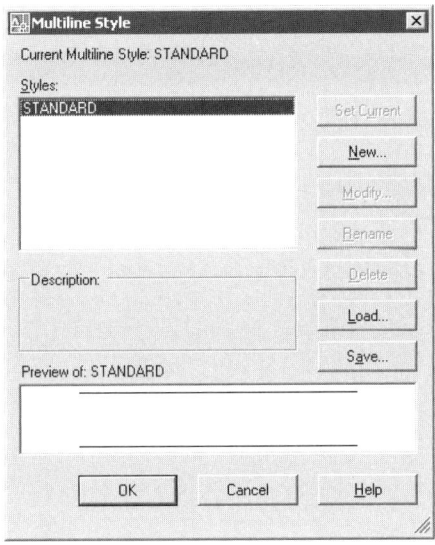

MUTLILINE STYLE DIALOG BOX

Set Current sets the selected multiline style name as the working style.

New creates new multiline styles; displays the Create New Multiline Style dialog box.

Modify changes the properties of existing multiline styles.

Rename changes the name of the selected multiline style.

Delete removes the multiline style from the Styles list.

Load loads styles from *.mln* files; displays the Load Multiline Styles dialog box.

Save saves a multiline style or renames a style; displays Save Multiline Style dialog box.

CREATE NEW MULTILINE STYLE DIALOG BOX

New Style Name specifies the name of the new multiline style. *Warning!* The name must contain no spaces.

Start With copies the multiline style from an existing style.

Continue displays the New Multiline Style dialog box.

NEW/MODIFY MULTILINE STYLE DIALOG BOX

Caps options

☐ **Line** draws a straight line start and/or end cap.

☐ **Outer Arc** draws an arc to cap the outermost pair of lines.

☐ **Inner Arcs** draws an arc to cap all inner pairs of lines.

Angle specifies the angle for straight line caps.

Fill options

Fill Color lists common colors, and displays the Select Color dialog box.

☐ **Display Joints** toggles the display of joints (miters) at vertices; affects all multiline segments.

Elements options

Add adds an element (line).

Delete deletes an element.

Offset specifies the distance from origin to element; the origin is often the center line.

Color specifies the element color; displays Select Color dialog box.

Linetype specifies the element linetype; displays Select Linetype dialog box.

LOAD MULTILINE STYLES DIALOG BOX

File selects an *.mln* multiline definition file.

RELATED COMMANDS

MlEdit edits multilines.

MLine draws up to 16 parallel lines.

RELATED SYSTEM VARIABLES

CMlJust specifies the current multiline justification.

CMlScale specifies the current multiline scale factor.

CMlStyle specifies the current multiline style name.

RELATED FILE

acad.mln is a multiline style definition file.

TIPS

- Use the MlEdit command to create (or close up) gaps for door and window symbols in multiline walls.

- The multiline scale factor has the following effect on the look of a multiline:

Scale	Meaning
1.0	Specifies the default scale factor (normal size).
0.5	Draws multiline half as wide.
2.0	Draws multiline twice as wide, not twice as long.
-1.0	Flips multiline about its origin.
0.0	Collapses multiline to a single line.

- The *.mln* file describes multiline styles in a DXF-like format.

- You cannot change the properties of the Standard style.

- Once multilines are placed in drawings, you cannot change their properties or their style. Once drawn, that's it! The workaround is to erase, change styles, and then reapply the MLine command.

Model

2000   Switches to Model tab.

Command	Alias	Ctrl+	F-key	Alt+	Menu Bar	Dashboard
model	...	...	...	...	...	...

Command: model
Switches to the model tab.

COMMAND LINE OPTIONS
None.

RELATED COMMANDS
Layout creates layouts.

MSpace switches to model space.

RELATED SYSTEM VARIABLE
Tilemode switches between Model and Layout tabs.

TIPS
- This command automatically sets TileMode to 1.

- As an alternative to this command, you can select the Model tab:

...or click the button or icon on the status bar :

- The Model tab replaces the TILE button on the status bar of AutoCAD Release 13 and 14.

- *Historical Note:* The system variable is named "tile"mode, because model space can only display *tiled* viewports. (Paper space, or layout mode, can display overlapping viewports.) Turning off tiled-viewport mode meant AutoCAD was switching to paper space, where viewports no longer had to be tiled.

  Going back further, it was a graphic board manufacturer, Control Systems, that first figured out how to make AutoCAD display four tiled viewports at once. Autodesk added the feature to AutoCAD Release 10.

  All of which leads to a question I cannot answer: Why can't viewports be tiled in model space? Technical editor Bill Fane attempts an answer: "Because that's the way we've always done it!" Seriously, though...

  The idea is that the model resides in model space, and then is looked at through viewports in paper space. Because multi-view drawings can be created, scaled, and plotted more easily in paper space, there is no need for floating viewports in model space.

Move

V. 1.0 Moves one or more objects.

Command	Alias	Ctrl+	F-key	Alt+	Menu Bar	Dashboard
move	m	...	...	MV	Modify ⌖Move	2D Draw

Command: move
Select objects: *(Select one or more objects.)*
Select objects: *(Press Enter to end object selection.)*
Specify base point or [Displacement] <Displacement>: *(Pick a point.)*
Specify second point of displacement or <use first point as displacement>: *(Pick a point, or press Enter.)*

1. Select Object(s)

2. Specify Base Point

3. Specify Second Point of Displacement

Intersection: 3'-0 7/16" < 0d0'

COMMAND LINE OPTIONS

Select objects selects the objects to copy.

Specify base point indicates the starting point for the move.

Displacement specifies relative x,y,z displacement when you press Enter at the next prompt.

Specify second point of displacement indicates the distance to move; you can use absolute (x,y), relative (@x,y), and polar (x<angle) coordinates.

RELATED COMMANDS

Copy copies the selected objects.

MlEdit moves the vertices of a multiline.

PEdit moves the vertices of a polyline.

TIPS

- When you press Enter at the 'Specify Second Point of Displacement' prompt, AutoCAD uses the first point as the "displacement."

  For example, when you enter **4,3** at the 'Specify base point' prompt, and press Enter at the second prompt, AutoCAD moves the objects 4 units in the x direction and 3 units in the y.

- You can move objects without using the Move command: select object, and then drag it (without grabbing a grip). After a brief pause, the object begins to move.

↷ ⋅ MRedo

Reverses the effect of the Undo command (short for Multiple REDO).

Command	Alias	Ctrl+	F-key	Alt+	Menu Bar	Dashboard
mredo	...	Y	...	...	...	...

Command: mredo
Enter number of actions or [All/Last]: *(Enter an option.)*

COMMAND LINE OPTIONS
Enter number of actions redoes the specified number of steps.

All redoes all commands undone.

Last redoes the last command.

RELATED COMMANDS
Redo redoes a single undo.

U undoes a single command.

Undo undoes one or more commands.

TIPS
■ The MRedo button on the Standard Annotation toolbar lists the redoable actions:

■ This command allows you to undo several undoes, but does not allow you to skip over actions.

MSlide

<u>Ver.2.0</u> Saves the current viewport as a *.sld* slide file on disk (short for Make SLIDE).

Command	Alias	Ctrl+	F-key	Alt+	Menu Bar	Dashboard
mslide	...	...	...	...	...	...

Command: mslide

*Displays Create Slide File dialog box. Specify a file name, and then click **Save**.*

COMMAND LINE OPTIONS
None.

RELATED COMMANDS
Save saves the current drawing as a DWG-format drawing file.

SaveImg saves the current view as a TIFF, Targa, or GIF-format raster file.

VSlide displays an SLD-format slide file in AutoCAD.

RELATED FILES
.sld files store slides created by this command.

.slb files store libraries of slides.

RELATED AUTODESK PROGRAM
SlideLib.exe compiles a group of slides into an SLB-format slide library file.

TIPS
- You view slides with the VSlide command.

- The slide was a predecessor to viewing raster and vector images inside AutoCAD.

- Slide files are used to create the images in the (rarely used) image dialog boxes.

MSpace

<u>Rel.11</u> Switches the drawing from paper space to model space (short for Model SPACE).

Command	Alias	Ctrl+	F-key	Alt+	Menu Bar	Dashboard
mspace	ms	...	...	...	...	...

Command: mspace

In model tab, AutoCAD complains, '** Command not allowed in Model Tab **'.

In model space of a layout tab, AutoCAD complains, 'Already in model space.'

In a layout tab (paper space), AutoCAD switches to model space in layout mode, and highlights a viewport:

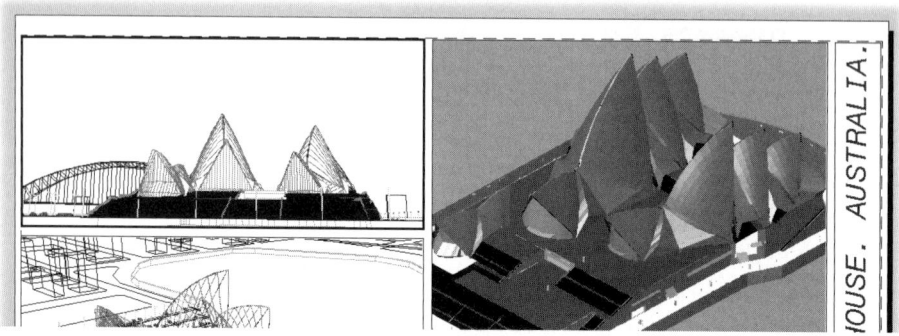

COMMAND LINE OPTIONS
None.

RELATED COMMANDS
PSpace switches from model space to paper space.

Model switches from layout mode to model mode.

Layout switches from model mode to layout mode.

RELATED SYSTEM VARIABLES
MaxActVp specifies the maximum number of viewports with visible objects; default=64.

TileMode specifies the current setting of tiled viewports.

TIPS
- To switch quickly between paper space and model space, click the Model and Paper icons on the status bar:

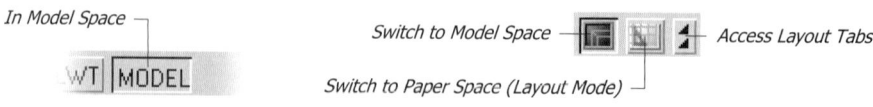

- AutoCAD clears the selection set when moving between paper space and model space.

- Note that there is a difference between model/layout *tabs* and the model/paper *button*.

- Double-click the viewport to switch from paper to model space; to return, double-click outside the viewport.

MTEdit

Rel.13 Edits mtext objects (short for Multiline Text EDITor).

Command	Alias	Ctrl+	F-key	Alt+	Menu Bar	Dashboard
mtedit	...	...	...	...	...	...

Command: mtedit
Select an MTEXT object: *(Pick an mtext object.)*

Displays Text Formatting toolbar; see MText command.

COMMAND LINE OPTION

Select an MTEXT object selects one paragraph text object for editing.

RELATED COMMANDS

DdEdit displays the text editor appropriate for the text object.

Properties changes the properties of an mtext object.

MtProp specifies the properties of an mtext object.

RELATED SYSTEM VARIABLE

MTextEd specifies the name of the external text editor to place and edit multiline text.

TIPS

- This command displays the same dialog box as the DdEdit command when an mtext object is selected.

- You can also invoke the mtext editor by double-clicking mtext.

A MText

<u>Rel.13</u> Creates multiline, or paragraph, text objects that fit the width defined by the boundary box (short for Multline TEXT).

Command	Alias	Ctrl+	F-key	Alt+	Menu Bar	Dashboard
mtext	t	...	...	DXXM	Draw	Text
	mt				⇘Text	2D Draw
					⇘Multiline Text	
-mtext	-t					

Command: mtext
Current text style: "Standard"   Text height: 0.20   Annotative: No
Specify first corner: *(Pick point 1.)*

AutoCAD displays the mtext bounding box:

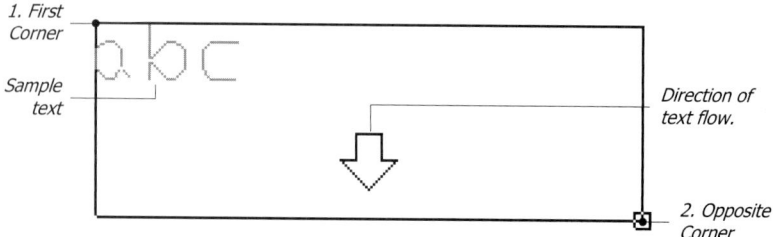

Specify opposite corner or [Height/Justify/Line spacing/Rotation/Style/Width/Columns]: *(Pick point 2, or enter an option.)*

See -MText command for description of the command-line options.

Displays toolbars and text editing area. Shown below is 2-column display, new to AutoCAD 2008:

TEXT FORMATTING TOOLBAR

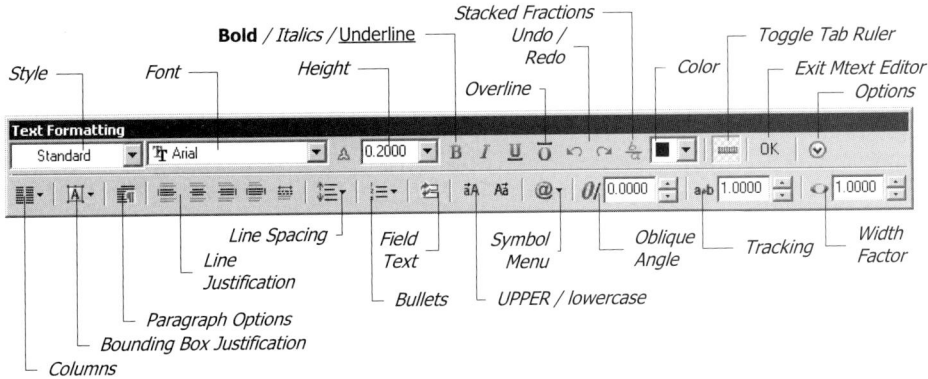

Style selects a predefined text style; see the Style command.

Font selects a TrueType (.*ttf*) or AutoCAD (.*shx*) font name (default=TXT).

Height specifies the height of the text in units (default = 0.2 units for Imperial drawings, 2.5 units for metric drawings).

Bold boldfaces the text, if permitted by the font.

Italic *italicizes* the text, if permitted by the font.

Underline underlines the text.

Overline draws a line over the selected text, the inverse of underline.

Undo undoes the last action.

Redo undoes the last undo.

Stack Fraction stacks a pair of characters separated by slash.

Color selects color for text; choose Other Color to display Select Color dialog box.

Ruler toggles the display of the tab and indent ruler.

OK closes the toolbar, and exits the MText command.

Options displays the shortcut menu of options.

Columns splits the bounding box into two or more static (fixed) or dynamic (as-needed) columns; choose Column Settings for the dialog box, detailed later (*new to AutoCAD 2008*).

Justification aligns the text within the bounding box:

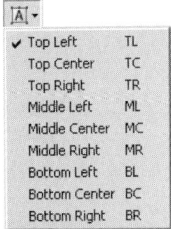

Paragraph displays the Paragraph dialog box, detailed later (*new to AutoCAD 2008*).

Alignment aligns the text horizontally against the left margin, centers it, aligns it to the right margin, forces all, but the last line to fit between margins, or forces all lines in the paragraph to fit.

Line Spacing places extra space between lines of text; click More for the Paragraph dialog box (*new to AutoCAD 2008*).

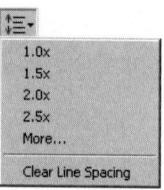

Bullets automatically applies bullet symbols, numbers (1., 2., 3., etc.), or letters (A., B., C., etc.) to selected text.

Field inserts field text from the Field dialog box; see the Field command.

Upper/lowercase changes selected text to all upper or all lowercase.

Symbol Menu displays a menu of symbols, detailed later.

Oblique Angle slants the text.

Tracking makes the spacing between characters tighter or looser.

Width Factor makes characters wider or narrower.

Ruler

Paragraph (formerly Indents and Tabs) displays Paragraph dialog box, detailed later (*new to AutoCAD 2008*).

Set MText Width displays Set MText Width dialog box, detailed later.

Set MText Height displays Set MText Height dialog box, detailed later (*new to AutoCAD 2008*).

PARAGRAPH DIALOG BOX

Formerly the Indents and Tabs dialog box; new to AutoCAD 2008:

Tab options
- ⊙ **Left** aligns text to the right of the tab position.
- ○ **Center** centers text about the tab position.
- ○ **Right** aligns text to the left of the tab position.
- ○ **Decimal** aligns text by the decimal marker.

Add adds tabs at the indicated distance.

Remove removes selected tabs.

Specify Decimal Style selects period, comma, or space.

Left Indent options
First Line specifies distance from left margin to the start of first line of text in a paragraph.

Hanging specifies distance from left margin to all other lines of text in the paragraph.

Right Indent options
Right specifies the distance from the right margin to the end of all text in the paragraph.

Paragraph Alignment options
- ⊙ **Left** aligns text against the left margin.
- ○ **Center** centers text between the two margins.
- ○ **Right** aligns text against the right margin.
- ○ **Justified** aligns text between both margins, except for the last line in the paragraph.
- ○ **Distributed** forces text to fit between margins, including the last line; see figure below
(new to AutoCAD 2008):

The Illustrated AutoCAD Quick Reference	The Illustrated AutoCAD Quick Reference

Left: Justified text.
Right: Distributed text.

Paragraph Spacing options

☑ **Paragraph Spacing** toggles spacing between paragraphs.

Before specifies line spacing before the selected paragraph.

After specifies line spacing after the selected paragraph.

Paragraph Line Spacing options

☑ **Paragraph Line Spacing** toggles spacing between lines of text.

Line Spacing specifies the amount of space between lines:

- **Exactly** specifies the spacing in inches or other units.
- **At least** specifies the minimum spacing.
- **Multiple** specifies the spacing as a factor of the text height, such as 1x.

At specifies the distance in units or factors.

COLUMNS

New to AutoCAD 2008:

No Columns switches the bounding box to a single column.

Dynamic Columns creates columns as required by the amount of text:

- **Auto Height** stretches the column to fit the text, until you press Alt+Enter or enter sufficient text for AutoCAD automatically to create another column.
- **Manual Height** displays the Column Height marker at the bottom of the bounding box. You manually adjust the height; if too small for the text, AutoCAD adds another column.

Static Columns creates a specific number of columns, two or more:

- **2, 3, 4, 5, 6** splits the bounding box into 2, 3, or more columns.
- **More** displays the Column Settings dialog box.

Insert Column Break forces the following text to start at the top of the next column; press Alt+Enter as a keystroke shortcut.

Column Settings diaplays the Column Settings dialog box:

Column Type specifies the type of column:

⦿ **Dynamic Columns** creates columns as required by the amount of text:

   ⦿ **Auto Height** stretches the column to fit the text.

   O **Manual Height** displays the Column Height marker at the bottom of the bounding box.

O **Static Columns** creates a specific number of columns.

O **No Columns** switches the bounding box to a single column.

Column Number specifies the number of static columns; range is 2 to 99 columns.

Height specifies the height of the bounding box.

Width specifies the width parameters of columns:

   Column specifies the width of each column, margin to margin.

   Gutter sets the distance between columns; the default is 5 times the text height.

   Total reports the total width of the columns and gutter(s).

SET MTEXT WIDTH & HEIGHT DIALOG BOXES

Alternatively, you can change the boundary box's size by dragging its right and bottom borders.

Width changes the width of the mtext bounding box.

Height changes the height of the bounding box.

SHORTCUT MENU

Right-click text to display menu, or enter shortcut keystrokes:

Select All (Ctrl+A) selects all text in the bounding box.

Cut (Ctrl+X) cuts the selected text to the Clipboard.

Copy (Ctrl+C) copies selected text to the Clipboard.

Paste (Ctrl+V) inserts text from the Clipboard, retaining most of its formatting.

Paste Special pastes text without formatting (*new to AutoCAD 2008*):

| Paste without Character Formatting |
| Paste without Paragraph Formatting |
| Paste without Any Formatting |

- **Paste Without Character Formatting** causes the text to take on the paragraph formatting of the surrounding mtext; formatting from original source is lost.
- **Paste Without Paragraph Formatting** causes the text to loose the formatting of its source.
- **Paste Without Any Formatting** pastes the text using the current mtext format.

Insert Field (Ctrl+F) displays the Field dialog box; select a field, and then click OK. See the Field command.

Symbol inserts symbols in the text (click Other for Character Map dialog box):

Degrees	%%d
Plus/Minus	%%p
Diameter	%%c
Almost Equal	\U+2248
Angle	\U+2220
Boundary Line	\U+E100
Center Line	\U+2104
Delta	\U+0394
Electrical Phase	\U+0278
Flow Line	\U+E101
Identity	\U+2261
Initial Length	\U+E200
Monument Line	\U+E102
Not Equal	\U+2260
Ohm	\U+2126
Omega	\U+03A9
Property Line	\U+214A
Subscript 2	\U+2082
Squared	\U+00B2
Cubed	\U+00B3
Non-breaking Space	Ctrl+Shift+Space
Other...	

Import Text imports text from ASCII (*.txt*) and RTF format files; displays the Open dialog box. *Warning!* Text files are limited to 64KB; spreadsheets are limited to 72 rows.

Paragraph Alignment (see Alignment, detailed earlier).

Paragraph displays the Paragraph dialog box (detailed earlier).

Bullets and Lists displays a submenu of options:

Columns displays a submenu of options, detailed earlier.

Find and Replace (Ctrl+R) displays the Find and Replace dialog box.

Change Case changes the case of the text:

- **UPPERCASE** (Ctrl+Shift+U) changes selected characters to uppercase.
- **lowercase** (Ctrl+Shift+l) changes selected characters to lowercase.

AutoCAPS places text as uppercase as it is typed.

Character Set selects alternate character sets, such as Western, Hebrew, and Thai.

Combine Paragraphs combines selected text into a single paragraph.

Remove Formatting removes bold, italic, and underlined formatting from selected text.

Background Mask displays the Background Mask dialog box, detailed later.

Editor Settings

- ✓ **Show Toolbar** toggles the display of the Text Formatting toolbar.
- ✓ **Show Options** toggles the display of the toolbar's lower half.
- ✓ **Show Ruler** toggles the display of the tab ruler.
- ✓ **Opaque Background** toggles the opacity of the bounding box.

  Text Heightlight Color specifies the color of highlighted text when selected; displays the Select Color dialog box.

Learn about MTEXT opens the New Features Workshop or accesses online help.

Cancel exits the mtext editor, like pressing ESC.

BACKGROUND MASK DIALOG BOX

☑ **Use background mask** toggles the display of the background mask. Note that the background mask applies to the full width of the mtext block, rather than the width of text.

Border offset factor determines the distance that the "margin" extends beyond the text. The *factor* is based on the text height: 1.0 means there is no offset; 1.5 means the offset distance is 1.5 times the text height. Maximum value = 5.0; minimum value = 1.0.

Use drawing background color:

☑ Uses the background color (usually white or black).

☐ Uses the specified color; for more colors, choose Select Color.

FIND AND REPLACE DIALOG BOX

Find what specifies the text to search for. This dialog box searches only text within the bounding box; to search for text in the entire drawing, use the Find command.

Replace with specifies the replacement text; leave blank to search only.

Match whole word only

☑ Matches the entire word(s).

☐ Matches parts of the word(s).

Match case

☑ Matches the case of the words.

☐ Ignores the word case.

Find Next finds the next occurrence of the word(s).

Replace replaces the found occurrence.

Replace All replaces all occurrences.

Cancel dismisses the dialog box.

. . . .

-MTEXT Command
Command: -mtext
Current text style: STANDARD. Text height: 0.20   Annotation: No
Specify first corner: *(Pick a point.)*
Specify opposite corner or [Height/Justify/Line spacing/Rotation/Style/ Width/Columns]: *(Pick another point.)*
MText: *(Enter text.)*
MText: *(Press Enter to end the command.)*

COMMAND LINE OPTIONS
Height specifies the height of UPPERCASE text *(default = 0.2 units)*.

Justify specifies a justification mode.

Line spacing specifies the distance between lines of text. AutoCAD prompts, 'Enter line spacing type [At least/Exactly] <At least>:" and then "Enter line spacing factor or distance <1x>:'.

. . . .

Rotation specifies the rotation angle of the boundary box.

Style selects the text style for multiline text (default = STANDARD).

Width sets the width of the boundary box; a width of 0 eliminates the boundary box.

Columns specifies the number of text columns (*new to AutoCAD 2008*); AutoCAD prompts:

Enter column type [Dynamic/Static/No columns] <Dynamic>: s
Specify total width: <16.0000>:
Specify number of columns: <2>:
Specify gutter width: <1.0000>:
Specify column height: <2.0000>:

RELATED COMMANDS

DdEdit edits mtext; alternatively, double-click the mtext to edit.

Properties changes all aspects of mtext.

MtProp changes properties of multiline text.

PasteSpec pastes formatted text from the Clipboard into the drawing.

Style creates a named text style from a font file.

RELATED SYSTEM VARIABLES

CenterMt determines how the bounding box is stretched.

MTextEd names the external text editor for placing and editing multiline text.

MTJigString specifies the sample text displayed when the bounding box is created; default = "abc".

TIPS

■ Use the MTextEd system variable to define a different text editor. To return to the old interface, enter the following system variables:

> MTextEd=oldeditor
> MTExtFixed=1

■ The Import Text option is limited to ASCII (unformatted) and RTF (rich text format) text files no more than 64KB in size.

■ To import Word documents, copy the text to the Clipboard, and then press **Ctrl+V** in the mtext editor. Most, but not all, formatting is retained.

■ To import formatted text, copy text from the word processor to the Clipboard, and then use AutoCAD's PasteSpec command.

■ To link text in the drawing with a word processor, use the InsertObj command. When the word processor updates, the linked text is updated in the drawing.

■ The mtext editor displays the diameter symbol as "%%c" and nonbreaking spaces as hollow rectangles, but these are displayed correctly in drawings.

■ Stacked text can be created on either side of the following symbols:
  • **Carat** (^) stacks text as left-justified tolerance values.
  • **Forward slash** (/) stacks text as center-justified fractional-style values; the slash is converted to a horizontal bar.
  • **Pound sign** (#) stacks text with a tall diagonal bar.

  Use the stack tool a second time to unstack stacked text.

MtProp

Rel.13 Changes the properties of multiline text (short for Multline Text PROPerties; an undocumented command).

Command	Alias	Ctrl+	F-key	Alt+	Menu Bar	Dashboard
mtprop	...	...	...	...	...	...

Command: mtprop
Select an MText object: *(Pick an mtext object.)*
   Displays the Text Formatting toolbar; see the MText command.

COMMAND LINE OPTIONS
   *See the **MText** command.*

RELATED COMMANDS
   DdEdit edits multiline text.
   MText places multiline text.
   Style creates a named text style from a font file.

TIP
- You can also invoke the mtext editor by double-clicking mtext.

Multiple

<u>**V. 2.5**</u> Repeats commands automatically that do not repeat on their own.

Command	Alias	Ctrl+	F-key	Alt+	Menu Bar	Dashboard
multiple	...	...	...	...	...	...

Command: multiple
Enter command name to repeat: *(Enter command name.)*

This command can also be used as a command modifier:
Command: multiple circle
3P/2P/TTR/<Center point>: *(Pick a point, or enter an option.)*
Diameter/<Radius>: *(Enter an option.)*
circle 3P3P/2P/TTR/<Center point>: *(Pick a point, or enter an option.)*
Diameter/<Radius>: *(Enter an option.)*
circle 3P3P/2P/TTR/<Center point>: *(Press Es to end command.)*

COMMAND LINE OPTIONS
Enter command name to repeat specifies the name of the command to repeat.

ESC stops the command from automatically repeating itself.

COMMAND INPUT OPTION
Spacebar repeats the previous command.

RELATED COMMANDS
Redo undoes an undo.

U undoes the previous command; undoes one multiple command at a time.

RELATED COMMAND MODIFIERS
' *(apostrophe)* allows the use of some commands within another command.

. *(period)* forces the use of undefined commands.

- *(dash)* forces the display of prompts on the command line for some commands.

+ *(plus)* prompts for the tab number of tabbed dialog boxes.

_ *(underscore)* uses the English command in international versions of AutoCAD.

(*(open parenthesis)* executes AutoLISP functions on the command line.

$(*(dollar and parenthesis)* executes Diesel functions on the command line.

RELATED SYSTEM VARIABLE
CopyMode toggles whether the Copy command repeats itself *(new to AutoCAD 2008)*.

TIPS
- Use the Multiple command to repeat commands that do not repeat on their own. This command does not cause options to repeat.

- *Warning!* Multiple U will undo all edits in the drawing.

- Multiple repeats the command name only; it does not repeat command options.

- Some commands automatically repeat, including Point and Donut.

MView

Rel.11 Creates and manipulates overlapping viewports (short for Make VIEWports).

Command	Alias	Ctrl+	F-key	Alt+	Menu Bar	Dashboard
mview	mv	...	...	...	...	...

Command: mview

In Model tab, AutoCAD complains, '** Command not allowed in Model Tab **'.

In a layout tab, AutoCAD prompts:

Specify corner of viewport or

[ON/OFF/Fit/Shadeplot/Lock/Object/Polygonal/Restore/LAyer/2/3/4]
<Fit>: *(Pick a point, or enter an option.)*

Specify opposite corner: *(Pick another point.)*

Regenerating drawing.

COMMAND LINE OPTIONS

Specify corner of viewport indicates the first point of a single viewport (default).

Fit creates a single viewport that fits the screen.

Shadeplot creates hidden-line or shaded views during plotting and printing.

Hideplot performs hidden-line removal during plotting and printing (*undocumented option*).

Lock locks the selected viewport's scale relative to model space.

Object converts a circle, closed polyline, ellipse, spline, or region into a viewport.

OFF turns off a viewport.

ON turns on a viewport.

Polygonal creates a multisided viewport of straight lines and arcs.

Restore restores a saved viewport configuration.

LAyer restores layer property overrides applied to viewports *(new to AutoCAD 2008)*; properties are reset to ByLayer:

Reset viewport layer property overrides back to global properties [Y/N]?: *(Type Y to reset changed properties to ByLayer.)*
Select viewports: *(Select one or more.)*
Select viewports: *(Press Enter to end viewport selection.)*

2 (Two Viewports) options
Enter viewport arrangement [Horizontal/Vertical] <Vertical>: *(Enter an option.)*
Specify first corner or [Fit] <Fit>: *(Pick a point, or enter an option.)*

Horizontal stacks two viewports.

Vertical places two viewports side-by-side (default).

3 (Three Viewports) options
[Horizontal/Vertical/Above/Below/Left/Right]<Right>: *(Enter an option.)*
Specify first corner or [Fit] <Fit>: *(Pick a point, or enter an option.)*

Horizontal stacks the three viewports.

Vertical places three side-by-side viewports.

Above places two viewports above the third.

Below places two viewports below the third.

Left places two viewports to the left of the third.

Right places two viewports to the right of the third *(default)*.

4 (Four Viewports) options
Specify first corner or [Fit] <Fit>: *(Pick a point, or enter an option.)*

Fit creates four identical viewports that fit the viewport.

First Point indicates the area of the four viewports (default).

Shadeplot options
Shade plot? [As displayed/Wireframe/Hidden/Visual style/Rendered] <As displayed>: *(Enter an option.)*

Left: Viewport set to "Wireframe" shadeplot.
Center: Viewport set to "Hidden" shadeplot.
Right: Viewport set to "Rendered" shadeplot.

As displayed plots the drawing as displayed.

Wireframe plots the drawing as a wireframe.

Hidden plots the drawing with hidden lines removed.

Visual style plots the drawing with specified visual style.

Rendered plots the drawing rendered.

Hideplot Options *(undocumented)*
Hidden line removal for plotting [ON/OFF]: on
Select objects: *(Select one or more viewports.)*

Hidden line removal for plotting toggles the removal of hidden lines for plotted output.

RELATED COMMANDS

Layout creates new layouts.

MSpace switches to model space.

PSpace switches to paper space before creating viewports.

RedrawAll redraws all viewports.

RegenAll regenerates all viewports.

VpLayer controls the visibility of layers in each viewport.

VPorts creates tiled viewports in model space.

Zoom zooms a viewport relative to paper space via the XP option.

RELATED SYSTEM VARIABLES

CvPort specifies the number of the current viewport.

MaxActVp controls the maximum number of active viewports; range is 2 to 64.

PsLtScale specifies linetype scaling in paper space.

TileMode controls the availability of overlapping viewports.

VpLayerOverride reports whether the current viewport has any overridden layers.

VpLayerOverrideMode toggles the plotting and display of overridden layers.

TIPS

- Drawings can have up to 32,767 viewports, but only a maximum of 64 that show content.

- Although the system variable MaxActVp limits the number of simultaneously-visible viewports, the Plot command plots all viewports.

- Snap, Grid, Hide, Shade, Ucs, and so on, can be set separately in each viewport.

- Press Ctrl+R to switch between viewports.

- Some of this command's options are also available from the Properties palette: select a viewport, right-click, and then select Properties from the shortcut menu.

- Any shape of viewports can be locked.
- The preset viewports created by the **MView** command have these shapes:

Fit Viewport

Fit   Creates a single viewport.

2 Viewports

Horizontal   Creates one viewport over another viewport.

Vertical   Creates one viewport beside another (default).

3 Viewports

Horizontal   Creates three viewports over each other.

Vertical   Creates three viewports side-by-side.

Above   Creates one viewport over two viewports.

Below   Creates one viewport below two viewports.

Left   Creates one viewport left of two viewports.

Right   Creates one viewport right of two viewports (default).

Four Viewports

4   Splits the current viewport into four viewports.

'MvSetup

<u>Rel. 11</u> Sets up a drawing quickly, complete with a predrawn border. Optionally, sets up multiple viewports, sets the scale, and aligns views in each viewport (short for Model View SETUP).

Command	Alias	Ctrl+	F-key	Alt+	Menu Bar	Dashboard
mvsetup	...	...	...	...	...	...

Command: mvsetup

When in model space:

Enable paper space? [No/Yes] <Y>: *(Type Y or N.)*

Command prompts in model tab (not paper space):

Enter units type [Scientific/Decimal/Engineering/Architectural/Metric]: *(Enter an option.)*

Enter the scale factor: *(Specify a distance.)*

Enter the paper width: *(Specify a distance.)*

Enter the paper height: *(Specify a distance.)*

Command prompts in layout mode (paper space):

Enter an option [Align/Create/Scale viewports/Options/Title block/Undo]: *(Enter an option.)*

COMMAND LINE OPTIONS

Align options

Pans the view to align a base point with another viewport.

Enter an option [Angled/Horizontal/Vertical alignment/Rotate view/Undo]: *(Enter an option.)*

Angled specifies the distance and angle from a base point to a second point.

Horizontal aligns views horizontally with a base point in another viewport.

Vertical alignment aligns views vertically with a base point in another viewport.

Rotate view rotates the view about a base point.

Undo undoes the last action.

Create options

Enter option [Delete objects/Create viewports/Undo] <Create>: *(Enter an option.)*

Delete objects erases existing viewports.

Create viewports creates viewports in these configurations:

Layout	Meaning
0	No layout.
1	Single viewport.
2	Standard engineering layout.
3	Array viewports along x and y axes.

Undo undoes the last action.

Scale Viewports options
Select the viewports to scale...
Select objects: *(Pick a viewport.)*
Select objects: *(Press Enter to end object selection.)*
Set the ratio of paper space units to model space units...
Enter the number of paper space units <1.0>: *(Enter a value.)*
Enter the number of model space units <1.0>: *(Enter a value.)*

Select objects selects one or more viewports.

Enter the number of paper space units scales the objects in the viewport with respect to drawing objects.

Enter the number of model space units scales the objects in the viewport with respect to drawing objects.

Options options
Enter an option [Layer/LImits/Units/Xref] <exit>: *(Enter an option.)*

Layer specifies the layer name for the title block.

Limits specifies whether to reset limits after title block insertion.

Units specifies inch or millimeter paper units.

Xref specifies whether title is inserted as a block or as an external reference.

Title Block options
Enter title block option [Delete objects/Origin/Undo/Insert] <Insert>: *(Enter an option.)*

Delete objects erases an existing title block from the drawing.

Origin relocates the origin.

Undo undoes the last action.

Insert displays the available title blocks.

RELATED SYSTEM VARIABLE

TileMode specifies the current setting of TileMode.

RELATED FILES

mvsetup.dfs is the default settings file for this command.

acadiso.dwg is a template drawing with ISO (international standards) defaults.

All *.dwt* template drawings can also be used as templates.

RELATED COMMANDS

LayoutWizard sets up the viewports via a "wizard."

TIPS

- When option **2 (Std. Engineering)** is selected at the Create option, the following views are created (counterclockwise from upper left):
  Top view.
  Isometric view.
  Front view.
  Right view.

- To create the title block, MvSetup searches the path specified by the AcadPrefix variable. If the appropriate drawing cannot be found, MvSetup creates the default border.

- MvSetup makes use the following predefined title blocks:

0:	None
1:	ISO A4 Size(mm)
2:	ISO A3 Size(mm)
3:	ISO A2 Size(mm)
4:	ISO A1 Size(mm)
5:	ISO A0 Size(mm)
6:	ANSI-V Size(in)
7:	ANSI-A Size(in)
8:	ANSI-B Size(in)
9:	ANSI-C Size(in)
10:	ANSI-D Size(in)
11:	ANSI-E Size(in)
12:	Arch/Engineering (24 x 36in)
13:	Generic D size Sheet (24 x 36in)

- The metric A0 size is similar to the Imperial E-size, while the metric A4 size is similar to A-size.

- This command provides the following preset scales (scale factor shown in parentheses):

Architectural Scales	Scientific Scales	Decimal Scales	Engineering Scales	Metric Scales
(480) 1/40"=1'	(4.0) 4 TIMES	(4.0) 4 TIMES	(120) 1"=10'	(5000) 1:5000
(240) 1/20"=1'	(2.0) 2 TIMES	(2.0) 2 TIMES	(240) 1"=20'	(2000) 1:2000
(192) 1/16"=1'	(1.0) FULL	(1.0) FULL	(360) 1"=30'	(1000) 1:1000
(96) 1/8"=1'	(0.5) HALF	(0.5) HALF	(480) 1"=40'	(500) 1:500
(48) 1/4"=1'	(0.25) QUARTER	(0.25) QUARTER	(600) 1"=50'	(200) 1:200
(24) 1/2"=1'			(720) 1"=60'	(100) 1:100
(16) 3/4"=1'			(960) 1"=80'	(75) 1:75
(12) 1"=1'			(1200) 1"=100'	(50) 1:50
(4) 3"=1'				(20) 1:20
(2) 6"=1'				(10) 1:10
(1) FULL				(5) 1:5
				(1) FULL

- You can add your own title block with the Add option. Before doing so, create the title block as an AutoCAD drawing.

NetLoad

2005 Loads *.dll* files written with Microsoft's .Net programming interface.

Command	Alias	Ctrl+	F-key	Alt+	Menu Bar	Dashboard
netload	...	...	...	...	...	...

Command: netload

Displays the Choose .NET Assembly dialog box.

Select a .dll file, and then click Open.

COMMAND LINE OPTIONS

None.

TIP

- Some parts of AutoCAD written in .Net include the Layer dialog box and the migration utility.

New

Rel.12 Starts new drawings from template drawings, from scratch, or through step-by-step drawing setup "wizards."

Command	Alias	Ctrl+	F-key	Alt+	Menu Bar	Dashboard
new	...	N	...	FN	File	...
					⮑New	

Command: new

AutoCAD displays one of three interfaces, depending on the settings of the FileDia and Startup variables.

FileDia	Startup	New
1	1	Displays Startup wizard.
1	0	Displays Select Template dialog box.
0	1 or 0	Prompts for *.dwt* file at command line.

SELECT TEMPLATE DIALOG BOX

Open creates new drawings based on the selected *.dwt* template file.

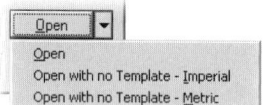

Open with no template - Imperial creates new drawings with default Imperial values.

Open with no template - Metric creates new drawings with default metric values.

STARTUP DIALOG BOX

The Startup wizard is displayed when Startup is set to 1 and when AutoCAD first starts.

The Create New Drawing wizard is similar, and is displayed when subsequent new drawings are opened.

 Open a Drawing page

(This page is not in the Create New Drawing wizard.)

Browse displays the Select File dialog box; see the Open command.

Start from Scratch page

⊙ **Imperial** creates a new drawing with AutoCAD's default values and InsUnits=1 (inches).

○ **Metric** creates a new drawing with AutoCAD's default values and InsUnits=4 (millimeters).

Use a Template page

Select a Template creates a new drawing based on the selected *.dwt* template file.
Browse displays the Select a Template File dialog box.

 Use a Wizard page

Select a Wizard

- **Advanced Setup** sets up a new drawing in several steps.
- **Quick Setup** sets up a new drawing in two steps.

QUICK SETUP WIZARD

This wizard can also be started by the undocumented QkSetupWiz command.

○ **Decimal** displays units in decimal (or "metric") notation (default): 123.5000.

○ **Engineering** displays units in feet and decimal inches: 10'-3.5000".

○ **Architectural** displays units in feet, inches, and fractional inches: 10' 3-1/2".

○ **Fractional** displays units in inches and fractions: 123 1/2.

○ **Scientific** displays units in scientific notation: 1.235E+02.

Cancel cancels the wizard, and returns to the previous drawing.

Back moves back one step.

Next moves forward one step.

Area page

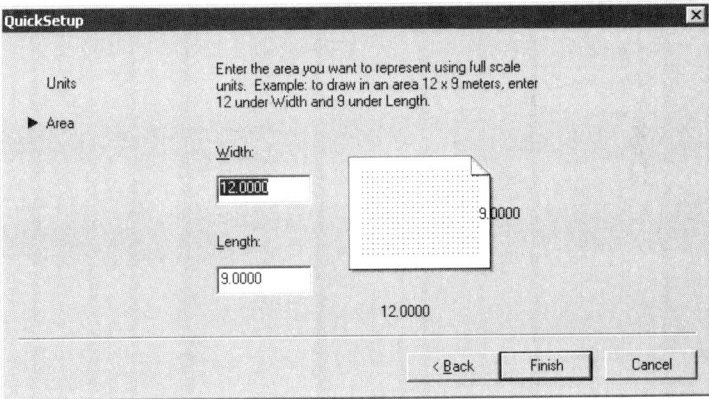

Width specifies the width of the drawing in real-world (not scaled) units; default = 12 units.

Length specifies the length or depth of the drawing in real-world units; default = 9 units.

ADVANCED SETUP WIZARD

This wizard can also be started by the undocumented AdvSetupWiz command.

⊙ **Decimal** displays units in decimal (or "metric") notation (default): 123.5000.

○ **Engineering** displays units in feet and decimal inches: 10'-3.5000".

○ **Architectural** displays units in feet, inches, and fractional inches: 10' 3-1/2".

○ **Fractional** displays units in inches and fractions: 123 1/2.

○ **Scientific** displays units in scientific notation: 1.235E+02.

○ **Precision** selects the precision of display up to 8 decimal places or 1/256.

Angle page

⊙ **Decimal Degrees** displays decimal degrees (default): 22.5000.

○ **Deg/Min/Sec** displays degrees, minutes, and seconds: 22 30.

○ **Grads** displays grads: 25g.

○ **Radians** displays radians: 25r.

○ **Surveyor** displays surveyor units: N 25d0'0" E.

Precision selects a precision up to 8 decimal places.

Angle Measure page

⊙ **East** specifies that zero degrees points East (default).

○ **North** specifies that zero degrees points North.

○ **West** specifies that zero degrees points West.

○ **South** specifies that zero degrees points South.

○ **Other** specifies any of the 360 degrees as zero degrees.

Angle Direction page

Counter-Clockwise measures positive angles counterclockwise from 0 degrees (default).

Clockwise measures positive angles clockwise from 0 degrees.

Area page

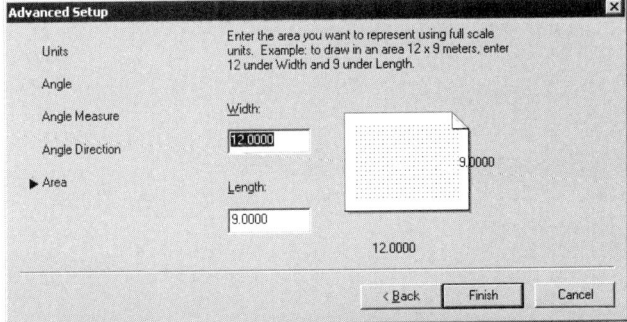

Width specifies the width of the drawing in real-world (not scaled) units; default = 12 units.

Length specifies the length or depth of the drawing in real-world units; default = 9 units.

COMMAND LINE SWITCHES

Switches used by Target field on Shortcut tab in the AutoCAD desktop icon's Properties dialog box:

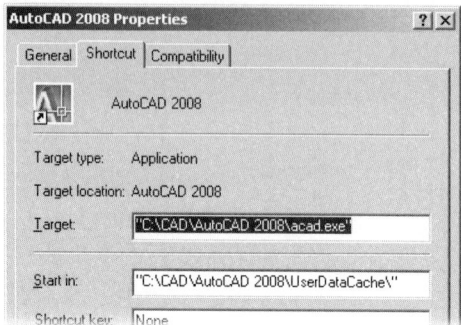

/b runs a script file after AutoCAD starts; uses the following format:

acad.exe "\acad 2008\drawing.dwg" /b "file name.scr"

/c specifies the path for alternative hardware configuration file; default = *acad2008.cfg*.

/layout specifies the layout to display *(undocumented)*.

/ld loads the specified ARx and DBx applications.

/nohardware prevents AutoCAD from using hardware-accelerated graphics, which can prevent AutoCAD from crashing in 3D modeling *(new to AutoCAD 2008)*.

/nologo suppresses the display of the AutoCAD logo screen.

/nossm prevents Sheet Set Manager palette from loading.

/p specifies a user-defined profile to customize AutoCAD's user interface.

/pl publishes a set of drawings defined by a *.dsd* file in the background.

/r restores the default pointing device.

/s specifies additional support folders; maximum is 15 folders, with each folder name separated by a semicolon.

/set specifies the *.dst* sheet set file to load.

/t specifies the *.dwt* template drawing to use.

/v specifies the named view to display upon startup of AutoCAD.

/w specifies the workspace to display first.

Command Line Options

When FileDia = 0, AutoCAD prompts you at the command line:

Command: new

Enter template file name or [. (for none)] *<default .dwt file path name>:*
(Enter the path and name of a .dwt , .dwg, or .dws file.)

Alternatively, enter the following options:

Enter accepts the default template drawing file.

. (period) eliminates use of a template; AutoCAD uses either *acad.dwt* or *acadiso.dwt*, depending on the setting of the MeasureInit system variable.

~ (tilde) forces the display of the Select Template dialog box.

RELATED COMMANDS

AdvSetupWiz starts the Advanced Setup wizard (an undocumented command).

QkSetupWiz starts the Quick Setup wizard (an undocumented command).

QNew starts a new drawing based on a predetermined template file.

SaveAs saves the drawing in *.dwg* or *.dwt* formats; creates template files.

RELATED SYSTEM VARIABLES

DbMod indicates whether the drawing has changed since being loaded.

DwgPrefix indicates the path to the drawing.

DwgName indicates the name of the current drawing.

FileDia displays prompts at the 'Command:' prompt.

InsUnits determines whether blocks are inserted in inches, millimeters, or other units.

Startup determines whether the dialog box or the wizard is displayed.

MeasureInit determines whether the units are imperial or metric.

RELATED FILES

wizard.ini names and describes template files.

***.dwt** are template files stored in *.dwg* format.

TIPS

- Until you name the drawing, AutoCAD names it *drawing1.dwg* (and subsquent drawings created in the same session as *drawing2.dwg, drawing3.dwg*, and so on).

- The default template drawing is the last one used, and not necessarily *acad.dwg*.

- Edit and save *.dwt* template drawings to change the defaults for new drawings.

- When you start a new drawing by pressing Ctrl+N, you will find that AutoCAD's behavior differs from that of Microsoft Office programs: these programs display new documents that take on the properties of the current document.

- The Startup system variable can no longer be set in the Options dialog box (*new to AutoCAD 2008*). The Today window was removed in AutoCAD 2004. It was replaced by the Communications Center.

NewSheetset

2005 Runs the New Sheet Set wizard.

Command	Alias	Ctrl+	F-key	Alt+	Menu Bar	Dashboard
newsheetset	...	...	...	TZN	Tools	...
					⌐Wizards	
					⌐New Sheet Set	

Command: newsheetset

Displays the Create Sheet Set dialog box.

CREATE SHEET SET DIALOG BOX

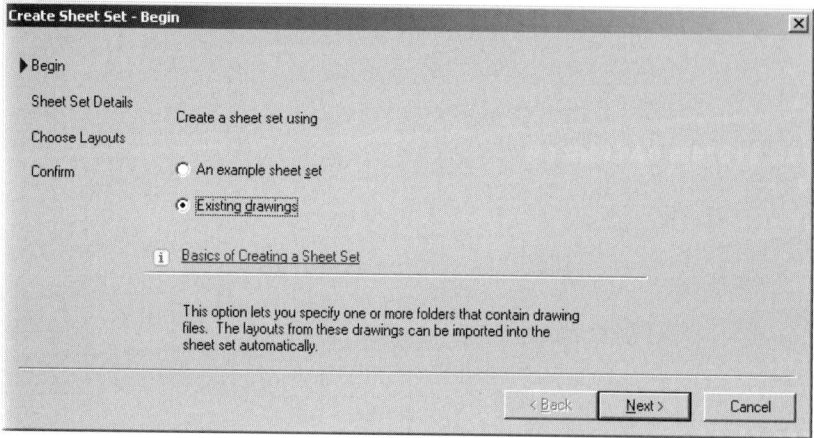

Create a sheet set using:

⊙ **An example sheet set** creates a new sheet set based on "templates."

○ **Existing drawings** selects one or more folders holding drawings, which are imported into the sheet set.

Click **Next** to continue.

Example Sheet Set

Sheet Set Example page

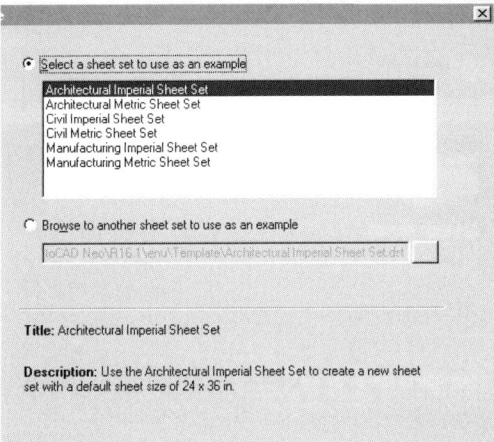

⊙ **Select a sheet set to use as an example** selects one of the sheet sets provided by Autodesk.

○ **Browse to another sheet set to use as an example** opens a *.dst* sheet set data file.

Opens the Browse for Sheet Set dialog box; select a *.dst* file, and then click Open.

Sheet Set Details page

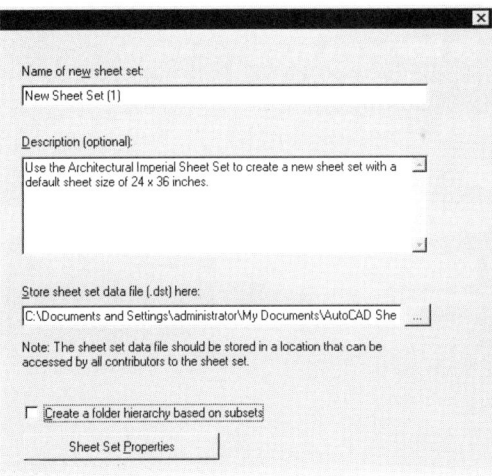

Name of new sheet set names the sheet set.

Description describes the sheet set.

Store sheet set data file here specifies the location of the *.dst* sheet set data file.

displays the Browse for Sheet Set Folder dialog box; select a folder, and then click Open.

Create a folder hierarchy based on subsets toggles creation of one or more folders to hold subset files.

Sheet Set Propeties displays the Sheet Set Properties dialog box.

Finishes with the Confirm page, illustrated below.

. . . .

Existing Drawings

Sheet Set Details page

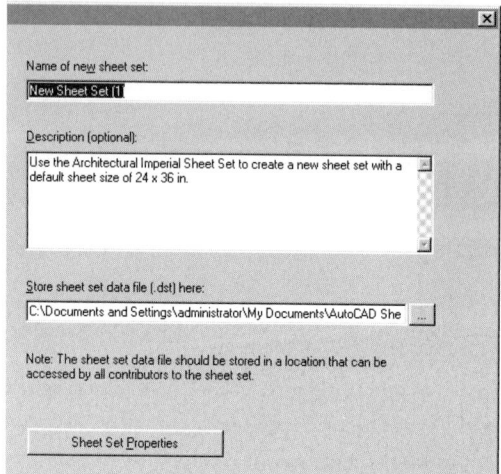

Name of new sheet set names the sheet set.

Description describes the sheet set.

Store sheet set data file here locates the sheet set data file.

▦ displays the Browse for Sheet Set Folder dialog box; select a folder, and then click Open.

Sheet Set Properties displays the Sheet Set Properties dialog box.

Choose Layouts page

Browse displays Browse for Folder dialog box; select a folder, and then click **OK**.

Import Options displays the Import Options dialog box.

Confirm page

Press CTRL+C to copy the settings, after selecting all text. (In a text editor, press Ctrl+V to paste the text in a document.)

Click **Back** to change and correct settings; click **Finish** to create the new sheet set.

SHEET SET PROPERTIES DIALOG BOX

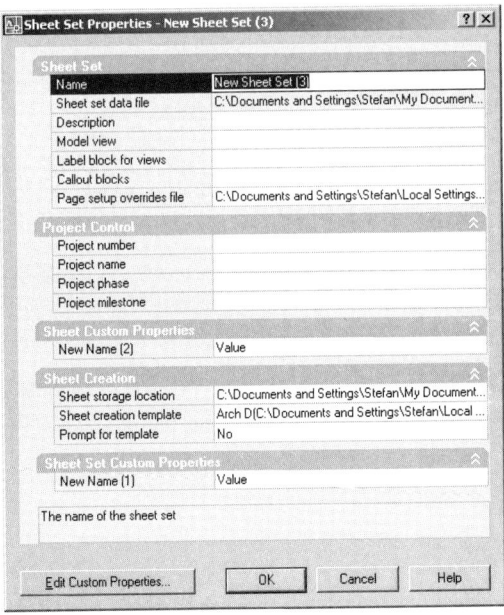

Edit Custom Properties displays the Custom Properties dialog box.

CUSTOM PROPERTIES DIALOG BOX

Add displays the Add Custom Property dialog box.

Delete removes the selected custom property without warning; click Cancel to undo erasure.

ADD CUSTOM PROPERTY DIALOG BOX

Name names the custom property.

Default value specifies the default value.

⦿ **Sheet set** indicates that the custom property belongs to the sheet set.

○ **Sheet** indicates that the custom property belongs to the sheet.

IMPORT OPTIONS DIALOG BOX

☑**Prefix sheet titles with file name** tags sheet set names with file names.

☐**Create subsheets based on folder structure** generates sheets based on folder names.

☐ **Ignore top level folder** ignores the topmost folder in sheet name generation.

RELATED COMMANDS

OpenSheetset opens existing sheet sets.

Sheetset displays the Sheet Set Manager palette.

. . . .

Moved Command

NewView is a shortcut to the New option of the View command; displays the New View dialog box; see the View command. NewView was introduced but undocumented in AutoCAD 2007, and then documented by Autodesk in AutoCAD 2008.

. . . .

'ObjectScale

2008 Adds and removes scale factors from annotatively-scaled objects.

Command	Alias	F-key	Alt+	Menu Bar	Dashboard
objectcsale	...	...	MOO⤵S	Modify	Annotation Scaling
				⤵Annotative Object Scale	
				⤵Add/Delete Scales	

-objectscale

Command: objectscale
Select annotative objects: *(Select one or more objects.)*
Select annotative objects: *(Press Enter.)*

Displays dialog box:

DIALOG BOX OPTIONS

Add adds displays the Add Scales to Object dialog box; detailed later.

Remove removes the selected scale factors.

ADD SCALES TO OBJECT DIALOG BOX

Select one or more scale factors, and then click OK.
To select more than one, hold down the Ctrl key.

Command: -objectscale
Select annotative objects: *(Select one or more objects.)*
Select annotative objects: *(Press Enter.)*
Enter an option [Add/Delete] <Add>: *(Type A or D.)*

COMMAND LINE OPTIONS

Add adds scales to the selected objects; AutoCAD prompts, 'Enter named scale to add or [?] '.

Remove removes scale factors; AutoCAD prompts, 'Enter named scale to delete or [?] '.

RELATED COMMANDS

AiObjectScaleAdd adds user-defined scales to annotative objects.

AiObjectScaleRemove removes scales from annotative objects.

AnnoReset resets the scale and location of selected annotative objects to the current setting.

AnnoUpdate updates annotative objects to match the properties of their respective styles.

ScaleListEdit edits the scale factors listed by this command.

RELATED SYSTEM VARIABLES

AnnoAllVisible toggles the display of annotative objects not at the current annotation scale.

AnnoAutoScale automatically adds scale factors to annotative objects when the annotation scale is changed.

CAnnoScale reports the name of the current annotation scale for the current space; a separate scale is used for model space and each paperspace viewport.

CAnnoScaleValue reports the value of the current annotation scale.

SaveFidelity toggles whether drawings are saved for earlier releases with visual fidelity.

SelectionAnnoDisplay toggles whether alternate scale representations are displayed faded when annotative objects are selected; amount of fading is specified by the XFadeCtl system variable.

TIPS

■ Use the Infobar to change annotation scales on the fly.

■ The ScaleListEdit command allows you to edit the scale factor used by drawings.

■ Each drawing contains its own list of scale factors used for annotative objects, viewport scaling, and plotting/publishing.

Offset

V. 2.5 Draws parallel lines, arcs, circles and polylines; repeats automatically until canceled.

Command	Alias	Ctrl+	F-key	Alt+	Menu Bar	Dashboard
offset	o	...	...	MS	Modify ↳ Offset	2D Draw

Command: offset
Specify offset distance or [Through/Erase/Layer] <1.0000>: *(Enter a distance or an option.)*
Select object to offset or [Exit/Undo]: *(Select an object.)*
Specify point on side to offset or[Exit/Multiple/Undo]: *(Pick a point.)*
Select object to offset or [Exit/Undo]: *(Select another object, or press Esc to end the command.)*

Offset objects, with original objects shown bold.

COMMAND LINE OPTIONS

Offset distance specifies the perpendicular distance to offset.

Through indicates the offset distance.

Erase erases the original object.

Layer specifies the layer on which to place offset objects.

Exit exits the command, as does Esc.

Undo undoes the last offset operation.

Multiple repeats the offset from the last object.

RELATED COMMANDS

Copy creates one or more copies of a group of objects.

MLine draws up to 16 parallel lines.

RELATED SYSTEM VARIABLES

OffsetDist specifies the current offset distance.

OffsetGapType determines how to close gaps created by offset polylines.

TIPS

■ Offsets of curved objects change their radii.

■ AutoCAD complains "Unable to offset" when curved objects are smaller than the offset distance.

OleConvert

Converts OLE objects, if possible (an undocumented command).

Command	Alias	Ctrl+	F-key	Alt+	Menu Bar	Dashboard
oleconvert	...	...	...	...	...	...

Select an OLE object before entering the command.
Command: oleconvert
Displays dialog box:

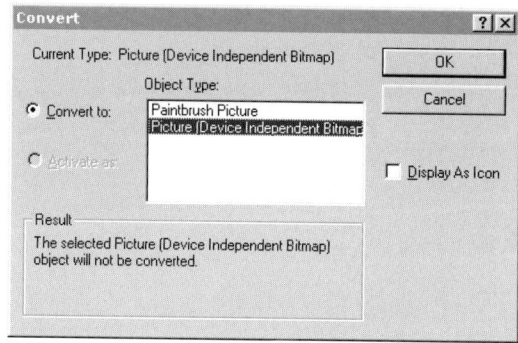

CONVERT DIALOG BOX

Object Type selects the type of object to convert to; list of types varies, depending on the object.

⊙ **Convert to** converts the OLE object to another type.

○ **Activate as** opens the OLE object with the source application.

☐ **Display as Icon** displays the OLE object as an icon.

RELATED COMMANDS

InsertObj places an OLE object in the drawing.

PasteSpec pastes objects from the Clipboard as linked objects in the drawing.

OleReset returns the OLE object to its original form.

RELATED SYSTEM VARIABLES

OleHide toggles the display of OLE objects.

OleQuality determines the plot quality of OLE objects.

OleStartup loads the source application for embedded OLE objects before plotting.

TIPS

- In many cases, this command is unable to convert the OLE objects.

- You can also access this command by: (1) selecting OLE object, (2) right-clicking, and (3) selecting OLE | Convert from the shortcut menu.

- OLE is short for "object linking and embedding." IBM invented this technology for its OS/2 operating system. OLE allows documents and objects from other applications to exist in "foreign" documents — documents that would not otherwise accept unknown objects.

OleLinks

Rel.13 Changes, updates, and cancels OLE links between the drawing and other Windows applications (short for Object Linking and Embedding LINKS).

Command	Alias	Ctrl+	F-key	Alt+	Menu Bar	Dashboard
olelinks	...	...	...	EO	Edit	...
					⌖OLE Links	

Command: olelinks

When no OLE links are in the drawing, the command does nothing.

When at least one OLE object is in the drawing, displays dialog box:

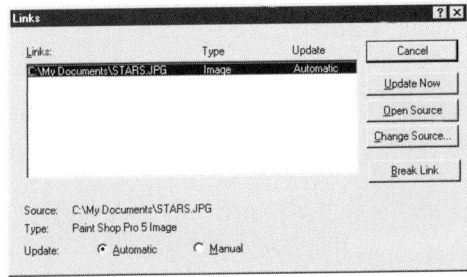

LINKS DIALOG BOX

Links displays a list of linked objects: source filename, type of file, and update mode — automatic or manual.

Update selects either automatic or manual updates.

Update Now updates selected links.

Open Source starts the source application program.

Break Link cancels the OLE link; keeps the object in place.

Change Source displays the Change Source dialog box.

RELATED COMMANDS

InsertObj places an OLE object in the drawing.

PasteSpec pastes objects from the Clipboard as linked objects in the drawing.

OleScale specifies the size, scale, and properties of selected OLE objects.

RELATED SYSTEM VARIABLES

MsOleScale specifies the scale of OLE objects in model space.

OleHide toggles the display of OLE objects.

OleQuality determines the plot quality of OLE objects.

OleStartup loads the source applications for embedded OLE objects before plotting.

RELATED WINDOWS COMMANDS

Edit | Copy copies objects from the source application to the Clipboard.

File | Update updates the linked object in the source application.

OleOpen, OleReset

2000 Opens OLE objects in their source applications; resets OLE objects (undocumented commands).

Command	Alias	Ctrl+	F-key	Alt+	Menu Bar	Dashboard
oleopen	...	...	...	...	...	...
olereset	...	...	...	...	...	...

Select an OLE object before entering the command.

Command: oleopen

Opens the OLE object in its source application. For example, selecting a text document opens it in the default word processor.

Command: olereset

Returns the OLE object to its original form.

RELATED COMMANDS

InsertObj places an OLE object in the drawing.

PasteSpec pastes objects from the Clipboard as linked objects in the drawing.

RELATED SYSTEM VARIABLES

MsOleScale specifies the scale of OLE objects in model space.

OleHide toggles the display of OLE objects.

OleQuality determines the plot quality of OLE objects.

OleStartup loads the source applications for embedded OLE objects before plotting.

TIPS

- You can also access these commands by: (1) selecting OLE objects, (2) right-clicking, and (3) selecting OLE | Open from the shortcut menu (or selecting OLE | Reset).

- If you do not first select an OLE object, AutoCAD complains, 'Unable to find OLE object. Object must be selected before entering the OleScale command.'

OleScale

<u>2000</u>   Modifies the properties of OLE objects.

Command	Alias	Ctrl+	F-key	Alt+	Shortcut Menu	Dashboard
olescale	...	...	...	...	...	...

Command: olescale

Displays dialog box:

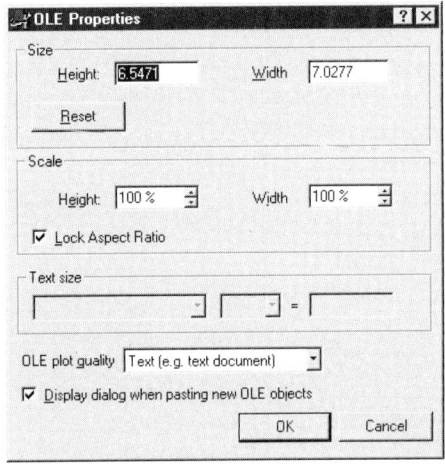

Alternatively, right-click a selected OLE object, and choose **OLE | Text Size** *from the shortcut menu.*
AutoCAD displays an abbreviated dialog box:

OLE PROPERTIES DIALOG BOX

The following options apply to text placed as OLE objects in drawings.

Size options

Height changes the height of the OLE object; displays the current height.

Width changes the width of the OLE object; displays the current width.

Reset resets the OLE object to its size when first inserted into the drawing.

Scale options

Height changes the height of the OLE object by a percentage of the original height.

Width changes the width of the OLE object by a percentage of the original width.

☑ **Lock Aspect Ratio** changes the Width size and ratio to match the Height, and vice versa.

Text Size options

Font displays the fonts used by the OLE object, if any.

Point Size displays the text height in point sizes; limited to the point sizes available for the selected font, if any (1 point = $^1/_{72}$ inch).

Text Height specifies the text height in drawing units.

OLE Plot Quality determines the quality of the pasted object when plotted:

> **Line Art** — text plotted as text; no colors or shading are preserved; some graphical images are not plotted, while others are plotted as monochrome images (black and white only, no shades of gray).
>
> **Text** — all text formatting preserved; text is plotted as graphics, which plots less cleanly than the Line Art setting; graphics are plotted less cleanly and at reduced colors than Graphics and Photograph settings.
>
> **Graphics** — graphics plotted at a reduced number of colors or fewer shades of gray (posterization); all text formatting is preserved; text is plotted as graphics, but more cleanly than with Text and Photograph settings.
>
> **Photograph** — graphics plotted at reduced resolution and colors; all text formatting is preserved; text is plotted as graphics.
>
> **High Quality** — graphics plotted at full resolution and colors.
>
> **Photograph** — all text formatting is preserved; text plots more cleanly than with Text and Graphics settings.

Display dialog when pasting new OLE object:

> ☑ Displays the OLE Properties dialog box automatically when inserting an OLE object.
>
> ☐ Does not display the dialog box.

RELATED SYSTEM VARIABLES

MsOleScale specifies the scale of OLE objects in model space.

OleHide toggles the display of OLE objects in the drawing and in plots.

OleQuality specifies the quality of displaying and plotting OLE objects.

OleStartup loads the source application of embedded OLE objects for plots.

RELATED COMMANDS

MsOleScale sets the scale for text in OLE objects.

InsertObj inserts an OLE object into the drawing.

OleLinks modifies the link between the object and its source.

PasteSpec allows you to paste an object with a link.

TIPS

- Change OleStartup to 1 to load the OLE source application, which may help improve the plot quality of OLE objects.

- The OLE Plot Quality list box determines the quality of the pasted object when plotted. I recommend the Line Art setting for text, unless the text contains shading and other graphical effects.

Oops

<u>V. 1.0</u> Restores the last-erased group of objects; restores objects removed by the **Block** and **-Block** commands.

Command	Alias	Ctrl+	F-key	Alt+	Menu Bar	Dashboard
oops	...	...	...	...	...	...

Command: oops

COMMAND LINE OPTIONS
None.

RELATED COMMANDS
Block, **Erase**, and **WBlock** can use this command to return erased objects.

U undoes the most recent command.

TIP
- This command restores only the most-recently erased object; use the Undo command to restore earlier objects.

Open

Rel.12 Loads one or more drawings into AutoCAD.

Command	Alias	Ctrl+	F-key	Alt+	Menu Bar	Dashboard
open	openurl	O	...	FO	File	...
					↳Open	

Command: open

Displays dialog box:

Places List — Files List — Toolbar — Preview

SELECT FILE DIALOG BOX

Look in selects the network drive, hard drive, or folder (subdirectory).

Preview displays a preview image of AutoCAD drawings.

☐ **Select initial view** selects a named view if the drawing has saved views. (After drawing is opened, AutoCAD displays the Select Initial View dialog box listing named views.)

M indicates a view created in model space.

P indicates a view created in paper space (layout mode).

File name specifies the name of the drawing.

Files of type specifies the type of file:

- **Drawing (*.dwg)** AutoCAD drawing file.
- **Standard (*.dws)** drawing standards file; see the Standards command.
- **DXF (*.dxf)** drawing interchange file; see the DxfIn command.
- **Drawing Template File (*.dwt)** template drawing file; see the **New** command.

Open opens the selected drawing file(s); to open more than one drawing at a time:

- In the files list, hold down the Shift key to select a contiguous range of files:

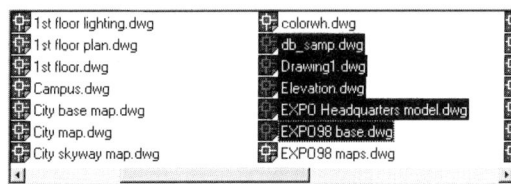

- Hold down the Ctrl key to select two or more non-contiguous files:

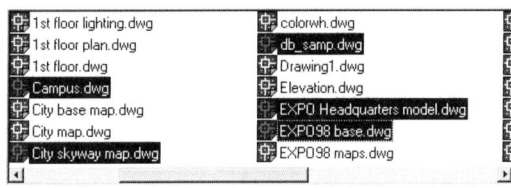

Cancel dismisses the dialog box without opening a file.

Open Button

Open opens the drawing.

Open as read-only loads the drawing, but you cannot save changes to the drawing except under another file name. AutoCAD displays "Read Only" on the title bar.

Partial Open loads selected layers or named views; displays the Partial Open dialog box; not available for *.dxf* and template files.

Partial Open Read-Only partially loads the drawing in read-only mode.

PARTIAL OPEN DIALOG BOX

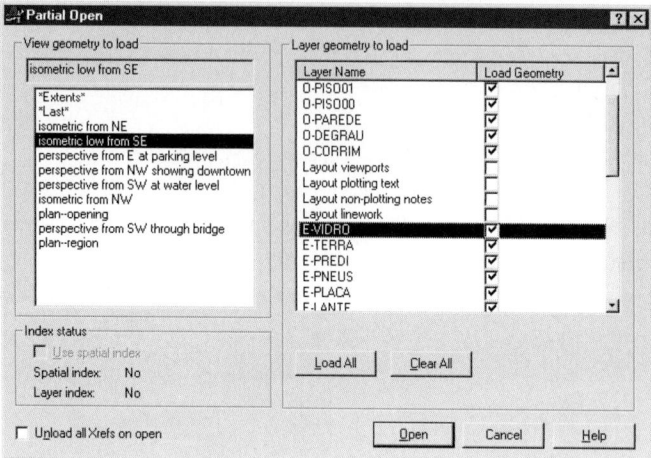

View geometry to load selects the model space views to load; paper space views are not available for partial loading.

Layer Geometry to Load

Load Geometry selects the layers to load.

Load All loads all layers.

Clear All deselects all layer names.

Index Status

Available only when the drawing was saved with spatial indices.

Use spatial index determines whether to use the spatial index for loading, if available.

Spatial Index indicates whether the drawing contains the spatial index.

Layer Index indicates whether the drawing contains the layer index.

☐ **Unload all xrefs on open** loads externally-referenced drawings when opening the drawing.

Open opens the drawing, and then partially loads the geometry.

DIALOG BOX TOOLBAR

Back returns to the previous folder (keyboard shortcut Alt+1).

Up moves up one level to the next folder or drive (Alt+2).

Search the Web displays the Browse the Web window (Alt+3); see Browser command.

Delete removes the selected file(s); does not delete folders or drives (Del).

Create New Folder creates new folders (Alt+5).

Views provides display options:

- **List** displays the file and folder names only.
- **Details** displays file and folder names, type, size, and date.
- **Thumbnails** displays thumbnail images of *.dwg* files.
- **Preview** toggles display of the preview window.

Tools provides file-oriented tools:

- **Find** displays the Find dialog box for searching files.
- **Locate** searches for the file along AutoCAD's search paths.
- **Add/Modify FTP Locations** displays a dialog box for storing the logon names and passwords for FTP (file transfer protocol) sites.
- **Add Current Folder to Places** adds the selected folders to the Places sidebar.
- **Add to Favorites** adds the selected files and folders to the Favorites list.

PLACES LIST

History displays files opened by AutoCAD during the last four weeks.

My Documents displays files and folders in the *my documents* folder.

Favorites displays files and folders in the *favorites* folder.

Desktop displays the contents of the *desktop* folder.

FTP displays the FTP Locations list.

Buzzsaw.com goes to the www.autodesk.com/buzzsaw Web site.

(The Point A and RedSpark destinations were removed from AutoCAD 2004.)

SHORTCUT MENUS

Places List menu

Right-click icons in the Places list:

Remove removes a folder from the list.

Add Current Folder adds the selected folder to the list; you can also drag a folder from the file list into the Places list.

Add displays the Add Places Item dialog box. (It's much easier simply to drag a folder onto the Places list.)

Properties displays the Places Item Properties dialog box, which is identical to the Add Places Item dialog box, and is available only for items you've added.

Restore Standard Folders restores the folders shown above.

File List

Right-click the File list without selecting a file or folder:

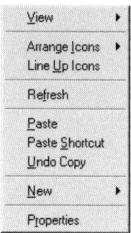

View switches between filename views: large icons, small icons, list, thumbnails, and details.

Arrange Icons arranges icons by name, type, size, and date.

Line Up Icons places icons in an orderly pattern.

Refresh updates the folder listing.

Paste pastes a file from the Clipboard.

Paste Shortcut pastes a file from the Clipboard as a shortcut.

Undo Copy undoes the copy-paste operation; available only after a copy or paste.

New creates a new folder (subdirectory) or shortcut.

Properties displays the Properties dialog box of the folder selected by the Look in list.

Right-click a file or folder name:

Select opens the drawing in AutoCAD.

Open opens the drawing in AutoCAD also.

Print does not work, even though it is on the menu.

Send To copies the file to another drive; may not work with some software.

Cut cuts the file to the Clipboard.

. . . .

Copy copies the file to the Clipboard.

Create Shortcut creates a shortcut icon for the selected file.

Delete erases the file; displays a warning dialog box.

Rename renames the file; displays a warning dialog box if you change the extension.

Properties displays the Properties dialog box in read-only mode; use the DwgProps command within AutoCAD to change the settings.

. . . .

Command Line Options

When FileDia = 0:

Command: open

Enter name of drawing to open <.>: *(Enter name of drawing file.)*

Opening an AutoCAD format file

Enter name specifies the (optional) path and name of the *.dwg, .dxf,* or *.dws* file.

If you leave out the extension, AutoCAD assumes a *.dwg* file.

Enter tilde (~) to display the Select File dialog box.

RELATED SYSTEM VARIABLES

TaskBar displays a button on the Windows taskbar for every open drawing.

DbMod indicates whether the drawing has been modified.

DwgCheck checks if the drawing was last edited by AutoCAD.

DwgName contains the drawing's filename.

DwgPrefix contains the drive and folder of the drawing.

DwgTitled indicates whether the drawing has a name other than *drawing1.dwg.*

FileDia toggles the interface between dialog boxes and the command-line.

FullOpen indicates whether the drawing is fully or partially opened.

RELATED COMMANDS

FileOpen opens drawings without the dialog box.

Import opens drawing files in other formats: 3DS, DGN, SAT, and WMF.

SaveAs saves drawings with new names.

PartiaLoad loads additional portions of partially-opened drawings.

TIPS

- When drawing file names are dragged from Windows Explorer into open drawings, they are inserted as blocks; when dragged to AutoCAD's title bar, they are opened as drawings.

- You can double-click *.dwg* files in Windows Explorer to open them in AutoCAD. If AutoCAD is not yet running, it starts with the last-used profile and vertical application.

- DXF and template files cannot be partially opened.

- After a drawing is partially opened, use PartiaLoad to load additional parts of the drawing.

- When a partially-opened drawing contains a bound xref, only the portion of the xref defined by the selected view is bound to the partially-open drawing.

- OpenUrl opened drawings over the Internet in earlier releases of AutoCAD.

. . . .

Alaised Command

OpenUrl was removed from AutoCAD 2000; it now activates the Open command.

. . . .

OpenDwfMarkup

Opens *.dwf* markup files in AutoCAD, and then loads the Markup Set Manager palette.

Command	Alias	Ctrl+	F-key	Alt+	Menu Bar	Dashboard
opendwfmarkup	...	...	...	FK	**File**	...
					↳ **Load Markup Set**	

Command: opendwfmarkup

Displays the Open Markup DWF dialog box. Select a .dwf file, and then click Open.

If the .dwf file contains no markup data, AutoCAD complains:

Filename.dwf does not contain any markup data. Would you like to open this DWF file in the viewer?

Click Yes to open the file in DesignReview (nee DWF Composer); click No to cancel.

COMMAND OPTIONS

See the Markup command.

RELATED COMMANDS

DwfAttach attaches DWF files as underlays, including un-marked-up files.

Markup displays the Markup Set Manager, and opens *.dwf* files containing markup data.

MarkupClose closes the Markup Set Manager palette opened by this command

RELATED SYSTEM VARIABLE

MsmState reports whether the Markup Set Manager palette is open.

TIPS

- MSM is short for "markup set manager." DWF is short for "drawing Web format."

- This command's purpose is to work with marked-up DWF files created by the DWF DesignReview software, available free from Autodesk at www.autodesk.com/designreview.

- This command replaces the RmlIn command, whose purpose was to import now-obsolete *.rml* redline markup files created by the now-obsolete Volo View.

OpenSheetset

<u>2005</u>   Loads sheet sets into the current drawing.

Command	Alias	Ctrl+	F-key	Alt+	Menu Bar	Dashboard
opensheetset	...	...	...	FE✎	File	...
					✎Open Sheet Set	
-opensheetset						

Command: opensheetset

Displays Open Sheet Set dialog box. Select a .dst file, and then click Open.

AutoCAD displays the Sheet Set Manager. See Sheetset command.

. . . .

-OPENSHEETSET Command
Command: -opensheetset

Enter name of sheet set to open: *(Enter .dst file name.)*

Enter the path and name of a *.dst* file, or enter tilde (~) to display the file dialog box.

RELATED COMMANDS

NewSheetset creates new sheet sets.

Sheetset opens the Sheet Set Manager palette.

SheetsetClose closes the Sheet Set Manager palette.

RELATED SYSTEM VARIABLES

SsFound records path and file name of the sheet set.

SsLocate toggles whether AutoCAD opens sheet sets associated with the drawing.

SsmAutoOpen toggles whether AutoCAD displays the Sheet Set Manager palette when the drawing is opened.

SsmState reports whether the Sheet Set Manager is active.

FileDia determines whether this command displays a dialog box or command-line prompts.

RELATED COMMAND LINE SWITCHES

These switches are used by the Target field on Shortcut tab in AutoCAD's desktop icon's Properties dialog box:

/nossm prevents Sheet Set Manager palette from loading.

/set specifies the *.dst* sheet set file to load.

TIPS

- The Sheetset command automatically opens the previously-opened sheet set file.

- SSM is short for "sheet set manager."

Options

Sets system and user preferences.

Commands	Aliases	Ctrl+	F-key	Alt+	Menu Bar	Dashboard
options	op	...	...	TN	Tools	...
	gr, ddgrips				⤷Options	
	preferences					
	ddselect					

+options

Command: options
Displays dialog box.

OPTIONS DIALOG BOX
OK applies the changes, and closes the dialog box.

Apply applies the changes, and keeps the dialog box open.

Files tab
Search paths, file names, and file locations specifies folders and files used by AutoCAD.

Browse displays dialog box for selecting folders or files.

Add adds an item below the selected path or file name.

Remove removes the selected item without warning; click Cancel to undo the removal.

Move Up/Move Down moves the selected item above or below the preceding item; applies to search paths only.

Set Current makes current the selected project names and spelling dictionaries.

Display tab

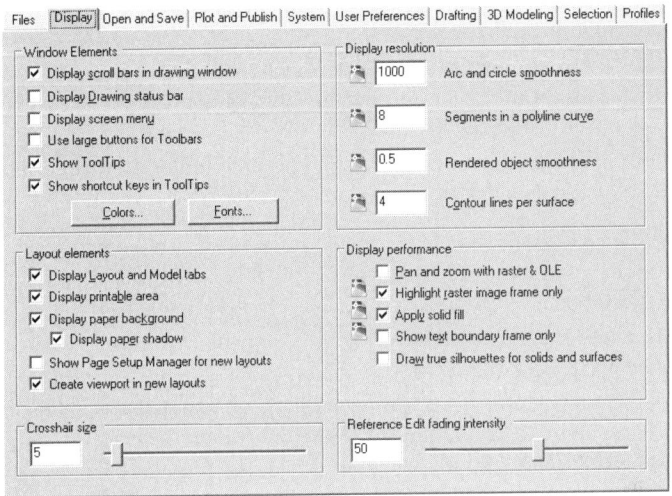 *— indicates setting is saved to a related system variable in the current drawing.*

Window Elements options

- ☐ **Display scroll bars in drawing window** toggles the presence of the horizontal and vertical scroll bars.
- ☐ **Display Drawing Status Bar** toggles the Infobar *(new to AutoCAD 2008)*.
- ☐ **Display screen menu** toggles the presence of the screen menu.
- ☐ **Use large buttons for Toolbars** switches toolbar buttons between normal and large size.
- ☑ **Show Tooltips** toggles the display of tooltips.
- ☑ **Show shortcut keys in Tooltips** toggles the display of shortcut keystrokes in tooltips.

Colors specifies the colors for the graphics and text windows; displays Color Options dialog box.

Fonts specifies the font for text on the command line; displays the Command Line Window Font dialog box, detailed later.

Layout Elements options

- ☑ **Display Layout and Model tabs** toggles the presence of the Model and Layout tabs.
- ☑ **Display printable area** toggles the display of dashed margin lines in layout modes.
- ☑ **Display paper background** toggles the presence of the page in layout modes.
  - ☑ **Display paper shadow** toggles the presence of the drop shadow under the page in layout modes.
- ☐ **Show Page Setup Manager for new layouts** specifies whether the Page Setup dialog box is displayed when you create a new layout. Use this dialog box to set options related to paper and plot settings.
- ☑**Create viewport in new layouts** toggles the automatic creation of a single viewport for new layouts.

Crosshair size specifies the size of the crosshair cursor; range is 1% to 100% of the viewport (stored in system variable CursorSize); default = 5%.

Display Resolution options

Arc and circle smoothness controls the displayed smoothness of circles, arcs, and other curves; range is 1 to 20000 (ViewRes); default = 1000.

Segments in a polyline curve specifies the number of line segments used to display polyline curves; range is -32767 to 32767 (SplineSegs); default = 8.

Rendered object smoothness controls the displayed smoothness of shaded and rendered curves; range is 0.01 to 10 (FaceTRes) default = 0.5.

Contour lines per surface specifies the number of contour lines on solid 3D objects; range is 0 to 2047 (IsoLines); default = 4.

Display Performance options

☐ **Pan and zoom with raster and OLE** toggles the display of raster and OLE images during realtime pan and zoom (RtDisplay).

☑ **Highlight raster image frame only** highlights only the frame, and not the entire raster image, when on (ImageHlt).

☑ **Apply solid fill** toggles the display of solid fills in multilines, traces, solids, solid fills, and wide polylines; this option does not come into effect until you click OK, and then use the **Regen** command (FillMode).

☐ **Show text boundary frame only** toggles the display of rectangles in place of text; this option does not come into effect until you click OK, and then use the Regen command (QTextMode).

☐ **Show silhouettes in wireframe** toggles the display of silhouette curves for 3D solid objects; when off, tesselation lines are drawn when hidden-line removal is applied to the 3D object (DispSilh).

Reference fading intensity specifies the amount of fading during in-place reference editing; range is 0% to 90% (XFadeCtl); default = 50%.

Open and Save tab

File Save options

Save as specifies the default file format used by the Save and SaveAs commands; default = "AutoCAD 2007 Drawing (*.dwg)."

☐ **Maintain Visual Fidelity for Annotative Objects** places annotatively-scaled objects on their own layers, when exported to AutoCAD 2007 and earlier (*new to AutoCAD 2008*).

Thumbnail Preview Settings displays Thumbnail Preview Settings dialog box.

Incremental save percentage indicates the percentage of wasted space allowed in a drawing file before a full save is performed; range is 0% to 100% (ISavePercent); default = 50.

File Safety Precautions options

☑ **Automatic save** saves the drawing automatically at prescribed time intervals (SaveFile and SaveFilePath).

Minutes between saves specifies the duration between automatic saves (SaveTime); default = 10 minutes.

☑ **Create backup copy with each save** creates backup copies when drawings are saved; (ISavBak).

☐ **Full-time CRC validation** performs cyclic redundancy check (CRC) error-checking each time an object is read into the drawing.

☐ **Maintain a log file** saves the Text window text to a log file (LogFileMode); default = off.

File extension for temporary files specifies the filename extension for temporary files created by AutoCAD (NodeName); default = .ac$.

Security Options displays the Security Options dialog box; see the SecurityOptions command.

☑ **Display digital signature information** displays digital signature information when opening files with valid digital signatures (SigWarn).

File Open options

Number of recently-used files to list specifies the number of recently-opened filenames to list in the Files menu; minimum = 0; default and maximum = 9.

☑ **Display full path in title** displays the drawing file's path in AutoCAD's titlebar.

External References (Xrefs) options

Demand load Xrefs specifies the style of demand loading of externally-referenced drawings a.k.a. xrefs (XLoadCtl):

- **Disabled** turns off demand loading.
- **Enabled** turns on demand loading to improve performance, but the drawing cannot be edited by another user; default.
- **Enabled with copy** turns on demand loading; loads a copy of the drawing so that another user can edit the original.

☐ **Retain changes to Xref layers** saves changes to properties for xref-dependent layers (VisRetain).

☑ **Allow other users to Refedit current drawing** allows another user to edit the current drawing when referenced by another drawing (XEdit).

ObjectARX Applications options

Demand load ObjectARX apps demand-loads an ObjectARx application when the drawing contains proxy objects (DemandLoad):

- **Disable load on demand** turns off demand loading.
- **Custom object detect** demand-loads applications when drawings contain proxy objects.

- **Command invoke** demand-loads applications when their commands are invoked.
- **Object detect and command invoke** demand-loads the application when the drawing contains proxy objects, or when one of the application's commands is invoked; default.

Proxy images for custom objects specifies how proxy objects are displayed:
- **Do not show proxy graphics** does not display proxy objects.
- **Show proxy graphics** displays proxy objects.
- **Show proxy bounding box** displays a rectangle instead of the proxy object.
- ☑ **Show Proxy Information dialog box** displays a warning dialog box when a drawing contains proxy objects (ProxyNotice).

Plot and Publish tab

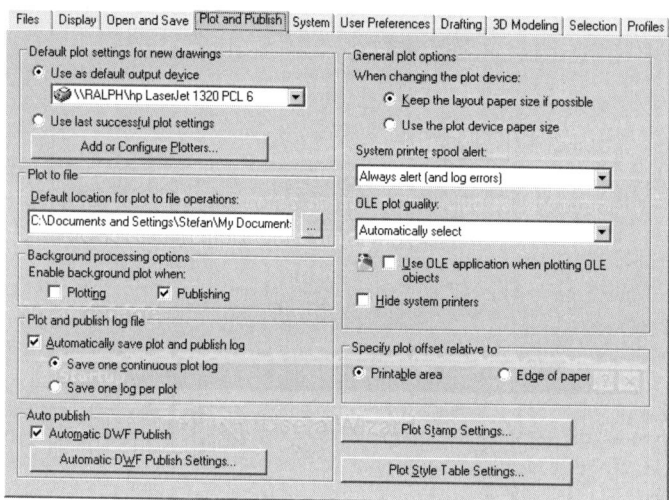

Default Plot Settings for New Drawings options
⊙ **Use as default output device** selects the default output device.

○ **Use last successful plot settings** reuses the plot settings from the last successful plot.

Add or configure plotters displays Plotter Manager window; see PlotterManager command.

Plot to File options
Default location for plot to file operations specifies the folder in which to store plot files; default=*documents and settings\username\my documents*. (**...** displays dialog box for selecting folder.)

Background Processing options
☐ **Plotting** executes the Plot command in the background (BackgroundPlot).

☑ **Publishing** executes the Publish command in the background (BackgroundPlot).

Plot and Publish Log File options
☑ **Automatically save plot and publish log file:**

⊙ **Save one continuous plot log** saves all log data in a single file.

○ **Save one log per plot** saves log data in separate files.

AutoPublish options

New to AutoCAD 2008:

☐ **Automatic DWF Publish** exports drawings as DWF files when saved or closed (AutoDWFPublish); see the AutoPublish command.

Automatic DWF Publish Settings displays a dialog box of the same name, as detailed by the AutoPublish command.

General Plot Options options

When changing the plot device:

⦿ **Keep the layout paper size if possible** uses the paper size specified by the Page Setup dialog box's Layout Settings tab, provided the output device can handle the paper size (PaperUpdate); default = on.

○ **Use the plot device paper size** uses the paper size specified by the *.pc3* plotter configuration file (PaperUpdate); default = off.

System printer spool alert displays an alert when a spooled drawing has a conflict:

- **Always alert (and log errors)** displays alert, and logs the error message.
- **Alert first time only (and log errors)** displays the alert once, but logs all error messages.
- **Never alert (and log first error)** does not display an alert, but logs the first error message.
- **Never alert (do not log errors)** neither displays an alert, nor logs any error messages.

OLE plot quality determines the quality of OLE objects when plotted (OleQuality); default = Automatically Select; see the OleScale command.

☐ **Use OLE application when plotting OLE objects** launches the application that created the OLE object when plotting a drawing with an OLE object (OleStarup).

☐ **Hide system printers** hides the names of Windows system printers not specific to CAD.

Specify Plot Offset Relative To options

⦿ **Printable area** measures offsets to plotter margins.

○ **Edge of paper** measures offsets to paper edges.

Plot Stamp Settings displays the Plot Stamp dialog box; see PlotStamp command.

Plot Style Table Settings displays the Plot Style Table Settings dialog box.

System tab

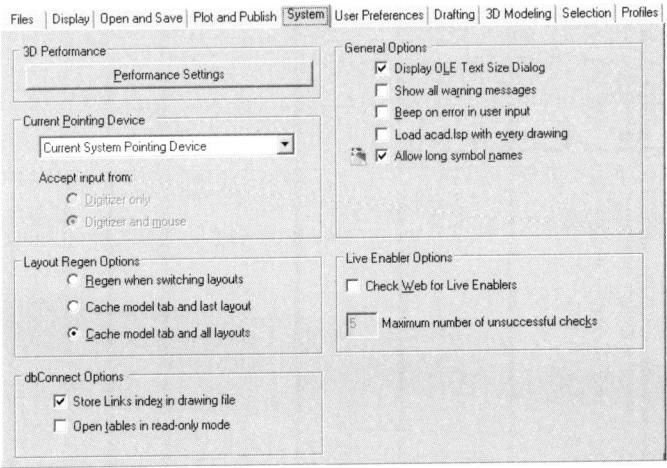

3D Performance options

Performance Settings displays the Adaptive Degradation and Performance Tuning dialog box; see the 3dConfig command.

Current Pointing Device options

Current Pointing Device selects the pointing device driver:

- **Current System Pointing Device** selects the pointing device used by Windows.
- **Wintab Compatible Digitizer** selects a Wintab-compatible digitizer driver.

Accept input from *(available only when a digitizing tablet is attached):*

○ **Digitizer only** reads input from the digitizer, and ignores the mouse.

◉ **Digitizer and mouse** reads input from the digitizer and the mouse.

Layout Regen options

○ **Regen when switching layouts** regenerates the drawing each time layouts are switched.

○ **Cache model tab and last layout** saves display lists of the model tab and last layout accessed.

◉ **Cache model tab and all layouts** saves the display list of the model tab and all layouts.

dbConnect Options options

☑ **Store links index in drawing file** stores the database index in the drawing file.

☐ **Open tables in read-only mode** opens database tables in read-only mode.

General Options options

☑ **Display OLE Text Size dialog** displays the OLE Properties dialog box after an OLE object is inserted in the drawing; see the OleScale command.

☐ **Show all warning messages** displays all dialog boxes with the Don't Display This Warning Again option.

☐ **Beep on error in user input** beeps the computer when AutoCAD detects a user error.

☐ **Load acad.lsp with every drawing** loads the *acad.lsp* file with every drawing (AcadLspAsDoc).

☑ **Allow long symbol names** allows symbol names — such as layers, dimension styles, and blocks, — to be up to 255 characters long, and to include letters, numbers, blank spaces, and most punctuation marks; when off, names are limited to 31 characters, and spaces may not be used (ExtNames).

The SDI and Startup options were removed from AutoCAD 2008.

Live Enabler Options options

☐ **Check Web for Live Enablers** checks whenever an Internet connection is present.

Maximum number of unsuccessful checks limits how often AutoCAD attempts to call home; default = 5.

User Preferences tab

Windows Standard Behavior options

☑ **Double click editing** toggles editing objects by double-clicking them; see Cui command.

☑ **Shortcut menus in drawing area:** toggles right-click action in drawing area: when on, displays shortcut menus; when off, equivalent to pressing the Enter key (ShortCutMenu).

Right-click Customization displays the Right-Click Customization dialog box, detailed later (ShortCutMenu).

Insertion Scale options

Source content units specifies the default units when an object is inserted into the drawing from AutoCAD DesignCenter; Unspecified-Unitless means the object is not scaled when inserted (InsUnitsDefTarget); default = inches or mm.

Target drawing units specifies the default units when "insert units" are not specified by the InsUnits system variable (InsUnitsDefTarget); default = inches or mm.

Field options

☑ **Display background of fields** displays gray behind field text (FieldDisplay).

Field Update Settings displays the Field Update Settings dialog box; see UpdateField command.

The Hidden Line Settings button was removed from AutoCAD 2007.

Lineweight Settings displays the Lineweight Settings dialog box; see Lineweight command.

Edit Scale List displays the Edit Scale List dialog box; see ScaleListEdit command.

Priority for Coordinate Data Entry options

○ **Running object snap** means that osnap overrides coordinates entered at the keyboard (OSnapCoord).

○ **Keyboard entry** means that coordinates entered at the keyboard override osnaps (OSnapCoord).

⊙ **Keyboard entry except scripts** means that coordinates entered at the keyboard override running object snaps, except when coordinates are provided by a script (OSnapCoord).

Associative Dimensioning options

☑ **Make new dimension associative** means that dimensions are associated with objects; when off, dimensions are associated with defpoints (DimAssoc).

Hyperlink options

☑ **Display hyperlink cursor and shortcut menu** displays hyperlink cursor (looks like chain links and the planet earth, when the cursor passes over objects containing hyperlinks), and adds the Hyperlink option to shortcut menus when right-clicking objects containing hyperlinks. See the HyperlinkOptions command.

Undo/Reo options

☑ **Combine zoom and pan commands** combines all sequential zooms and pans into a single undo or redo.

Drafting tab

AutoSnap Settings options

☑ **Marker** displays the object snap icon (AutoSnap).

☑ **Magnet** turns on the AutoSnap magnet (AutoSnap).

☑ **Display AutoSnap tooltip** displays the AutoSnap tooltip (AutoSnap).

☐ **Display AutoSnap aperture box** displays the AutoSnap aperture box (ApBox).

Colors specifies the color of the AutoSnap icons; default = color #31. Displays the Drawing Window Colors dialog box.

AutoSnap marker size sets the size for the AutoSnap icon; range is 1 to 20 pixels.

Object Snap Options options

☑ **Ignore hatch objects** prevents AutoCAD from snapping to hatch objects.

☐ **Replace Z value with current elevation**.

☑ **Ignore negative Z object snaps for Dynamic UCS** (OsOptions).

AutoTrack Settings options
☑ **Display polar tracking vector** displays the Polar Tracking vectors at specific angles (TrackPath).

Polar tracking vector (dotted line) and AutoTrack tooltip.

☑ **Display full-screen tracking vector** displays the tracking vectors (TrackPath).

☑ **Display AutoTrack tooltip** displays the AutoTrack tooltip (AutoSnap).

Full-screen tracking vector and AutoTrack tooltip.

Alignment Point Acquisition options
⊙ **Automatic** displays tracking vectors automatically when aperture moves over an object snap.

○ **Shift to acquire** displays tracking vectors when pressing Shift and moving the aperture over an object snap.

Aperture Size options
Aperture Size sets the size for the aperture; range is 1 to 50 pixels (Aperture); default = 10.

Drafting Tooltip Settings displays the Tooltip Appearance dialog box; see DSettings command.

Light Glyphs Settings displays the Light Glyphs Appearance dialog box, detailed later

Camera Glyphs Settings displays the Camera Glyphs Appearance dialog box, detailed later.

3D Modeling tab

3D Crosshairs options
☑ **Show Z Axis in Crosshairs** toggles display of z axis in crosshair cursor.

☐ **Label Axes in Standard Crosshairs** toggles whether labels (X, Y, and Z) are displayed with the crosshair cursor.

☐ **Show Labels for Dynamic UCS** displays labels during dynamic UCS, even when the axis labels are off.

☐ **Crosshair Labels:**
⊙ **Use X, Y, Z** labels the axes with X, Y, and Z.

○ **Use N, E, z** labels the axes with N (north), E (east), and z.

○ **Use Custom Labels** labels each axis with up to eight user-defined characters.

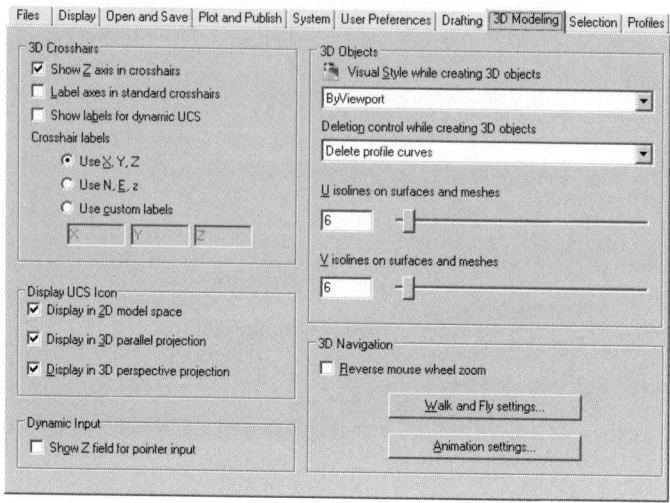

Display UCS Icon options

☑ **Display in 2D Model Space** toggles display of the UCS icon in 2D Wireframe visual style.

☑ **Display in 3D Parallel Projection** toggles display of the UCS icon in parallel projection with non-2D visual styles.

☑ **Display in 3D Perspective Projection** toggles display of the UCS icon in perspective projection with non-2D visual styles.

Dynamic Input options

☐ **Show Z Field for Pointer Input** toggles display of the z coordinate during dynamic input.

3D Objects options

Visual Style While Creating 3D Objects specifies the visual style while dragging 3D primitives into shape (DragVs system variable):
- **By Viewport** uses the viewport's setting (see the MView command).
- 2D Wireframe.
- 3D Hidden.
- 3D Wireframe.
- Conceptual.
- Realistic.

Deletion Control While Creating 3D Objects (DelObj):
- Retain defining geometry.
- Delete profile curves.
- Delete profile and path curves.
- Prompt to delete profile curves.
- Prompt to delete profile and path curves.

U and **V Isolines on Surfaces and Meshes** specifies the number of isolines in the u and v directions (SurfU and SurvV).

3D Navigation options

☐ **Reverse Mouse Wheel Zoom** switches the direction of zoom when the mouse wheel is rolled (ZoomWheel).

Walk and Fly Settings displays the Walk and Fly Settings dialog box; see the WalkFlySettings command.

Animation Settings displays the Animation Settings dialog box; see the AniPath command.

Selection tab

This tab is displayed when the DdGrips command is entered.

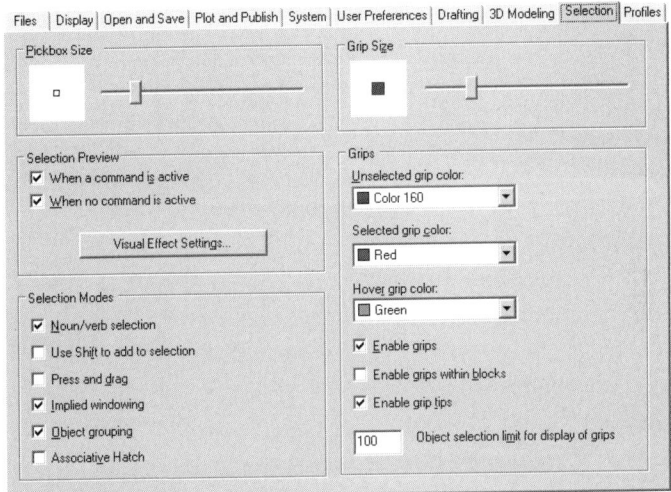

Pickbox Size specifies the size of the pickbox; range is 1 to 20 pixels (PickBox); default = 3.

Grip size specifies the size of grips; range is from 1 to 20 pixels (GripSize); default = 5.

Selection Preview options

☑ **When a command is active** previews object selection during commands.

☑ **When no command is active** previews object selection as cursor passes over objects.

Visual Effect Settings displays the Visual Effects Settings dialog box.

Selection Modes options

☑ **Noun/verb selection** allows you to select an object before executing an editing command (PickFirst).

☐ **Use Shift to add to selection** requires you to press Shift to add or remove objects from the selection set, a la Windows (PickAdd).

☐ **Press and drag** allows you to create the selection window by dragging (PickDrag).

☑ **Implied windowing** creates a selection window when you pick a point in the drawing that does not pick an object (PickAuto).

☑ **Object grouping** selects the entire group when an object in the group is selected (PickStyle).

☐ **Associative hatch** selects boundary objects, along with the associative hatch patterns (PickStyle).

Grips options

Unselected grip color specifies color of unselected (cold) grips (GripColor); default = blue.

Selected grip color specifies the color of selected (hot) grip (GripHot); default = red.

Hover grip color specifies the color of the grip when the cursor hovers over it; default = green.

☑ **Enable grips** displays grips on selected objects (Grips).

☐ **Enable grips within blocks** displays all grips for every object in the selected block; when off, a single grip at the block's insertion point is displayed (GripBlock).

☑ **Enable grip tips** toggles the display of grip tips on custom objects (GripTips).

Object selection limit for display of grips limits the number of selected objects that display grips (GripObjLimit); default = 100.

Profiles tab

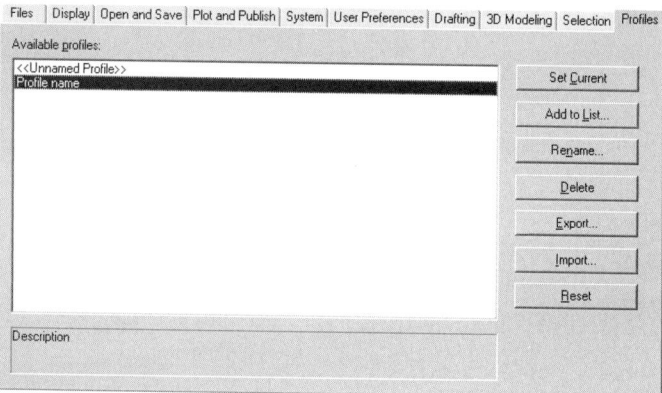

Set Current sets the selected profile as the current profile.

Add to List displays the Add Profile dialog box; allows you to enter a name and description for the new profile.

Rename displays the Change Profile dialog box; allows you to change the name and the description of the selected profile.

Delete erases the selected profile; the current profile cannot be erased.

Export exports profiles as *.arg* files.

Import imports *.arg* profiles into AutoCAD.

Reset resets the selected profile to AutoCAD's default settings.

COMMAND LINE WINDOW FONT DIALOG BOX

In the Display tab, click the Font button:

Font selects the font for the command line palette.

Font Style changes the style.

Size specifies the size of the text.

COLOR OPTIONS DIALOG BOX

In the Display tab, click the Colors button:

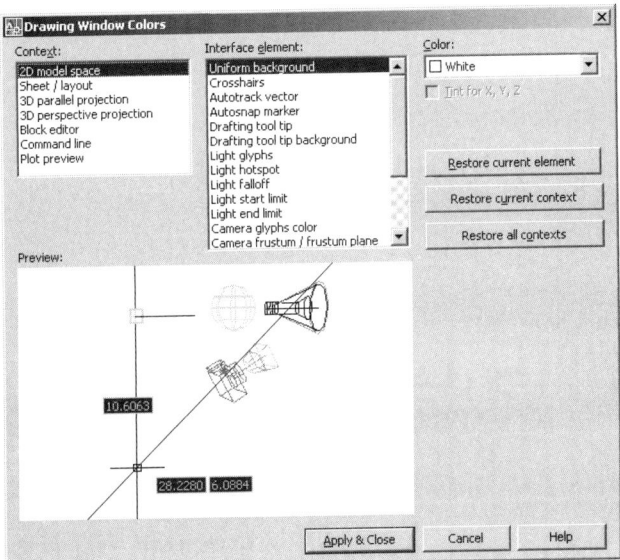

Context selects the user interface.

Interface element selects the individual interface item.

Color selects the color for the interface element.

☐ **Tint for X, Y, Z** selects colors for the crosshair cursor and 3D UCS icon's x, y, z axes.

Restore current element sets the selected item back to its default color.

Restore current context sets the selected UI back to its default colors.

Restore all contexts sets all UIs back to their default colors.

FIELD UPDATE SETTINGS DIALOG BOX

In the User Preferences tab, click the Field Update Settings button:

☑ Specifies which commands force AutoCAD to update field text.

PLOT STYLE TABLE SETTINGS DIALOG BOX

In the Plot and Publish tab, click the Plot Style Table Settings button:

Default Plot Style Behavior for New Drawings (PStylePolicy).

⊙ **Use Color Dependent Plot Styles** uses color-dependent plot styles in new drawings; plotted colors are defined by a number between 1 and 255.

○ **Use Named Plot Styles** uses named plot styles in new drawings; objects are plotted based on plot style definitions.

Current Plot Style Table Settings options

Default Plot Style Table selects the plot style table to attach to new drawings.

Default Plot Style for Layer 0 specifies the default plot style for Layer 0 in new drawings (DefLPlStyle).

Default Plot Style for Objects specifies the default plot style assigned to new objects (DefPlStyle). The list displays BYLAYER, BYBLOCK, Normal styles, and any plot styles defined in the currently-loaded plot style table.

Add or Edit Plot Style Tables displays the Plot Style Manager window; see the StylesManager command.

VISUAL EFFECT SETTINGS DIALOG BOX

In the Selection tab, click the Visual Effect Settings button:

○ **Dash** changes selected object lines to dashes.

○ **Thicken** thickens selected object lines.

⊙ **Both** changes selected object lines to thick dashes.

Advanced Options displays the Advanced Options dialog box.

☑ **Indicated selection area** toggles filled selection areas.

Window selection color selects the color for Window selection mode.

Crossing selection color selects the color for Crossing selection mode.

Selection area opacity specifies the "see thru ness" of the selection area; 100=opaque.

ADVANCED OPTIONS DIALOG BOX

☑ **Exclude objects on locked layers** ignores objects on locked layers, which cannot be edited.

Exclude excludes specific object types.

THUMBNAIL PREVIEW SETTINGS DIALOG BOX

In the Open and Save tab, click the Thumbnail Preview Settings button:

☑ **Save a Thumbnail Preview Image** saves the current view of the drawing as a bitmap preview image in the *.dwg* file (RasterPreview).

☑ **Generate Sheet, Sheet View and Model View Thumbnails** updates thumbnails.

Performance updates thumbnail previews only when UpdateThumbsNow is executed.

Accuracy updates thumbnail previews with every drawing save.

RIGHT-CLICK CUSTOMIZATION DIALOG BOX

In the User Preferences tab, click the Right Click Customization button:

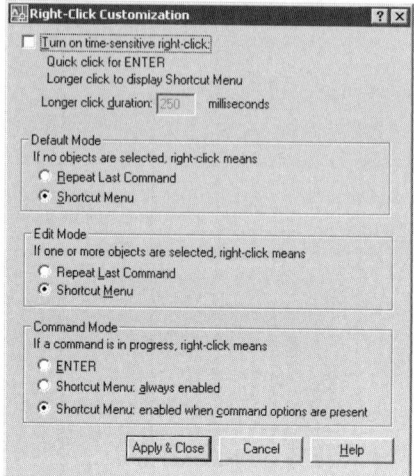

☐ **Turn on Time-Sensitive Right-Click** toggles Enter-or-shortcut menu.

Default Mode:

○ **Repeat Last Command** repeats the previous command, like pressing Enter.

⊙ **Shortcut Menu** displays the appropriate shortcut menu.

Edit Mode:

○ **Repeat Last Command** repeats the previous command, like pressing Enter.

⊙ **Shortcut Menu** displays the appropriate shortcut menu.

Command Mode during a command:

○ **ENTER** disables the shortcut menu, and executes Enter.

○ **Shortcut Menu: Always Enabled** enables the shortcut menu, and disables Enter.

⊙ **Shortcut Menu: Enabled When Command Options Are Present** executes Enter only when no options are available.

LIGHT AND CAMERA GLYPH APPEARANCE DIALOG BOXES

In the Drafting tab, click the Light Glyph Settings or Camera Glyph Settings button:

 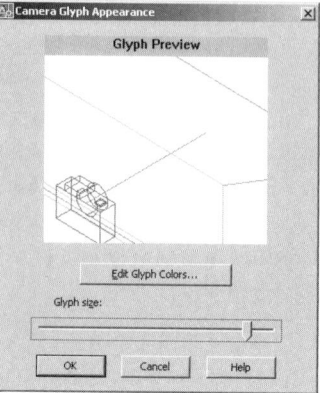

Edit Glyph Colors changes the color(s) of the glyphs; displays the Drawing Windows Color dialog box.

Glyphy size changes the size of the glyph.

Light Glyph Appearance options

Show:

⊙ **Point** glyph for point lights (PointLight command).

○ **Spot** glyph for point lights (SpotLight command).

- -

+OPTIONS COMMAND

Command: +options

Tab index <0>: *(Enter a digit between 0 and 9.)*

COMMAND LINE OPTION

Tab index specifies the tab to display:

0 — Files tab.

1 — Display tab.

2 — Open and Save tab.

3 — Plot and Publish tab.

4 — System tab.

5 — User Preferences tab.

6 — Drafting tab.

7 — 3D Modeling tab.

8 — Selection tab.

9 — Profiles tab.

TIPS

- *Grips* are small squares that appear on an object when the object is selected at the 'Command' prompt. In other Windows applications, grips are known as *handles*.

GripBlock = 0.   GripBlock = 1.

- When an object is first selected, the grips are blue, and are called *unselected* or *cold* grips. When a grip is selected, it turns into a solid red square; this is called a *hot* grip. Press Esc to turn off unselected grips; press Esc twice to turn off hot grips.

- A larger pickbox makes it easier to select object generally, but also easier to select objects accidentally.

- The first time you use the DrawOrder command, it turns on all object sort method options.

- It's best to choose a crosshair color that is different from your drawing colors.

- -

'Ortho

V. 1.0 Constrains drawing and editing commands to the vertical and horizontal directions only (short for ORTHOgraphic).

Command	Alias	Ctrl+	F-key	Alt+	Status Bar	Dashboard
ortho	...	L	F8	...	ORTHO	...

Command: ortho
Enter mode [ON/OFF] <OFF>: *(Enter ON or OFF.)*

COMMAND LINE OPTIONS
OFF turns off ortho mode.
ON turns on ortho mode.

STATUS BAR OPTIONS
Ortho mode can be toggled on and off by clicking ORTHO on the status bar:

10.1046, 0.9275 , 0.0000	SNAP GRID ORTHO POLAR OSNAP

RELATED COMMANDS
DSettings toggles ortho mode via a dialog box.
Snap rotates the ortho angle.

RELATED SYSTEM VARIABLES
OrthoMode stores the current ortho modes.
PolarMode specifies polar and object snap tracking.
SnapAng specifies the rotation angle of the ortho cursor.

TIPS
- Use ortho mode when you want to constrain your drawing and editing to right angles.

- This command has been effectively replaced by the Polar command, which is not restricted to 90-degree increments.

- Rotate the angle of ortho with the Snap command's Rotate option.

- In isoplane mode, ortho mode constrains the cursor to the current isoplane.

- AutoCAD ignores ortho mode when you enter coordinates from the keyboard and in perspective mode; ortho is also ignored by object snap modes.

- Ortho is not necessarily horizontal or vertical; its orientation is determined by the current UCS and snap alignment.

'-OSnap

V. 2.0 Sets and turns on and off object snap modes at the command line (short for Object SNAP.

Command	Alias	Ctrl+	F-key	Alt+	Status Bar	Dashboard
-osnap	-os	...	F3	...	OSNAP	...

The OSnap command displays the Drafting Settings dialog box; see the DSettings command.

Command: -osnap
Current osnap modes: Ext
Enter list of object snap modes: *(Enter one or more modes separated by commas.)*

COMMAND LINE OPTIONS

You only need enter the first three letters as the abbreviation for each option:

APParent snaps to the intersection of two objects that don't physically cross, but appear to intersect on the screen, or would intersect if extended.

CENter snaps to the center point of arcs and circles.

ENDpoint snaps to the endpoint of lines, polylines, traces, and arcs.

EXTension snaps to the extension path of objects.

FROm extends from a point by a given distance.

INSertion snaps to the insertion point of blocks, shapes, and text.

INTersection snaps to the intersection of two objects, to a self-crossing object, or to objects that would intersect if extended.

MIDpoint snaps to the middle of lines and arcs.

NEArest snaps to the object nearest the crosshair cursor.

NODe snaps to a point object.

NONe turns off all object snap modes temporarily.

OFF turns off all object snap modes.

PARallel snaps to a parallel offset.

PERpendicular snaps perpendicularly to objects.

QUAdrant snaps to the quadrant points of circles and arcs.

QUIck snaps to the first object found in the database.

TANgent snaps to the tangent of arcs and circles.

STATUS BAR OPTIONS

Right-click OSNAP on the status bar:

On turns on previously-set object snap modes.

Off turns off all running object snaps.

Settings displays the Object Snap tab of the Drafting Settings dialog box.

SHORTCUT MENU

Hold down the Ctrl key, and then right-click anywhere in the drawing:

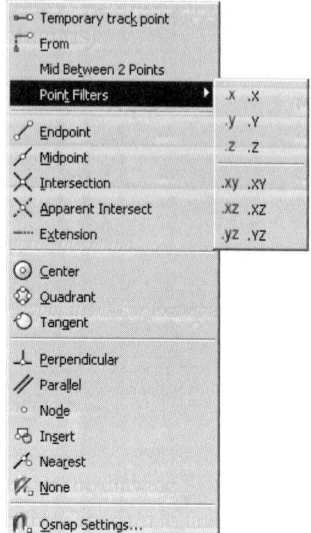

Temporary track point invokes Tracking mode; see the Tracking command.

From locates temporary points during drawing and editing commands. AutoCAD prompts:

From point: from
Base point: *(Pick a point.)*
<Offset>: *(Pick another point.)*

Mid Between 2 Points locates a point between two pick points.

From point: m2p
First point of mid: *(Pick a point.)*
Second point of mid: *(Pick another point.)*

Point filters invokes point filter modes:

.x — **need YZ:** *(Enter value for y and z.)*
.y — **need XZ:** *(Enter value for x and z.)*
.z — **need XY:** *(Enter value for x and y.)*
.xy — **need Z:** *(Enter value for z.)*
.xz — **need Y:** *(Enter value for y.)*
.yz — **need X:** *(Enter value for x.)*

Osnap Settings displays the Object Snap tab of the Drafting Settings dialog box; see **DSettings** command.

RELATED SYSTEM VARIABLES

Aperture specifies the size of the object snap aperture in pixels.

AutoSnap controls the display of AutoSnap (default = 63).

DgnOsnap toggles snapping to elements in attached MicroStation V8 DGN underlays.

OsnapNodeLegacy determines whether the NODe object snap snaps to text insertion points.

OsOptions toggles between using the z coordinate and the current elevation setting.

OsnapCoord overrides object snaps when entering coordinates at 'Command' prompt.

OsnapZ determines whether osnaps uses the z coordinate or the elevation.

OsMode stores the current object snap mode(s).

TempOverrides toggles temporary override keys.

TEMPORARY OVERRIDE KEYS

These keystrokes temporarily override object snaps during commands:

Override	Left Side	Right Side
Disables all osnaps and tracking	SHIFT+D	SHIFT+L
Enables osnap enforcement	SHIFT+S	SHIFT+;
Overrides CENter osnap	SHIFT+C	SHIFT+,
Overrides ENDpoint osnap	SHIFT+E	SHIFT+P
Overrides MIDpoint osnap	SHIFT+V	SHIFT+M
Toggles osnap mode	SHIFT+A	SHIFT+'

TIPS

- The Aperture command controls the drawing area through which AutoCAD searches.

- If AutoCAD finds no snap matching the current modes, then the pick point is selected.

- The m2p modifier can also be entered as mtp.

- The APPint and INT object snap modes should not be used together.

- To turn on more than one object snap at a time, use a comma to separate mode names:

    Enter list of object snap modes: int,end,qua

- The location of object snaps:

PageSetup

2000 Sets up model and layout views in preparation for plotting drawings.

Command	Alias	Ctrl+	F-key	Alt+	Menu Bar	Dashboard
pagesetup	...	...	...	FG	File	...
					⤷ Page Setup	

Command: pagesetup

Displays dialog box:

PAGE SETUP MANAGER DIALOG BOX

Set Current selects the page setup to be used by the drawing.

New displays the New Page Setup dialog box.

Modify displays the Page Setup dialog box.

Import displays the Select Page Setup From File dialog box; select a *.dwg*, *.dwt*, or *.dxf* file, and then click Open; displays the Import Page Setups dialog box.

☐ **Display when creating a new layout** displays this dialog box the first time you click a layout tab in new drawings.

NEW PAGE SETUP DIALOG BOX

New Page Setup Name specifies the name of the setup.

Start With selects the default settings:

<None> — no page setup is used as the template.

<Default Output Device> — selects the plotter specified as the default output device in the Options dialog box.

<Previous Plot> — uses settings from a previous plot job.

Layout Name — selects an existing layout.

OK displays the Page Setup dialog box.

PAGE SETUP DIALOG BOX

See Plot command.

IMPORT PAGE SETUPS DIALOG BOX

Name lists the names of page setups found in the drawing, if any.

Location indicates the location of the page setups: model or layout.

OK returns to the Page Setup Manager dialog box.

SHORTCUT MENU

Right-click a page setup name:

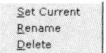

Set Current selects the page setup to use for the drawing.

Rename changes the name of the page setup.

Delete erases the page setup from the drawing, without warning.

RELATED COMMANDS

Layout creates new layouts.

LayoutWizard creates layouts and page setups.

Plot plots drawings based on the settings of the PageSetup command.

TIPS

■ The sheet set icon shows when this command is activated from the Sheet Set Manager palette.

■ Right-click Model or Layout tabs, and then select Page Setup Manager.

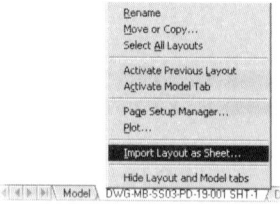

■ Page setups cannot be applied to entire sheet sets after they are created.

'Pan

V. 1.0 Moves the view in the current viewport.

Commands	Aliases Ctrl+	F-key	Alt+	Menu Bar	Dashboard
pan	p ...	...	...	View	2D Navigate
	rtpan			⤷Pan	
	3dpantransparent			⤷Realtime	
-pan	-p		VPP	View	
				⤷Pan	
				⤷Point	

Command: pan
Press Esc or Enter to exit, or right-click to display shortcut menu. *(Move cursor to pan, and then press esc to exit command.)*

Enters real-time panning mode, and displays hand cursor:

Drag the hand cursor to pan the drawing in the viewport.
Press Enter or Esc to return to the 'Command' prompt.

SHORTCUT MENU OPTIONS

During real-time pan mode, right-click the drawing to display shortcut menu:

Exit exits real-time pan mode; returns to the 'Command' prompt.

Pan switches to real-time pan mode, when in real-time zoom mode.
Zoom switches to real-time zoom mode; see the Zoom command.
3D Orbit switches to 3D orbit mode; see the 3dOrbit command.

Zoom Window prompts you, 'Press pick button and drag to specify zoom window.'
Zoom Original returns to the view when you first started the Pan command.
Zoom Extents displays the entire drawing.

. . . .

-PAN Command

Command: -pan
Displacement: *(Pick a point, or enter x,y-coordinates.)*
Second point: *(Pick another point, or enter x,y-coordinates.)*

. . . .

Left: *Before panning to the left.*
Right: *After panning to the left.*

COMMAND LINE OPTIONS

Enter or **Esc** exits real-time panning mode.

Displacement specifies the distance and direction to pan the view.

Second point pans to this point.

RELATED COMMANDS

DsViewer displays the Aerial View palette, which pans in an independent palette.

RegenAuto determines how regenerations are handled.

View saves and restores named views.

Zoom pans with the Dynamic option.

3dOrbit pans during perspective mode.

RELATED SYSTEM VARIABLES

MButtonPan determines the action of a mouse's third button or wheel.

RtDisplay toggles the display of raster and OLE images during realtime panning.

VtDuration specifies the duration of view transitions.

VtEnable turns on smooth view transitions.

VtFps specifies the speed of view transitions.

TIPS

- With MButtonPan set to 1 (default), pan by holding down the wheelbutton and moving the mouse.

- You can pan each viewport independently.

- Use 'Pan to start drawing objects in one area of the drawing, pan over, and then continue working in another area of the drawing.

- You cannot use transparent pan under the following conditions: perspective mode, VPoint, DView, another Pan command, View, Zoom, or 3dOrbit.

- When the drawing no longer moves during real-time panning, you have reached the panning limit; AutoCAD changes the hand icon to show the limit. (Use Regen to increase the limit.)

- As an alternative, use the horizontal and vertical scroll bars to pan the drawing.

PartialCui

<u>**2006**</u> Loads *.cui* files into AutoCAD.

Command	Alias	Ctrl+	F-key	Alt+	Menu Bar	Dashboard
partialcui	...	...	...	TC	Tools	...
					↳Customize	
					↳Import Customizations	

Command: partialcui

Displays Select Customization File dialog box.

Select a .cui, .mnu, or .mns file, and then click Open.

TIPS

- The U command reverses the effect of this command.

- This command is meant to replace the MenuLoad command.

PartiaLoad

2000 Loads additional views and layers of a partially-loaded drawing; this command works only with drawings that have been partially loaded.

Commands	Alias	Ctrl+	F-key	Alt+	Menu Bar	Dashboard
partiaload	...	...	...	FL	File	...
					⤷Partial Load	

-partiaload

Command: partiaload

If the current drawing has not been partially opened through the Open command's Partial Open option, AutoCAD complains, 'Command not allowed unless the drawing has been partially opened.'

Otherwise, displays dialog box:

PARTIAL LOAD DIALOG BOX

View geometry to load selects the model space views to load. (Paper space views are not available for partial loading.)

⬛ **Pick a Window** button dismisses the dialog box temporarily to allow you to specify an area that becomes the view to load. AutoCAD prompts you:

Specify first corner: *(Pick a point.)*
Specify opposite corner: *(Pick a point.)*

When the Partial Load dialog box returns, *New View* is listed by the View Geometry to Load list.

Layer Geometry to Load options

Load Geometry selects the layers to load.

Load All loads all layers.

Clear All deselects all layer names.

-PARTIALOAD Command

Command: -partiaload

Specify first corner or [View]: *(Pick a point, or type V.)*

Enter view to load or [?] <*Extents*>: *(Enter a view name, or type ?.)*

Enter layers to load or [?] <none>: *(Enter a layer name, press Enter for no layers, or type ?.)*

COMMAND LINE OPTIONS

Specify first corner specifies a corner to create a new view.

Specify opposite corner specifies the second corner; causes geometry in the new view to be loaded into the drawing.

View options

Enter view to load loads the geometry found in the view into the drawing.

? lists the names of views in the drawing:

Enter view name(s) to list <*>: *(Enter a view name, or type *.)*

Sample display:

> Saved model space views:
> View name"1"
> "Plot"

Extents loads all geometry into the drawing.

Enter layers to load loads geometry on the layers into the drawing.

? lists the names of layers in the drawing:

Enter layers to list <*>: *(Press Enter.)*

Sample display:

> Layer names:
> "0"
> "3"

None loads no layers.

RELATED COMMANDS

Open partially opens a drawing via a dialog box.

PartialOpen partially opens a drawing via the command line.

RELATED SYSTEM VARIABLE

FullOpen indicates whether the drawing is fully or partially opened.

-PartialOpen

2000   Opens a drawing and loads selected layers and views.

Command	Alias	F-key	Alt+	Menu Bar	Dashboard
-partialopen	partialopen	...	...	...	...

Command: -partialopen
Enter name of drawing to open <filename.dwg>: *(Enter a file name.)*
Enter view to load or [?]<*Extents*>: *(Enter a view name, or type ?.)*
Enter layers to load or [?]<none>: *(Enter a layer name, or type ?.)*
Unload all Xrefs on open? [Yes/No] <N>: *(Type Y or N**N**.)*

COMMAND LINE OPTIONS

Enter name of drawing to open specifies the name of the DWG file.

Enter view to load loads the geometry found in the view into the drawing.

? lists the names of the views in the drawing.

Extents loads all geometry into the drawing.

Enter layers to load loads geometry on the layers into the drawing.

? lists the names of layers in the drawing.

None loads no layers.

Unload all Xrefs on open:

    Yes does not load externally-referenced drawings.

    No loads all externally-referenced drawings.

RELATED COMMANDS

Open displays the Select Drawing dialog box; includes the Partial Open option.

PartiaLoad loads additional views or layers of a partially-opened drawing.

RELATED SYSTEM VARIABLES

FullOpen indicates whether the drawing is fully or partially opened.

TIPS

- PartialOpen is an alias for the -PartialOpen command.

- As an alternative, you may use the Partial Open option of the Open command's Select File dialog box.

PasteAsHyperlink

<u>**2000**</u> Pastes object as hyperlinks in drawings; works only if the Clipboard contains appropriate data.

Command	Alias	Ctrl+	F-key	Alt+	Menu Bar	Dashboard
pasteashyperlink	...	...	...	EH	Edit	...
					⸖Paste as Hyperlink	

Command: pasteashyperlink
Select objects: *(Select one or more objects.)*
Select objects: *(Press Enter.)*

COMMAND LINE OPTION

Select objects selects the objects to which the hyperlink will be pasted.

RELATED COMMANDS

CopyClip copies a hyperlink to the Clipboard.

Hyperlink adds hyperlinks to selected objects.

SelectURL highlights all objects with hyperlinks.

RELATED SYSTEM VARIABLES

None.

PasteBlock

<u>**2000**</u> Pastes objects as blocks in drawings; works only when the Clipboard contains AutoCAD objects.

Command	Alias	Ctrl+	F-key	Alt+	Menu Bar	Dashboard
pasteblock	...	Shift+V ...		EK	Edit	...
					↳ **Paste as Block**	

Command: pasteblock
Specify insertion point: *(Pick a point.)*

COMMAND LINE OPTION
Specify insertion point specifies the position for the block.

RELATED COMMANDS
Insert inserts blocks in the drawing.

CopyClip copies a block to the Clipboard.

RELATED SYSTEM VARIABLES
None.

TIPS
- The block is pasted onto the current layer.

- This command does not work when the Clipboard contains data that cannot be pasted as a block.

- This command pastes any AutoCAD object as a block, generating a generic block name similar to "A$C65D94228."

- If the Clipboard contains a block, the block is nested by this command.

- As of AutoCAD 2004, you can use the Ctrl+Shift+V shortcut.

PasteClip

<u>**Rel.13**</u>   Places objects from the Clipboard in drawings (short for PASTE CLIPboard).

Command	Alias	Ctrl+	F-key	Alt+	Menu Bar	Dashboard
pasteclip	...	V	...	EP	Edit	...
					⌐ Paste	

Command: pasteclip

When the Clipboard contains AutoCAD objects, the following prompt is displayed:
Specify insertion point: *(Pick a point.)*

COMMAND LINE OPTION

Specify insertion point specifies the lower-left corner of the pasted object.

SHORTCUT MENU

Right-click the pasted object to display a shortcut menu; not all options appear with all pasted objects.

OLE displays submenu:

Open opens the object in its source application; does not work with static ActiveX objects.

Reset resets the object to its original settings.

Text Size displays the Text Size dialog box; available only when the object contains text.

Convert converts (or, in most cases, does not convert) the object to another format; displays dialog box:

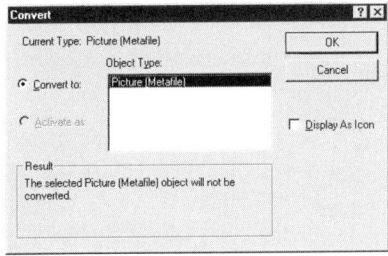

RELATED COMMANDS

CopyClip copies selected objects to the Clipboard.

Insert inserts an AutoCAD drawing in the drawing.

InsertObj inserts an OLE object in the drawing.

PasteSpec places the Clipboard object as pasted or linked objects.

RELATED SYSTEM VARIABLE

OleHide toggles the display of the OLE object (1 = off).

TIPS

- Graphical objects are placed in the drawing as OLE objects.

- Text is usually — but not always — placed in the drawing as Mtext objects.

- Use the PasteBlock command to paste objects as an AutoCAD block.

PasteOrig

2000 Pastes AutoCAD objects from the Clipboard into the current drawing at the objects' original locations (short for PASTE ORIGinal).

Command	Alias	Ctrl+	F-key	Alt+	Menu Bar	Dashboard
pasteorig	...	...	...	ED	Edit	...
					⤷ Paste to Original Coordinates	

Command: pasteorig

COMMAND LINE OPTIONS
None.

RELATED COMMANDS
CopyClip copies a drawing to the Clipboard.

Insert inserts an AutoCAD drawing in the drawing.

PasteBlock pastes AutoCAD objects as a block at a user-specified insertion point.

RELATED SYSTEM VARIABLES
None.

TIPS
- Use this command to copy objects from one drawing to another.

- This command cannot be used to paste objects into the drawing from which they originate.

PasteSpec

Rel.13 Pastes Clipboard objects in the current drawing as embedded, linked, pasted, or AutoCAD objects (short for PASTE SPECial).

Command	Alias	Ctrl+	F-key	Alt+	Menu Bar	Dashboard
pastespec	pa	...	...	ES	Edit	...
					⮡ Paste Special	

Command: pastespec

Displays dialog box:

Left: *Paste.*
Right: *Paste Link.*

PASTE SPECIAL DIALOG BOX

⊙ **Paste** pastes the object as an embedded object.

○ **Paste Link** pastes the object as a linked object.

☐ **Display as Icon** displays the object as an icon from the originating application.

Change Icon allows you to select the icon.

RELATED COMMANDS

CopyClip copies the drawing to the Clipboard.

InsertObj inserts an OLE object in the drawing.

OleLinks edits the OLE link data.

PasteClip places the Clipboard object as a pasted object.

RELATED SYSTEM VARIABLE

OleHide toggles the display of OLE objects (1 = off).

TIP

- As of AutoCAD 2008, you can paste spreadsheet files directly into the drawing as tables. Follow these steps:
    1. In the spreadsheet program, copy a range of cells to the Clipboard (use **Ctrl+C**).
    2. In AutoCAD, start the **PasteSpec** command.
    3. From the **As** list, select "AutoCAD Entities."
    4. Choose the **Paste Link** option.
    5. Click **OK**, and follow the remaining prompts.

PcInWizard

2000 Converts PCP and PC2 plot configurations to PC3 format.

Command	Alias	Ctrl+	F-key	Alt+	Menu Bar	Dashboard
pcinwizard	...	...	...	TZI	Tools	...
					⤷Wizard	
					⤷Import Plot Settings	

Command: pcinwizard

Displays dialog box.

IMPORT PCP OR PC2 PLOT SETTINGS DIALOG BOX

Back returns to the previous step.

Next continues to the next step.

Cancel cancels the wizard.

Browse File page

PCP or PC2 filename specifies the name of the *.pcp* or *pc2* file.

Browse displays the Import dialog box.

Finish page

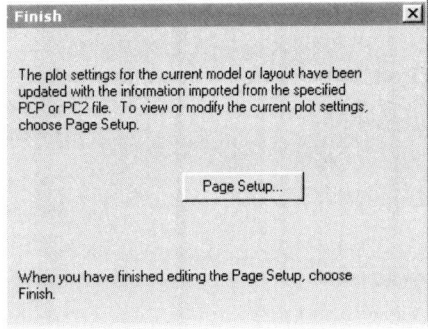

Page Setup displays the Page Setup dialog box; see the PageSetup command.

Finish completes the importation.

RELATED COMMANDS

PageSetup creates and modifies page setups.

Plot uses PC3 files to control the plotter configuration.

RELATED SYSTEM VARIABLES

None.

TIPS

- This command imports *.pcp* and *.pc2* files, and applies them to the current layout or model tab.

- Only use this command when have you created *.pcp* or *.pc2* files with earlier versions of AutoCAD.

- PCP is short for "plotter configuration parameters":
  - *.pcp* files are used by AutoCAD Release 13.
  - *.pc2* files are used by AutoCAD Release 14.
  - *.pc3* files are used by AutoCAD 2000 to now.

- AutoCAD imports the following information from *.pcp* and *.pc2* files: paper size, plot area, plot scale, plot origin, and plot offset.

- To import color-pen mapping, use the Plot Style Table wizard, run by the *StyShWiz.Exe* program.

- To import the optimization level and plotter connection, use the Add-a-Plotter wizard, run by the *AddPlWiz.Exe* program.

PEdit

V. 2.1 Edits 2D polylines, 3D polylines, or 3D meshes — depending on the selected object (short for Polyline EDIT).

Command	Alias	Ctrl+	F-key	Alt+	Menu Bar	Dashboard
pedit	pe	...	...	MO⬥P	Modify	...
					⬥Object	
					⬥Polyline	

Command: pedit

Options vary, depending on whether a 2D polyline, 3D polyline, or polymesh is picked.

. . . .

2D Polyline Operations
Select polyline or [Multiple]: *(Pick a 2D polyline, or type M.)*
Enter an option [Open/Join/Width/Edit vertex/Fit/Spline/Decurve/Ltype gen/Undo]: *(Enter an option.)*

Original Polyline · PEdit Close · PEdit Curve · PEdit Width · PEdit Spline · PEdit Edit vertex

COMMAND LINE OPTIONS
Multiple selects more than one polyline to edit.

Close closes an open polyline by joining the two endpoints with a single segment.

Decurve reverses the effects of Fit and Spline operations.

Edit vertex options are listed below.

Fit fits a curve to the tangent points of each vertex.

Ltype gen specifies the linetype generation style.

Join adds other polylines to the current polyline.

Open opens a closed polyline by removing the last segment.

Spline fits a splined curve along the polyline.

Undo undoes the most recent PEdit operation.

Width changes the width of the entire polyline.

eXit exits the PEdit command.

. . . .

Edit Vertex options
Enter a vertex editing option
[Next/Previous/Break/Insert/Move/Regen/Straighten/Tangent/Width/eXit]
<N>: *(Enter an option.)*

Note: It is much easier to use grips to edit polyline vertices than this Edit vertex option:

Break removes a segment or breaks the polyline at a vertex:

    Next moves the x-marker to the next vertex.

    Previous moves the x-marker to the previous vertex.

    Go performs the break.

    eXit exits the **Break** sub-submenu.

Insert inserts another vertex.

Move relocates a vertex; more easily accomplished with grips editing.

Next moves the x-marker to the next vertex.

Previous moves the x-marker to the previous vertex.

Regen regenerates the screen to show the effect of the PEdit command.

Straighten draws a straight segment between two vertices:

    Next moves the x-marker to the next vertex.

    Previous moves the x-marker to the previous vertex.

    Go performs the straightening.

    eXit exits the Straighten sub-submenu.

Tangent shows the tangent to the current vertex.

Width changes the width of a segment.

eXit exits the Edit vertex submenu.

- -

3D Polyline Operations
Command: pedit
Select polyline or [Multiple]: *(Pick a 3D polyline, or type M.)*
Enter an option [Open/Edit vertex/Spline curve/Decurve/Undo]: *(Enter an option.)*

COMMAND LINE OPTIONS

    Multiple selects more than one 3D polyline to edit.

    Close closes an open polyline.

    Decurve reverses the effects of the Fit curve and Spline curve operations.

    Open removes the last segment of a closed polyline.

- -

Spline curve fits a splined curve along the polyline.

Undo undoes the most recent PEdit operation.

eXit exits the PEdit command.

Edit Vertex options
**Enter a vertex editing option
[Next/Previous/Break/Insert/Move/Regen/Straighten/eXit] <N>:** *(Enter an option.)*

Break removes a segment or breaks the polyline at a vertex:

Next moves the x-marker to the next vertex.

Previous moves the x-marker to the previous vertex.

Go performs the break.

eXit exits the Break sub-submenu.

Insert inserts another vertex.

Move relocates a vertex; more easily accomplished with grips editing.

Next moves the x-marker to the next vertex.

Previous moves the x-marker to the previous vertex.

Regen regenerates the screen to show the effect of PEdit options.

Straighten draws a straight segment between two vertices:

Next moves the x-marker to the next vertex.

Previous moves the x-marker to the previous vertex.

Go performs the straightening.

eXit exits the Straighten sub-submenu.

eXit exits the Edit vertex submenu.

. . . .

3D Mesh Operations
Select polyline or [Multiple]: *(Pick a 3D mesh, or type **M**.)*
**Enter an option [Edit vertex/Smooth surface/Desmooth/Mclose/Nclose/
Undo]:** *(Enter an option.)*

Original 3D surface

Pedit Mclose

Pedit Smooth surface

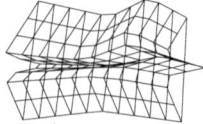
Pedit Nclose

. . . .

COMMAND LINE OPTIONS

Multiple selects more than one 3D mesh to edit.

Desmooth reverses the effect of the Smooth surface options.

Mclose closes the mesh in the m-direction.

Mopen opens the mesh in the m-direction.

Nclose closes the mesh in the n-direction.

Nopen opens the mesh in the n-direction.

Smooth surface smooths the mesh with a Bezier-spline.

Undo undoes the most recent PEdit operation.

eXit exits the PEdit command.

Edit Vertex Options
Current vertex (0,0).
Enter an option [Next/Previous/Left/Right/Up/Down/Move/REgen/eXit]
<N>: *(Enter an option.)*

Down moves the x-marker down the mesh by one vertex.

Left moves the x-marker along the mesh by one vertex left.

Move relocates the vertex to a new position; more easily accomplished with grips editing.

Next moves the x-marker along the mesh to the next vertex.

Previous moves the x-marker along the mesh to the previous vertex.

REgen regenerates the drawing to show the effects of the PEdit command.

Right moves the x-marker along the mesh by one vertex right.

Up moves the x-marker up the mesh by one vertex.

eXit exits the Edit vertex submenu.

RELATED COMMANDS

Break breaks 2D polylines at any position.

Chamfer chamfers all vertices of 2D polylines.

Convert converts older polylines to the new lwpolyline format.

EdgeSurf draws 3D meshes.

Fillet fillets all vertices of 2D polylines.

Join joins like objects, as well as lines, arcs, and polylines into a single polyline.

PLine draws 2D polylines.

RevSurf draws 3D meshes of revolution.

RuleSurf draws 3D ruled meshes.

TabSurf draws 3D tabulated meshes.

3D draws 3D surface objects.

3dPoly draws 3D polylines.

RELATED SYSTEM VARIABLES

SplFrame determines the visibility of a polyline spline frame:

0 — Does not display control frame (default).

1 — Displays control frame.

SplineSegs specifies the number (-32768 to 32767) of lines or arcs used to draw a splined polyline (default = 8); when the number is negative, arc segments are used, when positive, line segments.

SplineType determines the Bezier-spline smoothing for 2D and 3D polylines:

5 — Quadratic Bezier spline.

6 — Cubic Bezier spline.

SurfType determines the smoothing using the Smooth option:

5 — Quadratic Bezier spline.

6 — Cubic Bezier spline.

7 — Bezier surface.

TIPS

- During vertex editing, move the x-marker by pressing the spacebar, the Enter key, right-clicking, or pressing mouse button #2. The marker is moved to the next or previous vertex, whichever was used last.

- It's easier to edit the position of vertices with grips, than with this command.

PFace

Rel.11 Draws multisided 3D meshes; intended for use by AutoLISP and ARx programs (short for Poly FACE).

Command	Alias	Ctrl+	F-key	Alt+	Menu Bar	Dashboard
pface	...	...	...	...	...	...

Command: pface
Specify location for vertex 1: *(Pick point 1.)*
Specify location for vertex 2 or <define faces>: *(Pick point 2, or press Enter.)*
Face 1, vertex 1:
Enter a vertex number or [Color/Layer]: *(Type a number, or enter an option.)*
Face 1, vertex 2:
Enter a vertex number or [Color/Layer] <next face>: *(Type a number, or enter an option.)*
   ...and so on.

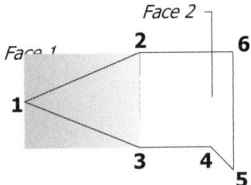

COMMAND LINE OPTIONS

Vertex defines the location of a vertex.

Face defines the faces, based on vertices.

Color gives the face a different color.

Layer places the face on a different layer.

RELATED COMMANDS

3dFace draws three- and four-sided 3D meshes.

3dMesh draws a 3D mesh with polyfaces.

RELATED SYSTEM VARIABLES

PFaceVMax specifies the maximum number of vertices per polyface (default = 4).

SplFrame controls the display of invisible faces (default = 0, not displayed).

TIPS

■ The 3dFace command creates 3- and 4-sided meshes, while this command creates meshes of an arbitrary number of sides, and allows you to control the layer and the color of each face.

■ The maximum number of vertices in the m- and n-direction is 256 when entered from the keyboard, and 32767 when entered from DXF files or from programs.

■ Make an edge invisible by entering a negative number for the beginning vertex of the edge.

■ This command is meant for programmers. To draw 3D mesh objects, use these commands instead: 3D, 3dMesh, RevSurf, RuleSurf, EdgeSurf, or TabSurf.

Plan

Rel.10 Displays the plan view of the WCS, the current UCS, or a named UCS.

Command	Alias	Ctrl+	F-key	Alt+	Menu Bar	Dashboard
plan	...	...	...	V3P	View	...
					⮩3D Views	
					⮩Plan View	

Command: plan
Enter an option [Current ucs/Ucs/World] <Current>: *(Enter an option.)*

Left: *Example of a 3D view.*
Right: *After using the Plan World command.*

COMMAND LINE OPTIONS

Current UCS shows the plan view of the current UCS.

Ucs shows the plan view of the named UCS.

World shows the plan view of the WCS.

RELATED COMMANDS

UCS creates new UCS views.

VPoint changes the viewpoint of 3D drawings.

RELATED SYSTEM VARIABLES

UcsFollow displays the plan view for UCS or WCS automatically.

ViewDir contains the x,y,z coordinates of the current view direction.

TIPS

- Entering VPoint **0,0,0** is an alternative to using this command.

- This command turns off perspective mode and clipping planes.

- Plan does not work in paper space; AutoCAD complains, "** Command only valid in Model space **."

- This command is an excellent method for turning off perspective mode.

PlaneSurf

2007 Draws flat rectangular surfaces, or converts 2D closed objects into surfaces (short for PLANar SURFace).

Command	Alias	Ctrl+	F-key	Alt+	Menu Bar	Dashboard
planesurf	...	...	...	DMF	Draw ↳Modeling ↳Planar Surface	3D Make

Command: planesurf
Specify first corner or [Object] <Object>: *(Pick a point, or type O.)*
Specify other corner: *(Pick another point.)*

COMMAND LINE OPTIONS

Specify first corner picks the first corner to make a rectangular surface.

Object converts the selected closed 2D object into a surface; choose a line, circle, arc, ellipse, elliptical arc, 2D polyline, planar 3D polyline, or spline.

Specify other corner specifies the other corner of the rectangular surface.

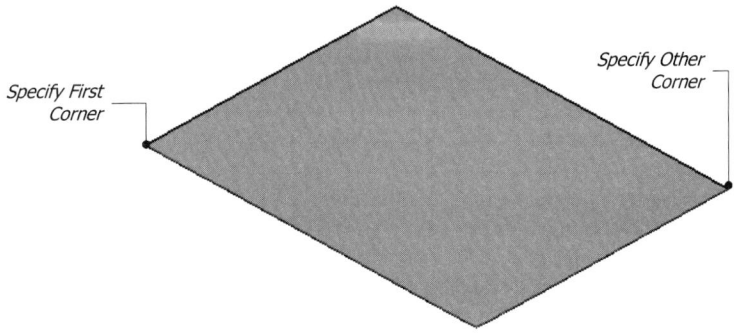

RELATED COMMAND

ConvertToSurface converts non-planar objects to surfaces.

RELATED SYSTEM VARIABLE

Elevation determines the z-height of planar surfaces.

TIPS

■ Use this command to create "floors."

■ Or ceilings, when Elevation is set to 96".

PLine

<u>**V. 1.4**</u> Draws complex 2D lines made of straight and curved sections of constant and variable width; treated as a single object (short for Poly LINE).

Command	Alias	Ctrl+	F-key	Alt+	Menu Bar	Dashboard
pline	pl	...	...	DP	Draw ⮡ Polyline	2D Draw

Command: pline
Specify start point: *(Pick a point.)*
Current line-width is 0.0000
Specify next point or [Arc/Halfwidth/Length/Undo/Width]: *(Pick a point, or enter an option.)*
Specify next point or [Arc/Close/Halfwidth/Length/Undo/Width]: *(Pick a point, or enter an option.)*

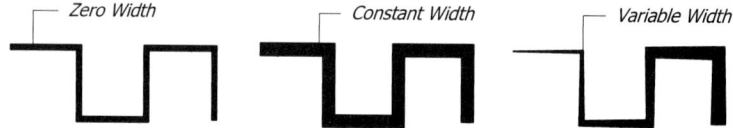
Zero Width   Constant Width   Variable Width

COMMAND LINE OPTIONS

Specify start point indicates the start of the polyline.

Close closes the polyline with a line segment.

Halfwidth indicates the half-width of the polyline.

Length draws a polyline tangent to the last segment.

Undo erases the last-drawn segment.

Width indicates the width of the polyline; start and end widths can be different.

Endpoint of line indicates the endpoint of the polyline.

Arc options
Specify endpoint of arc or
[Angle/CEnter/CLose/Direction/Halfwidth/Line/Radius/Second pt/Undo/
Width]: *(Enter an option.)*

Endpoint of arc indicates the arc's endpoint.

Angle indicates the arc's included angle.

CEnter indicates the arc's center point.

CLose uses an arc to close a polyline.

Direction indicates the arc's starting direction.

Halfwidth indicates the arc's half width.

Line switches back to the menu for drawing lines.

Radius indicates the arc's radius.

Second pt draws a three-point arc.

Undo erases the last drawn arc segment.

Width indicates the width of the arc.

RELATED COMMANDS

Boundary draws a polyline boundary.

Donut draws solid-filled circles as polyline arcs.

Ellipse draws ellipses as polyline arcs when PEllipse = 1.

Explode reduces a polyline to lines and arcs with zero width.

Fillet fillets polyline vertices with a radius.

PEdit edits the polyline's vertices, widths, and smoothness.

Polygon draws polygons as polylines of up to 1024 sides.

Rectang draws a rectangle out of a polyline.

Sketch draws polyline sketches, when SkPoly = 1.

Xplode explodes a group of polylines into lines and arcs of zero width.

3dPoly draws 3D polylines.

RELATED SYSTEM VARIABLES

PLineGen specifies the style of linetype generation:

0 — Vertex to vertex (default).

1 — End to end (compatible with Release 12).

PLineType controls the conversion of old (pre-R14) polylines and the creation of lwpolyline objects by the PLine command:

0 — Old polylines not converted; old-format polylines created.

1 — Not converted; lwpolylines created.

2 — Polylines in pre-R14 drawings converted on open;

PLine command creates lwpolyline objects (default).

PLineWid sets the current width of polyline (default = 0.0).

TIPS

- **Boundary** uses a polyline to outline a region automatically; use the List command to find its area.

- If you cannot see a linetype on a polyline, change system variable PlineGen to 1; this regenerates the linetype from one end of the polyline to the other.

- If the angle between a joined polyline and polyarc is less than 28 degrees, the vertex is chamfered; at greater than 28 degrees, the vertex is not chamfered.

- Use the object snap modes INTersection or ENDpoint to snap to the vertices of a polyline.

Plot

V.1.0 Creates hard copies of drawings on vector, raster, and PostScript plotters; also plots to files on disk.

Commands	Alias	Ctrl+	F-key	Alt+	Menu Bar	Dashboard
plot	print	P	...	FP	File	...
	dwfout				⌐ **Plot**	
-plot						

Command: plot

Displays dialog box::

PLOT DIALOG BOX

Page Setup options

Name selects or imports a previous plot setup.

Add prompts for the name of a new plot setup; displays the Add Page Setup dialog box.

Printer/Plotter Configuration options

Name selects a system printer, a *.pc3* named plotter configuration, or previous plot setting.

Properties displays the Plotter Configuration Editor dialog box; see PlotterManager command.

☐ **Plot to file** plots the drawing to a *.plt* file.

Paper Size option

Paper size displays paper sizes supported by the selected output device.

Plot Area options

What to plot:

- **Layout** plots all parts of the drawing within the margins of the specified paper size; the origin is calculated from 0,0 in the layout (not available in model space).
- **Display** plots all parts of the drawing within the current viewport.
- **Extents** plots the entire drawing.
- **Limits** plots the drawing in the Model tab within a rectangle defined by the Limits command.
- **View** plots a view saved with the View command (not available when no views are named).
- **Window** plots all parts of the drawing within a picked rectangle.

Plot Offset options

Center the plot centers the plot on the paper.

X specifies the plot origin offset from x = 0; the origin varies, depending on the plotter model.

Y specifies the plot origin offset from y = 0.

☑ **Center the plot** centers the plot between the page edge or printer margins (not available when the Layout option is selected, above).

Number of Copies option

Number of copies specifies the number of copies to plot.

Plot Scale options

☑ **Fit to paper** scales the drawing to fit within the plotter's margins (not available when the **Layout** option selected).

Scale specifies the plotting scale; ignored by layouts.

- **Custom** specifies user-defined scale; these can be edited by the ScaleListEdit command.

Inches / mm specifies the number of inches (or millimeters) on the plotted page.

Unit specifies number of units in drawing.

☐ **Scale lineweights** scales lineweights proportionately to the plot scale.

Preview displays the preview window; see the **Preview** command.

Apply to layout applies setting to layout.

OK starts the plot.

⊘ EXPANDED DIALOG BOX

Plot Style Table (Pen Assignments) options
Droplist selects a .ctb plot style table file; see the PlotStyle command.

New displays the Add Color-Dependent Plot Style Table wizard; see the StylesManager command.

Edit button displays the Plot Style Table Editor dialog box; see the StylesManager command.

Shaded Viewport Options options
Available only in model space; in layouts, right-click the viewport border, and then select Properties.

Shade Plot plots the viewport as-displayed, in a visual style, or rendered.

Quality sets print quality as draft (wireframe), preview (quarter-resolution, up to 150dpi), normal (half resolution, up to 300dpi), presentation (maximum printer resolution up to 600dpi), maximum (maximum printer resolution), or custom (as specified by DPI text box).

DPI specifies the resolution for the output device; DPI is short for "dots for inch."

Plot Options options
☐ **Plot in background** plots drawings as a background process.

☑ **Plot object lineweights** plots objects with lineweights (available only when plot style table turned off).

☑ **Plot with plot styles** plots objects using the plot styles.

☑ **Plot paperspace last** plots the model space objects first, followed by paper space objects (available in layouts only).

☐ **Hide paperspace objects** removes hidden lines before plotting.

☐ **Plot Stamp On** stamps the plot; when on, the Plot Stamp Settings button appears. See PlotStamp command.

☐ **Save changes to layouts** saves changes made in this dialog box to the layout.

Draw Orientation options
⊙ **Portrait** positions the paper vertically.

○ **Landscape** positions the paper horizontally.

☐ **Plot Upside Down** plots the drawing upside down.

-PLOT Command

Command: -plot
Detailed plot configuration? [Yes/No] <No>: *(Type Y or N.)*

Brief Plot Configuration options
Enter a layout name or [?] <Model>: *(Enter a layout name, or type ?.)*
Enter a page setup name <>: *(Enter a page setup name, or press Enter for none.)*
Enter an output device name or [?] <default printer>: *(Enter a printer name, or type ? for a list of printers.)*
Write the plot to a file [Yes/No] <N>: *(Type Y or N.)*
Save changes to page setup [Yes/No]? <N> *(Type Y or N.)*
Proceed with plot [Yes/No] <Y>: *(Type Y or N.)*

Detailed Plot Configuration options
Enter a layout name or [?] <Model>: *(Enter a layout name, or type ?.)*
Enter an output device name or [?] <default printer>: *(Enter a printer name, or type ? for a list of printers.)*
Enter paper size or [?] <Letter 8 ½ x 11 in>: *(Enter a paper size, or type ?.)*
Enter paper units [Inches/Millimeters] <Inches>: *(Type I or M.)*
Enter drawing orientation [Portrait/Landscape] <Landscape>: *(Type P or L.)*
Plot upside down? [Yes/No] <No>: *(Type Y or N.)*
Enter plot area [Display/Extents/Limits/View/Window] <Display>: *(Enter an option.)*
Enter plot scale (Plotted Inches=Drawing Units) or [Fit] <Fit>: *(Enter a scale factor, or type F.)*
Enter plot offset (x,y) or [Center] <0.00,0.00>: *(Enter an offset, or type C.)*
Plot with plot styles? [Yes/No] <Yes>: *(Type Y or N.)*
Enter plot style table name or [?] (enter . for none) <>: *(Enter a name, or enter an option.)*
Plot with lineweights? [Yes/No] <Yes>: *(Type Y or N.)*
Enter shade plot setting [As displayed/Wireframe/Hidden/Visual styles/Rendered] <As displayed>: *(Enter an option.)*

Write the plot to a file [Yes/No] <N>: *(Type Y or N.)*
Save changes to page setup? Or set shade plot quality? [Yes/No/Quality] <N>: *(Enter an option.)*
Proceed with plot [Yes/No] <Y>: *(Type Y or N.)*

COMMAND OPTIONS

Enter a layout name specifies the name of a layout; enter Model for model view or ? for a list of layout names.

Enter an output device name specifies the name of a printer, or enter ? for a list of print devices.

Enter paper size specifies the name of a paper size, or enter ? for a list of paper sizes supported by the print device.

Enter paper units specifies Inches (imperial) or Millimeters (metric) units.

Enter drawing orientation specifies Portrait (vertical) or Landscape (horizontal) orientation.

Plot upside down:
- **Yes** plots the drawing upside down.
- **No** plots the drawing normally.

Enter plot area specifies that the current Display, Extents of the drawing, Limits defined by the **Limits** command, View name, or window area be plotted.

Enter plot scale specifies the scale using the Plotted Inches=Drawing Units format; or enter F to fit the drawing to the page.

Enter plot offset specifies the distance between the lower left corner of the paper and the drawing; or enter **C** to center the drawing on the page.

Plot with plot styles:
- **Yes** prompts you to specify the name of a plot style.
- **No** ignores plot styles, and uses color-dependent styles.

Plot with lineweights:
- **Yes** uses lineweight settings to draw thicker lines.
- **No** ignores lineweights.

Enter shade plot setting specifies whether the plot should be rendered As displayed, Wireframe, Hidden lines removed, named Visual style, or Rendered.

Write the plot to a file:
- **Yes** sends the plot to a file.
- **No** sends the plot to the plot device.

Save changes to page setup:
- **Yes** saves the settings you specified during the previous set of questions.
- **No** does not save the settings.
- **Quality** specifies the quality of the plot.

Proceed with plot:
- **Yes** plots the drawing.
- **No** cancels the plot.

Quality options
Enter shade plot quality [Draft/Preview/Normal/pResentation/Maximum/Custom] <Normal>: *(Select an option.)*

Draft, Preview, Normal, pResentation, and Maximum apply preset dpi (dots per inch) settings to the plot. The higher the dpi, the better the quality, but the slower the plot.

Custom allows you to specify the dpi setting; default = 150dpi.

RELATED COMMANDS
PageSetup selects one or more plotter devices.
Preview goes directly to the plot preview screen.
Publish plots multi-page drawings.
PublishToWeb plots drawings to HTML format.
PsOut saves drawings in *.eps* format.
ViewPlotDetails reports on succussful and unsuccessful plots.

RELATED SYSTEM VARIABLES

BackgroundPlot toggles background plotting.

BgrndPlotTimeout specifies the time AutoCAD waits for plotter to respond.

FullPlotPath sends either the full path or just the drawing name to the plot spooler software.

RasterPercent specifies the percentage of system memory to allocate to plotting raster images; default = 20%.

RasterDpi specifies the scaling for raster devices.

RasterThreshold specifies the maximum amount of physical memory to assign to plotting raster images; plot fails when memory is insufficient. Default = 20MB.

TextFill toggles the filling of TrueType fonts.

TextQlty specifies the quality of TrueType fonts.

RELATED FILES

.pc3 holds plotter configuration parameter files.

.plt holds plot files created with this command.

TIPS

- AutoCAD R12 replaced PrPlot with the Plot command. R13 removed -p freeplot (plotting without using up a network license). AutoCAD 2005 removed the partial preview.

- AutoCAD cannot plot perspective views to-scale, only to-fit.

- Background plotting displays an icon in the tray; click it for options.

 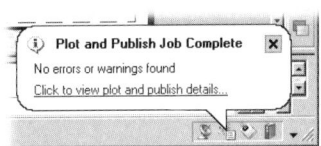

- You must wait for one background plot to complete before beginning the next one.

- Turn off BackgroundPlot when plotting drawings with scripts.

- As of AutoCAD 2008, the Plot command takes into account these 3dConfig command settings: Emulate Unsupported Hardware Effects, and Texture Compression.

'PlotStamp

2000i Adds information about drawings to hard copy plots.

Commands	Alias	F-key	Alt+	Menu Bar	Dashboard
plotstamp	ddplotstamp	...	...	...	...
-plotstamp					

Command: plotstamp

Displays dialog box:

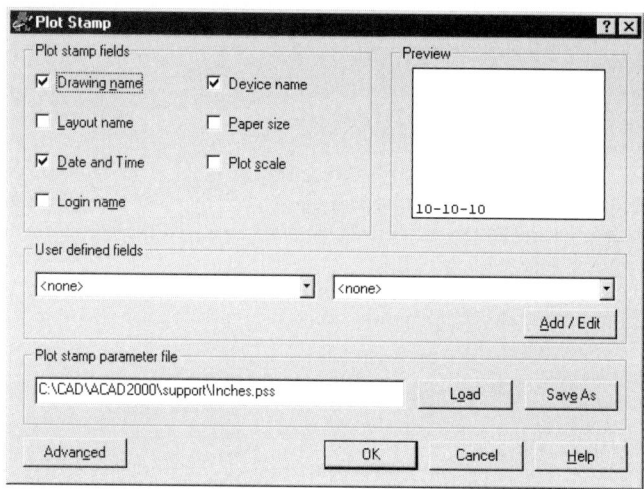

PLOT STAMP DIALOG BOX

Plot Stamp Fields options

☑ **Drawing name** adds the path and name of the drawing.

☐ **Layout name** adds the layout name.

☑ **Date and time** adds the date (short format) and time of the plot.

☐ **Login name** adds the Windows login name, as stored in the LogInName system variable.

☑ **Device name** adds the plotting device's name.

☐ **Paper size** adds the paper size, as currently configured.

☐ **Plot scale** adds the plot scale.

User Defined Fields options

Add/Edit displays the User Defined Fields dialog box.

Plot Stamp Parameter File options

Load/Save displays the Plotstamp Parameter File Name dialog box for opening (or saving) *.pss* files.

Advanced displays the Advanced Options dialog box.

ADVANCED OPTIONS DIALOG BOX

Location and Offset options

Location specifies the location of the plot stamp: Top Left, Bottom Left (default), Bottom Right, or Top Right — relative to the orientation of the drawing on the plotted page.

Stamp upside down:

☑ Draws the plot stamp upside down.

☐ Draws the plot stamp rightside up.

Orientation rotates the plot stamp relative to the page: Horizontal or Vertical.

X offset specifies the distance from the corner of the printable area or page; default = 0.1.

Y offset specifies the distance from the corner of the printable area or page; default = 0.1

⊙ **Offset relative to printable area.**

○ **Offset relative to paper border.**

Text Properties options

Font selects the font for the plot stamp text.

Height specifies the height of the plot stamp text.

Single line plot stamp:

☑ Plotstamp text is placed on a single line.

☐ Plotstamp text is placed on two lines.

Plot Stamp Units option

Units selects either inches, millimeters, or pixels.

Log File Location options

☑ **Create a log file** saves the plot stamp text to a text file.

Browse displays the Log File Name dialog box.

-PLOTSTAMP Command

Command: -plotstamp

Current plot stamp settings:

Displays the current setting of nearly 20 plot stamp settings.

Enter an option [On/OFF/Fields/User Fields/Log file/LOCation/Text properties/UNits]: *(Enter an option.)*

COMMAND LINE OPTIONS

On turns on plot stamping.

OFF turns off plot stamping.

Fields specifies plot stamp data: drawing name, layout name, date and time, login name, plot device name, paper size, plot scale, comment, write to log file, log file path, location, orientation, offset, offset relative to, units, font, text height, and stamp on single line.

User fields specifies two user-defined fields.

Log file writes the plotstamp data to a file instead of the drawing.

LOCation specifies the location and orientation of the plotstamp.

Text properties specifies the font name and height.

UNits specifies the units of measurement: inches, millimeters, or pixels.

RELATED COMMAND

Plot plots the drawing.

RELATED FILES

*.log is the plotstamp log file; stored in ASCII text format.

*.pss holds the plotstamp parameters; stored in binary format.

TIPS

- When the options of the Plot Stamp dialog box are grayed out, or when the -Plotstamp command reports 'Current plot stamp file or directory is read only,' this means that the *Inches.pss* or *Mm.pss* file in the *Support* folder is read-only. To change, in Explorer: (1) right-click the file, (2) select **Properties**, (3) uncheck **Read-only**, and (4) click **OK**.

- You can access the Plot Stamp dialog box via the Plot command's More Options dialog box:

- *Warning!* Too large of an offset value positions the plotstamp text beyond the plotter's printable area, which causes the text to be cut off. To prevent this, use the **Offset relative to printable area** option.

PlotStyle

2000 Selects and assigns plot styles to objects.

Commands	Alias	Ctrl+	F-key	Alt+	Menu Bar	Dashboard
plotstyle	...	...	...	OY	Format	Object Properties
					⤳Plot Style	
-plotstyle						

Command: plotstyle

When no objects are selected, displays the Current Plot Style dialog box.

When one or more objects are selected, displays the Select Plot Style dialog box.

SELECT PLOT STYLE DIALOG BOX

Displayed when one or more objects are selected:

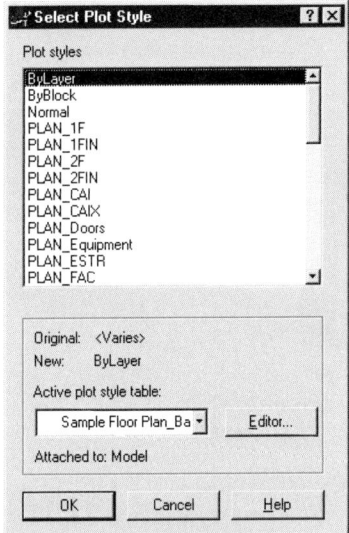

Plot styles lists the plot styles available in the drawing.

Active plot style table selects the plot style table to attach to the current drawing.

Editor displays the Plot Style Table Editor dialog box; see the StylesManager command.

CURRENT PLOT STYLE DIALOG BOX

Displayed when no objects are selected:

Plot styles lists the plot styles available in the drawing.

Active plot style table selects the plot style table to attach to the current drawing.

Editor displays the Plot Style Table Editor dialog box; see the StylesManager command.

. . . .

-PLOTSTYLE Command

Command: -plotstyle
Current plot style is "Default"
Enter an option [?/Current] : *(Type ? or C.)*
Set current plot style : *(Enter a name.)*
Current plot style is "*plotstylename*"
Enter an option [?/Current] : *(Press Enter.)*

COMMAND LINE OPTIONS

Current prompts you to change plot styles.

Set current plot style specifies the name of the plot style.

ENTER exits the command.

? displays plot style names; sample output:

```
        Plot Styles:
        -----------------
        ByLayer
        ByBlock
        Normal
        ARCH_Dimensions for stairway
        Current plot style is "PLAN_ESTR"
```

. . . .

RELATED COMMANDS

Plot plots drawings with plot styles.

PageSetup attaches plot style tables to layouts.

StylesManager modifies plot style tables.

Layer specifies plot styles for layers.

RELATED SYSTEM VARIABLES

CPlotStyle specifies the plot style for new objects; defined values include "ByLayer," "ByBlock," "Normal," and "User Defined."

DefLPlStyle specifies the plot style name for layer 0.

DefPlStyle specifies the default plot style for new objects.

PStyleMode indicates whether the drawing is in Color-Dependent or Named Plot Style mode.

PStylePolicy determines whether object color properties are associated with plot style.

TIPS

- This command does not operate until a plot style table has been created for the drawing; the PlotStyle command displays this message box:

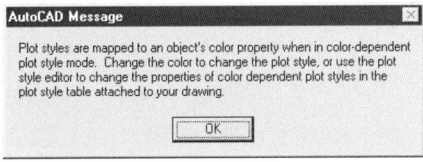

- A plot style can be assigned to any object and to any layer.

- A plot style can override the following plot settings: color, dithering, gray scale, pen assignment, screening, linetype, lineweight, end style, join style, and fill style.

- Plot styles are useful for plotting the same layout in different ways, such as emphasizing objects using different lineweights or colors in each plot.

- Plot style tables can be attached to the Model tab and layout tabs, and attach different plot style tables to layouts, to create different looks for plots.

- By default, drawings do not use plot styles. To turn on plot styles, follow the tutorial.

- Before you can apply plot styles to a new drawing, you must turn on plot styles; notice that Plot Style Control in the Object Properties toolbar is grayed out.

Follow these steps to turn on plot style:

1. From the **Tools** menu, select **Options**.

2. In the Options dialog box, select the **Plot and Publish** tab, and then click the **Plot Style Table Settings** button.

3. In the Plot Style Table Settings dialog box, select the **Use Named Plot Styles** option.

4. From the **Default plot style** list, select any *.stb* plot style.

5. Click **OK** to dismiss both dialog boxes.

The Plot Style Control in the Properties toolbar and Layer Properties Manager dialog box is not available until the *next* new drawing you start, curiously enough.

PlotterManager

__2000__ Displays the Plotters window, the Add-A-Plotter wizard, and PC3 configuration Editor.

Command	Alias	Ctrl+	F-key	Alt+	Menu Bar	Dashboard
plottermanager	...	...	...	FM	File	...
					⮡ Plotter Manager	

Command: plottermanager

Displays window:

PLOTTERS WINDOW

Add-a-Plotter Wizard adds a plotter configuration; double-click to run the Add Plotter wizard.

*.pc3 specifies parameters for creating plotted output; double-click to display the Plotter Configuration Editor; see the StylesManager command.

ADD PLOTTER WIZARD

The following steps create PC3 plotter configurations for plotting drawings to generic HP plotters; the steps are similar to creating .pc3 files for other types of plotters:

Next displays the next dialog box.
Cancel dismisses the dialog box.

Begin page

My Computer configures printers and plotters connected to your computer.

Network Plotter Server configures printers and plotters connected to other computers on networks.

System Printer configures the default Windows printer.

Plotter Model page

Manufacturer selects a brand name of plotter.

Model selects a specific model number.

Have Disk selects plotter drivers provided by the manufacturer; displays the Open dialog box.

Import Pcp or Pc2 page

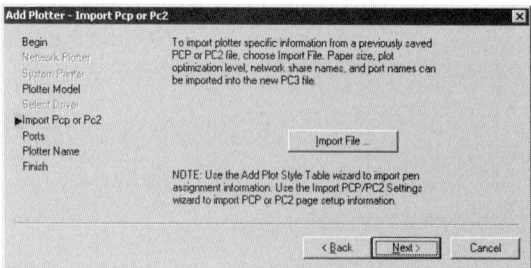

Import File imports *.pcp* and *.pc2* plotter configuration files from earlier versions of AutoCAD.

Ports page

⦿ **Plot to a port** sends the plot to an output port.

○ **Plot to File** plots the drawing to a file with user-defined file names; default file name is the same as the drawing name with the *.plt* extension; PostScript plot files are given the *.eps* extension.

○ **AutoSpool** plots the drawing to files with the file name generated by AutoCAD, and then executes the command specified in the Option dialog box's Files tab. Enter the name of the program that should process the AutoSpool file in Print Spool Executable. You may include these DOS command-line arguments:

Argument	Meaning
%s	Substitutes path, spool filename, and extension.
%d	Specifies the path, AutoCAD drawing name, and extension.
%c	Describes the device.
%m	Returns the plotter model.
%n	Specifies the plotter name.
%p	Specifies the plotter number.
%h	Returns the height of the plot area in plot units.
%w	Returns the width of the plot area in plot units.
%i	Specifies the first letter of the plot units.
%l	Specifies the login name (*LogInName system variable*).
%u	Specifies the user name.
%e	Specifies the equal sign (=).
%%	Specifies the percent sign (%).

Port options

Port lists the virtual ports defined by the Windows operating system:

Port	Meaning
USB	Universal serial bus.
COM	Serial port.
LPT	Parallel port.
HDI	Autodesk's Heidi Device Interface.

Description describes the type of port:

Description	Meaning
Local Port	Printer is connected to your computer.
Network Port	Printer is accessible through the network.

Printers describes the brand name of the printer connected to the port.

Configure Port displays the Configure Port dialog box; allows you to specify parameters specific to the port, such as timeout and protocol.

What is AutoSpool? displays an explanatory dialog box:

☐ **Show all system ports and disable I/O port validation** prevents AutoCAD from checking whether the port is valid.

Plotter Name page

Plotter name specifies a user-defined name for the plotter configuration; you may have many different configurations for a single plotter.

Finish Page

Edit Plotter Configuration displays the Edit Plotter Configuration dialog box.

Calibrate Plotter displays the Calibrate Plotter wizard.

PLOTTER CONFIGURATION EDITOR DIALOG BOXES

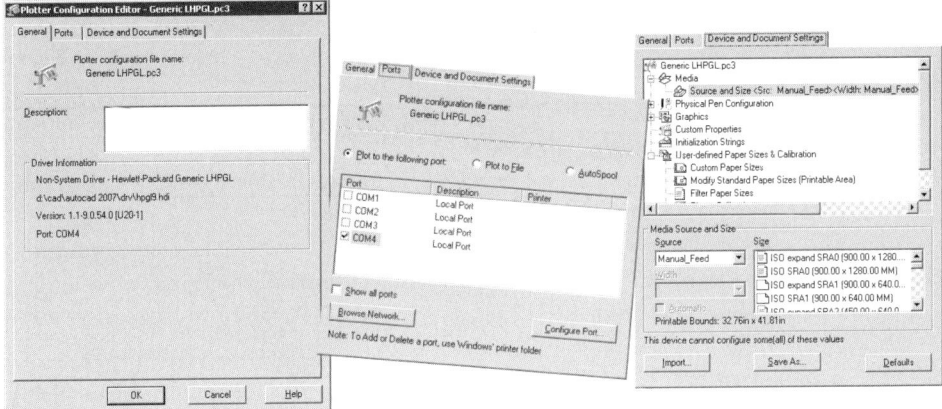

General tab

Description allows you to provide a detailed description of the plotter configuration.

Ports tab

⊙ **Plot to the following port** sends the plot to an output port.

○ **Plot to File** plots the drawing to a file with a user-defined file name; default file name is the same as the drawing name, with a *.plt* extension; PostScript plot files are given the *.eps* extension.

○ **AutoSpool** plots the drawing to a file with a filename generated by AutoCAD, and then executes the command specified in the Option dialog box's Files tab.

☐ **Show all ports** lists all ports on the computer.

Browse Network displays the Browse for Printer dialog box; selects a printer on the network.

Configure Port displays the Configure Port dialog box; allows you to specify the parameters specific to the port, such as timeout and protocol.

Device and Document Settings tab

Media specifies the paper source, paper size, type, and destination.

Physical Pen Configuration specifies the physical pens in the pen plotter (for pen plotters only).

Graphics specifies settings for plotting vector and raster graphics and TrueType fonts.

Custom Properties specifies settings specific to the plotter, printer, or other output device.

Initialization Strings specifies the control codes for pre-initialization, post-initialization, and termination (for non-system plotters only).

Calibration Files and Paper Sizes calibrates the plotter by specifying the *.pmp* file; adds and modifies custom paper sizes; see the Calibrate Plotter wizard.

Import imports *.pcp* and *.pc2* plotter configuration files from earlier versions of AutoCAD.

Save As saves the plotter configuration data to a *.pc3* file.

Defaults resets the plotter configuration settings to the previously-saved values.

RELATED COMMANDS

Plot plots drawings with plot styles.

PageSetup attaches plot style tables to layouts.

StylesManager modifies plot style tables.

AddPlWiz.Exe runs the Add Plotter wizard.

RELATED SYSTEM VARIABLES

PaperUpdate toggles the display of a warning before AutoCAD plots a layout with a paper size different from the size specified by the plotter configuration file:

0 — Displays warning dialog box (default).

1 — Changes paper size to match the size specified by the plotter configuration file.

PlotRotMode controls the orientation of plots:

0 — Aligns rotation icon with media at the lower left for 0 degrees; calculates x and y origin offsets relative to lower-left corner.

1 — Aligns the lower-left corner of plotting area with lower-left corner of the paper.

2 — Same as 0, except x and y origin offsets relative to the rotated origin position.

TIPS

- You can create and edit *.pc3* plotter configuration files without AutoCAD. From the Start button on the Windows toolbar, select **Settings | Control Panel | Autodesk Plotter Manager**.

- The *dwf classic.pc3* file specifies parameters for creating *.dwf* files via the Release 14-compatible DwfOut command.

- The *dwf eplot.pc3* file specifies parameters for creating *.dwf* files via the Plot command.

PngOut

2004 Exports the current view in PNG raster format.

Command	Alias	Ctrl+	F-key	Alt+	Menu Bar	Dashboard
pngout	...	...	...	...	...	...

Command: pngout

*Displays Create Raster File dialog box. Specify a filename, and then click **Save**.*

Select objects or <all objects and viewports>: *(Select objects, or press Enter to select all objects and viewports.)*

COMMAND LINE OPTIONS

Select objects selects specific objects.

All objects and viewports selects all objects and all viewports, whether in model space or in layout mode.

RELATED COMMANDS

BmpOut exports drawings in BMP (bitmap) format.

Image places raster images in the drawing.

JpgOut exports drawings in JPEG (joint photographic expert group) format.

Plot exports drawings in PNG and other raster formats.

TifOut exports drawings in TIFF (tagged image file format) format.

TIPS

- The rendering effects of the VsCurrent command are preserved, but not those of the Render command.

- PNG files are used as a royalty-free alternative to JPEG files.

- PNG is short for "portable network graphics."

▪ **Point**

<u>**V. 1.0**</u>   Draws a 3D point.

Command	Alias	Ctrl+	F-key	Alt+	Menu Bar	Dashboard
point	po	...	...	DO	Draw ⮡**Point**	2D Draw

Command: point
Current point modes: PDMODE=0 PDSIZE=0.0000
Specify a point: *(Pick a point.)*

COMMAND LINE OPTION
Point positions a point, or enters a 2D or 3D coordinate.

RELATED COMMANDS
DdPType displays a dialog box for selecting PsMode and PdSize.

Regen regenerates the display to see the new point mode or size.

RELATED SYSTEM VARIABLES
PDMode determines the appearance of a point:

0	1	2	3	4

32	33	34	35	36

64	65	66	67	68

96	97	98	99	100

PDSize determines the size of a point:
  0 — 5% of height of the ScreenSize system variable (default).
  1 — No display.
  -10 — *(Negative)* Ten percent of the viewport size.
  10 — *(Positive)* Ten pixels in size.

TIPS
▪ The size and shape of points are determined by PdSize and PdMode; changing these changes the appearance and size of *all* points in the drawing with the next regeneration (Regen).

▪ Entering only x,y coordinates places the point at the z coordinate of the current elevation; setting Thickness to a value draws points as lines in 3D space.

▪ Prefix coordinates with * (such as *1,2,3) to place points in the WCS, rather than in the UCS. Use the object snap mode NODe to snap to points.

▪ Points plot according to the setting of PdMode.

Aliased Commands
The **SpotLight** and **PointLight** command options are accessed from the Light command.

Polygon

<u>**V. 2.5**</u> Draws 2D polygons of between three 3 and 1024 sides.

Command	Alias	Ctrl+	F-key	Alt+	Menu Bar	Dashboard
polygon	pol	...	...	DY	Draw ↳Polygon	2D Draw

Command: polygon
Enter number of sides <4>: *(Enter a number.)*
Specify center of polygon or [Edge]: *(Pick a point, or type E.)*
Enter an option [Inscribed in circle/Circumscribed about circle] <I>: *(Type I or C.)*
Specify radius of circle: *(Enter a value, or pick two points.)*

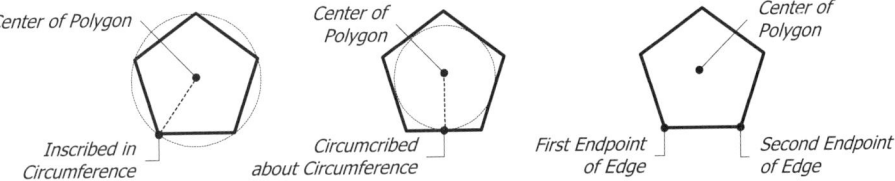

COMMAND LINE OPTIONS
Center of polygon indicates the center point of the polygon; then:

C (*Circumscribed*) fits the polygon outside an indicated circumference.

I (*Inscribed*) fits the polygon inside an indicated circumference.

Edge draws the polygon based on the length of one edge.

RELATED COMMANDS
PEdit edits polylines, including polygons.

Rectang draws squares and rectangles.

RELATED SYSTEM VARIABLE
PolySides specifies the most recently-entered number of sides (default = 4).

TIPS
- Polygons are drawn as a polyline; use PEdit to edit the polygon, such as its width.

- The pick point determines the polygon's first vertex; polygons are drawn counterclockwise.

- Use the system variable PolySides to preset the default number of polygon sides.

- Use the Snap command to place the polygon precisely; use INTersection or ENDpoint object snap modes to snap to the polygon's vertices.

Aliased Command
Preferences was removed from AutoCAD 2000; it now is an alias for the **Options** command.

PolySolid

2007 Draws 3D solid "walls" made of straight and curved sections in a manner similar to polylines.

Command	Alias	Ctrl+	F-key	Alt+	Menu Bar	Dashboard
polysolid	psolid	...	...	DM⬥	Draw	3D Make
					⬥Modeling	
					⬥Polysolid	

Command: polysolid
Specify start point or [Object/Height/Width/Justify] <Object>: *(Pick a point, or enter an option.)*
Specify next point or [Arc/Undo]: *(Pick a point, press Enter, or enter an option.)*

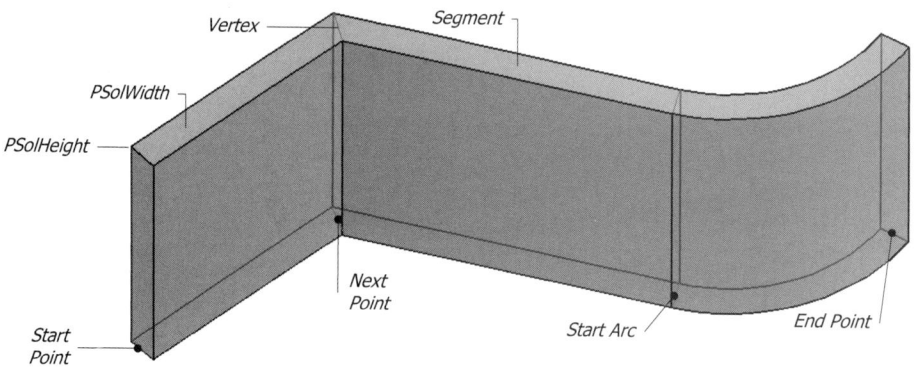

COMMAND LINE OPTIONS

Specify start point indicates the starting point by picking a point or entering x,y coordinates.

Object selects the line, arc, circle, or 2D polyline to convert into a solid.

Height specifies the height of the polysolid; default value is stored in the PSolHeight system variable.

Width specifies the width; default value is found in the PSolWidth system variable.

Justify determines whether the base of the polysolid is drawn to the left, on the center, or to the right of the pick points.

Specify the next point indicates the vertex of the wall; press Enter to stop drawing the polysolid.

Arc switches to arc-drawing mode with prompts similar to that of the PLine command.

Undo undraws the last segment.

Close closes the polysolid by joining the endpoint with the start point.

RELATED SYSTEM VARIABLES

PSolHeight specifies the default height of polysolids.

PSolWidth specifies the default width of polysolids.

PressPull

<u>**2007**</u> Extends and shortens the faces of 3D solids.

Command	Alias	Ctrl+	F-key	Alt+	Menu Bar	Dashboard
presspull	...	...	...	...	...	3D Make

Command: presspull
Click inside bounded areas to press or pull. *(Pick a face, and then drag.)*

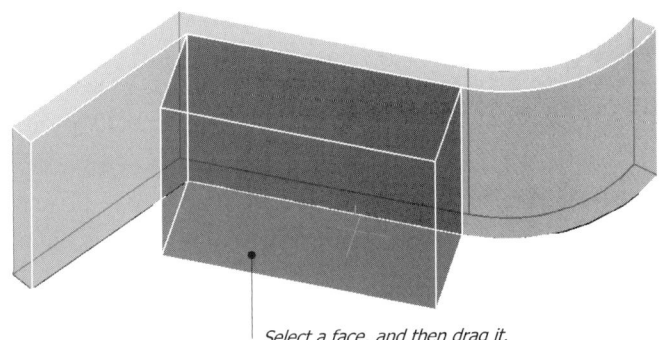

Select a face, and then drag it.

COMMAND LINE OPTION

Click inside bounded areas lets you drag the following faces:

- Areas that can be hatched by picking a point with zero gap tolerance, such as circles.
- Areas enclosed by crossing coplanar linear geometry — includes edges and geometry in blocks.
- Closed polylines, regions, 3D faces, and 2D solids with coplanar vertices.
- Areas created by geometry (including edges on faces) drawn coplanar to any face of 3D solids.

RELATED COMMANDS

Extrude converts 2D areas into 3D solids with optional tapered sides.

Thicken converts surfaces to 3D solids.

TIPS

- This command can be used to lengthen and shorten the faces of 3D solids.
- When used on 2D objects, like circles, the circle is turned into a 3D solid.
- When pushpulling "holes" (cylinders subtracted from solids), the result may be unexpected.
- I recommend using this command only for creating solids from closed boundaries and regions.
- Use Ctrl+pick to select a face, and then use grips editing to press and pull the face.

Preview

Rel.13 Displays plot preview; bypasses the Plot command.

Command	Alias	Ctrl+	F-key	Alt+	Menu Bar	Dashboard
preview	pre	...	...	FV	File 🖑Plot Preview	...

Command: preview
Press ESC or ENTER to exit, or right-click to display shortcut menu.
Displays preview screen:

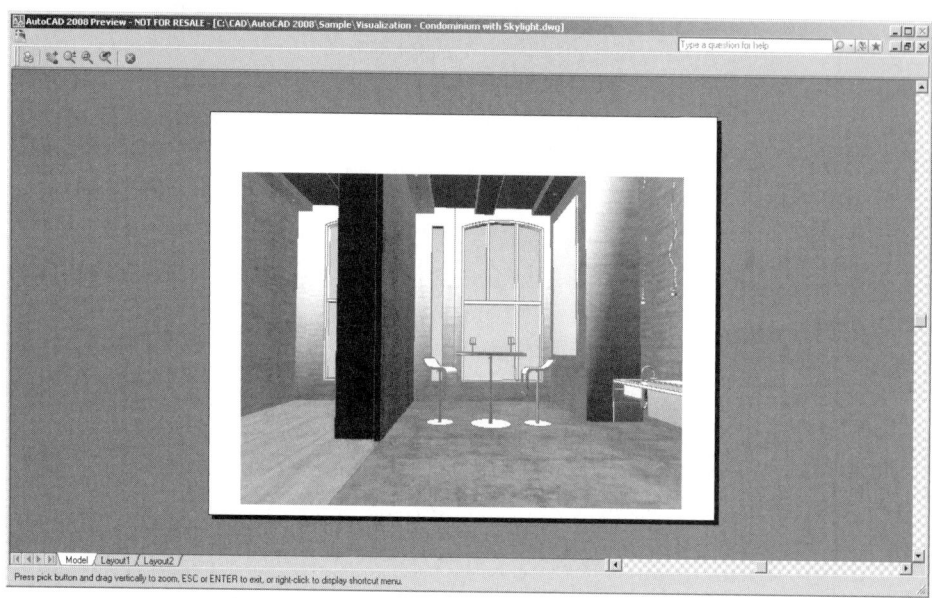

COMMAND LINE OPTION

Esc or **Enter** returns to the drawing window.

Preview Toolbar

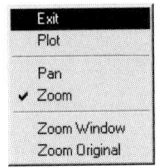

Exit exits preview mode.

Plot plots the drawing; goes immediately to plotting, bypassing the plot dialog box.

Pan pans the preview image in real time.

Zoom enlarges and shrinks the preview image in real time.

Zoom Window zooms into a windowed area.

Zoom Original returns to the orignal size.

RELATED COMMANDS

Plot plots the drawing.

PageSetup enables plot preview once a plotter is assigned to the layout.

TIPS

- This command does not operate when no plotter is assigned; use the PageSetup command to assign a plotter to the Model and layout tabs.

- Press ESC to exit preview mode.

- Partial preview mode was removed with AutoCAD 2005.

- To zoom or pan, select the mode, and then hold down the left mouse button. The image zooms or pans as you move the mouse.

- To change the background color of the preview screen, go to Tools | Options | Display | Colors, and then select "Plot Preview Background" from the Window Element droplist.

'Properties, 'PropertiesClose

2000 Opens and closes the Properties palette for examining and modifying properties of selected objects.

Command	Aliases	Ctrl+	F-key	Alt+	Menu Bar	Dashboard
properties	ch	1	...	TP	Tools	...
	mo				⤷Palettes	
	pr, props				⤷Properties	
	ddchprop			MP	Modify	
	ddmodify				⤷Properties	
propertiesclose	prclose	1	...	TP	Tools	...
					⤷Properties	

You can right-click any object and select Properties from the shortcut menu. Better yet, double-click objects to bring up this palette.

Command: properties

Displays palettte with different content, depending on the objects selected. When no objects are selected:

- Title Bar
- Parameters
- Description
- Resize
- Close Palette
- Toolbar
- Change Values
- Collapse Section
- AutoHide
- Options

Palette toolbar

- Select Objects
- Quick Select
- Toggle PickAdd

Toggle Value of PickAdd Variable controls selections when the Shift key is held down:

- **Off** adds objects to the selection set.
- **On** replaces objects in the selection set.

Select Objects prompts, 'Select objects:'; you can also just pick objects without this button.

Quick Select displays the Quick Select dialog box; see the QSelect command.

RELATED COMMANDS

ChProp changes an object's color, layer, linetype, and thickness.

Style creates and changes text styles.

RELATED SYSTEM VARIABLES

OpmState reports whether the Properties palette is open.

CeColor specifies the current color.

CeLtScale specifies the current linetype scale factor.

CeLtype specifies the current linetype.

CLayer specifies the current layer.

CAnnoScale reports the name of the current annotation scale for the current space; a separate scale can be used for model space and paperspace viewports.

CMaterial reports the current material.

CMleaderStyle reports the current multileader style.

Elevation specifies the current elevation in the z-direction.

Thickness specifies the current thickness in the z-direction.

TIPS

- To count objects in the drawing, press Ctrl+A to select all objects, and then click the Selection list:

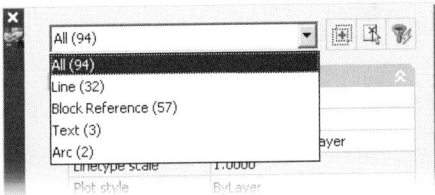

- As an alternative to entering the PropertiesClose command, you can click the **x** button on the palette's title bar.

- Double-click the title bar to dock and undock the palette within the AutoCAD window.

- The Properties palette can be dragged larger and smaller by its edges.

- When an item is displayed by gray text, it cannot be modified.

- This command can only be used transparently after it has been used at least once non-transparently.

. . . .

Removed Command

The **PsDrag** command was discontinued with AutoCAD 2000i. It has no replacement.

. . . .

PSetupIn

2000 Imports user-defined page setups into the current drawing layout (short for Page SETUP IN).

Commands	Alias	Ctrl+	F-key	Alt+	Menu Bar	Dashboard
psetupin	...	...	...	...	...	...
-psetupin						

Command: psetupin

Displays the Select File dialog box. Select a .dwg, .dwt, or .dxf file, and then click **Open.**
AutoCAD displays dialog box:

IMPORT PAGE SETTINGS DIALOG BOX

Name lists the names of page setups.

Location lists the location of the setups.

OK closes the dialog box, and loads the page setup.

. . . .

-PSETUPIN Command

Command: -psetupin
Enter filename: *(Enter .dwg file name.)*
Enter user defined page setup to import or [?]: *(Enter a name, or type **?**.)*

COMMAND LINE OPTIONS

Enter filename enters the name of a drawing file.

Enter user defined page setup to import enters the name of a page setup.

? lists the names of page setups in the drawing.

RELATED COMMANDS

PageSetup creates a new page setup configuration.

Plot plots the drawing.

. . . .

PsFill

Rel.12 Fills 2D polyline outlines with raster PostScript patterns (short for Post-Script FILL; an undocumented command).

Command	Alias	Ctrl+	F-key	Alt+	Menu Bar	Dashboard
psfill	...	...	...	...	...	...

Command: psfill
Select polyline: *(Pick an outline.)*
Enter PostScript fill pattern name (. = none) or [?] <.>: *(Enter pattern name, or type ?.)*

COMMAND LINE OPTIONS

Select polyline selects the closed polyline to fill.

PostScript pattern specifies the name of the fill pattern.

. (dot) selects no fill pattern.

? lists the available fill patterns.

* specifies that the pattern should not be outlined with a polyline.

RELATED COMMANDS

BHatch fills an area with vector hatch patterns and solid colors.

PsOut exports drawings as PostScript files.

RELATED SYSTEM VARIABLE

PSQuality specifies the display options for PostScript images:

      75 — *(Positive)* Displays filled image at 75dpi (default).

      0 — Displays bounding box and filename; no image.

      -75 — *(Negative)* Displays image outline at 75dpi; no fill.

TIPS

■ These fill patterns are available:

Grayscale	RGBcolor
AIlogo	Lineargray
Radialgray	Square
Waffle	Zigzag
Stars	Brick
Specks	

■ You cannot see fill patterns in AutoCAD drawings until the drawings are exported with the PsOut command.

. . . .

Removed Command

PsIn command was removed with AutoCAD 2000i; there is no replacement.

. . . .

PsOut

Rel.12 Exports the current drawing as an encapsulated PostScript file (an undocumented command).

Command	Alias	Ctrl+	F-key	Alt+	Menu Bar	Dashboard
psout	...	...	...	FE	File	...
					↳Export	

Command: psout

Displays the Create PostScript File dialog box.

Select Tools | Options to display dialog box of options:

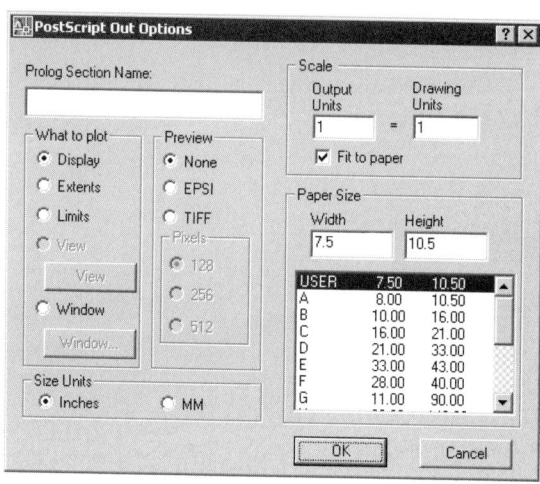

POSTSCRIPT OUT OPTIONS DIALOG BOX

Prolog Section Name specifies the name of the optional prolog section read from the *acad.psf* file. It customizes the PostScript output.

What to Plot

⊙ **Display** selects the display in the current viewport.

○ **Extents** selects the drawing extents.

○ **Limits** selects the drawing limits.

○ **View** selects a named view.

○ **Window** picks two corners of a window.

Preview

⊙ **None** specifies no preview image (default).

○ **EPSI** specifies Macintosh preview image format.

○ **TIFF** specifies Tagged Image File Format (recommended).

Pixels

128 specifies a preview image size of 128x128 pixels (default).

256 specifies a preview image size of 256x256 pixels.

512 specifies a preview image size of 512x512 pixels (recommended).

Size Units

Inches specifies the plot parameters in inches.

MM specifies the plot parameters in millimeters.

Scale

Output Units scales the output units.

Drawing Units specifies the drawing units.

Fit to Paper forces the drawing to fit paper size.

Paper Size

Width enters a width for the output size.

Height enters a height for the output size.

RELATED COMMAND

Plot exports the drawing in a variety of formats, including raster EPS and PDF.

RELATED SYSTEM VARIABLE

PSProlog specifies the PostScript prolog information.

RELATED FILE

*.eps extension of file produced by PsOut.

TIPS

- The "screen preview image" is only used for screen display purposes, since graphics software generally cannot display PostScript graphic files.

- When you select the Window option, AutoCAD prompts you for the window corners *after* you finish selecting options.

- Although Autodesk recommends using the smallest screen preview image size (128x128), even the largest preview image (512x512) has a minimal effect on file size and screen display time. Some software programs, such as those from Microsoft, might reject an *.eps* file when the preview image is larger than 128x128. The screen preview image size has no effect on the quality of the PostScript output.

- If you're not sure which screen preview format to use, select TIFF.

PSpace

Rel.11 Switches from model space to paper space in layout mode (short for Paper SPACE).

Command	Alias	Ctrl+	F-key	Alt+	Menu Bar	Dashboard
pspace	ps	...	...	...	...	...

This command does not operate in model space.
Command: pspace

COMMAND LINE OPTIONS
None.

RELATED COMMANDS
MSpace switches from paper space to model space.

MView creates viewports in paper space.

UcsIcon toggles the display of the paper space icon.

Zoom scales paper space relative to model space with the XP option.

RELATED SYSTEM VARIABLES
MaxActVP specifies the maximum number of viewports displaying an image.

PsLtScale specifies the linetype scale relative to paper space.

TileMode allows paper space when set to 0.

TIPS
- Use paper space to layout multiple views of a single drawing.

- In layout mode, you can switch to paper space by clicking the word MODEL on the status bar; switch to a layout's model space by clicking the word PAPER.

| 4.0679, 0.4813, 0.0000 | SNAP GRID ORTHO POLAR OSNAP OTRACK LWT MODEL |

| 4.0679, 0.4813, 0.0000 | SNAP GRID ORTHO POLAR OSNAP OTRACK LWT PAPER |

- When a drawing is in paper space, AutoCAD displays **PAPER** on the status line and the paper space icon:

- Paper space is known as "drawing composition" in some other CAD packages.

'Publish

2004 Outputs multiple layout sheets from one or more drawings as a single, multi-page *.dwf* file or hardcopy plot.

Command	Alias	Ctrl+	F-key	Alt+	Menu Bar	Dashboard
publish	...	...	...	FH	File ⤷Publish	...
-publish						

Command: publish

Displays dialog box:

PUBLISH DIALOG BOX

List of Drawing Sheets options

Sheet Name concatenates the drawing name and the layout name with a dash (-); edit sheet names with the Rename option.

Page Setup lists the named page setup for each layout; click to select other setups.

Status displays a message as layouts are published.

☐ **Include plot stamp** adds plot stamp data to the edge of each plot; see the PlotStamp command. (Click the Plot Stamp Settings button to access the Plot Stamp dialog box.)

Number of copies allows you make multi-copy prints (available only when plotting).

Publish To options

⊙ **Plotters named in page setups** generates hardcopy plots, or to-file plots.

○ **DWF file** generates multi-sheet *.dwf* files.

Include when Adding Sheets options

Model Tab determines whether the model layout is included with the list of sheets.

Layout Tabs includes all layouts.

Publish Options displays the Publish Options dialog box.

Show Details expands the dialog box to provide added information about the selected sheet.

Publish generates the *.dwf* file or plot; displays Now Plotting dialog box.

PUBLISH OPTIONS DIALOG BOX

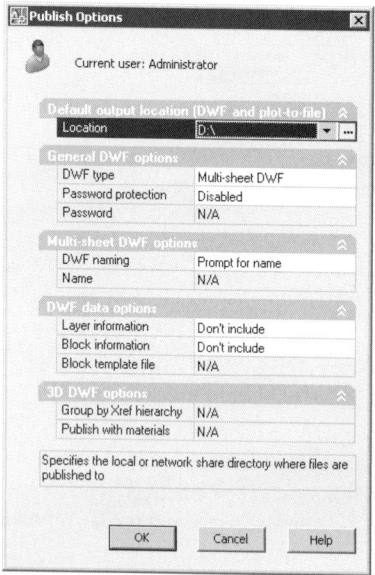

Default Output Location options

Location specifies the folder in which to save the *.dwf* and *.plt* plot files.

... displays the Select a Folder for Generated Files dialog box; select a folder, and click Select.

General DWF options

DWF Type:

- **Single-Sheet DWF** generates one *.dwf* file for each sheet.
- **Multi-Sheet DWF** generates a single *.dwf* file for all sheets.

Password Protection:

- **Disabled** leaves out the password.

- **Prompt for Password** prompts you for the password during publishing. Passwords are case-sensitive, and may consist of letters, numbers, punctuation, and non-ASCII characters. *Warning!* Passwords cannot be recovered.
- **Specify Password** specifies the password.

Password specifies the password.

Multi-Sheet DWF options
DWF Naming:
- **Prompt for Name** prompts you for the file and folder names during publishing.
- **Specify Name** specifies file and folder names for the multisheet *.dwf* file.

Name:
... displays the Select DWF File dialog box; select a file name, and then click Select.

DWF Data options
Layer information toggles whether layers are included in *.dwf* files.

Block information toggles whether blocks are included in *.dwf* files.

Block template file specifies the file name of the *.dxe* or *.blk* files used to format data.

... displays the Publish Block Template dialog box; specifies which blocks and attributes to include in *.dwf* files.

3D DWF options
Grouped by Xref Hierarchy displays objects by xref in the DWF Viewer.

Publish with Materials displays objects with materials in the DWF Viewer.

SHORTCUT MENU
Right-click in a sheet name the Sheets to Publish area:

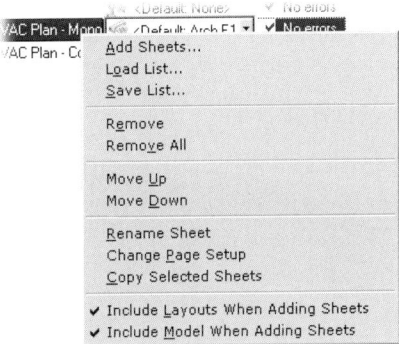

Add Sheets displays the Select Drawings dialog box; choose one or more drawings, and then click Select.

Load List displays the Load List of Sheets dialog box; select a *.dsd* drawing set description or *.bp3* batch plot list file, and then click Load.

Save List displays the Save List As dialog box; enter a file name, and then click Save.

Remove removes selected sheets from the list, without warning.

Remove All removes all sheets.

Move Up moves the selected sheets up the list.

Move Down moves the selected sheets down the list.

Rename Sheet renames the selected sheet.

Change Page Setup displays the Page Setup droplist.

Copy Selected Sheets copies selected sheets, and adds the "-Copy*n*"suffix to the sheet name.

Include Layouts when Adding Sheets includes all layouts in the sheet set.

Include Model when Adding Sheets includes model space with the layouts.

Dialog Box Toolbar

Preview displays the plot preview window; see the Preview command.

Add Sheets displays the Select Drawings dialog box; choose one or more drawings, and then click **Select**.

Remove Sheets removes selected sheets from the list, without warning.

Move Sheet Up moves the selected sheets up the list.

Move Sheet Down moves the selected sheets down the list.

Load Sheet List displays the Load List of Sheets dialog box; select a *.dsd* drawing set description or *.bp3* batch plot list file, and then click Load.

Save Sheet List displays the Save List As dialog box; enter a file name, and then click Save.

Plot Stamp Settings displays the Plot Stamp dialog box; see PlotStamp command.

. . . .

-PUBLISH Command

Command: -publish

> *If drawing is not saved, displays Drawing Modified - Save Changes dialog box. Click* **OK**.
>
> *Displays Select List of Sheets dialog box. Select a .dsd file, and then click* **Select**.
>
> *Immediately plots the drawing set, and generates a log file.*

RELATED COMMANDS

AutoPublish automatically publishes drawings as DWF files upon saving or closing (*new to AutoCAD 2008*).

Plot outputs drawings as *.dwf* files.

PublishToWeb coverts drawings to Web pages.

3dDwf exports model space drawings in 3D DWF format.

RELATED SYSTEM VARIABLES

AutoDwfPublish toggles the AutoPublish feature.

PublishCollate toggles whether a set of sheets is published as a single print job (1).

. . . .

RELATED STARTUP SWITCH

/pl specifies the *.dsd* file to publish in the background when AutoCAD starts up.

TIPS

- The Model tab is included only when the Include Model When Adding Sheets setting is turned on.

- DWF passwords are case sensitive. If the password is lost, it cannot be recovered, and the *.dwf* file cannot be opened. Erase the *.dwf* file, and create a new set.

- The -Publish command is good for generated sheets when a *.dsd* (drawing set description) file exists.

- To have AutoCAD publish a set of drawings unattended, follow these steps:

  1. Open the drawings in AutoCAD, and then use the Publish command to create a *.dsd* file. (Click the Save Sheet list button.)

  2. Exit AutoCAD.

  3. Modify the Target of AutoCAD's desktop item's Properties:

     "D:\CAD\AutoCAD 2008\acad.exe" **/pl "C:\path name\file name.dst"**

  4. Close the Properties dialog box.

  5. Double-click the icon, and AutoCAD starts plotting.

- This command can only be used transparently after it has been used at least once non-transparently.

'PublishToWeb

<u>2000i</u>   Exports drawings as DWF, JPG, or PNG images embedded in pre-format-
ted Web pages.

Command	Alias	Ctrl+	F-key	Alt+	Menu Bar	Dashboard
publishtoweb	...	...	...	FW	File	...
					⮑Publish to Web	

Before starting this command, save drawings.
Command: publishtoweb
Displays dialog box:

◉ **Create New Web Page** guides you through the steps to create new Web pages from drawings.

○ **Edit Existing Web Page** guides you through the steps to edit existing Web pages created earlier by this command.

Create Web Page page

Specify the Name of Your Web Page requires you to enter a file name. (AutoCAD uses the name for the files making up the Web page, which allows you later to edit the Web page.) The name also appears at the top of the Web page.

Provide a Description to Appear On Your Web Page provides a description appearing below the name on the Web page.

Select Image Type page

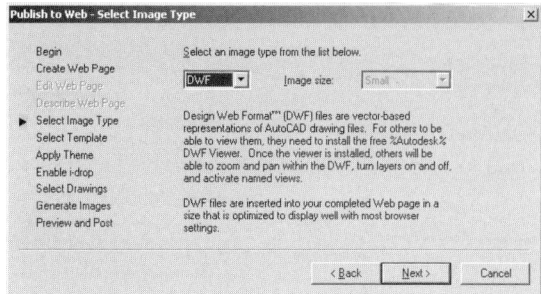

Select an Image Type From the List Below:

- **DWF** (drawing Web format) is a vector format that displays cleanly, and can be zoomed and panned; not all Web browsers can display DWF.
- **JPEG** (joint photographic experts group) is a raster format that all Web browsers display; may create *artifacts* (details that don't exist).
- **PNG** (portable network graphics) is a raster format that does not suffer the artifact problem, but some Web browsers do not display PNG.

Image size selects a size of raster image (available for JPEG and PNG only):

- **Small**— 789x610 resolution.
- **Medium** — 1009x780 resolution.
- **Large** — 1302x1006 resolution.
- **Extra Large** — 1576x1218 resolution.

Select Template page

Select a Template From the List Below selects one of the pre-designed formats for the Web page:

Apply Theme page

Select a Theme From the List Below selects one of the pre-designed themes (colors and fonts) for the Web page.

Enable i-drop page

☐ **Enable i-Drop** adds i-drop capability to the Web page. This allows blocks to be dragged from the Web page directly into AutoCAD and other Autodesk software.

Select Drawings page

Image Settings options

Drawing selects the drawing; the current drawing is the default.

Layout selects the name of a layout, or Model space.

Label names the image.

Description provides a description that appears with the drawing on the Web page.

Add adds the image setting to the image list.

Update changes the image setting in the image list.

Remove removes the image setting from the image list.

Move Up moves the image setting up the image list.

Move Down moves the image setting down the image list.

Generate Images page

⊙ **Regenerate Images For Drawings That Have Changed** updates images for those drawings that have been edited.

○ **Regenerate All Images** regenerates all images from all drawings; ensures all are up to date.

Preview and Post page

Preview launches the Web browser to preview the resulting Web page.

Post Now uploads the files (HTML, JPEG, PNG, DWF, and so on) to the Web site.

Send Email sends an email to alert others of the posted Web page.

Examples of resulting Web pages:

RELATED COMMANDS

Publish exports drawings as multi-sheet *.dwf* files.

Plot exports drawings as *.dwf* files via the ePlot option.

Hyperlink places hyperlinks in drawings.

RELATED FILES

*.ptw are PublishToWeb parameter files, stored in tab-delimited ASCII file.

*.js are JavaScript files.

*.jpg are joint photographic experts group (raster image) files.

*.png are portable network graphics (raster image) files.

*.dwf are drawing Web format (vector image) files.

TIPS

- Use the Regenerate all images option, unless you have an exceptionally slow computer or a large number of drawings to process. The Generate Images step can take a long time.

- After you click Preview to view the Web page (and after AutoCAD launches the Web browser), click the Back button to make changes, if the result is not to your liking.

- The Post Now option works only if you have correctly set up the FTP (file transfer protocol) parameters. If so, you can have AutoCAD upload the HTML files directly to a Web site. If not, use a separate FTP program to upload the files from the *\windows\applications data\autodesk* folder.

- You can customize the themes and templates by editing the *acwebpublish.css* (themes) and *acwebpublish.xml* (templates) files.

- This command can only be used transparently after it has been used at least once non-transparently.

- This command fails if you use the Options command to redirect templates to a folder other than *\template*, because AutoCAD expects to find the *\ptwtemplate* folder in *\template*.

Purge

<u>V. 2.1</u> Removes unused, named objects from the drawing: blocks, dimension styles, layers, linetypes, plot styles, shapes, text styles, table styles, application ID tables, and multiline styles.

Commands	Alias	Ctrl+	F-key	Alt+	Menu Bar	Dashboard
purge	pu	...	...	FUP	File	...
					⤷Drawing Utilities	
					⤷Purge	
-purge						

Command: purge

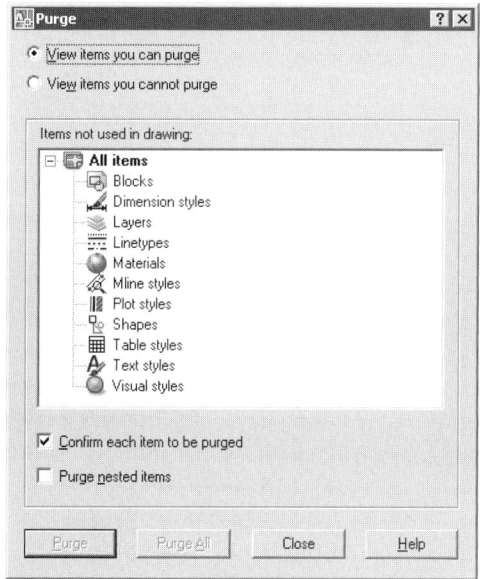

PURGE DIALOG BOX

⊙ **View items you can purge** lists objects that can be purged from the drawing.

○ **View items you cannot purge** lists objects that cannot be purged from the drawing.

☑ **Confirm each item to be purged** displays a confirmation dialog box for each object being purged.

☐ **Purge nested items** purges nested objects, such as unused blocks within unused blocks.

-PURGE Command

Command: -purge

Enter type of unused objects to purge
[Blocks/Dimstyles/LAyers/LTypes/MAterials/Plotstyles/SHapes/textSTyles/
Mlinestyles/Tablestyles/Visualstyles/Regapps/All]: *(Enter an option.)*

Sample response:
Purge linetype CENTER? <N> **y**
Purge linetype CENTER2? <N> **y**

COMMAND LINE OPTIONS

Blocks purges named but unused (not inserted) blocks.

Dimstyles purges unused dimension styles.

LAyers purges unused layers.

LTypes purges unused linetypes.

MAterials purges unused linetypes.

Plotstyle purges unused plot styles.

SHapes purges unused shape files.

textSTyles purges unused text styles.

Mlinestyles purges unused multiline styles.

Tablestyles purges unused table styles.

Visualstyles purges unused visual styles.

Regapps purges unused application ID tables of registered applications.

All purges all named objects, if possible.

RELATED COMMAND

-WBlock writes the current drawing to disk with the * option, and removes spurious information from the drawing.

TIPS

- It may be necessary to use this command several times; follow each purge with the Close command, then open the drawing, and purge again. Repeat until the Purge command reports nothing to purge.

- The View Items You Cannot Purge option lists items being used, as well as system items that can never be purged, as illustrated at right.

- A new standard style name is "Annotative," as of AutoCAD 2008.

Pyramid

<u>**2007**</u> Draws 3D solid pyramids with pointy or flat tops, and 3 to 32 sides.

Command	Alias	Ctrl+	F-key	Alt+	Menu Bar	Dashboard
pyramid	pyr	...	...	DMY	Draw ⤷Modeling ⤷Pyramid	3D Make

Command: pyramid
 4 sides Circumscribed
Specify center point of base or [Edge/Sides]: *(Pick a point, or enter an option.)*
Specify base radius or [Inscribed]: *(Pick a point, or type I.)*
Specify height or [2Point/Axis endpoint/Top radius]: *(Pick a point, or enter an option.)*

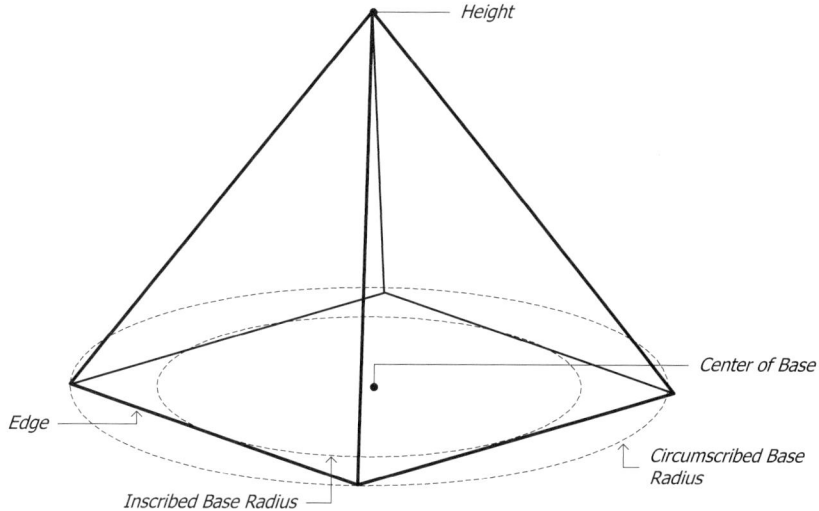

COMMAND LINE OPTIONS

The base is defined by a center point or edge length, and the number of sides:

Specify center point of base locates the center of the pyramid's base by defining x,y,z coordinates or by picking a point in the drawing.

Edge specifies the size of the base by the length of one edge; the base is always made from equilateral sides.

Sides specifies the number of sides; the range is 3 to 32, and the default is 4.

When the base is drawn like a polygon: circumscribed or inscribed:

Specify base radius specifies the radius of a "circle" that circumscribes the base.

Inscribed specifies the radius of a "circle" inscribed within the base.

Specify height indicates the height by specifying x,y,z coordinates or by picking a point in the drawing.

2Point specifies the height by picking two points.

Axis endpoint specifies the location and angle of the top point; this allows the pyramid to be drawn at an angle; the top and bottom are perpendicular to the axis.

Top radius specifies the size of the flat top to the pyramid.

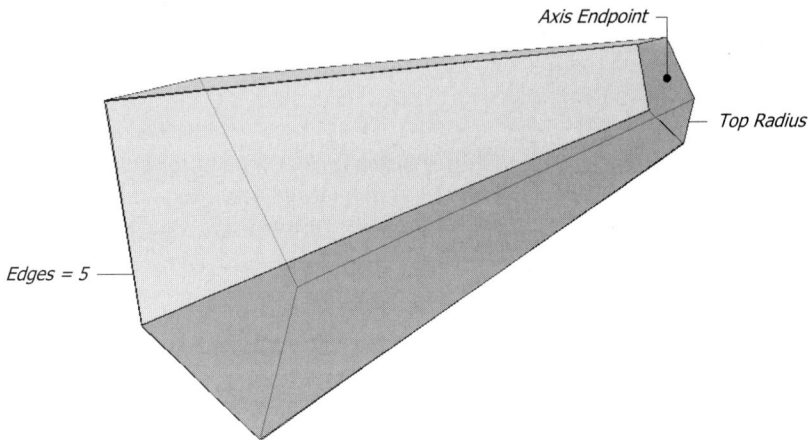

RELATED COMMANDS

Ai_Pyramid draws 3D mesh pyramids.

Cone draws 3D solid cones.

Extrude creates 3D objects with optional tapered sides.

RELATED SYSTEM VARIABLE

DragVs specifies the visual style while creating 3D solids.

TIPS

■ To create a pyramid-like object with a non-equilateral base, use the Extrude command.

■ After the pyramid is drawn, it can be edited with grips.

■ When the top radius equals the base radius, this command draws a prism.

QDim

2000 Dimensions entire objects (with continuous, baseline, ordinate, radius, diameter, and staggered dimensions) using just three picks (short for Quick DIMensioning).

Command	Alias	Ctrl+	F-key	Alt+	Menu Bar	Dashboard
qdim	...	...	...	N🖰	**Dimension**	**Dimensions**
					🖰**Quick Dimenion**	

Command: qdim
Select geometry to dimension: *(Select one or more objects; press Ctrl+A to select the entire drawing.)*
Select geometry to dimension: *(Press Enter to end object selection.)*
Specify dimension line position, or
[Continuous/Staggered/Baseline/Ordinate/Radius/Diameter/datumPoint/ Edit/seTtings] <Continuous>: *(Enter an option.)*

COMMAND LINE OPTIONS

Select geometry to dimension selects objects to dimension.

Specify dimension line position specifies the location of the dimension line.

Continuous draws continuous dimensions.

Staggered draws staggered dimensions.

Baseline draws baseline dimensions.

Ordinate draws ordinate dimensions relative to the UCS origin.

Radius draws radial dimensions; prompts 'Specify dimension line position:'.

Diameter draws diameter dimensions; prompts 'Specify dimension line position:'.

datamPoint sets a new datum point for ordinate and baseline dimensions; prompts 'Select new datum point:'.

Edit options
Indicate dimension point to remove, or [Add/eXit] <eXit>: *(Select a dimension, point, or enter an option.)*

Indicate dimension to remove selects the dimension point to remove from the dimension.

Add adds a dimension point to the dimension.

eXit returns to dimension drawing mode.

seTtings options
Associative dimension priority [Endpoint/Intersection] <Endpoint>: *(Enter an option.)*

Endpoint applies associative dimensions to endpoints over intersections.

Intersection applies associative dimensions to intersections over endpoints.

RELATED COMMANDS

DimStyle creates dimension styles, which specify the look of a dimension.

Dimxxx draws other kinds of dimensions.

QLeader draws leaders.

RELATED SYSTEM VARIABLES

DimStyle specifies the current dimension style.

DimAnno toggles annotative scaling for dimensions.

TIPS

- Example of continuous dimensions:

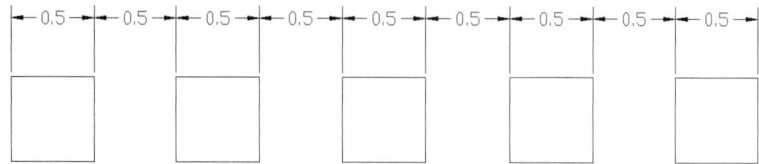

- Example of staggered dimensions:

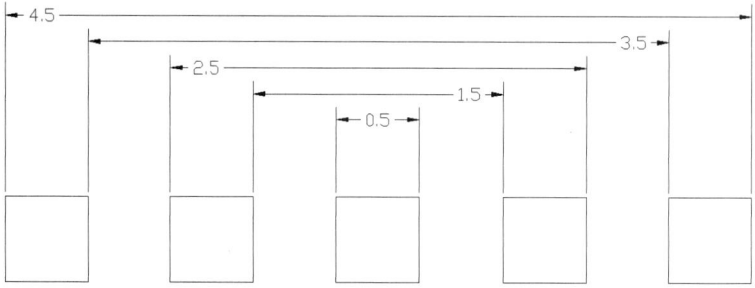

- Example of ordinate dimensions:

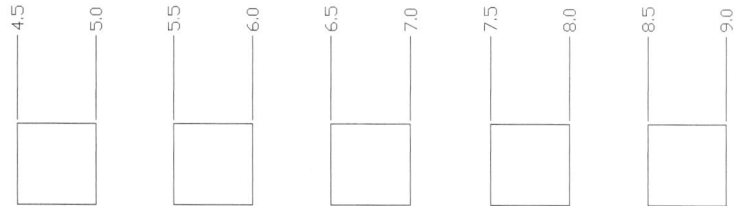

- As of AutoCAD 2004, dimensions created with QDim are fully associative. As of AutoCAD 2008, QDim supports annotative scaling.

- This command sometimes fails. AutoCAD complains, "Invalid number of dimension points found."

QLeader

<u>Rel.14</u> Draws leaders; dialog box specifies options for custom leaders and annotations (short for Quick LEADER).

Command	Alias	Ctrl+	F-key	Alt+	Menu Bar	Dashboard
qleader	le	...	...	...	...	...

Command: qleader
Specify first leader point, or [Settings] <Settings>: *(Pick point 1 for the arrowhead, or type S.)*

*When **S** is entered, displays dialog box:*

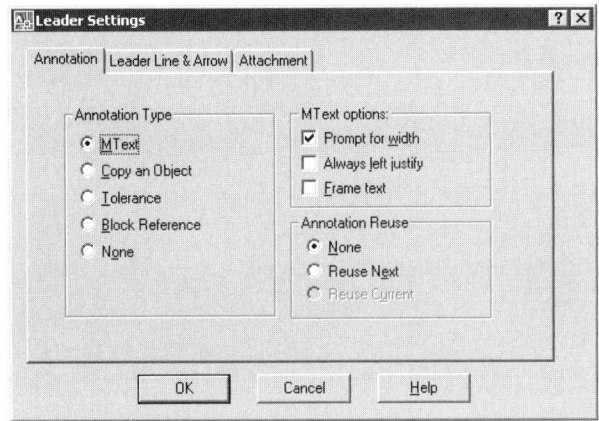

Click OK to continue with the command; the prompts vary, depending on the options selected in the dialog box:

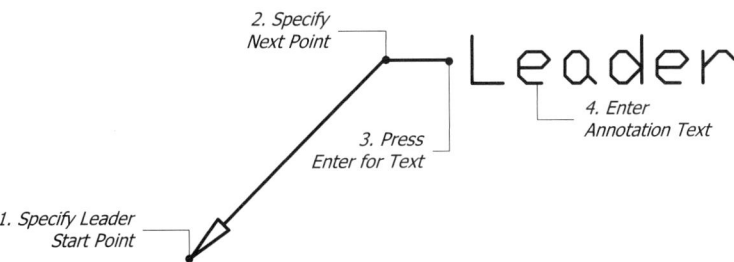

Specify next point: *(Pick point 2 for the leader shoulder.)*
Specify next point: *(Pick point, or press Enter for text options.)*
Specify text width <0.0000>: *(Enter a value, or press Enter.)*
Enter first line of annotation text <Mtext>: *(Enter text, or press Enter for the mtext editor.)*
Enter next line of annotation text: *(Press Enter to end command.)*

COMMAND LINE OPTIONS

Specify first leader point picks the location for the leader's arrowhead; press Enter to display tabbed dialog box.

Specify next point picks the vertices of the leader; press Enter to end the leader line.

Specify text width specifies the width of the bounding box for the leader text.

Enter first line of annotation text specifies the text for leader annotation; press Enter twice to end.

Enter next line of annotation text specifies more text; press Enter once to end.

LEADER SETTINGS DIALOG BOX

Annotation Type options

⊙ **MText** prompts you to enter text for the annotation.

○ **Copy an Object** attaches any object in the drawing as an annotation.

○ **Tolerance** prompts you to select tolerance symbols for the annotation.

○ **Block Reference** prompts you to select a block for the annotation.

○ **None** attaches no annotation.

MText Options options

☑ **Prompt for width** displays the 'Specify text width' prompt.

☐ **Always left justify** left justifies the text, even when the leader is drawn to the right.

☐ **Frame text** places a rectangle around the text.

Annotation Reuse options

⊙ **None** does not retain annotation for next leader.

○ **Reuse Next** remembers the current annotation for the next leader.

○ **Reuse Current** uses the last annotation for the current leader.

Leader Line & Arrow tab

Leader Line options

⊙ **Straight** draws the leader with straight lines.

○ **Spline** draws the leader as a spline curve.

Number of Points options

☐ **No limit** keeps prompting for leader vertex points until you press Enter.

Maximum stops the command prompting for leader vertex points; default=3.

Arrowhead options

Arrowhead selects the type of arrowhead, including Closed filled (default), None, and User Arrow.

Angle Constraints options

First Segment selects from Any angle (user-specified), Horizontal (0 degrees), 90, 45, 30, or 15-degree leader line, first segment.

Second Segment selects from Any angle (user-specified), Horizontal, 90, 45, 30, or 15-degree leader line, second segment.

Attachment tab

Multiline Text Attachment options

Left Side positions the annotation at one of several locations relative to the last leader segment, when the annotation is located to the left of the leader:

- ○ Top of top line.
- ○ Middle of top line.
- ⊙ Middle of multi-line text.
- ○ Middle of bottom line (default).
- ○ Bottom of bottom line.

Right Side positions the annotation at one of several locations relative to the last leader segment, when the annotation is located to the right of the leader.

☐ **Underline bottom line** underlines the last line of leader text.

RELATED COMMANDS

DdEdit edits leader text; see the MText command.

DimStyle sets dimension variables, including leaders.

MLeader draws multiline leaders (*new to AutoCAD 2008*).

Leader draws leaders without dialog boxes.

TIPS

- This command draws leaders, just like the Leader command. The difference is that it brings up a triple-tab dialog box for setting the leader options.

- Some options have interesting possibilities, such as using a mtext, text, block reference, or tolerance in place of the leader text.

- You may prefer to use the MLeader command, because of its greater flexibility over QLeader.

QNew

2004   Starts new drawings based on a default template file *(short for Quick NEW)*.

Command	Alias	Ctrl+	F-key	Alt+	Menu Bar	Dashboard
qnew	...	...	...	...	...	...

Command: qnew

Display depends on the value of the Startup system variable. See the New command.

RELATED COMMANDS

New starts new drawings.

SaveAs saves drawings in *.dwg* or *.dwt* format; creates template files.

RELATED SYSTEM VARIABLES

DbMod indicates whether the current drawing has changed since being loaded.

DwgPrefix indicates the path to the current drawing.

DwgName indicates the name of the current drawing.

FileDia displays file prompts at the 'Command' prompt.

RELATED FILES

wizard.ini holds the names and descriptions of template files.

***.dwt** are template files stored in *.dwg* format.

TIPS

- The New and QNew commands operate in an identical manner, except when a template drawing is specified by the Option command's Files tab:

- The toolbar icon executes the QNew command, while the File | New menu pick executes the New command.

⊞ QSave

Rel.12 Saves the current drawing without requesting a file name (short for Quick SAVE).

Command	Alias	Ctrl+	F-key	Alt+	Menu Bar	Dashboard
qsave	...	S	...	FS	File	...
					⸽Save	

Command: qsave

If the drawing has never been saved, displays the Save Drawing As dialog box.

COMMAND LINE OPTIONS

None.

RELATED COMMANDS

Quit ends AutoCAD, with or without saving the drawing.

Save saves the drawing, after requesting the filename.

RELATED SYSTEM VARIABLES

DBMod indicates whether the drawing has changed since it was loaded.

DwgName specifies the current drawing filename (default = *drawing1*).

DwgTitled specifies the status of drawing's filename:

0 — Drawing is named *drawing1* (default).

1 — Drawing was given another name.

TIPS

- When the drawing is unnamed, the QSave command displays the Save Drawing As dialog box to request a file name; see the SaveAs command.

- When the drawing file, its folder, or drive (such as a CD-ROM drive) is marked read-only, use the SaveAs command to save the drawing by another file name, or to another folder or drive.

- If the AutoDwfPublish system variable is turned on, then the drawing is simultaneously exported as a DWF file at the same time as saving (*new to AutoCAD 2008*). The DWF file is published in the background, if the BackgroundPlot system variable is turned on.

- Use the Options command's Files tab to specify the location of automatic backup files. I recommend that they be saved to another drive.

QSelect

2000 Creates selection sets of objects based on their properties (short for Quick SELECT).

Command	Alias	Ctrl+	F-key	Alt+	Menu Bar	Dashboard
qselect	...	...	...	TK	Tools	...
					↳Quick Select	

Command: qselect

Displays dialog box:

QUICK SELECT DIALOG BOX

Apply to applies the selection criteria to the entire drawing or current selection set; to create a selection set, click the Select Objects button.

Select Objects allows you to select objects; AutoCAD prompts: 'Select objects: '. Right-click or press Enter to return to this dialog box.

Object type lists the objects in the selection set; allows you to narrow the selection criteria to specific types of objects (default = Multiple).

Properties lists the properties valid for the selected object types; when you select more than one object type, only the properties in common are listed.

Operator lists logical operators available for the selected property; operators include:

Operator		Meaning
=	Equals	Selects objects equal to the property.
<>	Not Equal To	Selects objects different from the property.
>	Greater Than	Selects objects greater than the property.
<	Less Than	Selects objects less than the property.
*	Wildcard Match	Selects objects with matching text.

Value specifies the property value for the filter. If known values for the selected property are available, Value becomes a list from which you can choose a value. Otherwise, enter a value.

How to Apply

⊙ **Include in new selection set** creates a new selection set.

○ **Exclude from new selection set** inverts the selection set, excluding all objects that match the selection criteria.

Append to current selection set

☑ adds to the current selection set.

☐ replaces the current selection set.

RELATED COMMANDS

Select selects objects on the command line.

Filter runs a more sophisticated version of the QSelect command.

RELATED SYSTEM VARIABLES

None.

TIPS

- Use this command to select objects based on their properties; use the Select command to select objects based on their location in the drawing.

- You may select objects before entering the QSelect command, and then add or remove objects from the selection set with the Quick Select dialog box's options.

- Since this command is not transparent, you cannot use it within other commands; instead, use the 'Filter command.

- As of AutoCAD 2005, this command also selects block insertions and tables; as of AutoCAD 2008, this command includes Annotative as a property.

- This command works with the properties of proxy objects created by ObjectARX applications.

'QText

V. 2.0 Displays lines of text as rectangular boxes (short for Quick TEXT).

Command	Alias	Ctrl+	F-key	Alt+	Menu Bar	Dashboard
qtext	...	...	...	...	...	...

Command: qtext
Enter mode [ON/OFF] <OFF>: *(Type ON or OFF.)*

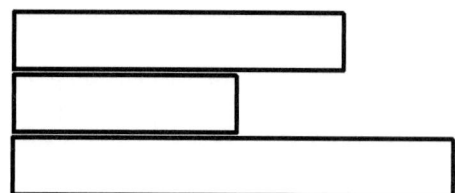

Left: Normally displayed text.
Right: Quick text after regeneration.

COMMAND LINE OPTIONS

ON turns on quick text after the next RegenAll command.

OFF turns off quick text after the next RegenAll command.

RELATED COMMAND

RegenAll regenerates the screen, which makes quick text take effect.

RELATED SYSTEM VARIABLE

QTextMode holds the current state of quick text mode.

TIPS

- A regeneration is required before AutoCAD displays text in quick outline form:

  Command: regenall
  Regenerating model.

- To reduce regen time, use QText to turn lines of text into rectangles, which redraw faster.

- The length of a qtext box does not necessarily match the actual length of text.

- Turning on QText affects text during plotting; qtext blocks are plotted as rectangles.

- To find invisible text (such as text made of spaces), turn on QText, thaw all layers, zoom to extents, and use the RegenAll command.

- The rectangles displayed by this command are affected by lineweights.

'QuickCalc, 'QcClose

<u>**2006**</u> Opens and closes the QuickCalc palette.

Command	Alias	Ctrl+	F-key	Alt+	Menu Bar	Dashboard
quickcalc	qc	8	...	TPQ	**Tools**	...
					⮡**Palettes**	
					⮡**Quick Calc**	
qcclose	...	8	...			

Command: quickcalc

Displays palette:

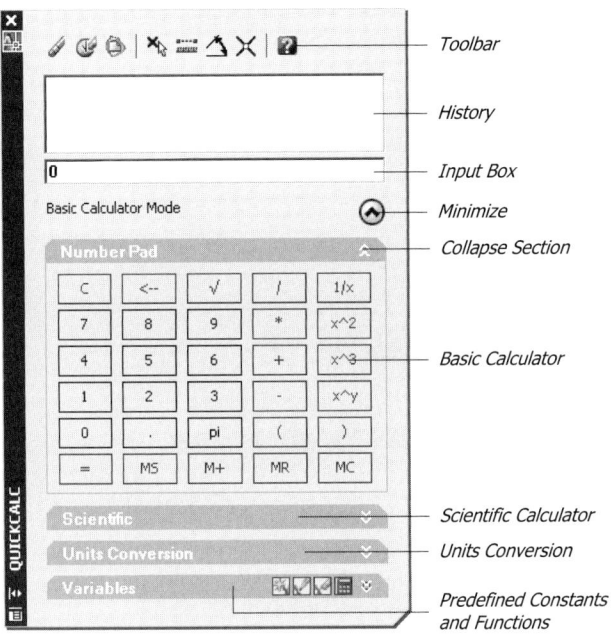

- Toolbar
- History
- Input Box
- Minimize
- Collapse Section
- Basic Calculator
- Scientific Calculator
- Units Conversion
- Predefined Constants and Functions

Command: qcclose

Closes the palette.

QUICKCALC PALETTE

Basic Calculator performs arithmetic and algebraic functions.

Scientific Calculator adds logarithmic and trigonometric functions (see below, left).

Units Conversion converts between units based on the *acad.unt* file, although does not include all conversion provided by the file, such as fortnights (see above, right).

Predefined Constants and Functions defines the values of variables.

Palette Toolbars

QuickCalc Toolbar

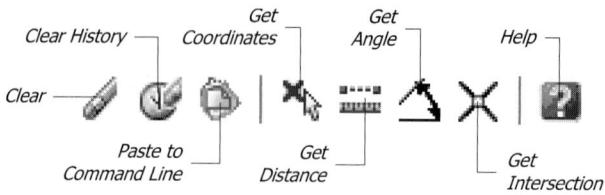

Clear clears the input area.

Clear History clears the history list.

Paste to Command Line places the value on the command line.

Get Coordinates returns the x,y,z coordinates of the pick point.

Get Distance returns the distance between two points picked in the drawing.

Get Angle returns the angle between two points relative to the x axis.

Get Intersection returns the x,y,z coordinates of the intersection of two lines defined by four pick points.

Help displays online help.

Variables Toolbar

New Variable displays the Variable Definition dialog box.

Edit Variable also opens the Variable Definition dialog box.

Delete erases the selected variable.

Return Variable to Input Area places variable on the input line.

VARIABLE DEFINITION DIALOG BOX

Variable Type:

Constant stores variables as constant values.

Function stores variables as functions.

Variable Properties:

Name specifies the name of the variable.

Group With stores the variable with a user-defined group.

Value or Expression defines the value of the constant, or expression for the variable.

Description describes the constant or variable.

RELATED COMMAND

Cal displays the geometry calculator function on the command line.

RELATED SYSTEM VARIABLES

CalcInput determines whether mathematical expressions are evaluated in dialog boxes.

QcState reports whether the QuickCalc palette is open or not.

RELATED FILE

acad.unt stores conversion units.

TIPS

- Unlike most calculators, QuickCalc does not calculate answers when you click a function; instead, it builds *expressions*. It expects you to compose expressions: enter the expression, edit it, and then click **=** or press Enter.

- Use the history area to modify expressions and recalculate them.

- This command can be used during other commands to provide numeric input. The QuickCalc icon appears in most dialog boxes that accept numeric input.   The QuickCalc icon appears in most dialog boxes that accept numeric input.

- You can enter expressions in fields of dialog boxes that expect numbers, such as the one illustrated below. Use the following syntax:

=expression (Press Alt+Enter)

- You can also use AutoLISP to perform calculations at the command line.

QuickCui

Opens the Customize User Interface dialog box in a partially-collapsed state.

Command	Alias	Ctrl+	F-key	Alt+	Menu Bar	Dashboard
quickui	qcui	...	...	...	...	...

Command: quickcui

Displays dialog box:

DIALOG BOX OPTION

⊙ opens the rest of the dialog box.

See CUI command for further details.

RELATED COMMAND

CUI opens the Customzie User Interface dialog box.

Quit

V. 1.0 Exits AutoCAD without saving changes made to the drawing after the most recent QSave or SaveAs command.

Command	Alias	Ctrl+	F-key	Alt+	Menu Bar	Dashboard
quit	exit	Q	Alt+F4	FX	File	...
					⇘Exit	

Command: quit
Displays dialog box:

AUTOCAD DIALOG BOX

Yes saves changes before leaving AutoCAD.

No does not save changes.

Cancel does not quit AutoCAD; returns to drawing.

RELATED COMMANDS

Close closes the current drawing.

SaveAs saves the drawing by another name and to another folder or drive.

RELATED SYSTEM VARIABLE

DBMod indicates whether the drawing has changed since it was loaded.

RELATED FILES

*.dwg are AutoCAD drawing files.

*.bak are backup files.

*.bkn are additional backup files.

TIPS

- You can change a drawing, yet preserve its original format:

    1. Use the SaveAs command to save the drawing by another name.

    2. Use the Quit command to preserve the drawing in its most recently-saved state.

- Even if you accidentally save over a drawing, you can recover the previous version — when AutoCAD is set to save backup files automatically. (See the Options command). Follow these steps:

    1. Use Windows Explorer to rename the drawing file.

    2. Use Windows Explorer to rename the *.bak* (backup) extension to *.dwg*.

- You cannot save changes to a read-only drawing with the Quit command; use the SaveAs command instead.

R14PenWizard

2000 Helps create color-dependent plot style tables (an undocumented command).

Command	Alias	Alt+	Menu Bar	Dashboard
r14penwizard	...	TZD	Tools	...
			⌊Wizard	
			⌊Add Color-Dependent Plot Style Table	

You can also access this command through the Windows Control Panel's Autodesk Plot Style Manager. (This command makes the Plot command compatible with versions of AutoCAD prior to 2000.)

Command: r14penwizard

Displays Add Color-Dependent Plot Style Table wizard.

ADD COLOR-DEPENDENT PLOT STYLE TABLE DIALOG BOX

⊙ **Start from scratch** creates a new color-dependent plot style table file.

○ **Use a CFG file** converts the plotter pen settings stored in AutoCAD Release 14 *acad.cfg* files.

○ **Use a PCP or PC2 file** converts the plotter pen settings stored in the plotter configuration parameter *.pcp* and *.pc2* files of earlier versions of AutoCAD.

Back goes back to the previous dialog box.

Next moves to the next dialog box.

Cancel exits the wizard.

File Name page

File name specifies the name of the file in which to store the new color-dependent plot style table.

Finish page

Plot Style Table Editor displays the Plot Style Table Editor dialog box; see StylesManager command.

☑ **Use this plot style table for the current drawing** applies the plot style table to the current drawing.

☐ **Use this plot style table for new and pre-AutoCAD 2008 drawings** applies the plot style table to all new drawings and drawings created by versions of AutoCAD prior to 2008.

RELATED COMMANDS

PcInWizard imports *.pcp* and *.pc2* configuration plot files into the current layout.

PlotterManager accesses the Add Plotter wizard and Assign Plot Style wizard.

Plot plots the drawing.

StylesManager displays the Plot Style Table Editor dialog box.

Ray

Rel.13 Draws semi-infinite construction lines.

Command	Alias	Ctrl+	F-key	Alt+	Menu Bar	Dashboard
ray	...	...	...	DR	Draw	...
					↳ **Ray**	

Command: ray
Specify start point: *(Pick a point.)*
Specify through point: : *(Pick another point.)*
Specify through point: : *(Press Enter to end the command.)*

Start point.

COMMAND LINE OPTIONS

Start point specifies the starting point of the ray.

Through point specifies the point through which the ray passes.

RELATED COMMANDS

Properties modifies rays.

Line draws finite line segments.

XLine draws infinite construction lines.

TIPS

- Ray objects have an endpoint at one end and are "infinite" in length at the other end.

- Rays display and plot, but do not affect the extents of the drawing.

- A ray has all the properties of a line (including color, layer, and linetype), and can be used as a cutting edge for the Trim command.

- The technical editor defines infinite: "You can type in coordinates for the ray's start and through points with restrictions. Scientific notation allows a maximum of e99. Maximum command line input is 255 characters. Thus, the start point can be -999...999E99, -999...999E99 and through point can be -999...999E99, -999...999E99 — where ... consists of enough 9s to total 255 characters."

. . . .

Removed Command

RConfig — render configuration — was removed from Release 14. It is replaced by Render.

. . . .

Recover

<u>Rel.12</u> Attempts to recover damaged drawings without user intervention.

Command	Alias	Ctrl+	F-key	Alt+	Menu Bar	Dashboard
recover	...	...	...	**FUR**	**File**	...
					⤷**Drawing Utilities**	
					⤷**Recover**	

Command: recover

Displays the Select File dialog box. Select a .dwg, .dxf, dws, or .dwt file, and then click Open.

Sample output:

```
Drawing recovery.
Drawing recovery log.
Scanning completed.

Validating objects in the handle table.
Valid objects 14526   Invalid objects 0
Validating objects completed.
16 error opening *Model_Space's layout.
Setting layout id to null.
16 error opening *Paper_Space's layout.
Setting layout id to null.

    Salvaged database from drawing.
A vertex was added to a 3D pline (3A18) which had only one vertex.
Auditing Header
Auditing Tables
Auditing Entities Pass 1
Pass 1 14500   objects audited
Auditing Entities Pass 2
Pass 2 14500   objects audited
Auditing Blocks
  10      Blocks audited
Total errors found 2 fixed 2
Erased 0 objects
```

RELATED COMMANDS

Audit checks a drawing for integrity.

DrawingRecovery displays a window listing recoverable drawings.

RecoverAll also recovers xrefs.

TIPS

- The Recover command does not ask permission to repair damaged parts of the drawing file; use the Audit command if you want to control the repair process.

- The Quit command discards changes made by the Recover command.

- If the Recover and Audit commands do not fix the problem, try using the DxfOut command, followed by the DxfIn command — assuming the file is not so damaged it cannot be opened at all.

RecoverAll

2008 Updates and, if necessary, repairs drawings and related xrefs.

Command	Alias	Ctrl+	F-key	Alt+	Menu Bar	Dashboard
recoverall	...	...	...	FUX	File	...
					⮡ Drawing Utilities	
					⮡ Recover Drawings and Xrefs	

Command: recoverall

May display an initial warning dialog box.

Displays the Select File dialog box. Select a .dwg, .dws, or .dwt file, and then click Open.

Reports drawing receovery process in the Drawing Recovering Log dialog box:

RELATED COMMANDS

Recover checks a drawing for integrity.

DrawingRecovery displays a window listing recoverable drawings.

TIPS

- The RecoverAll command works like this:

  1. The selected drawing and its related xrefs are opened.

  2. The files are recovered, and custom objects are updated (if the object enabler is available).

  3. The files are saved in AutoCAD 2008 format; the original files are saved as .bak files.

  4. The files are closed.

⬜ Rectang

Rel.12 Draws squares and rectangles with a variety of options.

Command	Aliases	Ctrl+	F-key	Alt+		Menu Bar	Dashboard
rectang	rec	...	...	DG		Draw	2D Draw
	rectangle					⬐Rectangle	

Command: rectang
Specify first corner point or [Chamfer/Elevation/Fillet/Thickness/Width]: : *(Pick a point, or enter an option)*
Specify other corner point or [Area/Dimensions/Rotation]: *(Pick another point.)*

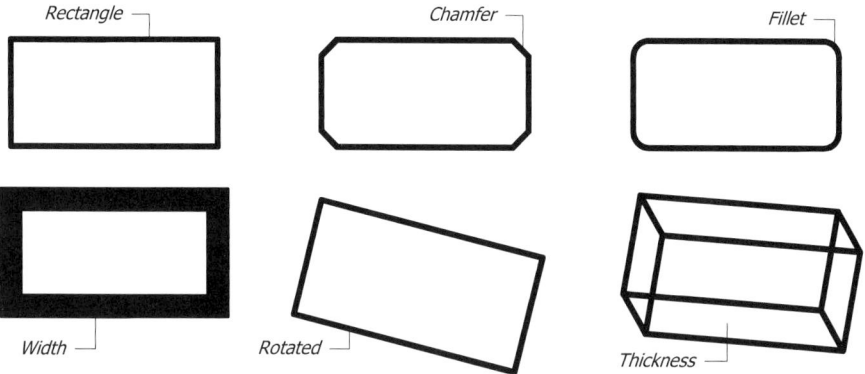

COMMAND LINE OPTIONS

Specify first corner point picks the first corner of the rectangle.

Specify other corner point picks the opposite corner of the rectangle.

Chamfer options
Specify first chamfer distance for rectangles <0.0>: : *(Enter a distance value.)*
Specify second chamfer distance for rectangles <0.0>: *(Enter another distance value.)*

First chamfer distance for rectangles sets the first chamfer distance for all four corners.

Second chamfer distance for rectangles sets the second chamfer distance for all four corners.

Elevation option
Specify the elevation for rectangles <0.0>: *(Enter an elevation value.)*

Elevation for rectangles sets the elevation *(height of the rectangle in the z direction)*.

Fillet option
Specify fillet radius for rectangles <1.0>: *(Enter a radius value.)*

Fillet radius for rectangles sets the fillet radius for all four corners of the rectangle.

Thickness option
Specify thickness for rectangles <0.0>: *(Enter a thickness value.)*

Thickness for rectangles sets the thickness of the rectangle's sides in the z direction.

Width option
Specify line width for rectangles <0.0>: *(Enter a width value.)*

Width for rectangles sets the width of all segments of the rectangle's four sides.

Area option
Enter area of rectangle in current units <100.0>: *(Enter an area value.)*
Calculate rectangle dimensions based on [Length/Width] <Length>: *(Type L or W.)*
Enter rectangle length <10.0>:

Length specifies the length of the rectangle; AutoCAD determines the width.

Width specifies the width of the rectangle; AutoCAD determines the length.

Dimensions options
Specify length for rectangles <0.0>: *(Enter a length value.)*
Specify width for rectangles <0.0>: *(Enter a width value.)*
Specify other corner point or [Area/Dimension/Rotation]: *(Pick a point.)*

Specify length specifies the length along the x axis.

Specify width specifies the length along the y axis.

Specify other corner point specifies the orientation of the rectangle.

Rotation options
Specify rotation angle or [Pick points] <0>: *(Enter an angle.)*

Rotation angle specifies the angle from the x axis to rotate about the first corner.

Pick points prompts to pick two points that show the angle.

RELATED COMMANDS

Donut draws solid-filled circles with polylines.

Ellipse draws ellipses with polylines, when PEllipse = 1.

PEdit edits polylines, including rectangles.

PLine draws polylines and polyline arcs.

Polygon draws polygons — 3 to 1024 sides — from polylines.

RELATED SYSTEM VARIABLE

PlineWid saves the current polyline width.

TIPS

- Rectangles are drawn from polylines; use the PEdit command to change the rectangle, such as the width of the polyline.

- The values you set for the Chamfer, Elevation, Fillet, Thickness, and Width options become the default for the next execution of the Rectangle command.

- The pick point determines the location of the rectangle's first vertex; rectangles are drawn counterclockwise.

- Use the Snap command and object snap modes to place the rectangle precisely.

- Use object snap modes ENDpoint or INTersection to snap to the rectangle's vertices.

- This command ignores the settings in the ChamferA, ChamferB, Elevation, FilletRad, PLineWid, and Thickness system variables.

Redefine

<u>Rel. 9</u> Restores the meaning of AutoCAD commands disabled by the Undefine command.

Command	Alias	Ctrl+	F-key	Alt+	Menu Bar	Dashboard
redefine	...	...	...	...	...	...

Command: redefine
Enter command name: *(Enter command name.)*

COMMAND LINE OPTION

Enter command name specifies the name of the AutoCAD command to redefine.

RELATED COMMANDS

All AutoCAD commands can be redefined.

Undefine disables the meaning of AutoCAD commands.

TIPS

- Prefix any command with a . *(period)* to redefine the undefinition temporarily, as in:

  Command: .line

- Prefix any command with an _ *(underscore)* to make an English-language command work in any linguistic version of AutoCAD, as in:

  Command: _line

- You must undefine a command with the Undefine command before using the Redefine command.

- *Warning!* If you undefine the Redefine command, it cannot be redefined with the Redefine command.

Redo

V. 2.5 Reverses the effect of the most-recent U and Undo commands.

Command	Alias	Ctrl+	F-key	Alt+	Menu Bar	Dashboard
redo	...	...	...	ER	Edit	...
					⇘Redo	

Command: redo

COMMAND LINE OPTIONS
None.

RELATED COMMANDS
MRedo redoes more than one undo.

Oops un-erases the most recently-erased objects.

U undoes the most recent AutoCAD command.

Undo undoes the most recent series of AutoCAD commands.

TIPS
■ The Redo command is limited to reversing a single undo, while the Undo and U commands can undo operations all the way back to the beginning of the editing session.

■ The Redo command must be used immediately following the U or Undo command.

'Redraw, 'RedrawAll

V. 1.0   Redraws viewports to clean up the screen.

Command	Alias	Ctrl+	F-key	Alt+	Menu Bar	Dashboard
redraw	r	...	...	...	...	...
redrawall	ra	...	...	VR	View ⇘Redraw	...

Command: redraw

Reraws the current viewport.

Command: redrawall

Redraws all viewports.

Left: *Before redraw, portions of the drawing are "missing."*
Right: *The drawing is clean after the redraw.*

COMMAND LINE OPTION

ESC stops the redraw.

RELATED COMMAND

Regen regenerates the current viewport.

RELATED SYSTEM VARIABLE

SortEnts controls the order of redrawing objects:
  0 — Sorts by the order in the drawing database.
  1 — Sorts for object selection.
  2 — Sorts for object snap.
  4 — Sorts for redraw.
  8 — Sorts for creating slides.
  16 — Sorts for regenerations.
  32 — Sorts for plotting.
  64 — Sorts for PostScript plotting.

TIPS

- Use this command to clean up the screen after a lot of editing; some commands automatically redraw the screen when they are done.

- Redraw does not affect objects on frozen layers.

RefClose

2000 Saves or discards changes made to reference objects (blocks and xrefs) edited in-place (short for REFerence CLOSE).

Command	Alias	Ctrl+	F-key	Alt+	Menu Bar	Dashboard
refclose	...	...	...	...	**Tools**	...
					⇗**Xref and Block In-place Editing**	
					⇘ **Close Reference**	
				...	**Tools**	
					⇗**Xref and Block In-place Editing**	
					⇘**Save Reference Edits**	

Command: refclose
Enter option [Save/Discard reference changes] <Save>: *(Enter an option.)*

COMMAND LINE OPTIONS

Save saves the editing changes made to the block or externally-referenced file.

Discard discards the changes.

RELATED COMMANDS

RefEdit edits blocks and externally-referenced files attached to the current drawing.

Insert inserts a block in the drawing.

XAttach attaches an externally-referenced drawing.

RELATED SYSTEM VARIABLE

RefEditName specifies the filename of the referenced file being edited.

TIPS

- AutoCAD prompts you with a warning dialog box to ensure you really want to discard or save the changes made to the reference:

- You can use this command only after the RefEdit command; otherwise AutoCAD reports, "** Command not allowed unless a reference is checked out with RefEdit command **."

RefEdit

__2000__   Edits-in-place blocks and externally-referenced files attached to the current drawing (short for REFerence EDIT).

Commands	Alias	Ctrl+	F-key	Alt+	Menu Bar	Dashboard
refedit	...	...	...	...	Tools	...
					⤷Xref and Block In-place Editing	
					⤷Edit Reference In-Place	
-refedit						

Command: refedit
Select reference: *(Pick an externally-referenced drawing or block.)*

Displays dialog box:

Click OK to continue with the command:

Select nested objects: *(Pick one or more objects.)*
Use REFCLOSE or the Refedit toolbar to end reference editing session.

Displays RefEdit toolbar.

COMMAND LINE OPTIONS

Select reference selects an externally-referenced drawing or an inserted block for editing.

Select nested objects selects objects within the reference — this becomes the selection set of objects that you may edit; you may select all nested objects with the All option (with the exception of OLE objects and objects inserted with the MInsert command, which cannot be refedited).

REFERENCE EDIT DIALOG BOX

Reference name lists a tree of the selected reference object and its nested references; a single reference can be edited at a time.

Preview displays a preview image of the selected reference.

⊙ **Automatically select all nested objects** selects all nested objects.

○ **Prompt to select nested objects** prompts you. 'Select nested objects.'

Settings Tab

Create unique layer, style, block names

☑ Prefixes layer and symbol names of extracted objects with **n**.

☐ Retains the names of layers and symbols, as in the reference.

Display attribute definitions for editing (available only when an xref contains attributes):

☑ Makes non-constant attributes invisible; attribute definitions can be edited.

☐ Attribute definitions cannot be edited.

Note: When edited attributes are saved back to the block reference, use the AttRedef or BAttMan commands to update attributes of the original references.

Lock objects not in working set options:

☑ Locks unselected objects in a manner similar to locking layers.

☐ Does not lock objects.

REFEDIT TOOLBAR

This toolbar appears automatically after you select nested objects to edit:

Edit block or xref executes the RefEdit command.

Add objects to working set executes the RefSet Add command.

Remove objects from working set executes the RefSet Remove command.

Discard changes to reference executes the RefClose Discard command.

Save back changes to reference executes the RefClose Save command.

-REFEDIT Command

Command: -refedit
Select reference: *(Pick an externally-referenced drawing, or a block.)*
Select nesting level [Ok/Next] <Next>: *(Type O or N.)*
Enter object selection method [All/Nested] <All>: *(Type A or N.)*
Display attribute definitions [Yes/No] <No>: *(Type Y or N.)*
Use REFCLOSE or the Refedit toolbar to end reference editing session.

Displays Refedit toolbar.

COMMAND LINE OPTIONS

Select reference selects an externally-referenced drawing or an inserted block for editing.

Select nesting level selects objects within the reference — this becomes the selection set of objects that you may edit; you may select all nested objects with All (with the exception of OLE objects and objects inserted with the MInsert command, which cannot be refedited).

Enter object selection method:

All selects all objects.

Nested selects only nested objects.

Display attribute definitions:

Yes makes non-constant attributes invisible; attribute definitions can be edited.

No means attribute definitions cannot be edited.

RELATED COMMANDS

RefSet adds and removes objects from a working set.

RefClose saves or discards editing changes to the reference.

AttRedef redefines blocks and updates attributes.

BAttMan controls all aspects of attributes in blocks.

BEdit edits blocks.

AttIPedit edits attributes in block definitions and insertions directly *(new to AutoCAD 2008)*.

RELATED SYSTEM VARIABLES

RefEditName stores the name of the externally-referenced file or block being edited.

XEdit determines whether the current drawing can be edited while being referenced by another drawing.

XFadeCtl specifies the amount of fading for objects not being edited in place.

TIPS

- You may find it more convenient to use the XEdit or BEdit commands.
- Double-clicking a block launches the Block Editor.
- In layouts, the drawing must be in model mode for you to select an xref.
- OLE objects and objects inserted with the MInsert command cannot be refedited.
- AutoCAD identifies the "working set" as those objects that you have selected to edit in-place.
- Objects *not* in a working set cannot be selected.

RefSet

<u>**2000**</u>   Adds and removes objects from working sets (short for REFerence SET).

Command	Alias	Ctrl+	F-key	Alt+	Menu Bar	Dashboard
refset	...	...	...	T	**Tools**	...
					↳**Xref and Block In-place Editing**	
					↳**Add to Working Set**	
				T	**Tools**	
					↳**Xref and Block In-place Editing**	
					↳**Remove from Working Set**	

Command: refset
Transfer objects between the Refedit working set and host drawing...
Enter an option [Add/Remove] <Add>: *(Type A or R.)*
Select objects: *(Pick one or more objects.)*

COMMAND LINE OPTIONS

Add prompts you to select objects to add to the working set.

Remove prompts you to select objects to remove from the working set.

Select objects selects the objects to be added or removed.

RELATED COMMANDS

RefEdit edits reference objects in place.

RefClose saves or discards editing changes to the reference.

RELATED SYSTEM VARIABLES

RefEditName stores the name of the xref or block being edited.

XEdit determines whether the current drawing can be edited while being referenced by another drawing.

XFadeCtl specifies the amount of fading for objects not being edited in place.

TIPS

- The purpose of this command is to add objects to — or remove them from — the "working set" of objects, while you are performing in-place editing of a block or an externally-referenced drawing.

- You cannot add or remove objects on locked layers. AutoCAD complains, "** *n* selected objects are on a locked layer.'

Regen, RegenAll

V. 1.0   Regenerates viewports to update the drawing.

Command	Alias	Ctrl+	F-key	Alt+	Menu Bar	Dashboard
regen	re	...	...	VG	View	...
					↳ Regen	
regenall	rea	...	...	VA	View	...
					↳ Regen All	

Command: regen
Regenerating model.

Regenerates the current viewport.

Command: regenall
Regenerating model.

Regenerates all viewports.

COMMAND LINE OPTION

Esc cancels the regeneration.

RELATED COMMANDS

Redraw cleans up the current viewport quickly.

RegenAuto checks with you before doing most regenerations.

ViewRes controls whether zooms and pans are performed at redraw speed.

RELATED SYSTEM VARIABLES

RegenMode toggles automatic regeneration.

WhipArc determines how circles and arcs are displayed:

   0 — Circles and arcs displayed as vectors (default).

   1 — Circles and arcs displayed as true circles and arcs.

TIPS

- Some commands automatically force a regeneration of the screen; other commands queue the regeneration.

- The Regen command reindexes the drawing database for better display and object selection performance.

- To save on regeneration time, freeze layers you are not working with, apply QText to turn text into rectangles, and place hatching on its own layer.

- Use the RegenAll command to regenerate all viewports.

'RegenAuto

V. 1.2 Prompts before performing regenerations, when turned off (short for REGENeration AUTOmatic).

Command	Alias	Ctrl+	F-key	Alt+	Menu Bar	Dashboard
regenauto	...	...	...	...	...	...

Command: regenauto
Enter mode [ON/OFF] <ON>: *(Type ON or OFF.)*

COMMAND LINE OPTIONS

OFF turns on "About to regen, proceed?" message.

ON turns off "About to regen, proceed?" message.

RELATED COMMANDS

Regen forces a regeneration in the current viewport.

RegenAll forces a regeneration in all viewports.

RELATED SYSTEM VARIABLES

Expert suppresses "About to regen, proceed?" message when value is greater than 0.

RegenMode specifies the current setting of automatic regeneration:

0 — Off.

1 — On (default).

TIPS

- If a regeneration is caused by a transparent command, AutoCAD delays it and responds with the message, "Regen queued."

- When RegenAuto is set to off, the following prompt is displayed with every command that causes a regeneration:

    Command: regen
    About to regen, proceed? <Y>: *(Press Enter.)*

- AutoCAD Release 12 reduces the number of regenerations by expanding the virtual screen from 16 bits to 32 bits.

- This command was useful when computers were so slow that a regeneration could take several minutes. The technical editor recalls that regnrations were timed to coincide with coffee and lunch breaks.

Region

Rel.11 Creates 2D regions from closed objects.

Command	Alias	Ctrl+	F-key	Alt+	Menu Bar	Dashboard
region	reg	...	...	DN	Draw	2D Draw
					⤷Region	

Command: region
Select objects: *(Select one or more closed objects.)*
Select objects: *(Press Enter to end object selection.)*
1 loop extracted.
1 region created.

COMMAND LINE OPTION
Select objects selects objects to convert to a region; AutoCAD discards unsuitable objects.

RELATED COMMANDS
All drawing commands.

RELATED SYSTEM VARIABLE
DelObj toggles whether objects are deleted during conversion by the Region command.

TIPS
- This command converts closed line sets, closed 2D and planar 3D polylines, and closed curves.
- This command rejects open objects, intersections, and self-intersecting curves.
- The resulting region is unpredictable when more than two curves share an endpoint.
- Polylines with width lose their width when converted to a region.
- Island are "holes" in regions.

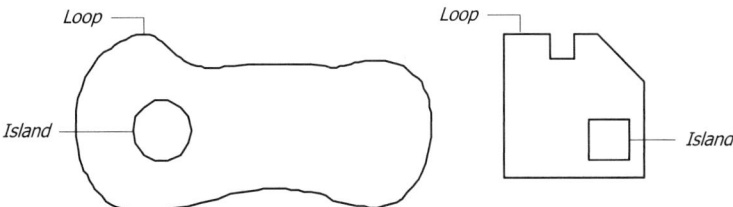

DEFINITIONS
Curve — an object made of circles, ellipses, splines, and joined circular and elliptical arcs.

Island — a closed shape fully within (not touching or intersecting) another closed shape.

Loop — a closed shape made of closed polylines, closed lines, and curves.

Region — a 2D closed area defined as a ShapeManager object.

Reinit

Rel.12 Reinitializes digitizers and input-output ports, and reloads the *acad.pgp* file (short for REINITialize).

Command	Alias	Ctrl+	F-key	Alt+	Menu Bar	Dashboard
reinit	...	...	...	...	...	...

Command: reinit

Displays dialog box:

RE-INITIALIZATION DIALOG BOX

I/O Port Initialization option+
 ☐ **Digitizer** reinitializes ports connected to the digitizer; grayed out if no digitizer is configured.

Device and File Initialization options
 ☐ **Digitizer** reinitializes the digitizer driver; grayed out if no digitizer is configured.
 ☐ **PGP File** reloads the *acad.pgp* file.

RELATED COMMAND
 CuiLoad reloads menu files.

RELATED SYSTEM VARIABLE
 Re-Init reinitializes via system variable settings.

RELATED FILES
 acad.pgp is the program parameters file in the hidden *documents and settings**<username>**application data**autodesk**autocad 2008**r17.1**enu**support* folder.
 ***.hdi** are device drivers specific to AutoCAD.

TIPS
- This command is almost unecessary now that most mice, printers, and plotters use USB ports, which do not need to be re-initialized.
- AutoCAD allows you to connect both the digitizer and the plotter to the same port, since you do not need the digitizer during plotting; use this command to reinitialize the digitizer after plotting.
- AutoCAD automatically reinitializes all ports and reloads the *acad.pgp* file each time a drawing is loaded.

Rename

V . 2.1 Changes the names of blocks, dimension styles, layers, linetypes, plot styles, text styles, table styles, UCS names, views, and viewports.

Commands	Aliases Ctrl+	F-key	Alt+	Menu Bar	Dashboard
rename	ren ...	...	OR	Format ⤷Rename	...
-rename	-ren				

Command: rename

Displays dialog box:

RENAME DIALOG BOX

Named Objects lists named objects in the drawing.

Items lists the names of named objects in the current drawing.

Old Name displays the current name of an object to be renamed.

Rename to allows you to enter a new name for the object.

. . . .

-RENAME Command

Command: -rename

Enter object type to rename
[Block/Dimstyle/LAyer/LType/Material/Plotstyle/textStyle/Tablestyle/Ucs/
VIew/VPort]: *(Enter an option.)*

Example usage:
[Block/Dimstyle/LAyer/LType/Material/Plotstyle/textStyle/Tablestyle/Ucs/
VIew/VPort]: b
Enter old block name: diode-2
Enter new block name: diode-02

. . . .

COMMAND LINE OPTIONS

Block changes the names of blocks.

Dimstyle changes the names of dimension styles.

LAyer changes the names of layers.

LType changes the names of linetypes.

Material changes the names of materials.

Plotstyle changes the names of plot styles.

Tablestyle changes the names of table styles.

textStyle changes the names of text styles.

Ucs changes the names of UCS configurations.

VIew changes the names of view configurations.

VPort changes the names of viewport configurations.

RELATED SYSTEM VARIABLES

CeLType names the current linetype.

CLayer names the current layer.

CMaterial names the current material.

CTableStyle names the current table style.

DimStyle names the current dimension style.

InsName names the current block.

TextStyle names the current text style.

UcsName names the current UCS view.

TIPS

- You cannot rename layer "0", anonymous blocks, groups, or linetype "Continuous." You can, however, rename the "Standard" text and dimension style.

- To rename a group of similar names, use * (the wildcard for "all") and ? (the wildcard for a single character).

- Names can be up to 255 characters in length.

- The Properties command does *not* allow you to rename blocks.

- The PlotStyle option is available only when plot styles are attached to the drawing.

- You cannot use this command during RefEdit.

Render

Rel.12 Renders 3D objects in an independent window.

Command	Alias	Ctrl+	F-key	Alt+	Menu Bar	Dashboard
render	rr	...	...	VER	View ⤷Render ⤷Render	Render

-render

Command: render

Displays window (illustrated below), or outputs the rendering in the current viewport, of a windowed selection, or of selected objects — depending on the setting in the RPref command.

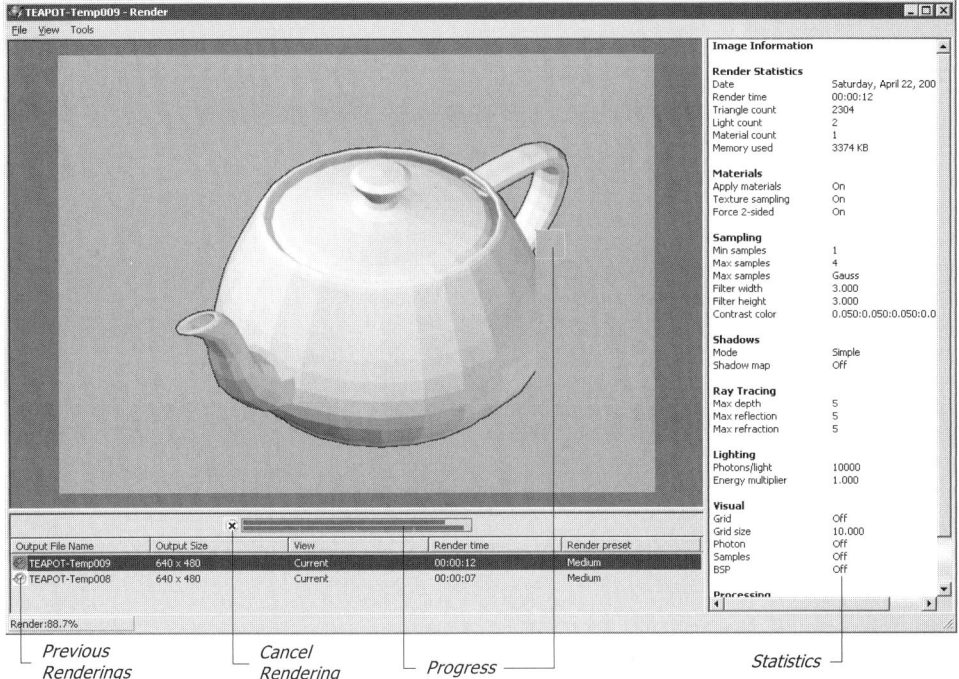

Previous Renderings Cancel Rendering Progress Statistics

RENDER WINDOW MENU BAR

File options

Save saves the rendered image to a file in BMP, TIFF, PNG, JPEG, Targa, or PCX formats. While the SaveImg command does not apply to this window, it has the same options as this command.

Save Copy saves a copy of the image by another name or to another folder.

Exit closes the Render window; reopen with the RenderWin command.

View options

Status Bar toggles the display of the status bar.

Statistics Pane toggles the display of the statistics pane.

Tools options

Zoom + makes the image up to 64x larger.

Zoom - makes the image up to 64x smaller.

Use the scroll bars or mouse wheel to pan the image. Hold down the Ctrl key and use the mouse wheel to zoom.

PROGRESS METERS

x cancels the rendering, as does pressing the ESC key.

Upper bar shows the progress within the current phase: translation, optional photon emission, optional final gather, and render.

Lower bar shows the progress of the entire rendering; also displayed by the status bar. The history pane reports the time each rendering took.

Status bar reports the name of the current rendering stage, and the current percentage of the total rendering.

SHORTCUT MENU

Right-click the history pane:

Render Again re-renders the selected history item.

Save saves the rendering to an image file.

Save Copy saves the rendering to a different folder.

Make Render Settings Current loads the rendering style associated with the history item.

Remove From The List removes the history item from the list.

Delete Output File erases the rendering from the image pane

-Render Command

(New to AutoCAD 2008.)

Command: -render

Specify render preset [Draft/Low/Medium/High/Presentation/Other] <Medium>: *(Enter a preset name.)*

Specify render destination [Render Window/Viewport] <Render Window>: *(Enter a destination.)*

Draft, **Low**, **Medium**, **High**, **Presentation** specify Autodesk-defined render parameters.

Other allows you to specifiy a user-defined rendering preset; AutoCAD prompts, 'Specify custom render preset [?]:'.

Render Window outputs the rendering in the Render window.

Viewport renders in the current viewport.

RELATED COMMANDS

RenderEnvironment sets up fog effects.

RenderPresets sets up rendering defaults for use with this command; displays the Render Presets Manager dialog box.

RPref sets up rendering styles; displays the Advanced Render Settings palette.

RenderWin acceses the Render window.

TIPS

■ Use the scoll bars or mouse wheel to pan the image. Hold down the Ctrl key, and use the mouse wheel to zoom.

■ Rendering is performed in four steps: translation, photon emission, final gather, and render. The photon emission and final gather steps occur only if specified in the render preferences.

■ Use the -Render command to batch renderings through scripts. It cannot set any rendering parameters, so preset them with the RPref or RenderPresets commands by creating named rendering presets.

RenderCrop

2007 Renders a windowed area of the current viewport.

Command	Alias	Ctrl+	F-key	Alt+	Menu Bar	Dashboard
rendercrop	rc	...	...	...	...	Render

Command: rendercrop
Pick crop window to render: *(Pick a point, or enter x,y coordinates.)*
please enter the second point: *(Pick another point.)*
Renders the windowed area:

COMMAND OPTIONS

Pick crop window to render picks the first point of the area to be rendered.

please enter the second point picks the other corner of the rectangle. (Note the unusual wording of the prompt.)

RELATED COMMANDS

RenderWin renders the drawing in an independent window.

RPref sets up rendering preferences.

TIPS

- Rendering a windowed area is faster than rendering the entire viewport.

- This command does not allow you to specify rendering preferences; before using this command, run the RPref command to set up preferences.

RenderEnvironment

<u>**2007**</u> Sets the parameters for fog effects in renderings; the command formerly known as Fog.

Command	Alias	Ctrl+	F-key	Alt+	Menu Bar	Dashboard
renderenvironment	fog	...	...	VEE	View Render ↳Render Environment	Render

Command: renderenvironment

Displays dialog box:

DIALOG BOX OPTIONS

Enable Fog toggles fog settings.

Color specifies the color of the fog; displays the Select Color dialog box.

Fog Background applies the fog effect to the background; see the View command.

Near Distance indicates the start of the fog effect, measured as a decimal (0.0 - 1.0) from the camera.

Far Distance indicates the distance from the camera where the fog effect ends.

Near Fog Percentage specifies the near intensity of the fog effect.

Far Fog Percentage specifies the far intensity of the fog effect.

RELATED COMMAND

Render creates renderings with the fog effect.

TIPS

- First set up the RenderEnvironment command, and then use Render to see the effect.
- The fog can be any color; here are some suggestions:

Color	RGB	HLS	Effect
White	1,1,1	0,1,0	Fog.
Black	0,0,0	0,0,0	Distance.
Green	0,1,0	0.33,0.5,0	Alien mist.

- The effect of using white as the fog color:

- It can be tricky getting the fog effect to work. Use the 3dClip command to set the back clipping plane at the back of the model, or where you want the fog to have its full effect. Then use RenderEnvironment to set up the following parameters:
  - Enable fog: On
  - Color: White
  - Fog Background: On or Off
  - Near distance: 0.70
  - Far distance: 1.00
  - Near fog percentage: 0.00
  - Far fog percentage: 100

RenderExposure

Modifies the global lighting of scenes rendered with photometric lighting.

Command	Alias	Ctrl+	F-key	Alt+	Menu Bar	Dashboard
renderexposure	...	...	...	...	...	Render

Command: renderexposure

When LightingUnits is set to 1 or 2, displays dialog box:

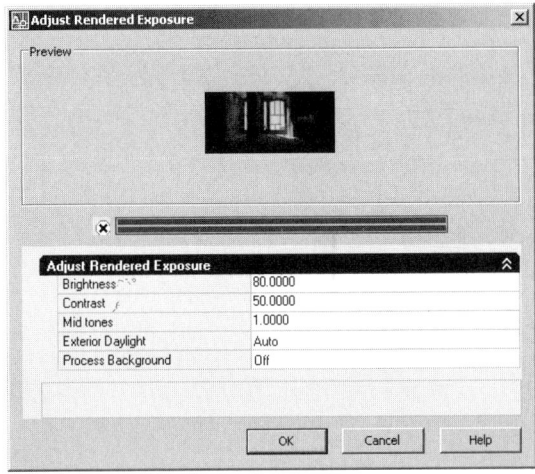

DIALOG BOX OPTIONS

Brightness ranges from from 0 to 200; default = 65.

Contrast ranges from 0 to 100; default = 100.

Mid tones ranges from 0 to 20; default = 1.

Exterior Daylight (aka sun) changes between auto, off, and on.

Process Background toggles between on and off.

RELATED SYSTEM VARIABLES

LogExpBrightness determines global brightness of photometric lighting.

LogExpContrast determines the global contrast of photometric lighting.

LogExpDaylight toggles global daylight of photometric lighting.

LogExpMidtones determines the global midtones of photometric lighting.

TIPS

- Use this command to interactively adjust the lighting of the current 3D drawing, such as making too-dark renderings lighter.

- This command does not work when the LightingUnits system variable is set to 0. Ensure that LightingUnits is set to 1 or 2.

RenderPresets

2007 Creates rendering styles; displays AutoCAD's longest dialog box.

Command	Alias	Ctrl+	F-key	Alt+	Menu Bar	Dashboard
renderpresets	rp	...	...	...	...	...
	rfileopt					

Command: renderpresets

Displays dialog box:

DIALOG BOX OPTIONS

See the RPref command for details of the parameters.

Set Current sets the selected rendering style as the default; the Render and RenderCrop commands use this rendering style.

Create Copy creates a copy of the parameters so that you can modify them for your needs.

Delete removes the named rendering style; you cannot delete the system styles.

TIPS

- AutoCAD comes preset with five styles: Draft, Low, Medium, High, and Presentation:

  - **Draft** rendering is very fast, but inaccurate and very coarse. Time to render the *teapot.dwg*: 3.5 seconds.
  - **Low** rendering is fast but inaccurate; time to render: 4.5 seconds.
  - **Medium** rendering is slightly slower but much more accurate; time to render: 10 seconds.
  - **High** rendering is slower and more accurate; time to render: 15 seconds.
  - **Presentation** rendering is very slow and very accurate; time to render: 55 seconds.

Left: *Teapot.dwg rendered in draft mode; notice the display errors.*
Right: *Drawing rendered in presentation mode; notice the material.*

- The complete contents of the Render Presets Manager look like this (see the RPref command for details):

- The teapot was used in the early days of computer rendering software as a complex test case to ensure the rendering was completed accurately.

Removed Command

RenderUnload was removed from Release 14.

RenderWin

<u>**2007**</u> Displays the Render window with the most recent rendering; formerly the RendScr command.

Command	Alias	F-key	Alt+	Menu Bar	Dashboard
renderwin	rendscr	...	...	...	Render
	rw				

Command: renderwin

Displays the Render window with the most recent rendering.

See Render command for details.

RELATED COMMANDS

Render generates renderings in the Render window.

RenderCrop generates renderings in a windowed area of the viewport.

RenderPresets sets the rendering parameters.

TIPS

- If this command fails to display the Render window, it's because the window is already open, but covered up by the AutoCAD window. Look for its button on the taskbar.

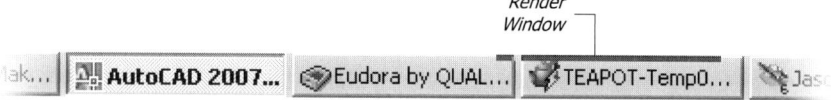

- If no rendering has occurred in this drawing, AutoCAD complains, "RENDERWIN Drawing contains no render history," and does not display the Render window.

. . . .

Removed Commands

The **Replay** command was removed from AutoCAD 2007.

The **RendScr** command is an alias of the RenderWin command as of AutoCAD 2007.

. . . .

ResetBlock

<u>2006</u> Resets dynamic block references to their default values.

Command	Alias	Ctrl+	F-key	Alt+	Menu Bar	Dashboard
resetblock	...	...	...	...	...	...

Command: resetblock
Select objects: *(Select one or more blocks.)*
Select objects: *(Press Enter to end object selection.)*
n **blocks reset.**

COMMAND LINE OPTION

Select objects selects the blocks to be reset.

RELATED COMMAND

BEditor creates and edits dynamic blocks.

TIP

- Dynamic blocks can be changed in many ways, according to the actions defined in the Block Editor. Use this command to reset blocks to their original definition.

'Resume

V. 2.0 Resumes script files paused by an error, or by pressing the Backspace or ESC keys.

Command	Alias	Ctrl+	F-key	Alt+	Menu Bar	Dashboard
resume	...	...	...	...	...	...

Command: resume

COMMAND LINE OPTIONS

BACKSPACE pauses the script file.

ESC stops the script file.

RELATED COMMANDS

RScript reruns the current script file.

Script loads and runs a script file.

RevCloud

2004   Draws revision clouds, and converts objects into revision clouds.

Command	Alias	Ctrl+	F-key	Alt+	Menu Bar	Dashboard
revcloud	...	...	...	DV	Draw	2D Draw
					ꀸRevision Cloud	

Command: revcloud
Minimum arc length: 0.5000   Maximum arc length: 0.5000   Style: Normal
Specify start point or [Arc length/Object/Style] <Object>: s
Select arc style [Normal/Calligraphy] <Normal>: c
Arc style = Calligraphy
Specify start point or [Arc length/Object/Style] <Object>: *(Pick a point, or enter an option.)*
Guide crosshairs along cloud path... *(Move cursor to create cloud.)*
Revision cloud finished.

Revision Cloud

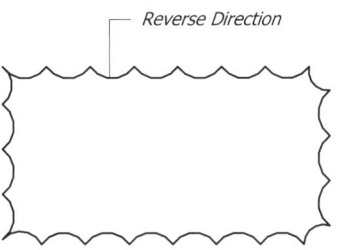
Reverse Direction

COMMAND LINE OPTIONS
   Specify start point specifies the starting point of the cloud.

   Guide crosshairs along cloud path outlines the cloud.

Arc Length Options
Specify minimum length of arc <0.5000>: *(Enter a minimum value.)*
Specify maximum length of arc <1.0000>: *(Enter a maximum value.)*

   Specify minimum length of arc specifies the minimum arc length.

   Specify maximum length of arc specifies the maximum arc length.

Object Options
Select object: *(Select one object to convert to a revision cloud.)*
Reverse direction [Yes/No] <No>: *(Type Y or N.)*

   Select object selects the object to convert into a revision cloud.

   Reverse direction turns the revision cloud inside-out.

Style Options

Select arc style [Normal/Calligraphy] <Normal>: *(Type N or C.)*

   Normal draws clouds with uniform-width arcs.

   Calligraphy draws clouds from variable-width polyarcs.

RELATED SYSTEM VARIABLE

   DimScale affects the size and width of the arcs.

TIPS

- When the cursor reaches the start point, the revision cloud closes automatically.

- The revision cloud is drawn as a polyline.

- To edit the revision cloud, select it, and then move the grips.

- The arc length is not available from a system variable, because it is stored in the Windows registry.

- The arc length is multiplied by the value stored in the DimScale system variable.

Revolve

<u>Rel.11</u>   Creates 3D solid objects by revolving closed 2D objects about an axis.

Command	Alias	Ctrl+	F-key	Alt+	Menu Bar	Dashboard
revolve	rev	...	...	DMR	Draw	3D Make
					↳ Modeling	
					↳ Revolve	

Command: revolve

Current wire frame density: ISOLINES=4

Select objects to revolve: *(Select one or more closed objects.)*

Select objects to revolve: *(Press Enter to end object selection.)*

Specify axis start point or define axis by [Object/X/Y/Z] <Object>: *(Pick a point, or enter an option.)*

Specify angle of revolution or [STart angle] <360>: *(Enter a value.)*

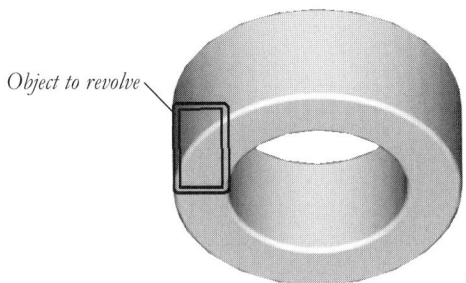

Object to revolve

The filleted rectangle was used to create the revolved object.

COMMAND LINE OPTIONS

Select objects to revolve selects a closed object to revolve: closed polyline, circle, ellipse, donut, polygon, closed spline, or a region.

Specify axis start point indicates the axis of revolution; you must specify the endpoint.

Object selects the object that determines the axis of revolution.

X uses the positive x axis as the axis of revolution.

Y uses the positive y axis as the axis of revolution.

Specify angle of rotation specifies the amount of rotation; full circle = 360 degrees.

STart angle specifies the starting angle.

RELATED COMMANDS

Extrude extrudes 2D objects into a 3D solid model.

RevSurf rotates open and closed objects, forming a 3D surface.

TIPS

- Revolving an open object creates surfaces; closed objects, solids.

- This command does not work with open objects or self-intersecting polylines.

RevSurf

Rel.10 Generates 3D meshes of revolution defined by a path curve and an axis (short for REVolved SURFace); does not draw surfaces.

Command	Alias	Ctrl+	F-key	Alt+	Menu Bar	Dashboard
revsurf	...	...	...	DMMM⬦	Draw	...
					⬦Modeling	
					⬦Meshes	
					⬦Revolved Surface	

Command: revsurf
Current wire frame density: SURFTAB1=6 SURFTAB2=6
Select object to revolve: *(Select an object.)*
Select object that defines the axis of revolution: *(Select an object.)*
Specify start angle <0>: *(Enter a value.)*
Specify included angle (+=ccw, -=cw) <360>: *(Enter a value.)*

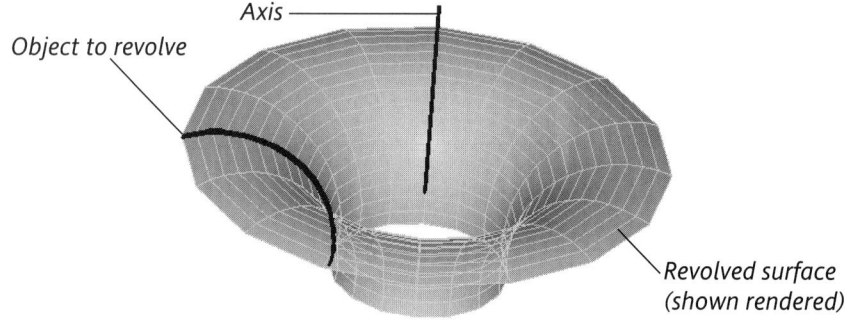

Axis

Object to revolve

Revolved surface
(shown rendered)

COMMAND LINE OPTIONS

Select object to revolve selects the single object that will be revolved about an axis.
Select object that defines the axis of revolution selects the axis object.
Start angle specifies the starting angle.
Included angle specifies the angle to rotate about the axis.

RELATED COMMANDS

PEdit edits revolved meshes.
Revolve revolves a 2D closed object into a 3D solid.

RELATED SYSTEM VARIABLES

SurfTab1 specifies the mesh density in the m-direction (default = 6).
SurfTab2 specifies the mesh density in the n-direction (default = 6).

TIPS

- This command works with open and closed objects.

- If a multi-segment polyline is the axis of revolution, the rotation axis is defined as the vector pointing from the first vertex to the last vertex, ignoring the intermediate vertices.

Rotate

<u>**V. 2.5**</u> Rotates objects about an axis perpendicular to the current UCS.

Command	Alias	Ctrl+	F-key	Alt+	Menu Bar	Dashboard
rotate	ro	...	...	MR	Modify ⬥Rotate	2D Draw

Command: rotate
Current positive angle in UCS: ANGDIR=counterclockwise ANGBASE=0
Select objects: *(Select one or more objects.)*
Select objects: *(Press Enter to end object selection.)*
Specify base point: *(Pick a point.)*
Specify rotation angle or [Copy/Reference]: *(Enter a value, or type R.)*

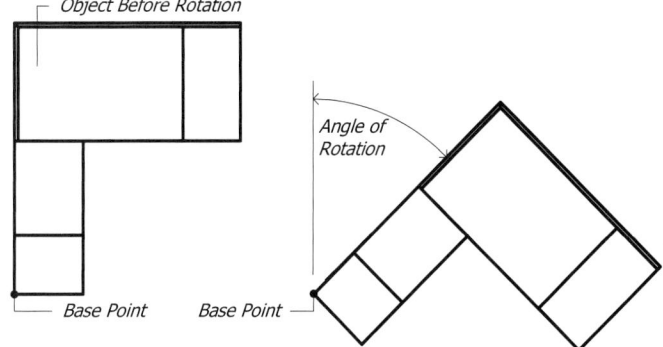

COMMAND LINE OPTIONS

Select objects selects the objects to be rotated.

Specify base point picks the point about which the objects will be rotated.

Specify rotation angle specifies the angle by which the objects will be rotated.

Reference allows you to specify the current rotation angle and new rotation angle.

Copy rotates a copy of the object, leaving the original in place.

RELATED COMMANDS

3dRotate rotates objects in 3D space using a 3D grip tool.

UCS rotates the coordinate system.

Snap command's Rotate option rotates the cursor.

TIPS

- AutoCAD rotates the selected object(s) about the base point.

- At the 'Specify rotation angle' prompt, you can show the rotation by moving the cursor. AutoCAD dynamically displays the new rotated position as you move the cursor.

- Use object snap modes, such as INTersection, to position the base point and rotation angle(s) accurately.

Rotate3D

<u>Rel.11</u> Rotates objects about any axis in 3D space with a user interface that's different from the 3dRotate command.

Command	Alias	Ctrl+	F-key	Alt+	Menu Bar	Dashboard
rotate3d	...	...	...	...	...	3D Make

Command: rotate3d
Current positive angle: ANGDIR=counterclockwise ANGBASE=0
Select objects: *(Select one or more objects.)*
Select objects: *(Press Enter to end object selection.)*

Specify first point on axis or define axis by
[Object/Last/View/Xaxis/Yaxis/Zaxis/2points]: *(Pick a point, or enter an option.)*
Specify second point on axis: *(Pick another point.)*
Specify rotation angle or [Reference]: *(Enter a value, or type R.)*

COMMAND LINE OPTIONS

Select objects selects the objects to be rotated.

Specify rotation angle rotates objects by a specified angle: relative rotation.

Reference specifies the starting and ending angle: absolute rotation.

Define Axis By options

Object selects object to specify the rotation axis.

Last selects the previous axis.

View makes the current view direction the rotation axis.

Xaxis makes the x axis the rotation axis.

Yaxis makes the y axis the rotation axis.

Zaxis makes the z axis the rotation axis.

2 points defines two points on the rotation axis.

RELATED COMMANDS

3dRotate rotates objects in 3D space using a 3D grip tool.

Rotate rotates objects in 2D space.

Align rotates, moves, and scales objects in 3D space.

Mirror3d mirrors objects in 3D space.

Removed Commands

The **RmlIn** command was removed from AutoCAD 2006; its replacement is Markup.

The **RMat** command was removed from AutoCAD 2007; its replacement is MaterialAttach.

RPref, RPrefClose

Rel.12 Specifies options for the Render and RenderCrop commands (short for Render PREFerences).

Command	Alias	Ctrl+	F-key	Alt+	Menu Bar	Dashboard
rpref	rpr	...	...	VED	View ↳Render ↳Advanced Render Settings	Render
rprefclose						

Command: rpref

Displays palette:

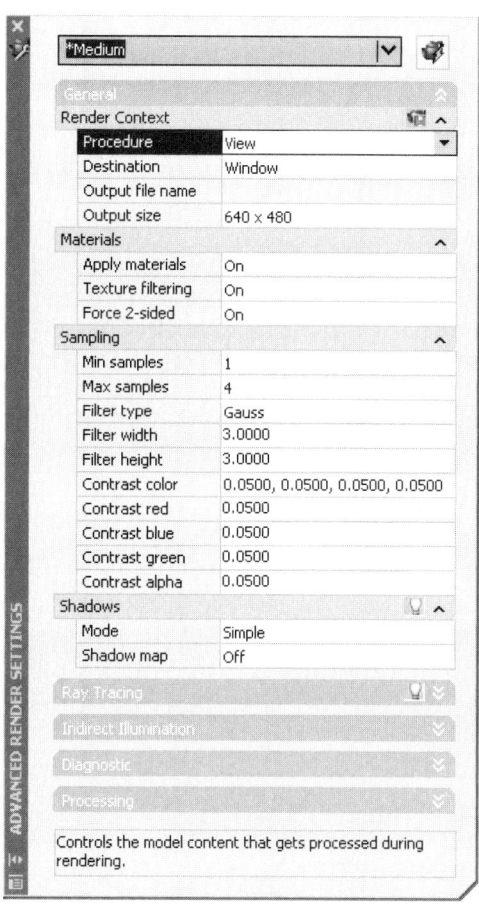

Command: rprefclose

Clsoes the palette.

Options are identical to the Render Presets Manager dialog box (RenderPresets command).

GENERAL

👷 performs the render operation.

Render Context options

🔳 toggles the Output File Name parameter for the Render command's Save option.

Procedure determines which part of the drawing is rendered:

- **View** — renders the entire viewport.
- **Crop** — prompts for you to pick two points, like the RenderCrop command.
- **Selected** — prompts you to select the objects to be rendered.

Destination determines where the model will be rendered:

- **Viewport** — model is rendered in the current viewport.
- **Window** — model is rendered in the Render window.

Output file name specifies the name and format of images saved to disc; see SaveImg command.

Output size specifies the size of the rendered image; smaller views render faster.

Materials options

Apply Materials toggles rendering of materials, if applied to the model. If not applied, the model takes on the look defined by the Global material. See Materials command.

Texture Filtering toggles the rendering of texture maps.

Force 2-Sided toggles the rendering of both sides of faces; single-sided is faster.

Sampling options

Min Samples specifies the minimum sample rate (number of samples per pixel). Range is 1/ 64 (one out of every 64 pixels is sampled) to 1024 (1,024 samples per pixel). More samples result in higher quality but slower rendering speed.

Max Samples specifies the maximum sample rate used when neighboring samples exceed the contrast limit (defined below).

Filter Type specifies how multiple samples are combined into one:

- **Box** — sums all samples in the filter area with equal weight (fastest).
- **Gauss** — weights samples using a bell curve.
- **Triangle** — weights the samples using a pyramid.
- **Mitchell** — weights the samples using a steeper bell curve.
- **Lanczos** — weights the samples using a steep bell curve that diminishes the effect of samples at the edge of the filter area.

Filter Width and **Filter Height** specify the size of the filtered area; range is 0 to 8. Larger values soften the image, but increase rendering time.

Contrast Color selects the threshold RGBA (red, green, blue, alpha) values; displays the Select Color dialog box. Range is from 0 to 1, where 0 is black and 1 is white.

Contrast Red, **Contrast Blue**, and **Contrast Green** indicate threshold values for the red, blue, and green components of samples.

Contrast Alpha indicates the threshold value for the alpha component, typically used for transparency.

Shadows options

    toggles shadow-casting.

    Mode specifies the type of shadow casting:

- **Simple** — casts random-order shadow.
- **Sort** — casts shadows from objects to lights.
- **Segments** — casts shadows along the light ray.

    Shadow Map Controls toggles shadow mapping:

    On — shadow-mapped shadows.

    Off — ray-traced shadows.

RAY TRACING

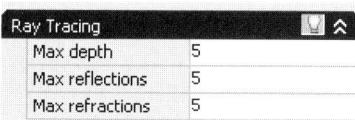

    toggles ray tracing.

Max Depth stops raytracing when the number of reflections reaches this number. If rays refract twice and reflect three times, the ray tracing is stopped when this value is set to 5.

Max Reflections stops ray tracing after this many *reflections* (light bounced off a surface). Turns off reflections when set to 0.

Max Refractions stops ray tracing after this many *refractions* (light distorted going through an object). Turns off refractions when set to 0.

INDIRECT ILLUMINATION

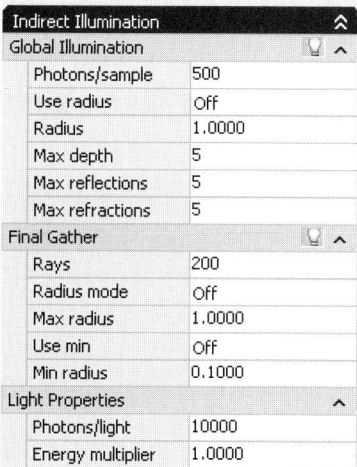

Global Illumination options

    toggles global illumination.

Photons/Samples specifies the number of photons used to compute the intensity of global illumination. Range is 1 to 2147483647; more photons make for less noise, more blur, and longer renderings.

Use Radius toggles the use of photon sizing. When off, photons are determined to be 1/10 the radius of the viewport.

Radius specifies the area in which photons are used.

Max Depth, Max Reflections, and **Max Refractions** settings are the same as for ray tracing.

Final Gather options

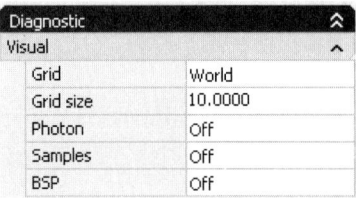 toggles gathering (additional global illumination calculations).

Rays specifies the number of rays used to compute indirect illumination. More rays make for less noise but longer renderings.

Radius Mode specifies the radius mode:
- **On** — Max Radius setting is used.
- **Off** — maximum radius is 10 percent of the maximum model radius.
- **View** — Max Radius setting in pixels.

Max Radius specifies the maximum radius used for final gathering. Lower values improve quality but increase rendering time.

Use Min Controls toggles the minimum radius setting.

Min Radius specifies the minimum radius used for final gathering. Higher values improve quality but increase rendering time.

Light Properties options

Affects how lights behave when calculating indirect illumination.

Photons/Light specifies the number of photons emitted by each light. Larger values increase accuracy but also increase the rendering time and RAM usage.

Energy Multiplier multiplies the global illumination intensity.

DIAGNOSTIC

Helps track down rendering peculiarities.

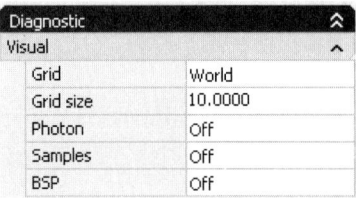

Visual options

Grid shows the coordinate space of:
- **Object** — local u,v,w coordinates for each object.
- **World** — world x,y,z coordinates for all objects.
- **Camera** — imposes a rectangular grid on the view.

Grid Size sets the size of the grid.

Photon renders the effect of a photon map, if present:

- **Density** — renders the photon map; red = high density; blue = low density.
- **Irradiance** — shades photons based on their irradiance; red = maximum irradiance; blue = low irradiance.

BSP (binary space partition) helps track down very slow renderings by visualizing the BSP raytrace parameters:

- **Depth** — illustrates the tree depth; red = top faces; blue = deep faces.
- **Size** — colors the leaf size in the tree.

Processing options

Tile Size specifies the rendering tile size; smaller values result in longer rendering times.

Tile Order specifies the tile render order:

- **Hilbert** — renders the next tile based on the difficulty of switching to the next one.
- **Spiral** — starts at the center and then spirals outward.
- **Left to Right** — bottom to top, left column to right.
- **Right to Left** — bottom to top, right column to left.
- **Top to Bottom** — right to left, top row to bottom.
- **Bottom to Top** — right to left, bottom row to top.

Memory Limit discards geometry to allocate memory for other objects, if this memory limit is reached.

RELATED COMMANDS

Render and **RenderCrop** use these settings to render 3D models.

RenderPresets creates rendering styles.

RScript

<u>**V. 2.0**</u>   Repeats script files (short for Repeat SCRIPT).

Command	Alias	Ctrl+	F-key	Alt+	Menu Bar	Dashboard
rscript	...	...	...	...	...	...

Command: rscript

COMMAND LINE OPTIONS
None.

RELATED COMMANDS
Resume resumes a script file after being interrupted.

Script loads and runs a script file.

TIP
- Placed at the end of a script file, this command causes the script file to run repeatedly until canceled with Backspace or Esc.

RuleSurf

<u>Rel.10</u> Draws 3D ruled meshes between two objects (short for RULEd SURFace); does not draw surfaces.

Command	Alias	Ctrl+	F-key	Alt+	Menu Bar	Dashboard
rulesurf	...	...	...	DMMR	Draw	...
					⤷Modeling	
					⤷Meshes	
					⤷Ruled Surf	

Command: rulesurf
Select first defining curve: *(Pick an object.)*
Select second defining curve: *(Pick another object.)*

Left to right: Ruled meshes created between point and line, line and line, and arc and line.

COMMAND LINE OPTIONS

Select first defining curve selects the first object for the ruled mesh.

Select second defining curve selects the second object for the ruled mesh.

RELATED COMMANDS

EdgeSurf draws a 3D mesh bounded by four edges.

RevSurf draws a 3D mesh of revolution.

TabSurf draws a 3D tabulated mesh.

RELATED SYSTEM VARIABLE

SurfTab1 determines the number of faces drawn.

TIPS

- This command uses these objects as the boundary curve: points, lines, arcs, circles, polylines, and 3D polylines.

- Pick order is important: the ruled mesh is drawn starting at the endpoint nearest the pick point. This can result in a twisted surface, as illustrated by the line-line example above.

- Both boundaries must either be closed or open; the exception is the point object.

- This command begins drawing its mesh as follows:
    Open object — from the object's endpoint closest to your pick.
    Circle — from the zero-degree quadrant.
    Closed polyline — from the last vertex.

- RuleSurf creates meshes with circles in the opposite direction from that of closed polylines.

Save, SaveAs

<u>**V. 2.0**</u>   Saves the current drawing to disk, after prompting for a file name.

Commands	Alias	Ctrl+	F-key	Alt+	Menu Bar	Dashboard
save	saveurl	...	...	...	...	...
saveas	...	Shift+S ...		FA	File ⮑Save As	...

Command: save *or* saveas

Displays dialog box.

SAVING DRAWING AS DIALOG BOX

Save in selects the folder, hard drive, or network drive in which to save the drawing.

☑ **Update sheet and view thumbnails now**.

File name names the drawing; maximum = 255 characters; default=*drawing1.dwg*.

Save saves the drawing, and then returns to AutoCAD.

Cancel does not save the drawing.

When a drawing of the same name already exists in the same folder, displays dialog box:

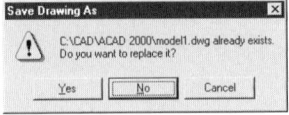

Files of type saves the drawing in a variety of formats:

Save as type	Saves drawings in format compatible with...
DWG formats:	
AutoCAD 2007 DWG	AutoCAD 2007/8.
AutoCAD 2004/LT2004 Drawing	AutoCAD 2004/5/6.
AutoCAD 2000/LT 2000 Drawing	AutoCAD 2000/i/2.
AutoCAD R14/LT98/LT97 Drawing	AutoCAD Release 14.
DWS and DWT formats:	
Drawing Standards	AutoCAD standards format.
Drawing Template	Templated drawings stired in the \template folder.
DXF formats:	
AutoCAD 2007 DXF	AC17 DXF, AutoCAD 2007 - 2008.
AutoCAD 2004/LT 2004 DXF	AC16 DXF, AutoCAD 2004 - 2006.
AutoCAD 2000/LT 2000 DXF	AC15 DXF, AutoCAD 2000 - 2002.
AutoCAD R12/LT 2 DXF	AC12 DXF format, also compatible with Release 11.

SAVEAS OPTIONS DIALOG BOX

From the dialog box's toolbar, select Tools | Options:

DWG Options tab

Save proxy images of custom objects option:

☑ Saves an image of the custom objects in the drawing file.

☐ Saves a frame around each custom object, instead of an image.

Index type saves indices with drawings; useful only for xrefed and partially-loaded drawings:

* **None** — creates no indices (default).
* **Layer** — loads layers that are on and thawed.
* **Spatial** — loads only the visible portion of clipped xrefs.
* **Layer and Spatial** — combines the above two options.

Save all drawings as selects the default format for saving drawings.

DXF Options tab

⊙ **ASCII** saves the drawing in ASCII format, which can be read by humans – as well as by common text editors – but takes up more disk space.

○ **BINARY** saves the drawing in binary format, which takes up less disk space, but cannot be read by some software programs.

☐ **Select objects** allows you to save selected objects to the DXF file; AutoCAD prompts you, 'Select objects:'.

☐ **Save thumbnail preview image** saves a thumbnail image of the drawing.

Decimal places of accuracy specifies the number of decimal places for real numbers: 0 to 16.

TEMPLATE DESCRIPTION DIALOG BOX

Displayed when drawing is saved as .dwt template file:

Description describes the template file; stored in *wizard.ini*.

Measurement selects English or metric as the default measurement system.

RELATED COMMANDS

Export saves the drawing in several of vector formats.

Plot saves the drawing in additional raster and vector formats.

Quit exits AutoCAD without saving the drawing.

QSave saves the drawing without prompting for a name.

AutoPublish simultaneously saves the drawing as a DWF file *(new to AutoCAD 2008)*.

RELATED SYSTEM VARIABLES

DBMod indicates that the drawing was modified since being opened.

DwgName specifies the name of the drawing; *drawing1.dwg* when unnamed.

AutoDwfPublish toggles simultaneous-saving as DWF files *(new to AutoCAD 2008)*.

SaveFidelity preserves annotative objects when exported to AutoCAD 2007 or earlier.

TIPS

- The Save and SaveAs commands perform exactly the same function.

- Use these commands to save drawings by other names; for example, saving read-only drawings.

- To control when AutoPublish operates, go to the Option command's Plot and Publish tab, and then click the Automatic DWF Publish Settings button; see the AutoPublish command.

- To save drawings without seeing the Save Drawing As dialog box, use the QSave command.

- All newer releases of AutoCAD can read drawing files produced by older releases, but AutoCAD 2004/5/6 and earlier cannot open drawings created by AutoCAD 2007/8. The Save command translates drawings to make them compatible with earlier releases. The problem is that incompatible objects are modified, as follows:

  - The Materials and ShadowDisplay properties are ignored and preserved by AutoCAD 2004/5/6. For example, apply a material to an object in AutoCAD 2008, export in 2004 DWG format. In AutoCAD 2006, the material is not displayed; you can modify the object, and when reopened in 2008, the original material definition is preserved.

  - When helixes are translated to DWG 2004/5/6 format, they appears as proxy entities. (Proxy entities are visual representations of ARx-defined objects or "custom objects" defined in other AutoCAD-based applications.) You can edit and change properties of proxies, such as move and rotate them — but not changing height or radius. The Explode command changes the helix to a series of arced polylines. A better solution is to use the Sweep command to convert the helix into a coil, a 3D solid entity that is represented accurately by AutoCAD 2004/5/6.

  - AutoCAD 2008 has extruded, lofted, planar, revolved, and swept surfaces, which become proxy entities in AutoCAD 2006 and earlier. When exploded, they become arcs, polylines, and circles. Use the ConvertToSolid command to convert surfaces to solids, which are acceptable to AutoCAD 2004/5/6.

  - When drawings with the SectionObject and multileader objects are exported to DWG 2004 format, they become proxy entities. When exploded, they becomes lines, polylines, and text.

  - Cameras and lights are represented by a new 3D icon called "glyphs." Glyphs are ignored but preserved by AutoCAD 2004/5/6. These materials specific to 2008 disappear: procedural maps, checkers, noise, speckles, tiles, and waves.

  - Tables created by the DataExtraction command cannot be edited/updated in earlier releases.

  - DGN underlays are not displayed in AutoCAD 2007 and earlier; imported DGN files are, because they have been converted to AutoCAD objects.

  - These dimension features are lost when opened in earlier releases: extension lines on radial and diameter dimensions; inspection text; and jogged linear dimensions. Breaks in dimension and extension lines survive back to 2004, but loose their associativity.

  - New paragraph features and fields that wrap across columns are lost when opened in earlier releases. Tables lose 2008-specific cell formatting.

  - VP layer overrides do not display the overridden properties in AutoCAD 2007 and earlier.

. . . .

Removed Command

SaveAsR12 was removed from AutoCAD Release 14. Use SaveAs instead, which saves drawings in earlier formats.

. . . .

SaveImg

Rel.12 Saves a rendering of the current viewport image as raster files on disk (short for SAVE IMAge).

Command	Alias	Ctrl+	F-key	Alt+	Menu Bar	Dashboard
saveimg	...	...	...	TYS	Tools	...
					⤷ Display Image	
					⤷ Save	

Command: saveimg

Displays the Render Output File dialog box.

Provide a file name, select a format, and then click Save.

One of the following dialog boxes appears:

IMAGE OPTIONS DIALOG BOXES

BMP Image Options dialog box

O **Monochrome** converts all lines to black.

O **8 Bits (256 Grayscale)** reduces the image to shades of gray.

O **8 Bits (256 Colors)** reduces the image to 256 colors.

⊙ **24 Bits (16.7 Million Colors)** maintains all colors in the image.

PCX Image Options dialog box

O **Monochrome** converts all lines to black.

O **8 Bits (256 Grayscale)** reduces the image to shades of gray.

⊙ **8 Bits Optimized (256 Colors)** reduces the image to the best 256 colors possible.

Targa Image Options dialog box

○ **8 Bits (256 Grayscale)** reduces the image to shades of gray.

○ **8 Bits (256 Colors)** reduces the image to 256 colors.

⊙ **24 Bits (16.7 Million Colors)** maintains all colors in the image.

○ **32 Bits (23 Bits + Alpha)** maintains all colors, plus transparency information (alpha).

☐ **Bottom Up** stores data in the file from the bottom to the top of the image.

TIFF Image Options dialog box

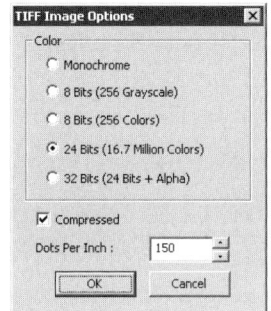

○ **Monochrome** converts all lines to black.

○ **8 Bits (256 Grayscale)** reduces the image to shades of gray.

○ **8 Bits (256 Colors)** reduces the image to 256 colors.

⊙ **24 Bits (16.7 Million Colors)** maintains all colors in the image.

○ **32 Bits (23 Bits + Alpha)** maintains all colors, plus transparency information (alpha).

☑ **Compressed** reduces the size of the file by compressing the data.

Dots Per Inch specifies the resolution of the image; range is 0 to 1000.

JPEG Image Options dialog box

Quality adjusts the amount of compression.

File size also adjusts the amount of compression; increase the compression for smaller file sizes. The sliders are linked, and cannot be set independently

PNG Image Options dialog box

○ **Monochrome** converts all lines to black.

○ **8 Bits (256 Grayscale)** reduces the image to shades of gray.

○ **8 Bits (256 Colors)** reduces the image to 256 colors.

⊙ **24 Bits (16.7 Million Colors)** maintains all colors in the image.

○ **32 Bits (23 Bits + Alpha)** maintains all colors, plus transparency information (alpha).

☑ **Progressive** displays the image more quickly by showing more of it, until all details are filled in.

Dots Per Inch specifies the resolution of the image; range is 0 to 1000.

RELATED COMMANDS

ImageAttach attaches raster images to drawings.

JpgOut saves selected objects in JPEG raster format.

PngOut saves selected objects in PNG raster format.

TifOut saves selected objects in TIFF raster format.

WmfOut saves selected objects in WMF vector format.

CopyClip copies selected objects to the Clipboard.

PRT SCR saves the entire screen to the Clipboard.

TIPS

- This command saves the entire viewport; in contrast, the JpgOut, PngOut, TifOut, and WmfOut commands prompt you to select objects.

- The GIF format, which is one of the two most common formats used to display images on the Internet (and was found in AutoCAD Release 12 and 13), was replaced by the BMP format in AutoCAD Release 14.

. . . .

Aliased Command

SaveUrl was removed from AutoCAD 2000; it now activates the Save.

. . . .

Scale

V. 2.5 Makes selected objects smaller or larger.

Command	Alias	Ctrl+	F-key	Alt+	Menu Bar	Dashboard
scale	sc	...	...	ML	Modify ⮡Scale	2D Draw

Command: scale
Select objects: *(Select one or more objects.)*
Select objects: *(Press Enter to end object selection.)*
Specify base point: *(Pick a point.)*
Specify a scale factor or [Copy/Reference]: *(Enter scale factor, or type C or R.)*

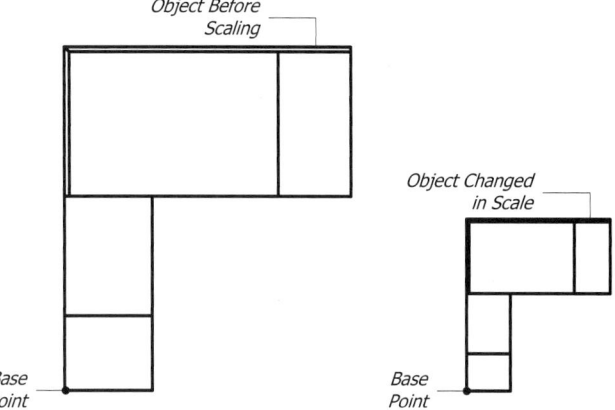

COMMAND LINE OPTIONS

Base point specifies the point from which scaling takes place.

Reference requests a distance, followed by a new distance.

Copy scales a copy of the selected objects, leaving the originals in place.

Scale factor indicates the scale factor, which applies equally in the x and y directions.

Scale Factor	Meaning
> 1.0	Enlarges object(s).
= 1.0	Makes no change.
> 0.0 *and* < 1.0	Reduces object(s).
= 0.0 *or negative*	Illegal values.

Reference options
Specify reference length <1>: *(Enter a base length.)*
Specify new length: *(Enter a new length.)*

Specify reference length specifies a baseline length; you can enter a value, or pick two points.

Specify new length specifies a new length that determines the scale factor.

RELATED COMMANDS

Align scales an object in 3D space.

Insert allows a block to be scaled independently in the x, y, and z directions.

Plot allows a drawing to be plotted at any scale.

TIPS

- You can interactively scale the object by moving the cursor to make the object larger and smaller.

- The objects scale from the base point.

- Points are the only object that cannot be scaled.

- Scale factors larger than 1.0 grow the object; smaller than 1.0 shrink the object.

- The Reference option is useful for:
  - Scaling raster images to match the size of a drawing.
  - Making one object the same size as another.
  - Changing the units in a drawing. For example, use 1:12 to change decimal feet into inches, or 2.54:1 to change cm to inches.

- This command changes the size in all dimensions (x,y,z) equally; to change an object in just one dimension, use the Stretch command.

'ScaleListEdit

2006 Edits lists of scale factors for plots, viewports, and annotative objects.

Command	Alias	Ctrl+	F-key	Alt+	Menu Bar	Dashboard
scalelistedit	...	...	...	OE	Format	...
					⮡Scale List	

-scalelistedit

Command: scalelistedit

Displays dialog box:

DIALOG BOX OPTIONS

Add adds scale factors to the list; displays the Add Scale Factor dialog box.

Edit changes the selected scale factor; displays the Edit Scale Factor dialog box.

Move Up moves the selected scale factor up the list.

Move Down moves the selected scale factor down the list.

Delete erases the selected scale factors from the list without warning; click Cancel or Restore to recover accidentally-erased scale factors.

Reset restores the list to its original status; AutoCAD warns custom scale factors will be lost.

Add Scale Factor dialog box

Name Appearing in Scale List provides descriptive names that appear in place of scale factors. These may be an actual scale factor, such as 1:10.

Paper Units specifies the size of objects on paper. This value is usually 1 for drawings larger than the paper.

Drawing Units specifies the size of objects in the drawing. This value is usually 1 for drawings smaller than the paper.

. . . .

-ScaleListEdit Command
Command: -scalelistedit
Enter option [?/Add/Delete/Reset/Exit] <Add>: *(Enter an option.)*
Enter name for new scale: *(Type a name.)*
Enter scale ratio: *(Type the scale ratio, such as 1:1.5.)*

COMMAND LINE OPTIONS
? lists the scale factors in the Text window.

Scale Name	Paper Units	Drawing Units	Effective Scale
1: 1:1	1.0000	1.0000	1.0000
2: 1:2	1.0000	2.0000	0.5000
3: 1:4	1.0000	4.0000	0.2500
4: 1:8	1.0000	8.0000	0.1250
5: 1:10	1.0000	10.0000	0.1000
6: 1:16	1.0000	16.0000	0.0625
7: 1:20	1.0000	20.0000	0.0500
8: 1:30	1.0000	30.0000	0.0333
9: 1:40	1.0000	40.0000	0.0250
10: 1:50	1.0000	50.0000	0.0200
11: 1:100	1.0000	100.0000	0.0100
12: 2:1	2.0000	1.0000	2.0000
13: 4:1	4.0000	1.0000	4.0000
14: 8:1	8.0000	1.0000	8.0000
15: 10:1	10.0000	1.0000	10.0000
16: 100:1	100.0000	1.0000	100.0000
17: 1/128" = 1'-0"	0.0078	12.0000	0.0007
18: 1/64" = 1'-0"	0.0156	12.0000	0.0013
19: 1/32" = 1'-0"	0.0313	12.0000	0.0026
20: 1/16" = 1'-0"	0.0625	12.0000	0.0052
21: 3/32" = 1'-0"	0.0938	12.0000	0.0078
22: 1/8" = 1'-0"	0.1250	12.0000	0.0104
23: 3/16" = 1'-0"	0.1875	12.0000	0.0156
24: 1/4" = 1'-0"	0.2500	12.0000	0.0208
25: 3/8" = 1'-0"	0.3750	12.0000	0.0313
26: 1/2" = 1'-0"	0.5000	12.0000	0.0417
27: 3/4" = 1'-0"	0.7500	12.0000	0.0625
28: 1" = 1'-0"	1.0000	12.0000	0.0833
29: 1-1/2" = 1'-0"	1.5000	12.0000	0.1250
30: 3" = 1'-0"	3.0000	12.0000	0.2500
31: 6" = 1'-0"	6.0000	12.0000	0.5000
32: 1'-0" = 1'-0"	12.0000	12.0000	1.0000

Add adds scale factors.

Delete erases scale factors.

Reset restores the list of scale factors to the original factory setting.

Exit exits the command.

TIPS
- This list of scale factors is displayed by the Plot command, the Properties command for viewports, and the ObjectScale command for annotative objects.

- Use this command to shorten the list of unused scale factors, and to add custom scales. It also affects the scale factors used for annotative scaling.

. . . .

ScaleText

Changes the height of text objects relative to their insertion points.

Command	Alias	Ctrl+	F-key	Alt+	Menu Bar	Dashboard
scaletext	...	...	...	MO🖑TS	Modify	...
					🖑Object	
					🖑Text	
					🖑Scale	

Command: scaletext
Select objects: *(Select one or more objects.)*
Select objects: *(Press Enter to end object selection.)*
Enter a base point option for scaling
[Existing/Left/Center/Middle/Right/TL/TC/TR/ML/MC/MR/BL/BC/BR]
<Existing>: *(Enter an option, or press Enter to keep insertion point as is.)*
Specify new height or [Paper height/Match object/Scale factor] <0.2>: *(Enter a value or option.)*

COMMAND LINE OPTIONS

Enter a base point option for scaling specifies the point from which scaling takes place.

Existing uses the existing insertion point for each text object.

Specify new height specifies the new height of the text; applies only to non-annotative text.

Paper height assigns the annotation property to the text *(new to AutoCAD 2008)*.

Match object matches the height of another text object.

Scale factor scales the text by a factor; cannot use 0 or a negative value.

Scale Factor Options
Specify scale factor or [Reference] <2.0000>: *(Enter a value, or type R.)*

Specify scale factor scales the text by a factor; see Scale command.

Reference supplies a reference value, followed by a new value.

RELATED COMMANDS

Justify changes the justification of text.

Properties changes all aspects of the text.

Style creates text styles.

TIPS

- *Warning!* Scaling text larger may make it overlap nearby text.

- For larger text, use a scale factor of more than 1.0; to reduce the text size, use a scale factor between 0.0 and 1.0. A factor of 1.0 leaves the text unchanged.

- To scale text automatically in model space, use text, dimension, and multileader styles with the annotative property turned on, as noted by the 🛆 icon *(new to AutoCAD 2008)*.

. . . .

Removed Command

The **Scene** command was removed from AutoCAD 2007.

. . . .

'Script

V. 1.4 Runs ASCII files containing sequences of AutoCAD commands and options.

Command	Alias	Ctrl+	F-key	Alt+	Menu Bar	Dashboard
script	scr	...	...	TR	Tools	...
					⇘Run Script	

Command: script

Displays the Select Script File dialog box. Select an .scr file, and then click Open.

Script file begins running as soon as it is loaded.

COMMAND LINE OPTIONS

BACKSPACE interrupts the script.

ESC stops the script.

RELATED COMMANDS

Delay specifies the delay in milliseconds; pauses the script before executing the next command.

Resume resumes a script after it is interrupted.

RScript repeats script files.

STARTUP SWITCH

/b specifies the script file name to run when AutoCAD starts up.

TIPS

- Since the Script command is transparent, it can be used during another command.

- Use the **/s** command-line switch to run a script when AutoCAD starts.

- Prefix the VSlide command with ***** to preload it; this results in a faster slide show:

  *vslide

- You can make a script file more flexible — such as pausing for user input, or branching with conditionals — by inserting AutoLISP functions.

Section

Rel.11 Creates 2D region objects from the intersection of a plane and a 3D solid.

Command	Alias	Ctrl+	F-key	Alt+	Menu Bar	Dashboard
section	sec	...	...	...	...	...

Command: section
Select objects: *(Select one or more objects.)*
Select objects: *(Press Enter to end object selection.)*
Specify first point on section plane by [Object/Zaxis/View/XY/YZ/ZX/ 3points]: *(Pick a point, or enter an option.)*

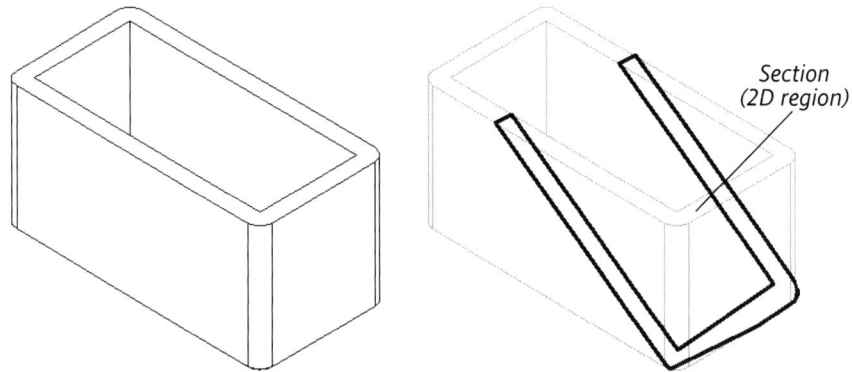

Section
(2D region)

COMMAND LINE OPTIONS

Select objects selects the 3D solid objects to be sectioned.

Object aligns the section plane with a selected object: circle, ellipse, arc, elliptical arc, 2D spline, or 2D polyline.

Zaxis specifies the normal (*z axis*) to the section plane.

View uses the current view plane as the section plane.

XY uses the x,y plane of the current view.

YZ uses the y,z plane of the current view.

ZX uses the z,x plane of the current view.

3 points picks three points to specify the section plane.

RELATED COMMAND

Slice cuts a slice out of a solid model, creating another 3D solid.

TIPS

■ Section regions are placed on the current layer, not the object's layer.

■ Regions are ignored.

■ One cutting plane can section multiple solids at a time.

SectionPlane

2007 Places "section plane" objects that visually cut 3D objects.

Command	Alias	Ctrl+	F-key	Alt+	Menu Bar	Dashboard
sectionplane	splane ...		...	DME	Draw	3D Make
					↳Modeling	
					↳Section Plane	

Command: sectionplane
Select face or any point to locate section line or [Draw section/Orthographic]:
(Pick a face or a point, or enter an option.)
Specify through point: *(Pick another point.)*

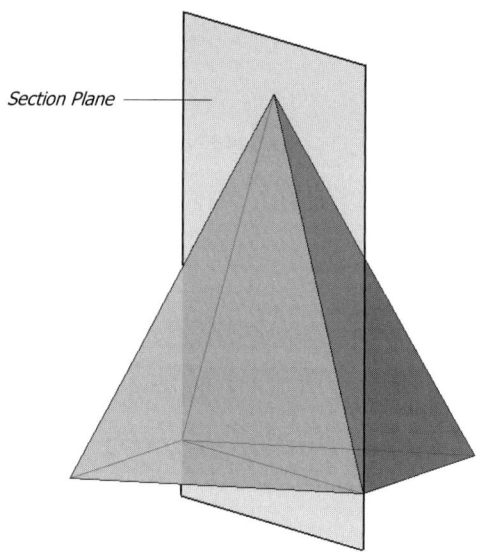

Section Plane

After the section plane is placed, double-click it to make it "live" (active). See the LiveSection command.

COMMAND LINE OPTIONS

Select face aligns the section plane with the face; you can pick a region or a face on a 3D solid. (There is no need to hold down the Ctrl key to select faces.)

Select any point specifies the starting point of the section plane; if this command insists on aligning the section plane with a face, override it with the Draw Section option.

Draw section places the section plane by picking points.

Orthographic places the section plane parallel to standard views: front, back, top, bottom, left, or right view.

Specify through point positions the section plane.

Draw Section options

It can be helpful to use ENDpoint and QUAdrant object snaps with these prompts.

Specify start point: *(Pick a point, 1.)*

Specify next point: *(Pick another point, 2.)*

Specify next point or ENTER to complete: *(Press Enter to skip to the next prompt, or pick a third point, 3.)*

Specify next point in direction of section view: *(Pick a point to align the section plane, 4.)*

Specify start point specifies the starting point of the section plane.

Specify next point allows you to draw a multi-segment section plane, as illustrated below.

Direction of section view specifies the cut-away portion.

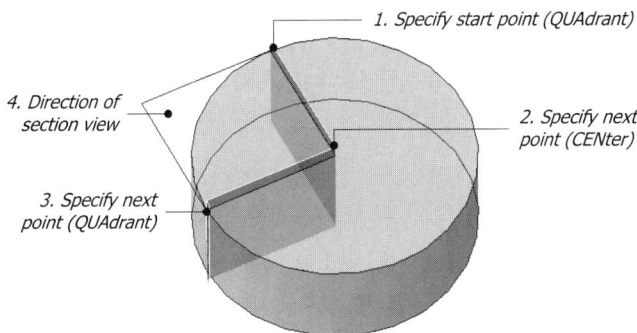

1. Specify start point (QUAdrant)

4. Direction of section view

2. Specify next point (CENter)

3. Specify next point (QUAdrant)

Orthographic option

Align section to: [Front/Back/Top/Bottom/Left/Right]: *(Enter an option.)*

Front, Back, Top, Bottom, Left, or Right aligns the section plane with current UCS. Live sectioning is turned on automatically (the forward portion of the sectioned objects being removed from view). See the LiveSection command.

RELATED COMMANDS

LiveSection edits section planes.

Section creates region slices from 3D models.

Interfere creates new objects from the interference of two or more 3D solids.

TIPS

■ This command is half of a two-part process in working with sections:
   1. Create the section plane with the SectionPlane command.
   2. Edit the section plane with the LiveSection command.

■ "Sectionplane" is a distinct object type introduced with AutoCAD 2007.

SecurityOptions

<u>**2004**</u> Sets passwords and digital signatures for securing drawings against unauthorized access.

Command	Alias	Ctrl+	F-key	Alt+	Menu Bar	Dashboard
securityoptions	...	...	...	...	...	...

Command: securityoptions

Displays dialog box:

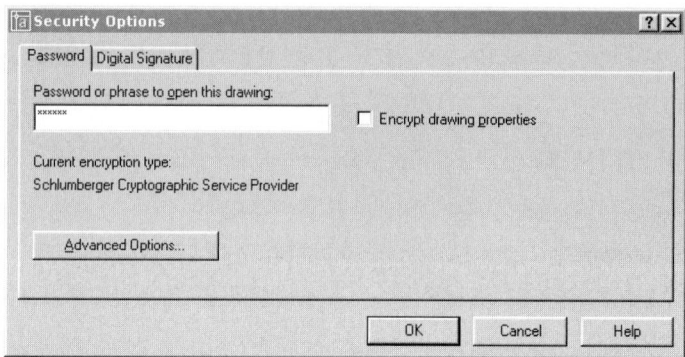

SECURITY OPTIONS DIALOG BOX

Password or phrase to open this drawing assigns a password to lock the drawing.

☐ **Encrypt drawing properties** encrypts drawing properties data; see the DwgProps command.

Advanced Options displays the Advanced Options dialog box; available only after a password is entered.

OK displays the Confirm Password dialog box:

Digital Signature tab

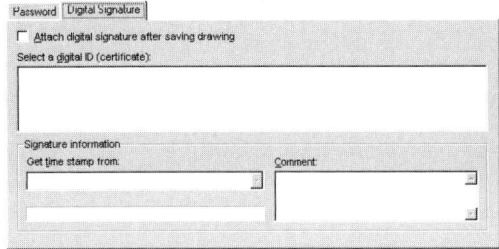

☐ **Attach digital signature after saving drawing** attaches the digital signature the next time the drawing is saved.

Select a digital ID (certificate) lists digital signature services.

Signature information

Get time stamp from lists time servers.

Comment provides additional comments to be included with the digital signature.

ADVANCED OPTIONS DIALOG BOX

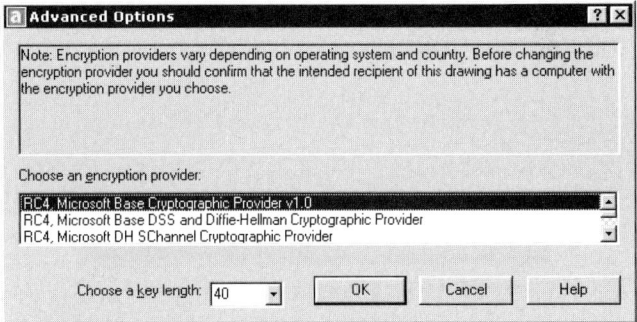

Choose an encryption provider selects from several encryption types.

Choose a key length selects 40, 48, or 56 bits of encryption.

RELATED COMMAND

SigValidate displays information about the digital signatures attached to drawings.

RELATED SYSTEM VARIABLE

SigWarn determines whether a warning alerts that drawings are opened with digital signatures.

TIPS

- When opening drawings protected by passwords, AutoCAD displays the Password dialog box for entering the password:

- When the password entered is correct, AutoCAD opens the drawing.

- When the password entered is incorrect, AutoCAD complains:

- To remove the password, in the Security Options dialog box, remove the password, and then click OK.

- The first time you assign a digital signature to the drawing, you are prompted to obtain one:

- *Warning!* If the password is lost, the drawing is not recoverable. Autodesk recommends that you (1) save a backup copy of the drawing not protected with any password; and (2) maintain a list of password names stored in a secure location.

- *Warning!* The encryption provider must be the same on the receiving computers; otherwise the drawings cannot be opened.

- To remove this feature from AutoCAD, install with the Custom option, and then unselect Drawing Encryption.

Select

<u>V. 2.5</u> Creates a selection set of objects based on their location in the drawing.

Command	Alias	Ctrl+	F-key	Alt+	Menu Bar	Dashboard
select	...	...	...	...	...	...

Command: select
Select objects: *(Select one or more objects, or enter an option.)*
Select objects: *(Press Enter to end object selection.)*

COMMAND LINE

pick selects a single object.

AU switches from pick mode to **C** or **W** mode, depending on whether an object is found at the initial pick point (short for AUtomatic).

ALL selects all objects in the drawing; press Ctrl+A to select all objects.

BOX goes into **C** or **W** mode, depending on how the cursor moves.

C selects objects in and crossing the selection box (Crossing).

CL selects application-specific objects (CLass).

CP selects all objects inside and crossing the selection polygon (Crossing Polygon).

F selects all objects crossing a selection polyline (Fence).

G selects objects contained in a named group (Group).

M makes multiple selections before AutoCAD scans the drawing; saves time in a large drawing (Multiple).

SI selects only a single set of objects before terminating the Select command (SIngle).

SU selects vertices, edges, and faces of 3D solids (SUbobject).

W selects all objects inside the selection box (Window).

WP selects objects inside the selection polygon (Windowed Polygon).

A continues to add objects after using the **R** option (Add).

L selects the last-drawn object still visible on the screen (Last).

O ends subobject selection mode (Object).

P selects previously-selected objects (Previous).

R removes objects from the selection set (Remove).

U removes the selected objects most-recently added (Undo).

Enter completes the Select command.

Esc abandons the Select command.

RELATED COMMANDS

All commands that prompt 'Select objects'.

Filter specifies objects to be added to the selection set.

QSelect selects objects via a dialog box interface based on their properties.

RELATED SYSTEM VARIABLES

PickAdd controls how objects are added to a selection set.

PickAuto controls automatic windowing at the 'Select objects' prompt.

PickDrag controls the method of creating a selection box.

PickFirst controls the command-object selection order.

TIPS

- Objects selected by the Select command become the Previous selection set, and may be selected by any subsequent 'Select objects:' prompt by responding **P** for "previous."

- To view a list of all selection options, enter any non-valid text at the prompt, such as "asdf":

  Command: select
  Select objects: asdf
  Invalid selection
  Expects a point or
  Window/Last/Crossing/BOX/ALL/Fence/WPolygon/CPolygon/Group/
  Add/Remove/Multiple/Previous/Undo/AUto/SIngle/SUbobject/Object
  Select objects:

- Pressing Ctrl+A selects all objects in the drawing without needing the Select command.

- All selects all objects in the drawing, except those on frozen and locked layers.

- Crossing selects objects within and crossing the selection rectangle.

- Window selects objects within the selection rectangle.

- Fence selects objects crossing the selection polyline.

- CPolygon selects objects within and crossing the selection polygon.

- WPolygon selects objects within the selection polygon.

- Selection preview highlights objects as the cursor passes over them. This can be useful for determining which objects belong together; for example, lines and arcs vs polylines, or blocks. However, it slows down the display speed with large, complex objects. In the figure below, the chair block is highlighted by the cursor:

You can change preview highlighting with the Options command: chose the Selection tab, and then click the Visual Effect Settings button.

SelectURL

Rel.14 Highlights all objects and areas that have hyperlinks attached to them (an undocumented command).

Command	Alias	Ctrl+	F-key	Alt+	Menu Bar	Dashboard
selecturl	...	...	...	...	...	...

Command: selecturl

Highlights hyperlinked objects and areas with grips.

If there are no hyperlinks in the drawing, AutoCAD reports:

No objects with hyperlinks found.

COMMAND LINE OPTIONS

None.

RELATED COMMANDS

AttachURL attaches URLs to objects and areas.

Hyperlink attaches URLs to objects via a dialog box.

TIPS

- Examples of URLs include:

  http://www.autodeskpress.com Autodesk Press Web site.
  news://adesknews.autodesk.com Autodesk news server.
  ftp://ftp.autodesk.com   Autodesk FTP server.
  http://www.upfrontezine.com   Author Ralph Grabowski's Web site.

- Do not delete layer URLLAYER. URL is short for "uniform resource locator," the universal file naming convention used by computers and the Internet.

- The URL is stored as follows:

Attachment	URL
One object	Stored as xdata (extended entity data).
Multiple objects	Stored as xdata in each object.
Area	Stored as xdata in rectangles on layer URLLAYER.

SetByLayer

2008 Forces overridden properties in viewports to ByLayer.

Command	Alias	Ctrl+	F-key	Alt+	Menu Bar	Dashboard
setbylayer	...	...	...	MB	Modify	...
					⮡Change to ByLayer	

Command: setbylayer
Current active settings: Color Linetype Lineweight Material Plot Style
Select objects or [Settings]: *(Select one or more objects, or type S.)*
Change ByBlock to ByLayer? [Yes/No] <Yes>: *(Type Y or N.)*
Include blocks? [Yes/No] <Yes>: *(Type Y or N.)*

COMMAND LINE OPTIONS

Select objects selects one or more objects; enter 'all' for all objects in the drawing.

Settings displays the Settings dialog box, detailed below.

Change ByBlock to ByLayer changes properties from ByBlock to ByLayer.

Include blocks toggles whether blocks should have their properties reset to Bylayer

SETTINGS DIALOG BOX

Displays dialog box:

Color changes the color of selected objects to ByLayer.

Linetype changes the linetype of selected objects to ByLayer.

Lineweight changes the lineweight of selected objects to ByLayer.

Material changes the material of selected objects to ByLayer.

Plot Style changes the plot style of selected objects to ByLayer.

RELATED COMMANDS

Change, ChProp, Properties, Color, Linetype, Lineweight, Materials, Plotstyle override properties.

Layer reports viewport overrides.

RELATED SYSTEM VARIABLES

SetByLayerMode specifies which properties are changed by this command.

VpLayerOverrides reports whether any layer in the current viewport has VP (viewport) property overrides.

VpLayerOverridesMode toggles whether viewport property overrides are displayed or plotted.

'SetIDropHandler

<u>2004</u>   Specifies how i-drop objects should be handled in drawings.

Command	Alias	Ctrl+	F-key	Alt+	Menu Bar	Dashboard
'setidrophandler	...	...	...	...	...	...

Command: setidrophandler

Displays dialog box:

SET DEFAULT I-DROP CONTENT TYPE DIALOG BOX

Choose the default i-drop content type specifies how i-drop objects are handled when placed in drawings.

TIPS

- When i-drop content is dragged from Web pages into drawings, AutoCAD treats the objects as blocks.

- Ever since this command was introduced, its only option has been "Block", so this command serves no purpose.

- For more information on i-drop (short for "Internet drag'n drop") technology, see www.autodesk.com/developidrop.

Removed Command

The **SetUV** command was removed from AutoCAD 2007; its replacement is MaterialMap.

'SetVar

V. 2.5 Lists the settings of system variables, and allows you to change variables that are not read-only (short for SET VARiable).

Command	Alias	Ctrl+	F-key	Alt+	Menu Bar	Dashboard
setvar	set	...	...	TQV	Tools ⬐Inquiry ⬐Set Variable	...

Command: setvar
Enter variable name or [?]: *(Enter a name, or type ?.)*

COMMAND LINE OPTIONS

Enter variable name names the system variable you want to access.

? lists the names and settings of system variables.

TIPS

- See Appendix C for a complete list of all system variables found in AutoCAD.

- Example usage of this command:

    Command: setvar
    Enter variable name or [?]: visretain
    Enter new value for VISRETAIN <0>: 1

- Almost all system variables can be entered without the SetVar command. For example:

    Command: visretain
    New value for VISRETAIN <0>: 1

- System variables marked "read only" cannot be changed:

    Command: _pkser
    _PKSER = "341-35000000" (read only)

. . . .

Replaced Commands

Shade was replaced in AutoCAD 2000 by the ShadeMode command's Flat+edges option.

ShadeMode was replaced by the -ShadeMode command in AutoCAD 2007; ShadeMode is now an alias for the VsCurrent command.

. . . .

-ShadeMode

2000 Generates renderings of 3D models quickly in a variety of modes (replaces the Shade and ShadeMode commands).

Command	Alias	Ctrl+	F-key	Alt+	Menu Bar	Dashboard
-shademode	...	...	...	...	...	...

Command: -shademode
Current mode: 2D wireframe
Enter option [2D wireframe/3D wireframe/Hidden/Flat/Gouraud/ fLat+edges/gOuraud+edges] <2D wireframe>: *(Enter an option.)*

COMMAND LINE OPTIONS

2D wireframe displays wireframe models in 2D space.

3D wireframe displays wireframe models in 3D space.

Hidden removes hidden faces.

Flat displays flat shaded faces.

fLat+edges displays flat shaded faces, with outlined faces of the background color.

Gouraud displays smooth shaded faces.

gOuraud+edges displays smooth shaded faces, with outlined faces of the background color.

RELATED COMMANDS

VsCurrent displays viewports in a variety of visual styles.

MView specifies hidden-line removal of individual viewports for plotting.

Render performs more realistic renderings than does -ShadeMode.

TIPS

- As an alternative to the -ShadeMode command, the Render command creates high-quality renderings of 3D drawings, but takes somewhat longer to do so.

- The smaller the viewport, the faster the shading.

- Drawing (at left, below) displayed in -ShadeMode's default 2D wireframe mode; note the standard 2D UCS icon. Wireframe colors are based on layer and object colors.

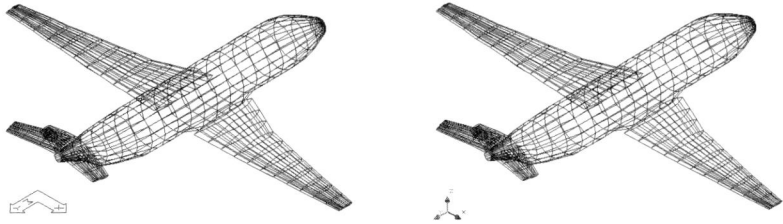

- Drawing (at right, above) displayed in 3D wireframe mode; note the 3D UCS icon. Wireframe colors are based on the color of the rendering materials, if any.

- Drawing (at left, below) displayed in Hidden mode; faces hidden by other faces are not displayed.

- Drawing (at right, above) displayed in Flat mode; each face is filled with shaded grays.
- Drawing (at left, below) displayed in Gouraud mode; faces are smoothed.

- Drawing (at right, above) displayed in fLat+edges mode; faces are outlined by the background color (*white*).
- Drawing (below) displayed in gOuraud+edges mode; each face is outlined.

Shape

<u>V. 1.0</u>   Inserts shapes into drawings.

Command	Alias	Ctrl+	F-key	Alt+	Menu Bar	Dashboard
shape	...	...	...	...	...	...

Command: shape
Enter shape name or [?]: *(Enter the name, or type ?.)*
Specify insertion point: *(Pick a point.)*
Specify height <1>: *(Specify the height.)*
Specify rotation angle <0>: *(Specify the angle.)*

COMMAND LINE OPTIONS

Enter shape name indicates the name of the shape to insert.

? lists the names of currently-loaded shapes.

Specify insertion point indicates the insertion point of the shape.

Specify height specifies the height of the shape.

Specify rotation angle specifies the rotation angle of the shape.

RELATED COMMANDS

Load loads SHX-format shape files into the drawings.

Insert inserts blocks into drawings.

Style loads *.shx* font files into drawings.

RELATED SYSTEM VARIABLE

ShpName specifies the default *.shp* file name.

TIPS

- Shapes are used to define the text and symbols found in complex linetypes.

Some electronic shapes.

- Shapes were an early alternative to blocks, but now are used primarily with complex linetypes. They take up extremely small amounts of memory, but are difficult to create.

- Shapes are defined by *.shp* files, which must be compiled into *.shx* files before they can be loaded by the Load command.

- Shapes must be loaded by the Load command before they can be inserted with the Shape command.

- Compile an *.shx* file into an *.shp* file with the Compile command.

Sheetset, SheetsetHide

2005 Displays and hides the Sheetset Manager palette.

Command	Alias	Ctrl+	F-key	Alt+	Menu Bar	Dashboard
sheetset	ssm	4	...	TPS	Tools	...
					⬫Palettes	
					⬫Sheetset Manager	
sheetsethide	...	4	...	TPS	Tools	...
					⬫Palettes	
					⬫✓Sheetset Manager	

Command: sheetset

Displays palette:

Command: sheetsethide

Closes the palette:

SHEETSET MANAGER PALETTE

x closes the palette; alternatively, use the SheetsetClose command.

Toolbar changes, depending on the tab selected.

Title bar docks against the side of AutoCAD, or can be dragged around.

Tabs change between Sheet List, View List, and Resource Drawings views.

Views lists sheets and sheet set names; right-click for shortcut menus.
- **Sheet List** displays names of sheets, which are organized into subsets.
- **Sheet Views** displays views for each sheet set; views can be organized into categories.
- **Model Views** displays folders and files used by the sheet set.

Open and select sheet sets displays a droplist of recently-opened sheet sets.

- **Recent** lists recently-opened *.dst* sheet set data files.
- **New Sheet Set** starts the Create Sheet Set wizard; see NewSheetset command.
- **Open** displays the Open Sheetset dialog box; select a *.dst* file, and then click Open. See the NewSheetset command.

Refresh Sheet Status (**F5**) updates the list of files and information stored in sheet set data *.dst* files, checks resource folders for new and removed drawings, and checks all drawings for new and removed model space views.

Details lists information about selected sheets and sheet sets.

Sheet Preview shows preview images of selected sheets.

AutoHide reduces palette to titlebar when cursor not over the palette.

Properties displays a shortcut menu.

Sheet List toolbar

SHORTCUT MENUS

 Right-click a sheet set name:

Close Sheet Set closes the entire sheet set; use the Open option to open another sheet set.

New Sheet displays the New Sheet dialog box; enter a number and name, and then click OK.
New Subset displays the Subsheet Properties dialog box; enter a name, and then click OK.
Import Layout as Sheet displays the Import Layout as Sheet dialog box; select a drawing and layouts, and then click **OK**.

Resave All Sheets saves sheet set information stored in each drawing file.
Archive displays the Archive as Sheet Set dialog box; see the Archive command.

Publish plots the sheet set as *.dwf* files or to a plotter; see the Publish command.
eTransmit collates the sheet set for transmittal by CD or email; see the eTransmit command.
Transmittal Setups displays the Transmittal Setup dialog box; see the eTransmit command.

Insert Sheet List Table creates a table listing all sheets in the sheet set, and adds it to a sheet.

Properties displays the Sheet Set Properties dialog box; make changes, and then click OK.

Subset

 Right-click a subset name:

Collapse hides the names of sheets.
Rename Subset changes the name of the subset.
Remove Subset erases the subset; available only when subset contains no sheets.
Properties displays the Subset Properties dialog box; make changes, and then click OK.

Sheet

 Right-click a sheet name:

Open opens the drawing in a new window.

Rename & Renumber displays the Rename & Renumber Sheet dialog box.

Remove Sheet removes the sheet from the set; does *not* erase the drawing or layout.

Properties displays the Sheet Properties dialog box; make changes, and then click OK.

Sheet Views tab

This tab was formerly called "View List."

 Right-click a sheet set name:

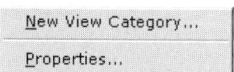

New View Category displays the View Category dialog box; enter info, and then click OK.

Properties displays the Sheetset Properties dialog box; make changes, and then click OK.

Category

 Right-click a category name:

Rename renames the category.

Remove Category erases the category; available only when category is empty.

Properties displays the View Category dialog box; make changes, and then click OK.

View

 Right-click a view name:

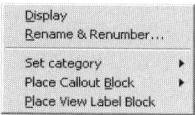

Open displays the drawing in a new window.

Rename & Renumber displays the Rename & Renumber View dialog box.

Set Category selects a view category, if defined. See the View command.

Place Callout Block:

- **Callout Bubble** places "Callout Bubble" blocks in drawings with view number (1) and sheet number (A-02) field text filled in; see the Insert command.

 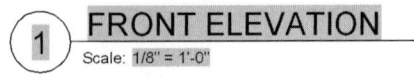

Place View Label Block places "Drawing Name" block in drawings, with view number (1), view name (Front Elevation), and viewport scale filled in as field text. (Gray background indicates field text; see the Field command.)

Model Views tab

This tab was formerly called "Drawing Resources."

Path

 Right-click a path name:

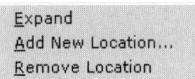

Expand displays folders, drawings, and views.

Add New Location displays the Browse for Folder dialog box; select a folder, and then click Open.

Remove Location deletes a location, with no warning.

Drawing

 Right-click a drawing name:

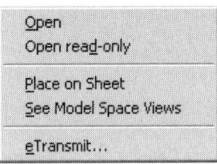

Open opens the drawing in a new window.

Open read-only opens the drawing read-only, so that changes cannot be saved.

Place on Sheet places the drawing in the current sheet; works only in layout mode.

See Model Space Views displays model space view icons, if any.

eTransmit packages drawings with support files; see the eTransmit command.

View

 Right-click a view name:

Open opens the view in a new window.

Place on Sheet places the drawing in the current sheet; works only in layout mode.

. . . .

NEW SHEET / SUBSET DIALOG BOX

Similar dialog boxes are used for properties of subsets and new sheets.

Number specifies the order number of the sheet; allows sheets to be placed in correct order.

Sheet Title names the sheet, usually with the name of the related layout tab.

File Name names the sheet file, usually consisting of the sheet number and sheet title.

SUBSET SHEET PROPERTIES DIALOG BOX

Subset name specifies the name of the subset.

Store new sheet DWG files in selects the folder for storing drawing files.

[...] displays Browser for Folder dialog box; select a folder, and then click Open.

Sheet Creation Template for Subset specifies the name of the *.dwt* template file to use.

[...] displays Browser for Folder dialog box; select a folder, and then click Open.

□ **Prompt for template** prompts for the template name later.

IMPORT LAYOUTS AS SHEETS DIALOG BOX

Select a drawing file containing layouts specifies the name of a drawing file.

[...] displays Browser for Folder dialog box; select a folder, and then click Open.

Select layouts to import as sheets lists the names of layouts found in the drawing file.

RELATED COMMANDS

NewSheetset creates new sheet sets.

OpenSheetset opens a specific sheet set.

RELATED SYSTEM VARIABLES

SsFound records the path and file name of current sheet set.

SsLocate toggles whether AutoCAD opens the sheet sets associated with drawings.

SsmAutoOpen toggles whether AutoCAD displays the Sheet Set Manager palette when drawings are opened.

SsmPollTime determines how often AutoCAD updates the Sheet Set Manager.

SsmSheetStatus specifies how data in sheetsets are updated.

SsmState reports whether the Sheet Set Manager palette is open.

STARTUP SWITCHES

/nossm (no Sheet Set Manager) prevents the Sheet Set Manager palette from opening at startup.

/set loads a specified *.dst* sheet set data file at startup.

RELATED FILES

.dst are sheet set data files.

TIPS

- The purpose of sheet sets is to create a hierarchy of drawings, arranged in order; as well, drawings can be organized into subsets (sheets) and categories (views).

- *SSM* is short for "sheet set manager."

- Sheet sets can be created with the NewSheetset command, based either on existing drawings or on another sheet set.

- You can add layouts to sheet sets easily: right-click a layout tab, and then select Import Layout as Sheet from the shortcut menu (*new to AutoCAD 2008*).

- Sheet sets, subsets, and sheets have properties, such as titles, descriptions, file paths, and custom properties. These can be accessed by the Field command.

- To publish sheets in reverse order, click the Publish button, and then select Publish in Reverse Order (*new to AutoCAD 2008*).

Shell

V. 2.5 Starts a new instance of the Windows command interpreter; runs external commands defined in *acad.pgp*.

Command	Alias	Ctrl+	F-key	Alt+	Menu Bar	Dashboard
shell	sh	...	...	...	...	...

Command: shell
OS Command: *(Enter a command.)*

Opens window:

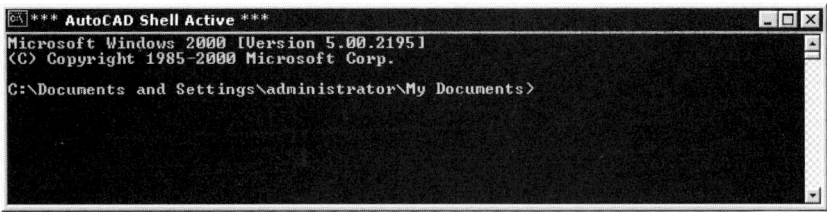

COMMAND LINE OPTIONS

OS Command specifies an operating system command, and then closes the new instance of the Windows command interpreter.

Enter remains in the operating system's command mode for more than one command.

Exit returns to AutoCAD from the OS Command mode.

RELATED COMMAND

Quit exits AutoCAD.

RELATED FILE

acad.pgp is the external command definition file.

TIP

- This command was an early solution to the problem of multi-tasking and task-switching lacking in DOS, before Windows. It is rarely used today.

. . . .

Aliased Command

ShowMat was removed from AutoCAD 2007; it now activates the List command.

. . . .

SigValidate

2004   Displays digital signature information stored in drawings.

Command	Alias	Ctrl+	F-key	Alt+	Menu Bar	Dashboard
sigvalidate	...	...	...	...	...	...

Command: sigvalidate

Displays dialog box:

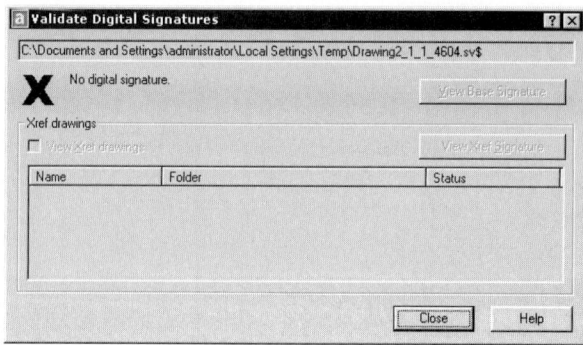

VALIDATE DIGITAL SIGNATURES DIALOG BOX

View Base Signature displays the Digital Signature Contents dialog box.

Xref Drawings options

View Xref Drawings

☑ Lists xref drawings attached to the drawing.

☐ Does not list xrefs.

View Xref Signature displays the Digital Signatures Contents dialog box for the selected xref; available only when the selected xref has a digital signature.

RELATED COMMAND

SecurityOptions attaches digital signatures to drawings.

TIPS

- Digital signatures validate the authenticity of drawings.

- Digital signatures indicate whether the drawing changed since being signed.

- Autodesk notes that digital signatures become invalid for these reasons: the drawing was corrupted when the digital signature was attached; the drawing was corrupted in transit; or the digital ID is no longer valid.

- This command requires that the drawing be saved first; if not, AutoCAD reports, "This drawing has been modified. The last saved version will be validated."

- If this command does not work, it was removed from AutoCAD; re-install with the Custom option, and then select Drawing Encryption.

Sketch

<u>V. 1.4</u> Allows freehand drawing as a series of lines or polylines.

Command	Alias	Ctrl+	F-key	Alt+	Menu Bar	Dashboard
sketch	...	...	...	...	...	...

Command: sketch
Record increment <0.1000>: *(Enter a value.)*
Sketch. Pen eXit Quit Record Erase Connect .
Click pick button to begin drawing:
<Pen down>
Click pick button again to stop drawing:
<Pen up>
Press Enter to record sketching and exit Sketch:
nnn **lines recorded.**

COMMAND LINE OPTIONS

Commands can be invoked by mouse and digitizer buttons:

Connect connects to last drawing segment (alternatively, press button #6 or C on the keyboard).

Erase erases temporary segments as the cursor moves over them (button #5 or E).

eXit records temporary segments, and exits the Sketch command (#3, R, Spacebar, or Enter).

Pen lifts and lowers the pen (button #1 or P).

Quit discards temporary segments, and exits the Sketch command (#4, Q, or Esc).

Record records the temporary segments as permanent (button #2 or R).

. (Period) connects the last segment to the current point (button #1).

RELATED COMMANDS

Line draws line segments.

PLine draws polyline and polyline arc segments.

RELATED SYSTEM VARIABLES

SketchInc specifies the current recording increment for the Sketch command (default = 0.1).

SKPoly controls the type of sketches recorded:

TIPS

■ During the Sketch command, the definitions of the pointing device's buttons change to those listed above, under Command Line Options.

■ Menus are unavailable during the Sketch command.

Slice

<u>Rel.11</u>   Cuts 3D solids with planes, creating one or two 3D solids.

Command	Alias	Ctrl+	F-key	Alt+	Menu Bar	Dashboard
slice	sl	...	...	M3S	Draw	3D Make
					⤷ 3D Operations	
					⤷ Slice	

Command: slice
Select objects: *(Select one or more solid objects.)*
Select objects: *(Press Enter to end object selection.)*
**Specify first point on slicing plane by [Object/Zaxis/View/XY/YZ/ZX/3points]
<3points>:** *(Pick a point, or enter an option.)*
Specify a point on desired side of the plane or [keep Both sides]: *(Pick a point, or type B.)*

Half a slice

COMMAND LINE OPTIONS

Select objects selects the 3D solid model to slice.

Slicing Plane options

Object aligns the cutting plane with a circle, ellipse, arc, elliptical arc, 2D spline, or 2D polyline.

View aligns the cutting plane with the viewing plane.

XY aligns the cutting plane with the x,y plane of the current UCS.

YZ aligns the cutting plane with the y,x plane of the current UCS.

Zaxis aligns the cutting plane with two normal points.

ZX aligns the cutting plane with the z,x plane of the current UCS.

3 points aligns the cutting plane with three points.

keep Both sides retains both halves of the cut solid model.

Specify a point on the desired side of the plane retains either half of the cut solid model.

TIPS

- This command cannot slice a 2D region, a 3D wireframe model, or other 2D shapes.
- It is helpful to use object snap when specifying points.

'Snap

<u>V. 1.0</u> Sets the drawing "resolution," the origin for the grid and hatches, isometric mode, and the angle of the grid, hatches, and orthographic mode.

Command	Alias	Ctrl+	F-key	Alt+	Status Bar	Dashboard
snap	sn	B	F9	...	SNAP	...

Command: snap
Specify snap spacing or [ON/OFF/Aspect/Style/Type] <0.5>: *(Enter a value, or an option.)*

COMMAND LINE OPTIONS

Snap spacing sets the snap increment.

Aspect sets separate x and y increments.

OFF turns off snap.

ON turns on snap.

Rotate rotates the crosshairs for snap and grid; a hidden option in AutoCAD 2008.

Style switches between standard and isometric style.

Type switches between grid or polar snap.

Aspect options
Specify horizontal spacing <0.5>: *(Enter a value.)*
Specify vertical spacing <0.5>: *(Enter a value.)*

Specify horizontal spacing specifies the spacing between snap points along the x-direction.

Specify vertical spacing specifies the spacing between snap points along the y-direction.

Rotate options
Specify base point <0.0,0.0>: *(Enter a value.)*
Specify rotation angle <0>: *(Enter a value.)*

Specify base point specifies the point from which snap increments are determined.

Specify rotation angle rotates the snap about the base point.

Style options
Enter snap grid style [Standard/Isometric] <S>: *(Type S or I.)*
Specify vertical spacing <0.5>: *(Enter a value.)*

Standard specifies rectangular snap.

Isometric specifies isometric snap.

Specify vertical spacing specifies the spacing between snap points along the y-direction.

Type options
Enter snap type [Polar/Grid] <Grid>: *(Type P or G.)*

Polar specifies polar snap.

Grid specifies rectangular snap.

Click SNAP *to turn snap on and off; right-click for shortcut menu:*

Polar Snap On turns on polar snap.

Grid Snap On turns on snap.

Off turns off snap.

Settings displays the Snap and Grid tab of the Drafting Settings dialog box.

RELATED COMMANDS

DSettings sets snap values via a dialog box.

Grid turns on the grid.

Isoplane switches to a different isometric drawing plane.

RELATED SYSTEM VARIABLES

SnapAng specifies the current angle of the snap rotation.

SnapBase specifies the base point of the snap rotation.

SnapIsoPair specifies the current isometric plane setting.

SnapMode determines whether snap is on.

SnapStyl determines the style of snap.

SnapUnit specifies the current snap increment in the x and y directions.

TIPS

- Setting the snap is setting the cursor resolution. For example, setting a snap distance of 0.1 means that when you move the cursor, it jumps in 0.1 increments. You can, however, still type in numerical values of greater resolution, such as 0.1234.

- There is no snap distance in the z-direction.

- The Aspect option is not available when the Style option is set to Isometric; you may, however, rotate the isometric grid.

- The Rotate option is no longer shown, as of AutoCAD 2007, but the option still works.

- The options of the Snap command affect several other commands:

Snap Option	Command	Effect
Style	Ellipse	Adds **Isocircle** option to the Ellipse command.
SnapBase	Hatch	Relocates the origin of the hatch.
Rotate	Hatch	Rotates the hatching angle.
Rotate	Grid	Rotates the grid display.
Rotate	Ortho	Rotates the ortho angle.

- You can toggle snap mode by clicking the word SNAP on the status bar.

SolDraw

<u>Rel.13</u> Creates profiles and sections in viewports created with the SolView command (short for SOLids DRAWing).

Command	Alias	Ctrl+	F-key	Alt+	Menu Bar	Dashboard
soldraw	...	...	...	DMUD	Draw	...
					⬎Modeling	
					⬎Setup	
					⬎Drawing	

Command: soldraw
Select viewports to draw ..
Select objects: *(Select one or more viewports.)*
Select objects: *(Press Enter to end object selection.)*

COMMAND LINE OPTION
Select objects selects a viewport; must be a floating viewport in paper space.

RELATED COMMANDS
SolProf creates profile images of 3D solids.

SolView creates floating viewports.

TIPS
- The SolView command must be used before this command.

- This command performs the following actions:

  1. Creates visible and hidden lines representing the silhouette and edges of solids in the viewport.
  2. Projects to a plane perpendicular to the viewing direction.
  3. Generates silhouettes and edges for all 3D solids and portions of solids behind the cutting plane.
  4. Crosshatches sectional views.

- Existing profiles and sections in the selected viewport are erased.

- All layers — except the ones needed to display the profile or section — are frozen in each viewport.

- The following layers are used by SolDraw, SolProf, and SolView: *viewname*-**VIS**, *viewname*-**HID**, and *viewname*-**HAT**.

- Hatching uses the values set in system variables HpName, HpScale, and HpAng.

Solid

V. 1.0 Draws solid-filled triangles and quadrilaterals; does *not* create 3D solids.

Command	Alias	Ctrl+	F-key	Alt+	Menu Bar	Dashboard
solid	so	...	...	DMM2	Draw	L8
					⤷Modeling	
					⤷Meshes	
					⤷2D Solid	

Command: solid
Specify first point: *(Pick a point.)*
Specify second point: *(Pick a point.)*
Specify third point: *(Pick a point.)*
Specify fourth point or <exit>: *(Pick a point, or press Enter to draw triangle.)*

  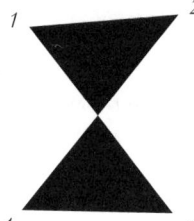

Three-point solid.   ³Pick order makes a difference for four-point solids.

COMMAND LINE OPTIONS

First point picks the first corner.

Second point picks the second corner.

Third point picks the third corner.

Fourth point picks the fourth corner; alternatively, press Enter to draw triangle.

Enter draws quadilaterials; ends the Solid command.

RELATED COMMANDS

Fill turns object fill off and on.

Hatch fills any shape with a solid fill pattern.

Trace draws lines with width.

PLine draws polylines and polyline arcs with width.

RELATED SYSTEM VARIABLE

FillMode determines whether solids are displayed filled or outlined.

SolidEdit

<u>2000</u> Edits the faces and edges of 3D solids (short for SOLids EDITor).

Command	Alias	Ctrl+	F-key	Alt+	Menu Bar	Dashboard
solidedit	...	...	...	MN	Modify	...
					⬐Solid Editing	

Command: solidedit
Solids editing automatic checking: SOLIDCHECK=1
Enter a solids editing option [Face/Edge/Body/Undo/eXit] <eXit>: *(Enter an option.)*

COMMAND LINE OPTIONS

Undo undoes the last editing actions, one at a time, up to the start of this command.

eXit exits body mode.

Face options
Enter a face editing option
[Extrude/Move/Rotate/Offset/Taper/Delete/Copy/coLor/Undo/eXit] <eXit>: *(Enter an option.)*

Extrude extrudes one or more faces to the specified distance, or along a path; a positive value extrudes the face in the direction of its *normal* (vector that points from its surface).

Move moves one or more faces the specified distance.

Rotate rotates one or more faces about an axis by a specified angle.

Offset offsets one or more faces by the specified distance, or through a specified point; a positive value increases the size of the solid, while a negative value decreases it.

Taper tapers one or more faces by a specified angle; a positive angle tapers in, while a negative angle tapers out.

Delete removes the selected faces; also removes attached chamfers and fillets.

Copy copies the selected faces as a 3D region or a 3D body object.

Color changes the color of the selected faces.

Edge options
Enter an edge editing option [Copy/coLor/Undo/eXit] <eXit>: *(Enter an option.)*

Copy copies the selected 3D edges as a line, arc, circle, ellipse, or spline.

coLor changes the color of the selected edges.

Body options
Enter a body editing option
[Imprint/seParate solids/Shell/cLean/Check/Undo/eXit] <eXit>: *(Enter an option.)*

Imprint imprints a selection set of arcs, circles, lines, 2D and 3D polylines, ellipses, splines, regions, bodies, and 3D solids on the face of a 3D solid. This function is also available through the Imprint command.

seParate solids separates 3D solids into independent 3D solid objects; the solid objects must have disjointed volumes; this option does *not* separate 3D solids that were joined by a Boolean editing command, such as Intersect, Subtract, and Union.

Shell creates a hollow, thin-walled solid of specified thickness; a positive thickness creates a shell on the inside of the solid, while a negative value creates the shell on the outside of the solid.

cLean removes redundant edges and vertices, imprints, unused geometry, shared edges, and shared vertices.

Check checks whether the object is a 3D solid; duplicates the function of the SolidCheck system variable.

RELATED SYSTEM VARIABLE

SolidCheck toggles solid validation on and off (default = on).

RELATED COMMAND

Imprint imprints the edges of objects onto 3D solid models.

TIP

- When working with this command, you can select a face, an edge, or an internal point on a face. Alternatively, use the **CP** (*crossing polygon*), **CW** (*crossing window*), or **F** (*fence*) object selection options.

SolProf

Rel.13 Creates profile images of 3D solids (short for SOLid PROFile).

Command	Alias	Ctrl+	F-key	Alt+	Menu Bar	Dashboard
solprof	...	...	...	DMUP	Draw	...
					↳Modeling	
					↳Setup	
					↳Profile	

Command: solprof
Select objects: *(Select one or more objects.)*
Select objects: *(Press* **ENTER** *to end object selection.)*
Display hidden profile lines on separate layer? [Yes/No] <Y>: *(Type Y or N.)*
Project profile lines onto a plane? [Yes/No] <Y>: *(Type Y or N.)*
Delete tangential edges? [Yes/No] <Y>: *(Type Y or N.)*
n **solids selected.**

COMMAND LINE OPTIONS

Select objects selects the objects to profile.

Display hidden profile lines on separate layer?

No specifies that all profile lines are visible; a block is created for the profile lines of every selected solid.

Yes generates just two blocks: one for visible lines and one for hidden lines.

Project profile lines onto a plane?

No creates profile lines with 3D objects.

Yes creates profile lines with 2D objects.

Delete tangential edges?

No does not display *tangential edges*, the transition line between two tangent faces.

Yes displays tangential edges.

RELATED COMMANDS

SolDraw creates profiles and sections in viewports.

SolView creates floating viewports.

TIPS

- The SolView command must be used before the SolProf command.

- Solids that share a common volume can produce dangling edges, if you generate profiles with hidden lines. To avoid this, first use the Union command.

- AutoCAD must display a layout, and a model space viewport must be active, before you can use the SolProf command.

SolView

<u>**Rel.13**</u> Creates floating viewports in preparation for the SolDraw and SolProf commands (short for SOLid VIEWs).

Command	Alias	Ctrl+	F-key	Alt+	Menu Bar	Dashboard
solview	...	...	...	DMUV	Draw	...
					⤷ Modeling	
					⤷ Setup	
					⤷ View	

Command: solview
Enter an option [Ucs/Ortho/Auxiliary/Section]: *(Enter an option.)*

Orthographic views created by the **SolView** command.

Top view.

Front view.

Side view.

COMMAND LINE OPTIONS

Ucs options
Enter an option [Named/World/?/Current] <Current>: *(Enter an option.)*

Named creates a profile view using the x,y plane of a named UCS.

World creates a profile view using the x,y plane of the WCS.

? lists the names of existing UCSs.

Current creates a profile view using the x,y-plane of the current UCS.

Ortho options
Specify side of viewport to project: *(Pick a point.)*
Specify view center: *(Pick a point.)*
Enter view name: *(Enter a name.)*

Pick side of viewport to project selects the edge of one viewport.

View center picks the center of the view.

Clip picks two corners for a clipped view.

View name names the view.

Auxiliary options
Specify first point of inclined plane: *(Pick a point.)*
Specify second point of inclined plane: *(Pick a point.)*
Specify side to view from: *(Pick a point.)*
Specify view center: *(Pick a point.)*
Enter view name: *(Enter a name.)*

> **Inclined plane's 1st point** picks the first point.
>
> **Inclined plane's 2nd point** picks the second point.
>
> **Side to view from** determines the view side.

Section options
Specify first point of cutting plane: *(Pick a point.)*
Specify second point of cutting plane: *(Pick a point.)*
Specify side to view from: *(Pick a point.)*
Enter view scale <5.9759>: *(Pick a point.)*
Specify view center <specify viewport>: *(Pick a point.)*
Specify first corner of viewport: *(Pick a point.)*
Specify opposite corner of viewport: *(Pick a point.)*
Enter view name: *(Enter a name.)*

> **Cutting plane 1st point** picks the first point.
>
> **Cutting plane 2nd point** picks the second point.
>
> **Side to view from** determines the view side.
>
> **Viewscale** specifies the scale of the new view.

> **eXit** exits the command.

RELATED COMMANDS

SolDraw creates profiles and sections in viewports.

SolProf creates profile images of 3D solids.

TIPS

- This command creates orthographic, auxiliary, and sectional views.

- This command must be used before the SolDraw command, because it creates the layers required by SolDraw.

- This command is useful for creating layouts for the SolProf command.

- The layers created by this command have the following prefixes:

Layer Name	View
viewname-**VIS**	Visible lines.
viewname-**HID**	Hidden lines.
viewname-**DIM**	Dimensions.
viewname-**HAT**	Hatch patterns for sectional views.

- *Warning!* Autodesk warns that "The information stored on these layers is deleted and updated when you run SolDraw. Do not place permanent drawing information on these layers."

'SpaceTrans

<u>2002</u> Converts distances between model and space units (short for SPACE
TRANSlation).

Command	Alias	Ctrl+	F-key	Alt+	Menu Bar	Dashboard
spacetrans	...	...	...	...	...	...

Command: spacetrans

This command does not operate in Model tab. In model view of a layout tab:

Specify paper space distance <1.000>: *(Enter a value.)*

If the layout contains just one viewport, it is selected automatically; otherwise, you are prompted:

Select a viewport: *(Pick a viewport.)*

Specify model space distance <1.000>: *(Enter a value.)*

COMMAND LINE OPTIONS

Specify paper space distance specifies the paper space length to be converted to its model space equivalent, usually the scale factor.

Select a viewport selects a paper space viewport.

Specify model space distance specifies the model space length to be converted to its paper space equivalent, usually the scale factor.

RELATED COMMANDS

Text places text in the drawing.

SolProf creates profile images of 3D solids.

TIPS

- This command is meant to be used transparently during other commands, and not at the 'Command:' prompt. It does not work in model tab.

- The purpose of this command is to convert lengths between model and paper space.

- Autodesk recommends using this command for converting model space text heights into paper space text heights.

'Spell

Checks the spelling of text in the drawing.

Command	Alias	Ctrl+	F-key	Alt+	Menu Bar	Dashboard
spell	sp	...	...	TE	Tools ⮡Spelling	Text

Command: spell

When unrecognized text is found, displays dialog box (changed in AutoCAD 2008):

When all text is recognized, or when spelling check is complete, displays dialog box:

CHECK SPELLING DIALOG BOX

Start starts spell checking all words in the current drawing, or as specified by the Settings dialog box (*new to AutoCAD 2008*).

Add to Dictionary adds the current word to the dictionary file.

Ignore ignores the word, and goes to the next word.

Ignore All ignores all words with this spelling.

Change changes the word to the suggested spelling.

Change All changes all words with this spelling.

Dictionaries selects a different dictionary; displays the Dictionaries dialog box.

Settings displays the Settings dialog box (*new to AutoCAD 2008*).

DICTIONARIES DIALOG BOX

Main dictionary selects a language for the dictionary.

Custom Dictionaries options

Directory specifies the drive, folder, and filename of the custom dictionary.

Content allows you to enter additonal words to the dictionary.

Add adds words (content).

Delete removes custom words from the dictionary.

Import imports custom dictionary CUS files; displays the Add Custom Dictionary File dialog box.

CHANGE DICTIONARIES DIALOG BOX
(New to AutoCAD 2008.)

Include options

☑ **Dimension Text** spell-checks text in dimensions.

☑ **Block Attributes** spell-checks text in attributes.

☐ **External References** spell-checks text in externally-referenced drawings.

Options options

☐ **Ignore Capitalized Words** ignores words that start with capital letters.

☐ **Ignore Words with Mix Cases** ignores words with mixed upper- and lowercase letters.

☐ **Ignore Words in Uppercase** ignores words that are in all uppercase letters.

☑ **Ignore Words with Numbers** ignores words that include digits.

☑ **Ignore Words Containing Punctuation** ignores words containing punctuation.

RELATED COMMANDS

DdEdit edits text.

MText places paragraph text.

Text places lines of text in the drawing.

AttDef defines attributes.

RELATED SYSTEM VARIABLES

DctCust names the custom spelling dictionary.

DctMain names the main spelling dictionary.

RELATED FILES

enu.dct is the dictionary word file.

*.cus are the custom dictionary files.

TIPS

- Spell checkers do not check your spelling, because words that are spelled correctly but used incorrectly (such as *its* and *it's*) are not flagged. Rather, spell checkers look for words that they do not recognize — namely words not in its dictionary file.

- As of AutoCAD 2002, the Spell command also checks words in blocks. As of AutoCAD 2008, this command zooms into the unrecognized word, as illustrated below:

- If you accidently add a misspelled word to the custom dictionary, use the Dictionaries button to remove it.

Sphere

Rel.11 Draws 3D spheres as solid models.

Command	Alias	Ctrl+	F-key	Alt+	Menu Bar	Dashboard
sphere	...	...	...	DMS	Draw	3D Make
					↳ Modeling	
					↳ Sphere	

Command: sphere
Specify center point or [3P/2P/Ttr]: *(Pick a point, or enter an option.)*
Specify radius or [Diameter] <1.0>: *(Enter a value, or type D.)*

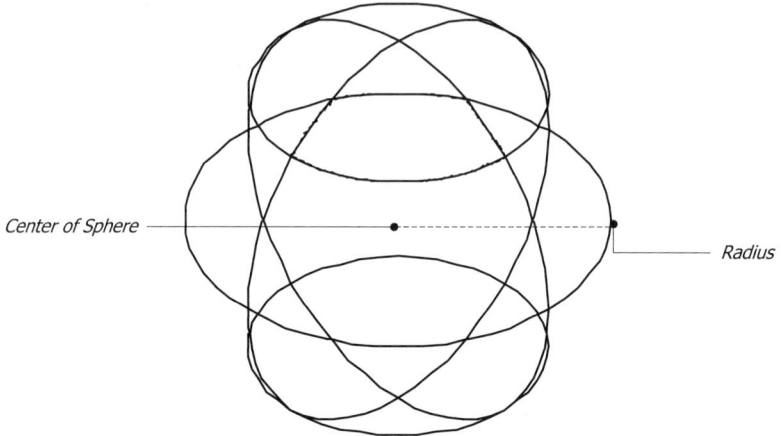

Center of Sphere

Radius

COMMAND LINE OPTIONS

Specify center point locates the center point of the sphere.

3P specifies three points on the outer circumference (equator) of the sphere.

2P specifies two points on the outer diameter of the sphere.

Ttr specifies two tangent points plus the radius of the outer circumference of the sphere.

Radius specifies the radius of the sphere.

Diameter specifies the diameter of the sphere.

RELATED COMMANDS

Cone draws solid cones.

Cylinder draws solid cylinders.

Torus draws solid tori.

Ai_Sphere draws surface meshed spheres.

RELATED SYSTEM VARIABLES

DragVs specifies the visual style during creation of solids.

DispSilh toggles the silhouette display of 3D solids.

· ·

IsoLines specifies the number of tessellation lines that define the surface of curved 3D solids; range is from 0 to 2047.

TIPS

- This command places the sphere's central axis parallel to the z axis of the current UCS, with the latitudinal isolines parallel to the x,y plane.

- Notice the effect of the DispSilh system variable, which toggles the silhouette display of 3D solids, after executing the Hide command:

- You must use the Regen command after changing the DispSilh and IsoLines system variables.

- Notice the effect of the IsoLines system variable, which controls the number of tessellation lines used to define the surface of a 3D solid:

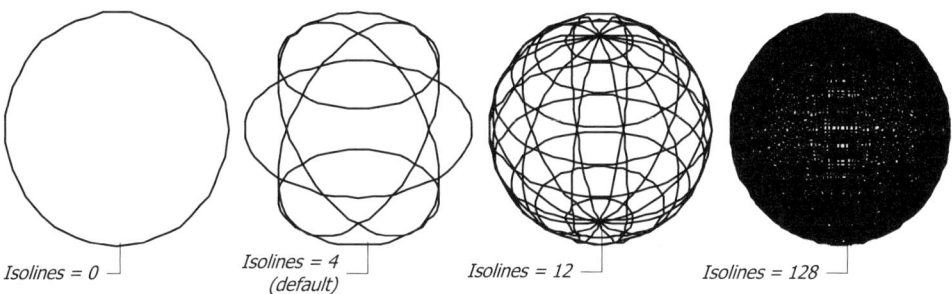

- Spheres can be edited using grips:

Spline

<u>Rel.13</u> Draws NURBS (Non-Uniform Rational Bezier Spline) curves.

Command	Alias	Ctrl+	F-key	Alt+	Menu Bar	Dashboard
spline	spl	...	...	DS	Draw ↳Spline	2D Draw

Command: spline
Specify first point or [Object]: *(Pick a point, or type O.)*
Specify next point: *(Pick a point.)*
Specify next point or [Close/Fit tolerance] <start tangent>: *(Pick a point, or enter an option.)*
Specify next point or [Close/Fit tolerance] <start tangent>: *(Press Enter to define tangency points.)*
Specify start tangent: *(Pick a point, or press Enter.)*
Specify end tangent: *(Pick a point, or press Enter.)*

Open spline   *Closed spline*

COMMAND LINE OPTIONS
Specify first point picks the starting point of the spline.
Object converts 2D and 3D splined polylines into a NURBS spline.
Specify next point picks the next tangent point.
Close closes the spline at the start point.
Fit changes the spline tolerance; 0 = curve passes through fit points.
Specify start tangent specifies the tangency of the starting point of the spline.
Specify end tangent specifies the tangency of the endpoint of the spline.

RELATED COMMANDS
PEdit edits splined polylines.
PLine draws splined polylines.
SplinEdit edits NURBS splines.

RELATED SYSTEM VARIABLE
DelObj toggles whether the original polyline is deleted with the Object option.

TIP
- A closed spline has the same start and end tangent.

SplinEdit

Rel.13   Edits NURBS splines.

Command	Alias	Ctrl+	F-key	Alt+	Menu Bar	Dashboard
splinedit	spe	...	...	MOS	Modify	...
					⤷ Object	
					⤷ Spline	

Command: splinedit
Select spline: *(Select one spline object.)*
Enter an option [Fit data/Close/Move vertex/Refine/rEverse/Undo]: *(Enter an option.)*

COMMAND LINE OPTIONS

Close closes the spline, if open.

Move vertex moves a control vertex.

Open opens the spline, if closed.

Refine adds a control point; changes the spline's order or weight.

rEverse reverses the spline's direction.

Undo undoes the most-recent edit change.

eXit exits the command.

Fit Dataoptions
Enter a fit data option [Add/Close/Delete/Move/Purge/Tangents/toLerance/eXit] <eXit>: *(Enter an option.)*

Add adds fit points.

Close closes the spline, if open.

Delete removes fit points.

Move moves fit points.

Open opens the spline, if closed.

Purge removes fit point data from the drawing.

Tangents edits the start and end tangents.

toLerance refits the spline with the new tolerance value.

eXit exits suboptions.

RELATED COMMANDS

PEdit edits a splined polyline.

PLine creates polylines, including splined polylines.

Spline draws a NURBS spline.

TIPS

■ The spline loses its fit data when you use these SplinEdit options:

    • **Refine.**

    • **Fit Purge.**

    • **Fit Tolerance** followed by **Fit Move.**

    • **Fit Tolerance** followed by **Fit Open** or **Fit Close.**

■ The maximum order for a spline is 26; once the order has been elevated, it cannot be reduced.

■ The larger the weight, the closer the spline is to the control point.

■ This command automatically converts a spline-fit polyline into a spline object, even if you do not edit the polyline.

. . . .

Aliased Command

Spotlight is also handled by the Light command.

. . . .

Standards

2002 Loads standards into the current drawing.

Command	Alias	Ctrl+	F-key	Alt+	Menu Bar	Dashboard
standards	sta	...	...	TSC	Tools ⤷CAD Standards ⤷Configure	...

Command: standards

Displays dialog box:

To add standards to the drawing, click the ✚ icon, and then select a .dws file.

CONFIGURE STANDARDS DIALOG BOX

✚ **Add** attaches a *.dws* standards file to the drawing (F3); displays the Select Standards File dialog box; more than one *.dws* file can be attached.

✕ **Remove Standards** removes the selected standards file (Del).

⤴ **Move up** moves the selected *.dws* standards file higher in the list; available only when two or more standards files are loaded.

⤵ **Move down** moves the selected standards file lower in the list.

Check Standards displays the Check Standards dialog box; see the CheckStandards command.

Settings displays CAD Standards Settings dialog box. See CheckStandards.

Once a standard has been attached to the drawing, the dialog box looks like this:

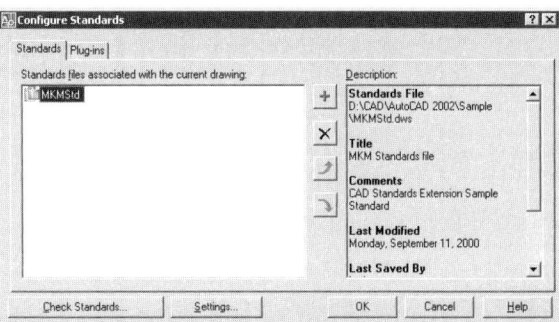

Plug-ins tab

Displays information about each item in the CAD standard:

☐ **Dimension Styles** checks if the drawing's dimension styles conform; click to turn off checking.

☐ **Layers** checks if the drawing's layer properties styles conform.

☐ **Linetypes** checks if the drawing's linetype definitions styles conform.

☐ **Text Styles** checks if the drawing's text styles conform.

RELATED COMMAND

CheckStandards checks the drawing against standards loaded by this command.

RELATED FILE

*.dws are AutoCAD standards files stored in DWG format.

. . . .

Removed Command

Stats was removed from AutoCAD 2007.

. . . .

'Status

<u>**V. 1.0**</u> Displays information about the current drawing and environment.

Command	Alias	Ctrl+	F-key	Alt+	Menu Bar	Dashboard
status	...	...	...	TQS	Tools	...
					⌐Inquiry	
					⌐Status	

Command: status

Example output:

```
223 objects in Drawing6.dwg
Model space limits are X:      0.0000   Y:      0.0000   (Off)
                        X:     12.0000   Y:      9.0000
Model space uses        X:     17.7037   Y:     20.3182
                        X:     38.1227   Y:     31.1948  **Over
Display shows           X:     16.4004   Y:     17.8401
                        X:     39.5708   Y:     34.9249
Insertion base is       X:      0.0000   Y:      0.0000   Z:      0.0000
Snap resolution is      X:      0.5000   Y:      0.5000
Grid spacing is         X:      0.5000   Y:      0.5000

Current space:          Model space
Current layout:         Model
Current layer:          "0"
Current color:          BYLAYER -- 7 (white)
Current linetype:       BYLAYER -- "Continuous"
Current material:       BYLAYER -- "Global"
Current lineweight:     BYLAYER
Current elevation:      0.0000  thickness:      0.0000
Fill on  Grid off  Ortho on  Qtext off  Snap off  Tablet off
Object snap modes:      Center, Endpoint, Insert, Intersection, Midpoint,
                        Nearest, Node, Perpendicular, Quadrant, Tangent,
                        Appint, Extension, Parallel
Free dwg disk (D:) space: 2344.3 MBytes
Free temp disk (C:) space: 3130.4 MBytes
Press ENTER to continue:
Free physical memory: 209.4 Mbytes (out of 1023.5M).
Free swap file space: 857.8 Mbytes (out of 1694.3M).
```

COMMAND LINE OPTIONS

Enter continues the listing.

F2 returns to the graphics screen.

RELATED COMMANDS

DbList lists information about all the objects in the drawing.

List lists information about the selected objects.

Properties lists information about selected objects in a palette.

DEFINITIONS

Model Space limits, Paper Space limits – the x,y coordinates stored in the LimMin and LimMax system variables; 'Off' indicates limits checking is turned off (LimCheck).

Model Space uses, Paper Space uses – the x,y coordinates of the lower-left and upper-right extents of objects in the drawing; 'Over' indicates drawing extents exceed the drawing limits.

Display shows – the x,y coordinates of the lower-left and upper-right corners of the current display.

Insertion base is – the x,y,z coordinates, as stored in system variable InsBase.

Snap resolution is – the snap settings, as stored in the SnapUnit system variable.

Grid spacing is – the grid settings, as stored in the GridUnit system variable.

Current space – the indication of whether model space or paper space is current.

Current layout – the name of the current layout.

Current layer, Current color, Current linetype, Current lineweight, Current plot style, Current elevation, Material Thickness – the current values for the layer name, color, linetype name, elevation, and thickness, as stored in system variables CLayer, CeColor, CeLType, CeLweight, CMaterial, CPlotSytle, Elevation, and Thickness.

Fill, Grid, Ortho, Qtext, Snap, Tablet – the current settings for the fill, grid, ortho, qtext, snap, and tablet modes, as stored in the system variables FillMode, GridMode, OrthoMode, TextMode, SnapMode, and TabMode.

Object Snap modes – the currently-set object modes, as stored in the system variable OsMode.

Free disk (dwg + temp = C) – the amount of free disk space on the drive storing AutoCAD's temporary files, as held by system variable TempPrefix.

Free physical memory – the amount of free RAM.

Free swap file space – the amount of free space in AutoCAD's swap file on disk.

TIPS

- When free disk space becomes close to zero, AutoCAD warns you.
- But when free temp space reaches zero, AutoCAD crashes without warning.

StlOut

Rel.12 Exports 3D solids and bodies in binary or ASCII SLA format (short for STereoLithography OUTput).

Command	Alias	Ctrl+	F-key	Alt+	Menu Bar	Dashboard
stlout	...	...	...	FE	File	...
				⬐STL	⬐Export	
					⬐Lithography	

Command: stlout
Select a single solid for STL output...
Select objects: *(Select one or more solid objects.)*
Select objects: *(Press Enter to end object selection.)*
Create a binary STL file? [Yes/No] <Y>: *(Type Y or N.)*
> *Displays the Create STL File dialog box; enter a name, and then click **Save**.*

COMMAND LINE OPTIONS

Select objects selects a single 3D solid object.

Y creates a binary-format *.sla* file.

N creates an ASCII-format *.sla* file.

RELATED COMMANDS

All solid modeling commands.

AcisOut exports 3D solid models to an ASCII SAT-format ACIS file.

AmeConvert converts AME v2.x solid models into ACIS models.

RELATED SYSTEM VARIABLE

FaceTRes determines the resolution of triangulating solid models.

RELATED FILE

***.stl** is the SLA-compatible file with STL extension created by this command.

TIPS

- The solid model must lie in the positive x,y,z-octant of the WCS.

- This command exports a single 3D solid; it does not export regions or any other AutoCAD object.

- Even though this command prompts you twice to 'Select objects', selecting more than one solid causes AutoCAD to complain, "Only one solid per file permitted."

- The resulting *.stl* file cannot be imported back into AutoCAD.

DEFINITIONS

STL — STereoLithography data file, a faceted representation of the model.

SLA — StereoLithography Apparatus.

Stretch

V. 2.5   Lengthens, shortens, or distorts objects.

Command	Alias	Ctrl+	F-key	Alt+	Menu Bar	Dashboard
stretch	s	...	...	MH	Modify ↳ Stretch	2D Draw

Command: stretch
Select objects to stretch by crossing-window or crossing-polygon...
Select objects: *(Type C or CP.)*
Specify first corner: *(Pick a point.)*
Specify opposite corner: *(Pick a point.)*
Select objects: *(Press Enter to end object selection.)*
Specify base point or displacement: *(Pick a point.)*
Specify second point of displacement or <use first point as displacement>:
(Pick a point.)

COMMAND LINE OPTIONS
First corner selects object; must be **CPolygon** or **Crossing** object selection.

Select objects selects other objects using any selection mode.

Base point indicates the starting point for stretching.

Second point makes the object larger or smaller.

RELATED COMMANDS
Lengthen changes the size of open objects.

Scale increases or decreases the size of objects.

TIPS
- The effect of Stretch on objects is not always obvious; be prepared to use the U command.

- Objects entirely within selection window are moved; objects crossing selection window are stretched.

- This command will not move a hatch pattern unless the hatch's origin is included in the selection set. This command does not stretch 3D solids, nor the width and tangents of polylines — although it does move them.

- As of AutoCAD 2008, selections made by crossing windows are cummulative.

'Style

V. 2.0 Creates and modifies text styles, which define the properties of fonts.

Commands	Aliases	Ctrl+	F-key	Alt+	Menu Bar	Dashboard
style	st	...	...	OS	Format	Text
	ddstyle				⌖Text Style	
-style						

Command: style

Displays dialog box (redesigned in AutoCAD 2008):

TEXT STYLE DIALOG BOX

To rename styles, right-click their name, and then select Rename from the shortcut menu.

Set Current sets the selected style as current.

New creates new text styles; displays dialog box:

Delete deletes text styles.

Apply applies the changes to the style without leaving the dialog box.

Styles option
- **All styles** lists all styles defined in the drawing.
- **Styles in use** lists those styles used by text in the drawing.

Font options
Font Name lists the names of AutoCAD SHX and TrueType TTF fonts installed in your computer's *\windows\fonts* folder.

Font Style selects available TTF font styles: **Bold**, *Italic*, and ***Bold Italic***.

☐ **Use Big Font** specifies the use of a big font file, typically for Asian alphabets.

Size options

(New to AutoCAD 2008.) When Annotative is off:

Height specifies the text height; 0 means users will be prompted for the height during the Text command and other text-related commands.

When Annotative is on:

☐ **Annotative** specifies that the text will be displayed at the correct height, based on the viewport scale factor.

    ☐ **Match Text Orientation to Layout** specifies that the text orientation in paper space viewports matches the orientation of the layout.

Paper Text Height specifies the height at which text is displayed in paperspace.

Effects options

Not available for all fonts:

☐ **Upside Down** draws text upside down:

☐ **Backwards** draws text backwards:

☐ **Vertical** draws text vertically.

Width Factor changes the width of characters.

Left: *Width factor = 0.5.* **Right**: *Width factor = 2.0.*

Oblique Angle slants characters forward or backward:

Left: *Oblique angle = 30.* **Right**: *Oblique angle = -30.*

. . . .

-STYLE Command

Command: -style
Enter name of text style or [?] <STANDARD>: *(Enter a name, or type ?.)*
Specify full font name or font filename (TTF or SHX) <txt>: *(Enter a name.)*
Specify height of text or [Annotative] <0.0000>: *(Enter a value or type A.)*
Specify width factor <1.0000>: *(Enter a value.)*
Specify obliquing angle <0>: *(Enter a value.)*
Display text backwards? [Yes/No] <N>: *(Type Y or N.)*
Display text upside-down? [Yes/No] <N>: *(Type Y or N.)*
Vertical? <N> *(Type Y or N.)*
"STANDARD" is now the current text style.

COMMAND LINE OPTIONS

Enter name of text style names the text style; maximum = 31 characters.

? lists the names of styles already defined in the drawing.

Specify full font name or font filename names the font file (SHX or TTF) from which the style is defined.

. . . .

Specify height of text specifies the height of the text.

Annotative specifies annotative scaling *(new to AutoCAD 2008)*.

Specify width factor specifies the width factor of the text.

Specify obliquing angle specifies the obliquing angle or slant of the text.

Display text backwards toggles mirror writing.

Display text upside-down toggles upside-down writing.

Vertical toggles vertical writing; not available for most fonts.

Annotative options
Create annotative text style [Yes/No] <Yes>: *(Enter Y or N.)*
Match text orientation to layout? [Yes/No] <Yes>: *(Enter Y or N.)*

Annotative specifies that the text will be displayed at the correct height in model space.

Match Text Orientation to Layout make the text face you, no matter how the viewport display is rotated.

REELATED COMMANDS

Properties changes the style assigned to selected text.

Purge removes unused text style definitions from drawings.

Rename renames text styles.

MText places paragraph text.

Text places a single line of text.

RELATED SYSTEM VARIABLES

TextStyle specifies the current text style.

TextSize specifies the current text height.

TextOutputFileFormat determines the Unicode text format for log files and plotting *(new to AutoCAD 2008)*.

RELATED FILES

*.shp is Autodesk's format for vector source fonts.

*.shx is Autodesk's format for compiled vector fonts; stored in *\autocad 2008\fonts* folder.

*.ttf are TrueType font files; stored in *\windows\fonts* folder.

TIPS

- The Style command affects the font used with the Text and MText commands, as well as with dimension and table styles.

- A width factor of 0.85 fits in 15% more text without sacrificing legibility.

- To use PostScript fonts in drawings, use Compile to convert PostScript *.pfb* files into *.shx* format.

- Text styles used in dimension styles should have a height of 0 to allow DimScale to work correctly.

StylesManager

2000 Displays the Plot Styles window.

Command	Alias	Ctrl+	F-key	Alt+	Menu Bar	Dashboard
stylesmanager	...	...	...	FY	File	...
					⮡Plot Style Manager	

You can also access this command through the Windows Control Panel: Autodesk Plot Style Manager.

Command: stylesmanager

Displays window, whose content varies:

PLOT STYLES WINDOW

*.ctb (color-table based) opens the Plot Style Table Editor dialog box.

* .stb (style-table based) also opens the Plot Style Table Editor dialog box.

Add-A-Plot Style Table Wizard opens Add Plot Style Table wizard; see the R14PenWizard command.

DEFINITIONS

StylesManager — the program that modifies plot styles stored in plot style tables.

Plot styles — when drawings have a plot style table attached to their model and layout tabs, any changes to the plot style changes the object using the plot style. Plot styles can be assigned by object or by layer. See the PlotStyle and Layer commands.

Color-table based (.cbt) — assigns colors to objects and layers, as used by older releases of AutoCAD. For example, the color of the object specifies the width of pen. This style of controlling the plot is now called "color dependent."

Style-table based (.stb) — newer releases of AutoCAD can control every aspect of the plot through "plot styles." By changing the *.stb* file attached to a layout tab, you immediately change the plot style for all objects and layers in the layout. This, for example, allows a quick change from monochrome to color plotting. A single drawing file can contain multiple plot styles.

PLOT STYLE TABLE EDITOR DIALOG BOX

General tab

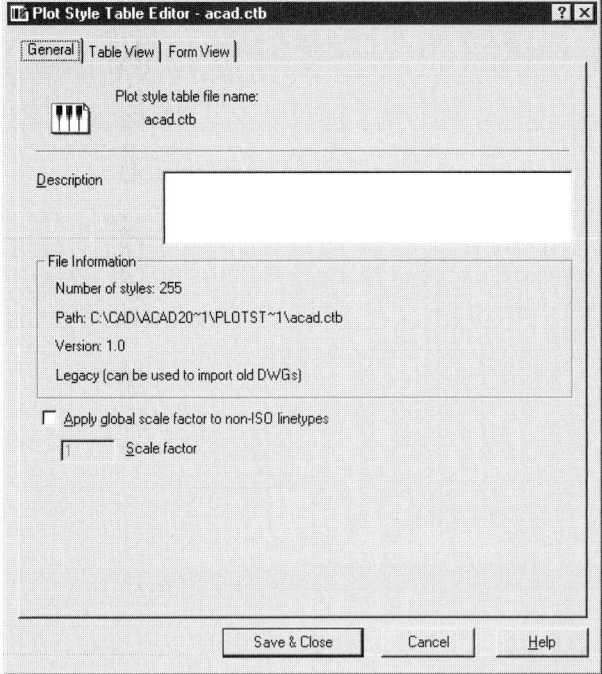

Description describes the plot style.

☐ **Apply global scale factor to non-ISO linetypes and fill patterns** applies the scale factor to all non-ISO linetypes and hatch patterns in the plot.

Scale factor specifies the scale factor.

Save and Close saves the changes, and closes the dialog box.

Cancel cancels the changes, and closes the dialog box.

CBT File Table View tab

See the following pages for options.

CBT File Form View tab

See the following pages for options.

STB File Table View tab

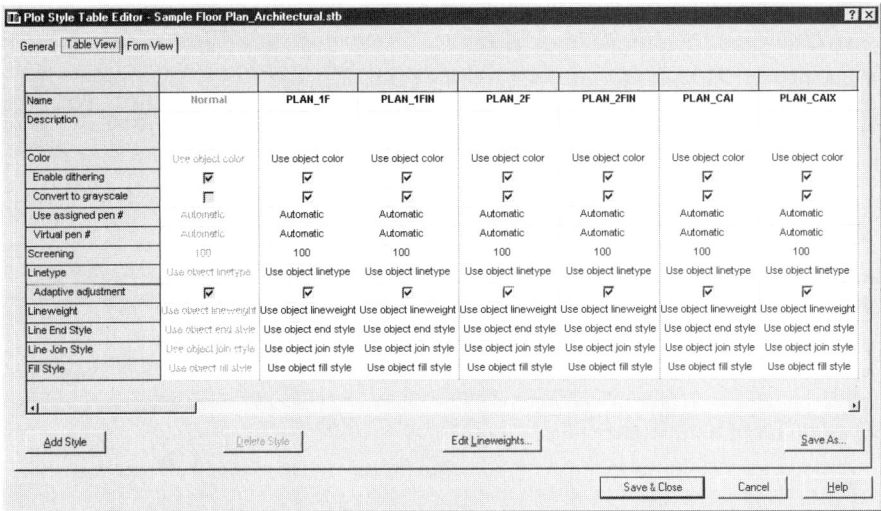

See the following pages for options.

STB File Form View tab

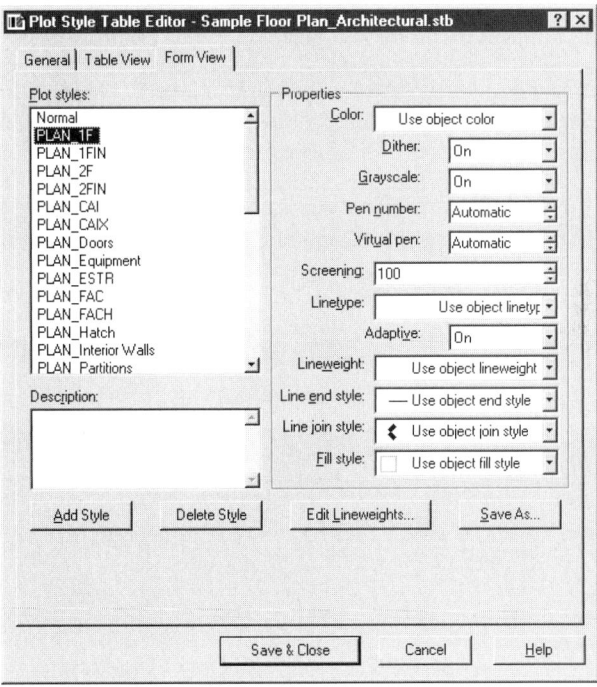

See the following pages for options.

Table View and Form View options are identical

Object color specifies the color of the object.

Description describes the plot style.

Color specifies the plotted color for the objects.

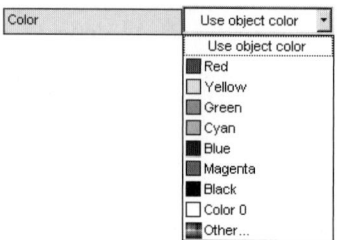

Enable dithering toggles dithering, if the plotter supports dithering, to generate more colors than the plotter is capable of; this setting should be turned off for plotting vectors, and turned on for plotting renderings.

Convert to grayscale converts colors to shades of gray, if the plotter supports gray scale.

Use assigned pen # specifies the pen number of pen plotters; range is 1 to 32.

Virtual pen number specifies the virtual pen number (default = Automatic); range is 1 to 255. A value of 0 (or Automatic) tells AutoCAD to assign virtual pens from ACI (AutoCAD Color Index); this setting is meant for non-pen plotters that can make use of virtual pens.

Screening specifies the intensity of plotted objects; range is 0 (plotted "white") to 100 (full density):

Linetype specifies the linetype with which to plot the objects:

Adaptive adjustment adjusts linetype scale to prevent the linetype from ending in the middle of its pattern; keep off when the plotted linetype scale is crucial.

Lineweight specifies how wide lines are plotted, in millimeters:

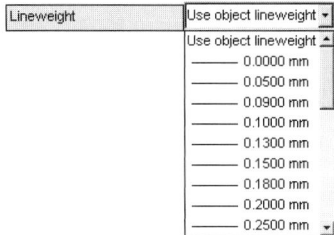

Line end style specifies how the ends of lines are plotted:

Line join style specifies how the intersections of lines are plotted:

Fill style specifies how objects are filled:

Add Style adds a plot style.

Delete Style removes a plot style.

Edit Lineweights displays the Edit Lineweights dialog box.

Save As displays the Save As dialog box.

TIPS

- To configure a printer or plotter for *virtual pens*:
  1. In the Device and Document Settings tab, open the PC3 Editor dialog box.
  2. In the Vector Graphics section, select 255 Virtual Pens from Color Depth.

- **CTB** is short for "color-dependent based" style table, which is compatible with earlier versions of AutoCAD.

- **STB** is short for "style-table based."

- You can attach a different *.stb* file to each layout and to the model tab in a drawing.

- You can create new *.stb* and *.ctb* files by copying and pasting them; double-click a file to open it for editing.

Subtract

<u>Rel.11</u>   Removes the volume of one set of 3D models or 2D regions from another.

Command	Alias	Ctrl+	F-key	Alt+	Menu Bar	Dashboard
subtract	su	...	...	MNS⬥	Modify ⬥Solids Editing ⬥Subtract	3D Make

Command: subtract
Select objects: *(Select one or more solid objects.)*
Select objects: *(Press Enter to end object selection.)*
1 solid selected.
Objects to subtract from them...
Select objects: *(Select one or more solid objects.)*
Select objects: *(Press Enter to end object selection.)*
1 solid selected.

*Left: Original objects. **Right:** Sphere subtracted from wedge.*

COMMAND LINE OPTION

Select objects selects the objects to be subtracted.

RELATED COMMANDS

Interfere find the volume common to two or more solids.

Intersect removes all but the intersection of two solid volumes.

Union joins two solids.

RELATED SYSTEM VARIABLE

ShowHist toggles the display of solids' history

TIPS

- AutoCAD subtracts the objects you select *second* from the objects you select *first*.

- You can use this commands on 3D solids and 2D regions.

- To subtract one region from another, both must lie in the same plane.

- When one solid is fully inside another, AutoCAD performs the subtraction, but reports, "Null solid created — deleted."

SunProperties, SunPropertiesClose

2007 Opens and closes the Sun Properties palette for controlling the sun light.

Command	Alias	Ctrl+	F-key	Alt+	Menu Bar	Dashboard
sunproperties	...	...	...	VELU	View	Lights
					⏷Render	
					⏷Light	
					⏷Sun Properties	

sunpropertiesclose						

Command: sunproperties

Displays palette:

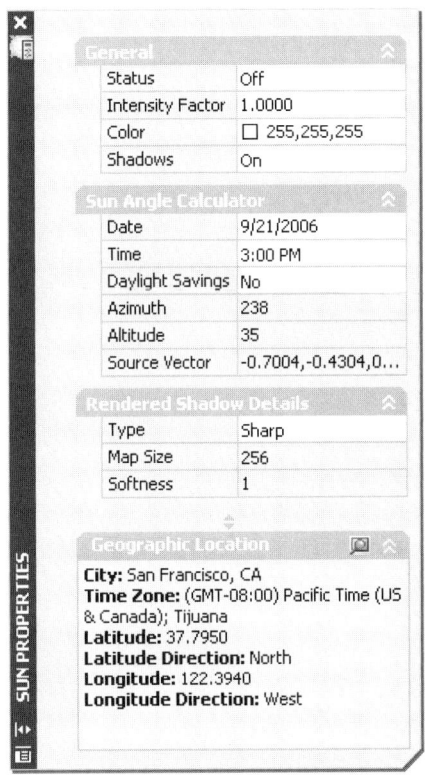

PALETTE OPTIONS

General options

Status toggles the sun light; inoperative when lighting is not enabled in the drawing.

Intensity Factor specifies the sun brightness; ranges from 0 (dark) to a very large number, like 9.999E+99 (bright).

Color specifies the color of the light.

Shadows toggles shadows generated by the sun.

Sun Angle Calculator options

Date specifies the day and month.

Time sets the time of day.

Daylight Saving toggles daylight savings time.

Azimuth reports the angle of the sun clockwise from due north; read-only.

Altitude reports the angle of the sun measured vertically from the horizon (90 = overhead); read-only.

Source Vector reports the direction of the sun; read-only.

Rendered Shadow Details options

Type selects the type of shadow, sharp or soft.

Map Size specifies the rendering area of soft shadows; range is 64 to 4096.

Softness specifies the appearance of shadow edges.

Geographic Location option

Edit Geographic Location displays the Geographic Location dialog box; see the GeographicLocation command.

RELATED COMMANDS

GeographicLocation displays the Geographic Location dialog box.

Render renders drawings, showing the effect of the sun light.

Light places lights in drawings.

RELATED SYSTEM VARIABLES

SunPropertiesState reports whether the Sun Properties palette is open.

SunStatus reports whether the sun is turned on or off.

TIPS

- Since lights can be difficult to place, the easy way to add lighting to 3D models is to turn on the sun light.

- Use the GeographicLocation command to specify the precise location of the sun, by date of year, time of day, and location on the surface of the earth.

- The View command's Background option simulates the sun and sky through an option added to AutoCAD 2008.

Sweep

<u>**2007**</u> Creates swept 3D solids and surfaces.

Command	Alias	Ctrl+	F-key	Alt+	Menu Bar	Dashboard
sweep	...	...	...	DMP↳	Draw ↳Modeling ↳Sweep	3D Make

Command: sweep
Current wire frame density: ISOLINES=4
Select objects to sweep: *(Select one or more objects.)*
Select objects to sweep: *(Press Enter.)*
Select sweep path or [Alignment/Base point/Scale/Twist]: *(Pick another object, or enter an option.)*

If the sweep object is closed (such as a circle), then this command produces a 3D solid; if open, such as an arc, then it generates a 3D surface.

COMMAND OPTIONS

Select objects to sweep selects the objects to be swept.

Select sweep path defines the path along which the objects are swept.

Alignment aligns the path normal (at 90 degrees) to the tangent direction of the path.

Base point relocates the sweep's base point.

Scale resizes the sweep object.

Twist specifies the twist angle.

Twist options
Enter twist angle or allow banking for a non-planar sweep path [Bank]<0.0>:
(Enter an angle, or type B.)

Twist angle specifies the angle through which the sweep twists, from one end to the other.

Bank causes the sweep to rotate along 3D curved paths: 3D polyline, 3D spline, or helix.

TIPS

- You can use the following objects as the sweep path: lines, arcs, circles, ellipses, elliptical arcs, 2D and 3D polylines, splines, helixes, and edges of surfaces and solids.

- The following objects can be swept: lines, traces, arcs, circles, ellipses, elliptical arcs, 2D polylines, 2D solids, splines, regions, flat 3D faces, flat solid faces, and planar surfaces.

- The figures below illustrate normal and twisted sweeps:

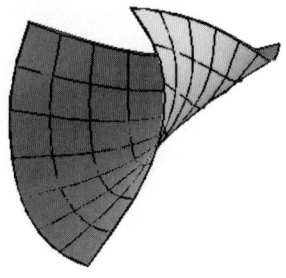

Left: Normal sweep.
Right: Twisted sweep.

- Apply this command to a helix to produce a spring:

- The Twist option does not work when the path is a closed loop.

SysWindows

Rel.13 Controls multiple windows (short for SYStem WINDOWS).

Command	Alias	Ctrl+	F-key	Alt+	Menu Bar	Dashboard
syswindows	...	F6	...	W	Windows	...
					⇘varies	

Command: syswindows
Enter an option [Cascade/tile Horizontal/tile Vertical/Arrange icons]: *(Enter an option.)*

AutoCAD displays drawings in four tiled windows.

COMMAND LINE OPTIONS

Arrange icons arranges icons in an orderly fashion.

Cascade cascades the window.

tileHorizontal tiles the window horizontally.

tileVertical tiles the window vertically.

TITLE BAR MENU

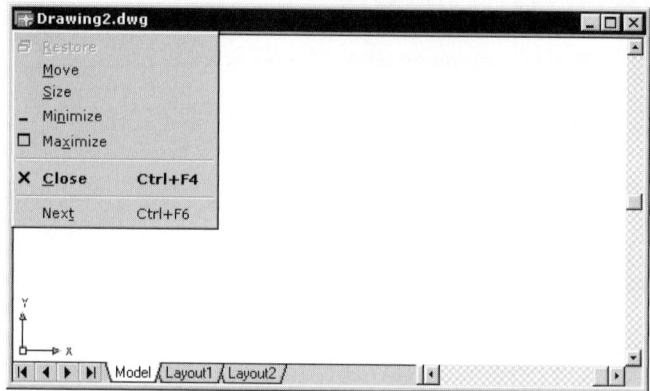

Restore restores the window to its "windowized" size.

Move moves the window.

Size resizes the window.

Minimize minimizes the window.

Maximize maximizes the window.

Close closes the window.

Next switches the focus to the next window.

RELATED COMMANDS

Close closes the current window.

CloseAll closes all windows.

Open opens one or more drawings, each in its own window.

MView creates paper space viewports in a window.

Vports creates model space viewports in a window.

RELATED SYSTEM VARIABLES

Taskbar determines whether each drawing opens in a single session of AutoCAD.

SDI forces AutoCAD to load just one drawing at a time.

TIPS

■ The SysWindows command had no practical effect until AutoCAD 2000, because AutoCAD Release 13 and 14 supported a single window only.

■ Press Ctrl+F6 to switch quickly between currently-loaded drawings.

■ Press Ctrl+F4 to close the current drawing

Table

2005 Inserts formatted tables in drawings; links to external data files.

Command	Alias	Ctrl+	F-key	Alt+	Menu Bar	Dashboard
table	tb	...	...	...	Draw	Tables
					⤷Table	2D Draw
-table						

Command: table

Displays dialog box (redesigned in AutoCAD 2008)

INSERT TABLE DIALOG BOX

Table Style options

▼ selects from a list of pre-formatted styles; new drawings contain one default style, Standard.

▧ displays the Table Style dialog box to edit or create new styles; see TableStyle command.

Insert Options options

New to AutoCAD 2008:

⊙ **Start from Empty Table** creates a blank table.

○ **From a Data Link** creates a table populated with data from an external spreadsheet file. All other options become unavailable in the dialog box.

▼ selects a named data link created previously with the DataLink command.

⊟ launches the DataLink command; see DataLink command.

⊙ **From Object Data in the Drawing (Data Extraction)** creates a table populated with data from block attributes, object properties, and drawing information; executes the DataExtraction command. All other options become unavailable in the dialog box.

Insertion Behavior options
⊙ **Specify Insertion Point** locates the table by its upper-left corner; after you click OK to exit the dialog box, AutoCAD prompts, 'Specify insertion point:'. Pick a point; AutoCAD inserts the table in the drawing:

○ **Specify window** fits the table to the window. After you click OK to dimiss the dialog box, AutoCAD prompts you to pick two points: 'Specify first corner:' and 'Specify second corner:'. Pick two points in the drawing; AutoCAD places the table at the first pick point, and fits the table to the second pick point.

Column & Row Settings options
Default size = 1 row x 5 columns, plus optional title and header rows.

Columns specifies the initial number of columns.

Column width specifies the initial width of columns.

Data Rows specifies the initial number of "data" rows; the table style will determine whether any header and title rows are added above the data rows.

Row height specifies the initial height of rows, in lines.

Set Cell Styles options
New to AutoCAD 2008:

First Row Style selects from data, header, or title.

Second Row Style selects from data, header, or title.

All Other Row Cell Styles selects from data, header, or title.

· ·

-Table Command
Command: -table
Current table style: Standard Cell width: 2.5000 Cell height: 1 line(s)
Enter number of columns or [Auto/from Style/data Link] <5>: *(Enter a number, or enter an option.)*
Enter number of rows or [Auto] <1>: *(Type A, or enter a number.)*
Specify insertion point or [Style/Width/Height]: *(Pick a point, or enter an option.)*

· ·

COMMAND OPTIONS

Auto creates columns and rows to fit the table.

from Style specifies table formatting, as detailed below *(new to AutoCAD 2008)*.

data Link prompts you for the name of a data link *(new to AutoCAD 2008)*.

Style selects the table style name; see TableStyle command.

Width specifies the initial width of the table.

Height specifies the initial height of the table.

From Style options
New to AutoCAD 2008:

Specify insertion point or [Style/Rows/Columns/Options]: *(Enter O.)*
Enter an insert option [Label text/Data text/Formulas/fIelds/data liNks/ Blocks/cell style Overrides] <exit>: *(Enter an option.)*

Label Text inserts rows with the Label cell style, including title and header cells styles.

Data Text inserts rows with the Data cell style, including data cell styles.

Formulas inserts formulas found in the table style.

Fields inserts fields found in the table style.

Data Links inserts data links in the table style.

Blocks inserts blocks found in the table style.

Cell Style Overrides retains cell style overrides found in the table style.

RELATED COMMANDS

MatchCell matches the style of one cell to other cells.

TableEdit edits text in table cells.

TableExport exports tables in CSV format.

TableStyle defines table styles.

DataLink creates named links with external spreadsheet files that can be turned into tables with the Table command.

DataExtraction extracts drawing data to tables.

RELATED SYSTEM VARIABLES

CTableStyle names the current table style.

TableIndicator toggles the display of row numbers and column letters.

TIPS

- To copy tables between drawings, use DesignCenter.

- Whether the table extends up or down from the insertion point depends on its style.

- In AutoCAD 2008, tables can be "live": when the spreadsheet data changes, the table can be updated by you. AutoCAD displays a warning baloon in the tray:

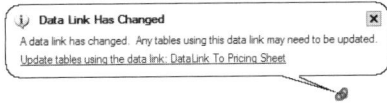

- Use the DataExtraction command to create tables of information extracted from the drawing.

- Tables created through the DataLink command display this icon when the cursor pauses over it.

- Linked tables are initially locked to avoid unintentional editing: The lock can be overriden by the TablEdit command's lock options; see TablEdit command.

- Unlike the Table command, the -Table command cannot set up data links; you first use the DataLink command to create a named link, and then you can use the -Table command to specify the name of the data link.

- AutoCAD shows the range of spreadsheet cells with green corner icons: .

- You can autofill cells by dragging an initial value across rows and columns, just as in spreadsheets.

- Tables can flow across multiple columns: enable the Table Breaks option in the Properties palette.

- To create new tables, you can bypass the Table command: copy a range of cells from a spreadsheet, and then paste them into the drawing. This workaround may only work with specific brands and versions of spreadsheet programs.

- Tables can be annotative: use an annotative text style, or else click the annotative icon in the text formating toolbar.

TablEdit

2005 Edits the text in cells; other methods edit the table itself.

Command	Alias	Ctrl+	F-key	Alt+	Menu Bar	Dashboard
tabledit	...	...	...	...	...	...

Command: tabledit

This command is substantially changed in AutoCAD 2008:

Pick a table cell: *(Pick inside a cell.)*

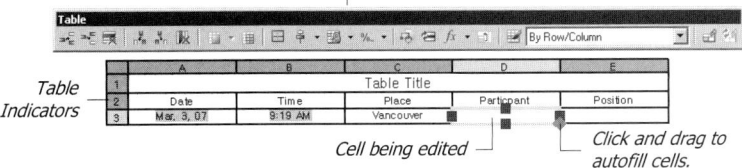

To edit other cells, press the arrow keys. To move to the "next" cell, press Tab.

As an alternative to the TablEdit command, double-click cells to edit their content.

TABLE TOOLBAR

New to AutoCAD 2008:

Right-click a cell for a shortcut menu listing similar options (illustrated at right).

From left to right:

Insert row above.

Insert row below.

Delete rows; deletes selected rows.

Insert column to the left.

Insert column to the right.

Delete columns; deletes selected columns.

Merge Cells: All, By Row, or By Column; operates only when two or more cells are selected.

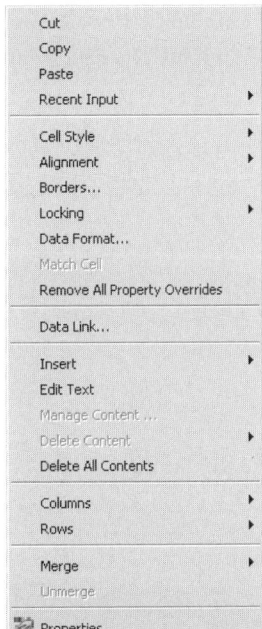

Cell Borders displays the Cell Border Properties dialog box; see the TableStyle command.

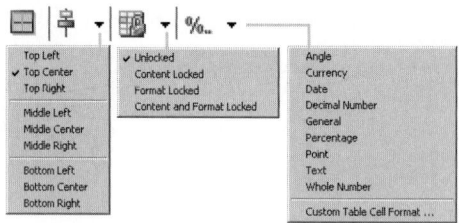

Alignment specifies top left, middle center, bottom right, and so on.

Locking specifies unlocked, content locked, format locked, or content and format locked.

Data Format specifies angle, currency, date, percentage, and so on; displays the Data Format dialog box; see the TableStyle command.

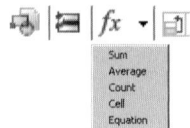

Insert Block: displays the Insert a Block in a Table Cell dialog box; see TInsert command.

Insert Field: displays the Field dialog box; see the Field command.

Insert Formula: provides sum (addition), average, count, cell (reference), or equation (displays the MText command's Text Formatting toolbar).

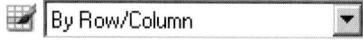

Match Cell: copies properties from one cell to others; see the MatchCell command.

Cell Styles: select from preset or user-defined cells; see the TableStyle command.

Link Cell: displays the Select a Data Link dialog box; see the DataLink command.

DownloadChanges from Source Cell; see the DataLinkUpdate command.

SHORTCUT MENU

Select a table, right-click, and then select an option from the shortcut menu:

Command specific to table editing:

Table Style displays a submenu (*new to AutoCAD 2008*):

- **Save as New table Style** saves the table's formatting as a style; displays the Save As New Table Style dialog box.
- **Save as Table in Current Table Style** adds the formatting to the current style.
- **Standard** chooses from a list of predefined table styles.

Size Columns Equally returns all columns to their original width.

Size Rows Equally adjusts the height of all rows to the height of the tallest row.

Remove All Property Overrides returns the table to its original style settings.

Export exports the table in CSV (comma separated values) format; see TableExport command.

Table Indicator Color changes the color of the row and column indicators; displays Select Color dialog box. See the Color command.

Update Table Data Links updates the table from data changed in external spreadsheet files (*new to AutoCAD 2008*).

Write Data Links to External Source updates changes in the table to the external files (*new to AutoCAD 2008*).

TABLE EDITING WITH GRIPS

To stretch columns and the table together, hold down the Ctrl key.

RELATED COMMANDS

MatchCell matches the style of one cell to other cells.

Table creates tables.

TableStyle defines table styles.

Properties changes table style names, and overrides style settings.

RELATED SYSTEM VARIABLES

CTableStyle names the current table style.

TableIndicator toggles the display of row numbers and column letters.

TableToolbar toggles the display of the toolbar when cells are selected in tables.

TableExport

<u>2005</u> Exports tables as CSV data files.

Command	Alias	Ctrl+	F-key	Alt+	Menu Bar	Dashboard
tableexport	...	...	...	...	...	...

Command: tableexport
Select a table: *(Select one table.)*

Displays the Export Data dialog box.

Enter a file name, select a folder, and then click Save.

TIPS

- *CSV* is short for "comma-separated values." Each table row is one line of data, and each cell is separated by a comma. This format can be read by spreadsheet and database programs.

- To import tables from a spreadsheet, copy and paste it into the drawing. AutoCAD converts the spreadsheet into a table object. Follow these steps:
  1. In the spreadsheet, copy the data to the Clipboard with Ctrl+C.
  2. In AutoCAD, use the Edit | Paste Special command to paste the data:

Format	Meaning
Picture (Metafile)	Pasted as an OLE object.
Device Independent Bitmap	Pasted as an OLE object.
AutoCAD Entities	Pasted as a table object.
Image Entity	Pasted as a raster image.
Text	Pasted as mtext.
Unicode Text	Pasted as mtext.

  To create a live table, ensure that the Paste Link option is selected.

- AutoCAD can import tables from CSV files via the DataLink command.

- To copy tables between drawings, use DesignCenter.

- You can also access this command by selecting the table, right-clicking, and then choosing Export from the shortcut menu.

TableStyle

2005 Creates and edits table styles.

Command	Alias	Ctrl+	F-key	Alt+	Menu Bar	Dashboard
tablestyle	ts	...	...	OB	Format	Table
					↳Table Style	

Command: tablestyle

TABLE STYLE DIALOG BOX

Styles lists the styles defined in the drawing; every new drawing has the "Standard" style.

List filters the style names:

- **All Styles** lists all styles.
- **Styles in use** lists the styles used by tables in the drawing.

Set Current sets the selected style as current.

New creates new table styles based on the settings of the current style; displays Create New Table Style dialog box, detailed below.

Modify changes the settings of the table style; displays the Modify Table Style dialog box.

Delete erases the style from the drawing; unavailable if the "Standard" table style is selected.

CREATE NEW TABLE STYLE DIALOG BOX

New Style Name names the new table style.

Start With selects the existing style upon which to base the new style.

Continue displays the New Table Style dialog box (identical to the Modify Table Style dialog box).

NEW/MODIFY TABLE STYLE DIALOG BOX

Substantially changed in AutoCAD 2008:

Starting Table options

Select Table to Start From selects the source for this style:

Selects a table in the drawing; dismisses the dialog box, and prompts you, 'Select a table:'. Pick a table, and the dialog box returns with the table as the source for the new style.

Removes the starting table from the table style; returns to the default style, as illustrated above.

General

Table Direction determines how the table grows relative to the insertion point:

- **Down** (insertion point is at upper-left corner).
- **Up** (insertion point is at lower-left corner).

Cell Styles options

New to AutoCAD 2008:

displays a list of cell format names; defaults are Data, Header, and Title.

Create a New Cell Style displays the Create New Cell Style dialog box for naming the new style.

Manage Cell Styles displays the Manage Cell Styles dialog box, detailed later.

General tab

Properties options
New to AutoCAD 2008:

Fill Color selects the color for cell backgrounds; default = None. Choose Select Color to access the Select Color dialog box; see the Color command.

Alignment specifies the alignment of objects in cells: top left, middle center, bottom right, and so on.

Format displays the Table Cell Format dialog box, which is identical to that used by the Field command.

Type selects either label or data cells; *label* cells are like titles and headers, while *data* cells are regular cells.

Margins options
Horizontal specifies the horizontal margin, distance between the cell content and its borders.

Vertical specifies the vertical margin, distance between the cell content and its borders.

☐ **Merge Cells on Row/column Creation** merges new rows and columns into one cell; this option is meant for creating title rows in tables.

Text tab

Properties options
Text Style selects a predefined text style.

... displays the Text Style dialog box; see the Style command.

Text Height specifies the height of the text; unavailable if the height is defined by text style.

Text Color specifies the color of the text; select Select Color for additional colors.

Text Angle specifies the angle of text in the cell, such as 30 degrees illustrated below (*new to AutoCAD 2008*). Either the text becomes smaller, or the cell expands to accomodate the text.

Table Title	
Date	Time
Mar. 3, 07	9:19 AM

BORDERS TAB

Properties options

Lineweight selects a lineweight from the droplist. Lineweights are applied to the center of the border lines, and some cases thick lineweights may overwrite part of the cell content.

Linetype selects the linetype for the cell borders; click Other to load additional linetypes (*new to AutoCAD 2008*).

Color chooses the color for the borders.

☐ **Double Line** toggles double-line borders, often used to emphasize cells or to surround the entire table (*new to AutoCAD 2008*).

Spacing specifies the spacing between double lines.

Properties are applied to all borders, or to those selected below:

All - Outer - Inner - Bottom - Left - Top - Right - None

MANAGE CELL STYLES DIALOG BOX

New names new cell styles; displays the Create New Cell Style dialog box.

Rename renames the cell style; standard styles cannot be renamed: Data, Header, and Title.

Delete removes cell styles from the drawing; in-use and standard styles cannot be erased.

RELATED COMMANDS

Table places tables in drawings.

MatchCell matches the style of one cell to other cells.

TableEdit edits text in table cells.

Rename renames table styles.

Purge purges unused table styles from drawings.

DesignCenter shares tables styles between drawings.

Properties changes table style names, and overrides style settings.

RELATED SYSTEM VARIABLE

CTableStyle names the current table style.

TIPS

- To copy table styles from other drawings, copy a table with Ctrl+C, and then paste it into the drawing with Ctrl+V.

- If you erase a table, the style remains in the drawing.

- To revert tables to original styles, select table, right-click, and select Remove All Property Overrides.

Tablet

<u>V. 1.0</u> Configures and calibrates digitizing tablets, and toggles tablet mode.

Command	Alias	Ctrl+	F-key	Alt+	Menu Bar	Dashboard
tablet	ta	T	F4	TBB✎	Tools	...
					✎Tablet	

Command: tablet

When no tablet is configured, AutoCAD complains, 'Your pointing device cannot be used as a tablet.'

When a digitizing tablet is configured, AutoCAD continues:

Enter an option [ON/OFF/CAL/CFG]: *(Enter an option.)*

COMMAND LINE OPTIONS

CAL calibrates the coordinates for the tablet.

CFG configures the menu areas on the tablet.

OFF turns off the tablet's digitizing mode.

ON turns on the tablet's digitizing mode.

RELATED SYSTEM VARIABLE

TabMode toggles use of the tablet:

   0 — Tablet mode disabled.

   1 — Tablet mode enabled.

RELATED FILES

acad.cui is the customization source code that defines the functions of tablet menu areas.

tablet.dwg is an AutoCAD drawing of the printed template overlay.

TIPS

- To change the tablet overlay, edit the *tablet.dwg* file, and then plot it to fit your digitizer.

- Tablet does not work if a digitizer has not been configured with the Options command.

- AutoCAD supports up to four independent menu areas; macros are specified by the Tablet Menu section of the *acad.cui* file.

- Menu areas may be skewed, but corners must form a right angle.

- Projective transformation is a limited form of "rubber sheeting": straight lines remain straight, but not necessarily parallel.

DEFINITIONS

Affine Transformation — requires three pick points; sets an arbitrary linear 2D transformation with independent x,y scaling and skewing.

Orthogonal Transformation — requires two pick points; sets the translation; the scaling and rotation angles remain uniform.

Residual Error — proves largest where mapping is least accurate.

Outcome of Fit — reports on the results of transformation types:

Outcome	Meaning
Exact	Enough points to transform data.
Success	More than enough points to transform data.
Impossible	Not enough points to transform data.
Failure	Too many colinear and coincident points.
Canceled	Fitting canceled during projective transformation.

Projective Transformation — maps a perspective projection from one plane to another plane.

RMS Error — specifies root mean square error; smaller is better; measures closeness of fit.

Standard Deviation —when near zero, residual error at each point is roughly the same.

TabSurf

Rel.10 Draws tabulated meshes (not surfaces) defined by path curves and direction vectors (short for TABulated SURFace).

Command	Alias	Ctrl+	F-key	Alt+	Menu Bar	Dashboard
tabsurf	...	...	...	DMMT	Draw	...
					⤷Modeling	
					⤷Meshes	
					⤷Tabulated Surface	

Command: tabsurf
Select object for path curve: *(Select an object.)*
Select object for direction vector: *(Select another object.)*

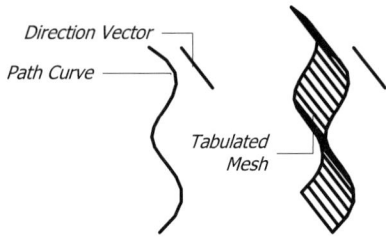

COMMAND LINE OPTIONS

Select object for path curve selects the object that defines the tabulation path.

Select object for direction vector selects the vector that defines the tabulation direction.

RELATED COMMANDS

Edge changes the visibility of 3D face edges.

PEdit edits 3D meshes, such as tabulated ones.

EdgeSurf draws 3D meshes between boundaries.

RevSurf draws revolved 3D meshes around an axis.

RuleSurf draws 3D meshes between open or closed boundaries.

RELATED SYSTEM VARIABLE

SurfTab1 defines the number of tesselations drawn by TabSurf in the *n*-direction.

TIPS

- The path curve can be an open or closed object: line, 2D polyline, 3D polyline, arc, circle, or ellipse.

- The direction in which you draw the *direction vector* determines the direction of the extrusion; the length of the vector determines the thickness of the extrusion.

- The number of *m*-direction tabulations is always 1, and lies along the direction vector. The number of *n*-direction tabulations is determined by SurfTab1 along curves only.

. . . .

Aliased Command

TargetPoint command is executed by the Light command.

. . . .

A⟂ Text

V. 1.0 Places text, one line at a time, in drawings.

Command	Alias	Ctrl+	F-key	Alt+	Menu Bar	Dashboard
text	dtext	...	...	DXS	Draw	Text
	dt				↳Text	
					↳Single Line Text	
-text						

Command: text
Current text style: "Standard" Text height: 0.2000 Annotative: No
Specify start point of text or [Justify/Style]: *(Pick a point, or enter an option.)*
Specify height <0.2000>: *(Enter a value.)*
Specify rotation angle of text <0>: *(Enter a value.)*

Displays in-drawing text box:

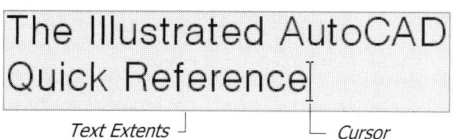

Text Extents ⌐   └ Cursor

Press Enter once to start a new line of text.

Press Enter twice to exit the command.

COMMAND LINE OPTIONS

Specify start point of text indicates the starting point of the text.

Justify selects a text justification.

Style specifies the text style; see the Style command.

Specify height indicates the height of the text; this prompt does not appear when the style sets the height to a value other than 0.

Specify paper height indicates the text height in model space; this prompt appears only when an annotative style is used *(new to AutoCAD 2008)*.

Specify rotation angle of text indicates the rotation angle of the text.

Enter continues text one line below the previously-placed text; press Enter twice to exit.

Justify options
Enter an option [Align/Fit/Center/Middle/Right/TL/TC/TR/ML/MC/MR/BL/BC/BR]: *(Enter an option.)*

Align aligns the text between two points with adjusted text height.

Fit fits the text between two points with fixed text height.

Center centers the text along the baseline.

Middle centers the text horizontally and vertically.

Right right-justifies the text.

TL top-left justifies the text.

TC top-center justifies the text.

TR top-right justifies the text.

ML middle-left justifies the text.

MC middle-center justifies the text.

MR middle-right justifies the text.

BL bottom-left justifies the text.

BC bottom-center justifies the text.

BR bottom-right justifies the text.

Style options

Enter style name or [?] <Standard>: *(Enter a name, or type* **?***.)*

Style name indicates a different style name.

? lists the currently-loaded styles.

SHORTCUT MENU

Right-click text to display shortcut menu:

Options specific to this command:

Editor Settings options

Opaque Background adds a gray background to the text editing area.

Highlight Color displays the Select Color dialog box; specifies the color of selected (highlighted) text *(new to AutoCAD 2008)*.

Insert Field inserts field text; displays the Field dialog box; see the Field command.

Find and Replace finds and optionally replaces text; see thd Find command.

Select All selects all text in the editing area.

Change Case options

You must select some text for these options to work:

Uppercase changes the case of selected text to all UPPERCASE; alternatively, press Ctrl+Shift+U.

Lowercase changes selected text to all lowercase; alternatively, press Ctrl+Shift+L.

SPECIAL SYMBOLS

%%c places the diameter symbol: Ø.

%%d places the degree symbol: °.

%%o starts and stops overlining.

%%p places the the plus-minus symbol: ±.

%%u starts and stops underlining.

. . . .

-Text command
Command: -text

This command displays prompts identical to those of the Text command, except that it ends after you enter one line of text. Also, no bounding box is displayed.

RELATED COMMANDS

DdEdit edits text.

Properties changes all aspects of text.

Style creates new text styles.

MText places paragraph text in drawings.

RELATED SYSTEM VARIABLES

DTextEd toggles between the in-place text editor and the old-style editor.

TextSize is the current height of text.

TextStyle is the current style of text.

TIPS

- You can enter any justification mode at the 'Start point' prompt.

- You can erase text by pressing Backspace, or by selecting text and then pressing the Del key.

- The spacing between lines of text does not necessarily match the snap spacing.

- Transparent commands and shortcut keystrokes do not work during the Text command.

- This command does not have an option for setting the annotative property, as does MText. The workaround is to specify a text style that has annotation turned on before using this command. Alternatively, you can use the Properties command to turn on annotation scaling afterwards.

- At the 'Enter text:' prompt, you can pick any point in the drawing to continue placing text. Press Shift+Tab to move to the previous line of text, or Tab to move to the next text line.

TextToFront

Places all text and dimensions visually on top of overlapping objects.

Command	Alias	Ctrl+	F-key	Alt+	Menu Bar	Dashboard
texttofront	...	...	...	TDF	Tools	...
					⬠Draw Order	
					⬠Bring Text and Dimensions to Front	

Command: texttofront
Bring to front [Text/Dimensions/Both] <Both>: *(Enter an option.)*
nnn **object(s) brought to front.**

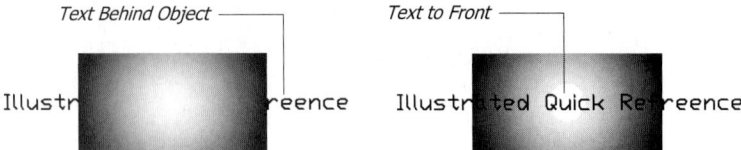

Text Behind Object · · · · · · · · · · · · · · · · Text to Front

Illustr[...]eence · · · · · Illustr[...]ted Quick Re[...]eence

COMMAND LINE OPTIONS

Text brings all text to the front.

Dimensions brings all dimensions to the front.

Both brings both text and dimensions to the front.

RELATED COMMANDS

DrawOrder controls the display order of selected objects.

BHatch controls the display of hatch and fill patterns relative to their boundaries and other objects.

TIPS

- You must apply this command separately to model space and each layout.

- Newer objects are drawn on top of older objects in drawings; to change this behavior, change the DrawOrderCtrl system variable.

'TextScr

V. 2.1 Switches from the AutoCAD window to the Text window (short for TEXT SCReen).

Command	Alias	Ctrl+	F-key	Alt+	Menu Bar	Dashboard
textscr	...	...	F2	VLT	View	...
					⤷ Display	
					⤷ Text Window	

Command: textscr

Displays the Text window:

EDIT MENU

Paste to CmdLine pastes text from the Clipboard to the command line; available only when the Clipboard contains text.

Copy copies selected text to the Clipboard.

Copy History copies all text to the Clipboard.

Paste pastes text from the Clipboard into the Text window; available only when the Clipboard contains text.

Options displays the Options dialog box; see the Options command.

COMMAND LINE OPTIONS

Command window navigation:

Key	Meaning
←	Moves the cursor left by one character.
→	Moves the cursor right by one character.
↑	Displays the previous line in the command history.
↓	Displays the next line in the command history.
Page Up	Displays the previous screen of text.
Page Down	Moves to the next screen of text.
Home	Moves the cursor to the start of the line.
End	Moves the cursor to the end of the line.
Insert	Toggles insert mode.
Delete	Deletes the character to the right of the cursor.
BACKSPACE	Deletes the character to the left of the cursor.

SHORTCUT MENU

*Right-click the **Text** window:*

Recent Command displays a list of ten recently-used commands.

Copy copies selected text to the Clipboard.

Copy History copies all text to the Clipboard.

Paste pastes text from the Clipboard into the Text window; available only when the Clipboard contains text.

Paste to CmdLine pastes text from the Clipboard to the command line; available only when the Clipboard contains text.

Options displays the Options dialog box; see the Options command.

RELATED COMMAND

GraphScr switches from the Text window to the AutoCAD drawing window.

RELATED SYSTEM VARIABLE

ScreenMode reports whether the screen is in text or graphics mode:
- 0 — Text screen.
- 1 — Graphics screen.
- 2 — Dual screen displaying both text and graphics.

· ·

Removed Command

TiffIn was removed from AutoCAD R14. Use the ImageAttach command instead.

· ·

Thicken

2007   Creates 3D solids by thickening surfaces.

Command	Alias	Ctrl+	F-key	Alt+	Menu Bar	Dashboard
thicken	...	...	...	M3T	Modify	3D Make
					⤷3D Operations	
					⤷Thicken	

Command: thicken
Select surfaces to thicken: *(Select one or more surfaces.)*
Select surfaces to thicken: *(Press Enter.)*
Specify thickness <0.0>: *(Type a number.)*

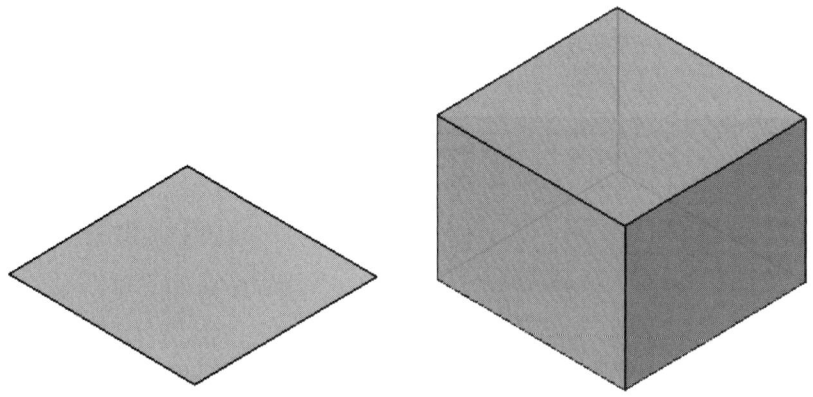

Left: Surface before thickening.
Right: Surface turned into 3D solid.

COMMAND LINE OPTIONS

Select surfaces to thicken selects one or more surfaces; this command does not work with meshes.

Specify thickness specifies the height of the thickened surface, such as 10. Enter a negative number, such as -10, to thicken the surface downwards.

RELATED COMMANDS

ConverToSolid converts circles and polylines with thickness to 3D solids.

PolySolid converts existing lines, 2D polylines, arcs, and circles to 3D solids.

RELATED SYSTEM VARIABLE

DelObj determines whether or not the surface is erased following the thickening.

TifOut

2004 Exports the current viewports in TIFF raster format.

Command	Alias	Ctrl+	F-key	Alt+	Menu Bar	Dashboard
tifout	...	...	...	...	...	...

Command: tifout

Displays Create Raster File dialog box. Specify a filename, and then click Save.

Select objects or <all objects and viewports>: *(Select objects, or press Enter to select all objects and viewports.)*

COMMAND LINE OPTIONS

Select objects selects specific objects.

All objects and viewports selects all objects and all viewports, whether in model space or in layout mode.

RELATED COMMANDS

BmpOut exports drawings in BMP (bitmap) format.

Image places raster images in the drawing.

JpgOut exports drawings in JPEG (joint photographic expert group) format.

Plot exports drawings in TIFF and other raster formats.

PngOut exports drawings in PNG (portable network graphics) format.

TIPS

- The rendering effects of the VsCurrent command are preserved, but not of the Render command.

- TIFF files are commonly used in desktop publishing, but the format produced by AutoCAD cannot be read by some graphics programs.

- TIFF is short for "tagged image file format," and was co-developed by Microsoft and Aldus, the forerunner of Adobe.

'Time

<u>V. 2.5</u> Displays time-related information about the current drawing.

Command	Alias	Ctrl+	F-key	Alt+	Menu Bar	Dashboard
time	...	...	...	TQT	Tools	...
					⮡ Inquiry	
					⮡ Time	

Command: time
Display/ON/OFF/Reset: *(Enter an option.)*

Example output:

```
Current time:          Wednesday, April 19, 2016  4:36:28:562 PM
Times for this drawing:
  Created:             Wednesday, April 19, 2016  3:19:27:359 PM
  Last updated:        Wednesday, April 19, 2016  3:19:27:359 PM
  Total editing time:  0 days 01:17:01:234
  Elapsed timer (on):  0 days 01:17:01:234
  Next automatic save in: <no modifications yet>
```

COMMAND LINE OPTIONS

Display displays the current time and date.

OFF turns off the user timer.

ON turns on the user timer.

Reset resets the user timer.

RELATED COMMAND

Status displays information about the current drawing and environment.

RELATED SYSTEM VARIABLES

CDate is the current date and time.

Date is the current date and time in Julian format.

SaveTime is the interval for automatic drawing saves.

TDCreate is the date and time the drawing was created.

TDInDwg is the time the drawing spent open in AutoCAD.

TDUpdate is the last date and time the drawing was changed.

TDUsrTimer is the current user timer setting.

TIPS

- The time displayed by the Time command is only as accurate as your computer's clock.

- The SaveAs command does not reset timings. The TdCreate system variable retains the original value, while the other timers continue accumulating seconds.

. . . .

Removed Command

Today was removed from AutoCAD 2004; it was replaced by the Communication Center.

. . . .

TInsert

<u>**2005**</u>   Inserts a block or a drawing in a table cell (short for Table INSERT).

Command	Alias	Ctrl+	F-key	Alt+	Menu Bar	Dashboard
tinsert	...	...	...	...	...	...

Command: tinsert
Pick a table cell: *(Select a single cell in a table.)*

Displays Insert a Block in a Table Cell dialog box.

If the block contains attributes, the Enter Attributes dialog box is displayed.

COMMAND LINE OPTION

Pick a table cell specifies the cell in which to place the block.

INSERT A BLOCK IN A TABLE CELL DIALOG BOX

Name specifies the name of the block; the droplist names all blocks found in the current drawing.

Browse displays the Select Drawing File dialog box; select a *.dwg* or *.dxf* file, and click Open.

Properties options

Cell Alignment specifies the position of the block within the cell, from top-left to bottom-right.

Scale sizes the block, and adjusts the cell to fit the block.

☑ **AutoFit** fits the block to the cell.

Rotation Angle specifies the angle at which to place the block.

RELATED COMMANDS

Table creates new tables.

Block creates blocks.

TIPS

- Alternatively, select a cell, right-click, and then choose Insert Block from the shortcut menu.

- A cell can contain at most one block.

⊕.1 Tolerance

Rel.13 Places geometric tolerancing symbols and text.

Command	Alias	Ctrl+	F-key	Alt+	Menu Bar	Dashboard
tolerance	tol	...	...	NT	Dimension ⮡Tolerance	Dimensions

Command: tolerance

Displays dialog box.

Fill in the dialog box, and then click **OK**. *AutoCAD prompts:*

Enter tolerance location: *(Pick a point.)*

GEOMETRIC TOLERANCE DIALOG BOX

Sym displays the Symbol dialog box.

Tolerance 1 and **Tolerance 2** specify the first and second tolerance values:

  Dia toggles the optional Ø (diameter) symbol.

  Value specifies the tolerance value.

  MC displays the Material Condition dialog box.

Datum 1, **Datum 2**, and **Datum 3** specify the datum references:

  Value specifies the datum value.

  MC displays the Material Condition dialog box.

Height specifies the projected tolerance zone value.

Projected Tolerance Zone toggles the projected tolerance zone symbol.

Datum Identifier creates the datum identifier symbol, such as -A-.

SYMBOL DIALOG BOX

Orientation Symbols

⊕   Position.

◎   Concentricity and coaxiality.

=   Symmetry.

//   Parallelism.

⊥   Perpendicularity.

∠   Angularity.

Form Symbols

⌭   Cylindricity.

▱   Flatness.

○   Circularity and roundness.

—   Straightness.

Profile Symbols

⌒   Profile of the surface.

⌒   Profile of the line.

↗   Circular runout.

↗↗   Total runout.

MATERIAL CONDITION DIALOG BOX

Ⓜ   specifies maximum material condition.

Ⓛ   specifies least material condition.

Ⓢ   specifies regardless of feature size.

RELATED FILES

gdt.shp is the tolerance symbol definition source file.

gdt.shx is the compiled tolerance symbol file.

TIPS

- You can use the DdEdit command to edit tolerance symbols and feature control frames.

- Tolerances can have the annotative property, which ensures that they display at the correct size in model space.

DEFINITIONS

Datum — a theoretically-exact geometric reference that establishes the tolerance zone for the feature. These objects can be used as a datum: point, line, plane, cylinder, and other geometry.

Material Condition — symbols that modify the geometric characteristics and tolerance values (modifiers for features that vary in size).

Projected Tolerance Zone — the height of the fixed perpendicular part's extended portion; changes the tolerance to positional tolerance.

Tolerance — the amount of variance from perfect form.

-Toolbar

Rel.13 Displays and hides toolbars via the command line.

Command	Alias	Ctrl+	F-key	Alt+	Menu Bar	Dashboard
-toolbar	...	...	...	...	...	...

Toolbar and TbConfig are aliases for the CUI command.

Command: -toolbar
Enter toolbar name or [ALL]: *(Enter a name, or type ALL.)*
Enter an option [Show/Hide]: *(Type S or H.)*

*Opening AutoCAD's toolbars with the **-Toolbar All Show** command.*

COMMAND LINE OPTIONS

Toolbar name specifies the name of the toolbar.

ALL applies the command to all toolbars; must be entered in all capital letters.

Show displays the toolbar.

Hide dismisses the toolbar.

RELATED COMMANDS

CUI customizes toolbars via a dialog box.

CuiLoad loads partial menu files, including toolbar definitions.

Tablet configures the tablet.

RELATED SYSTEM VARIABLE

ToolTips toggles the display of tooltips.

RELATED FILES

*.cui are user interface customization files.

*.bmp are bitmap files that define custom icon buttons.

TIP

- Right-click any toolbar for a shortcut menu that toggles the display of toolbars. Use this commadn when all toolbars are off.

'ToolPalettes, 'ToolPalettesClose

2004 Displays and closes the Tool palette.

Command	Alias	Ctrl+	F-key	Alt+	Menu Bar	Dashboard
toolpalettes	tp	3	...	TPTTools	...	
					⌐Palettes	
					⌐Tool Palettes	
toolpalettesclose		3				

Command: toolpalettes

Opens palette:

Command: toolpalettesclose

Closes the palette.

SHORTCUT MENUS

Tabs shortcut menu

Right-click an active tab:

Move up moves the tab up one position.

Move down moves the tab down one position.

View Options displays the View Options dialog box.

Paste pastes a tool previously copied from a palette.

New Tool Palette creates a new tab, and then prompts you for a name.

Delete Tool Palette removes the tab, after prompting you for confirmation.

Rename Tool Palette renames the tab.

Properties shortcut menu

Right-click the palette, or click the Properties button:

Move moves the palette.

Size changes the size of the palette.

Close closes the palette.

Allow Docking toggles whether the palette can be docked at the sides of the AutoCAD window.

Anchor Left < anchors (docks and minimizes) the palette at the left edge of AutoCAD.

Anchor Right > anchors the palette at the right edge of AutoCAD.

Auto-hide reduces the palette to the size of its title bar when the cursor is elsewhere.

Transparency displays the Transparency dialog box.

New Tool Palette creates a new tab, and then prompts you for a name.

Rename renames the selected tab.

Customize Palettes displays the Customize dialog box; see the Customize command.

Customize Commands displays the Customize User Interface dialog box; see the Cui command.

Palette shortcut menu

Right-click in the palette, away from any icon:

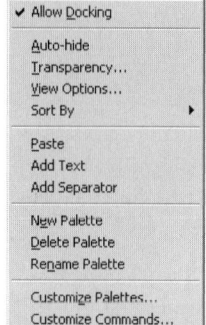

Allow Docking toggles whether the palette can be docked at the sides of the AutoCAD window.

Auto-hide reduces the palette to the size of its title bar when the cursor is elsewhere.

Transparency displays the Transparency dialog box.

View Options displays the View Options dialog box.

Sort by sorts the tools (doesn't seem to work correctly):

- **Name** of tool.
- **Type** of tool: Camera, Command, Face Settings, General, Insert, Materials Editor, Pattern, or Table.

Paste pastes a tool previously copied from a palette.

Add Text adds text to the palette.

Add Separator adds a horizontal line to the palette.

New Palette adds a new palette, names it "New Palette," and then encourages you to change the name.

Delete Palette removes the palette.

Rename Palette changes the palette's name, which appears on its associated tab.

Customize Palettes displays the Customize dialog box; see the Customize command.

Customize Commands displays the Customize User Interface dialog box; see the Cui command.

Icon shortcut menu
Right-click an icon:

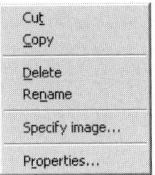

Cut copies the icon and its properties to the Clipboard, and then erases the icon.

Copy copies the icon and its properties, which can then be pasted to another palette.

Delete removes the icon.

Rename changes the icon's name.

Specify Image displays the Select Image File dialog box; choose a BMP, JPG, PNG, GIF, or TIF file *(new to AutoCAD 2008).*

Properties displays the Properties dialog box.

DIALOG BOXES

Transparency dialog box
Right-click the palette title bar, and then select Transparency. (Transparency is unavailable when AutoCAD uses hardware acceleration.)

Transparency Level changes the transparency of the palette from Less (none) to More.

☐ **Turn off window transparency** makes the palette opaque.

View Options dialog box
Right-click a tab, and then select View Options:

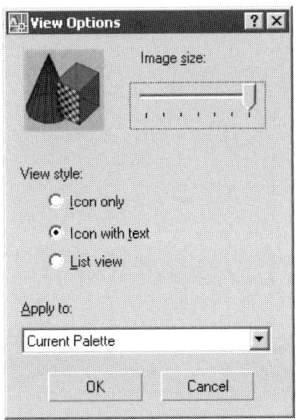

Image Size changes the size of icons from small (14 pixels square) to large (54 pixels).

View style toggles the display of icons and text:

○ **Icon only** displays icons only.

⊙ **Icon with text** displays icons with text below.

○ **List view** displays small icons with text beside.

Apply to determines if changes apply to the current palette or all palettes.

Properties dialog box
Right-click an icon, and then select Properties:

Image displays the icon associated with the tool. To change the icon, right-click it and then select **Specify Image**.

Name specifies the name that appears on the palette; it can be different from the Command String.

Description is the text that accompanies the icon on the palette; filling it in is optional.

The other items in this dialog box vary depending on the type of tool.

RELATED COMMANDS
TpNavigate displays the specified tool palette or palette group (*new to AutoCAD 2008*).

Customize exports and imports Tool palettes, and creates groups of palettes.

CUI customizes the content of tool palettes.

RELATED SYSTEM VARIABLES
PaletteOpaque determines whether the Tools palette can be transparent.

TpState notes whether the Tools palette is open.

RELATED FILES

.xtp stores the content of Tools palette in an XML-like format.

TIPS

- You can drag objects (while the right mouse button is held down) or toolbar buttons (while the Customize dialog box us open) onto palettes. Program code (macros, scripts, and so on) can also be stored on palettes by editing existing tools.

- To bring content from the DesignCenter into the Tool palette:

    1. In DesignCenter, right-click an item.

    2. From the shortcut menu, select Create Tool Palette.

- The *.xtp* files store content in XML (extended markup language) format.

- The Customize command allows you to create groups of palettes.

Torus

Rel.11 Draws 3D tori as solid models.

Command	Alias	Ctrl+	F-key	Alt+	Menu Bar	Dashboard
torus	tor	...	...	DMT	Draw	3D Make
					⤷ Modeling	
					⤷ Torus	

Command: torus
Specify center point or [3P/2P/Ttr]: *(Pick point 1.)*
Specify radius or [Diameter] <1.0>: *(Specify the radius, 2, or type D.)*
Specify tube radius or [2Point/Diameter]: *(Specify the radius, 3, or type D.)*

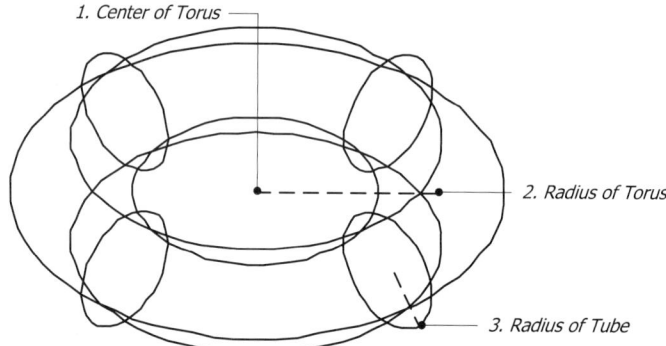

1. Center of Torus
2. Radius of Torus
3. Radius of Tube

COMMAND LINE OPTIONS

Center point indicates the center of the torus.

3P specifies three points on the outer circumference of the torus.

2P specifies two points on the outer diameter of the torus.

Ttr specifies two tangent points plus the radius of the outer circumference of the torus.

Specify radius indicates the radius of the torus.

Diameter indicates the diameter of the torus.

Tube radius indicates the radius of the tube.

2Point asks for two points to determine the diameter of the tube.

RELATED COMMANDS

Ai_Torus creates a torus from 3D polyfaces.

Cone draws solid cones.

Cylinder draws solid cylinders.

Sphere draws solid spheres.

RELATED SYSTEM VARIABLES

Isolines specifies the number of isolines to draw on the torus.

DragVs specifies the visual style during tori construction.

· ·

TIPS

- When the torus radius is negative, the tube radius must be a larger positive number.
- A negative torus radius creates a football shape, as illustrated below:

Football: *Hole-less torus:*

Specify a tube diameter larger than the torus diameter to create a hole-less torus.

- After it is constructed, the torus can be edited by grips.

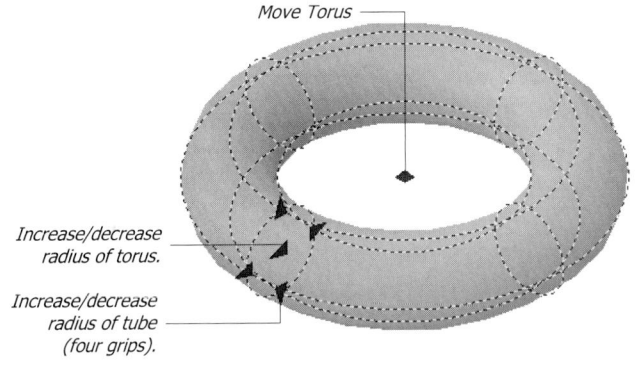

Move Torus

Increase/decrease radius of torus.

Increase/decrease radius of tube (four grips).

TpNavigate

<u>2008</u> Displays the specified tool palette or palette group (short for "Tool Palettes NAVIGATE").

Command	Alias	Ctrl+	F-key	Alt+	Menu Bar	Dashboard
tpnavigate	...	...	...	...	...	...

Command: tpnavigate
Specify tool palette to display or [palette Group]: *(Enter the name of a tool palette name, or type G.)*

COMMAND LINE OPTIONS

Specify Tool Palette to Display displays the specified tool palette.

Palette Group prompts, 'Specify palette group to display:'; enter the name of a valid tool palette group.

RELATED COMMAND

ToolPalettes opens the Tools palette.

RELATED SYSTEM VARIABLES

_AuthorPalettepath specifies the path to folder holding customized Tool palettes (undocumented).

_ToolPalettepath specifies the path to folder holding Tool palettes (undocumented).

TpState reports whether the Tools palette is open.

TIPS

- You can get the names of groups and palettes from the Customize command.

- This command is meant for use in macros and scripts.

- When you enter an unrecognized name, AutoCAD opens the Tools palette anyhow, displaying the last-accessed group.

Trace

V. 1.0   Draws line segments with width.

Command	Alias	Ctrl+	F-key	Alt+	Menu Bar	Dashboard
trace	...	...	...	...	...	...

Command: trace
Specify trace width <0.050>: *(Enter a value.)*
Specify start point: *(Pick a point.)*
Specify next point: *(Pick another point.)*
Specify next point: *(Press Enter to end the command.)*

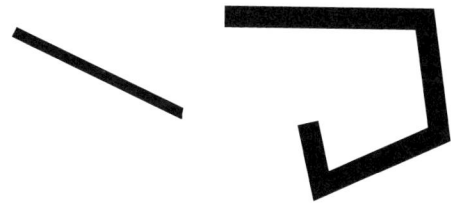

COMMAND LINE OPTIONS

Trace width specifies the width of the trace.

Start point picks the starting point.

Next point picks the next vertex.

ENTER exits the **Trace** command.

RELATED COMMANDS

Line draws lines with zero width.

MLine draws up to 16 parallel lines.

PLine draws polylines and polyline arcs with varying widths.

LWeight gives every object a width.

RELATED SYSTEM VARIABLES

FillMode toggles display of fill or outline traces (default = 1, on).

TraceWid specifies the current width of the trace (default = 0.05).

TIPS

- This command has largely been replaced by the PLine command; Trace was meant for drawing traces in printed circuit board designs.

- Traces are drawn along the centerline of the pick points. During drawing, the display of a trace segment is delayed by one pick point.

- During the drawing of traces, you cannot back up, because an Undo option is missing; if you require this feature, draw wide lines with the PLine command, setting the Width option.

- There is no option for controlling joints (always bevelled) or endcapping (always square); if you wish to modify these features, draw wide lines with the MLine command, after setting the solid fill, endcap, and joint options with the MlStyle command.

Tracking

Rel.14 Locates x and y points visually, relative to other points in the command sequence; *not* a command, but a option modifier.

Modifer	Alias	Ctrl+	F-key	Alt+	Menu Bar	Dashboard
tracking	tk	...	...	...	...	...
	track					

Example usage:
Command: line
Specify first point: *(Pick a point.)*
Specify next point or [Undo]: tk
Enters tracking mode:
First tracking point: *(Pick a point.)*
Next point (Press ENTER to end tracking): *(Pick a point.)*
Next point (Press ENTER to end tracking): *(Press Enter to end tracking.)*
Exits tracking mode:
Specify next point or [Undo]: *(Pick a point.)*

COMMAND LINE OPTIONS
First tracking point picks the first tracking point.

Next point picks the next tracking point.

ENTER exits tracking mode.

RELATED KEYBOARD MODIFIERS
Direct distance entry specifies an angle and relative distance to the next point.

from locates an offset point.

m2p finds a point midway between two picked points.

TIPS
- Tracking is not a command, but an option modifier. It is entered during commands without the ' transparent-command prefix.

- Tracking can be used in conjunction with direct distance entry.

- In tracking mode, AutoCAD automatically turns on ortho mode to constrain the cursor vertically and horizontally.

- If you start tracking in the x direction, the next tracking direction is y, and vice versa.

- You can use tracking as many times as you need to at the 'Specify first point' and 'Specify next point' prompts.

Transparency

Rel.14   Toggles the transparency of pixels in raster images.

Command	Alias	Ctrl+	F-key	Alt+	Menu Bar	Dashboard
transparency	...	...	...	MO⬢IT	Modify	...
					⬢Object	
					⬢Image	
					⬢Transparency	

Command: transparency
Select image(s): *(Select one or more images inserted with the Image command.)*
Select image(s): *(Press Enter to end object selection.)*
Enter transparency mode [ON/OFF] <OFF>: *(Type ON or OFF.)*

COMMAND LINE OPTIONS

Select image(s) selects the objects whose transparency to change.

ON makes the background pixels transparent.

OFF makes the background pixels opaque.

RELATED COMMANDS

ImageAttach attaches a raster image as an externally-referenced file.

ImageAdjust changes the brightness, contrast, and fading of a raster image.

TIPS

- This command works only with raster images placed in drawings with the Image command.

- This command is meant for use with raster file formats that allow transparent pixels.

- Transparent pixels allow graphics under the image to show through.

- For this command to work correctly, you must first specify the color of pixel to be transparent. Use a raster editor, such as PaintShop Pro: when saving the image, designate the color as transparent.

TraySettings

2004 Specifies settings for the icons that appear at the right end of the status bar (known as the "tray").

Command	Alias	Ctrl+	F-key	Alt+	Menu Bar	Dashboard
traysettings	...	...	...	...	...	...

Command: traysettings

Displays dialog box:

TRAY SETTINGS DIALOG BOX

Display icons from services:

☑ Displays "tray" at the right end of the status line; see below.

☐ Turns off the tray.

Display notification from services:

☑ Displays balloon notifications from a variety of services; see below.

☐ Turns off notifications:

　○ **Display time** specifies the duration that a notification balloon is displayed.

　⊙ **Display until closed** specifies that the notification balloon is displayed until closed by user.

RELATED SYSTEM VARIABLES

TrayIcons toggles the display of the tray on the status bar.

TrayNotify toggles whether notification balloons are displayed.

TrayTimeOut determines the length of time that notification balloons are displayed.

TIP

- Examples of notification balloons are illustrated below. Click <u>underlined</u> links for more information, or click **x** to dismiss.

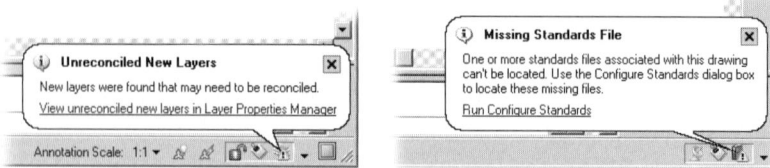

'TreeStat

<u>Rel.12</u> Displays the status of the drawing's spatial index, including the number and depth of nodes (short for TREE STATistics).

Command	Alias	Ctrl+	F-key	Alt+	Menu Bar	Dashboard
treestat	...	...	...	...	...	...

Command: treestat

Sample output:

```
Deleted objects: 12

Model-space branch
-------------------
Oct-tree, depth limit = 30
Subtree containing objects with defined extents:
    Nodes: 21    Objects: 8   Maximum depth: 23
    Average objects per node: 0.38
    Average node depth: 14.24   Average object depth: 17.00
    Objects at depth 14: 2    17: 5    23: 1
    Nodes with population 0: 18    1: 1    2: 1    5: 1
Total nodes: 24    Total objects: 8

Paper-space branch
-------------------
Quad-tree, depth limit = 20
Subtree containing objects with defined extents:
    Nodes: 1    Objects: 0
    Average objects per node: 0.00
    Average node depth: 5.00
    Nodes with population 0: 1
Total nodes: 4    Total objects: 0
```

RELATED SYSTEM VARIABLES

TreeDepth specifies the size of the tree-structured spatial index in *xxyy* format:

> **xx** — number of model space nodes (default = 30).
>
> **yy** — number of paper space nodes (default = 20).
>
> **-xx** — 2D drawing.
>
> **+xx** — 3D drawing (default).

TreeMax is the maximum number of nodes (default = 10,000,000).

TIPS

- Better performance occurs with fewer objects per oct-tree node. When redraws and object selection seem slow, increase the value of TreeDepth.

- Each node consumes 80 bytes of memory.

DEFINITIONS

Oct Tree — the model space branch of the spatial index, where all objects are either 2D or 3D. *Oct* comes from the eight volumes in the x,y,z coordinate system of 3D space.

Quad Tree — the paper space branch of the spatial index, where all objects are two-dimensional. *Quad* comes from the four areas in the x,y coordinate system of 2D space.

Spatial Index — objects indexed by oct-region to record their position in 3D space; it has a tree structure with two primary branches: oct tree and quad tree. Objects are attached to *nodes*; each node is a branch of the *tree*. It's not clear that anyone understands this.

Trim

<u>**V. 2.5**</u> Trims lines, arcs, circles, 2D polylines, and hatches to existing and projected cutting lines and views.

Command	Alias	Ctrl+	F-key	Alt+	Menu Bar	Dashboard
trim	tr	...	...	MT	Modify ↳Trim	2D Draw

Command: trim
Current settings: Projection=UCS Edge=None
Select cutting edges ...
Select objects or <select all>: *(Select one or more objects.)*
Select objects: *(Press Enter to end object selection.)*
Select object to trim or shift-select to extend or [Fence/Crossing/Project/Edge/eRase/Undo] *(Select object, or enter an option.)*
Select object to trim or shift-select to extend or [Fence/Crossing/Project/Edge/eRase/Undo] *(Press Enter to end command.)*

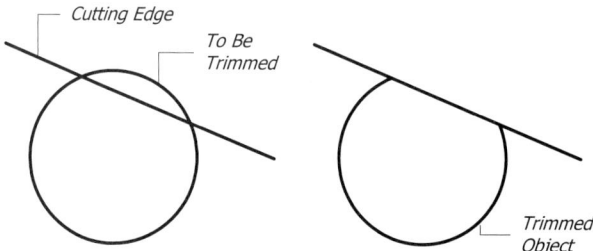

COMMAND LINE OPTIONS

Select objects selects the cutting edges.

Select all selects all objects in the drawing .

Select object to trim trims the object; picked portion is trimmed.

Undo untrims the last trim action.

Fence options

Specify first fence point locates the first point for fence selection mode.

Specify next fence point locates subsequent points; press Enter to exit selection mode.

Crossing options

Specify first corner locates the first corner for crossing selection mode.

Specify opposite corner locates the second corner.

Project options
Enter a projection option [None/Ucs/View] <Ucs>: *(Enter an option.)*

None uses only objects as cutting edges.

Ucs trims at the x,y plane of the current UCS.

View trims at the current view plane.

Edge options

Enter an implied edge extension mode [Extend/No extend] <No extend>:
(Type E or N.)

Extend extends the cutting edge to trim object.

No extend trims only at an actual cutting edge.

Erase option

Select objects to erase or <exit> selects objects to erase; press ENTER to exit object selection mode.

RELATED COMMANDS

Change changes the size of lines, arcs and circles.

Extend lengthens lines, arcs and polylines.

Lengthen lengthens open objects.

RELATED SYSTEM VARIABLES

EdgeMode determines whether this command projects cutting edges.

ProjMode determines how cutting edges are projected in 3D space.

TIPS

- This command also trims hatches.

- You can select all objects in the drawing by pressing Enter at the 'Select objects' prompt.

- To trim intersecting lines, it can be easier to use the Fillet command with Radius set to 0.

- To have this command extend objects (rather than trim them), hold down the Shift key while selecting objects to extend; the opposite occurs in the Extend command.

- Cutting edges can be trimmed, too.

U

V. 2.5 Undoes the most recent AutoCAD command (short for Undo).

Command	Alias	Ctrl+	F-key	Alt+	Menu Bar	Dashboard
u	...	Z	...	EU	Edit	...
					↳Undo	

Command: u

COMMAND LINE OPTIONS
None.

RELATED COMMANDS
Oops unerases the most-recently erased object.

Quit exits the drawing, undoing all changes.

Redo reverses the most-recent undo, if the prior command was U or Undo.

Undo allows more sophisticated control over undo than U.

RELATED SYSTEM VARIABLE
UndoCtl toggles the undo mechanism.

TIPS
- The U command is convenient for stepping back through the design process, undoing one command at a time.

- As of AutoCAD 2006, multiple zooms and pans are grouped into a single undo. This feature can be toggled through the Combine Zooms and Pans option found on the User Preferences tab of the Options dialog box.

- The U command is the same as the Undo 1 command; for greater control over the undo process, use the Undo command.

- The Redo command redoes the most-recent undo only; use MRedo otherwise.

- The Quit command, followed by the Open command, restores the drawing to its original state, if not already saved.

- Because the undo mechanism creates a mirror drawing file on disk, disable the Undo command with system variable UndoCtl (set to 0) when your computer is low on disk space.

- Commands that involve writing to file, plotting, and some display functions (such as Render) are not undone.

Ucs

Rel.10 Defines new coordinate planes, and restores existing UCSs (short for User-defined Coordinate System).

Command	Alias	Ctrl+	F-key	Alt+	Menu Bar	Dashboard
ucs	dducs	...	...	TW	Tools	...
					⤷New UCS	
				TH	Tools	
					⤷Orthographic UCS	

Command: ucs

For compatibility with earlier versions of AutoCAD, you may enter any of the older options at the first prompt.

Current ucs name: *WORLD*

Specify origin of UCS or [Face/NAmed/OBject/Previous/View/World/X/Y/Z/ZAxis] <World>: *(Enter an option.)*

COMMAND LINE OPTIONS

Specify origin of new UCS moves the UCS to a new origin.

Face aligns the UCS with the face of a 3D solid object.

NAmed saves, restores, and deletes named UCSs.

OBject aligns the UCS with a selected object; see tips.

Previous restores the previous UCS orientation.

World aligns the UCS with the WCS.

View aligns the UCS with the current view.

X rotates the UCS about the x axis.

Y rotates the UCS about the y axis.

Z rotates the UCS about the z axis.

ZAxis aligns the UCS with a new origin and z axis.

Face options

Select face of solid object: *(Pick a face.)*

Enter an option [Next/Xflip/Yflip] <accept>: *(Press Enter, or enter an option.)*

Select face selects a face to which the UCS will be applied.

Next selects the next face.

Xflip flips the x axis.

Yflip flips the y axis.

accept accepts the selected face (press Enter to accept.)

NAme options

Enter an option [Restore/Save/Delete/?]: *(Enter an option.)*

Restore restores a named UCS.

Save saves the current UCS by name.

Del deletes a saved UCS.

· ·

? lists the names of saved UCS orientations.

Hidden options

The following options also work with this command:

New defines a new user-defined coordinate system.

Move moves the UCS along the z axis, or specifies a new x,y-origin.

orthoGraphic selects a standard orthographic UCS: top, bottom, front, back, left, and right.

Apply applies the UCS setting to a selected viewport, or all active viewports.

3point aligns the UCS with a point on the positive x-axis and positive x,y plane.

STATUS BAR OPTION

Click DUCS to turn on dynamic UCS mode, where the UCS matches the orientation of the selected face:

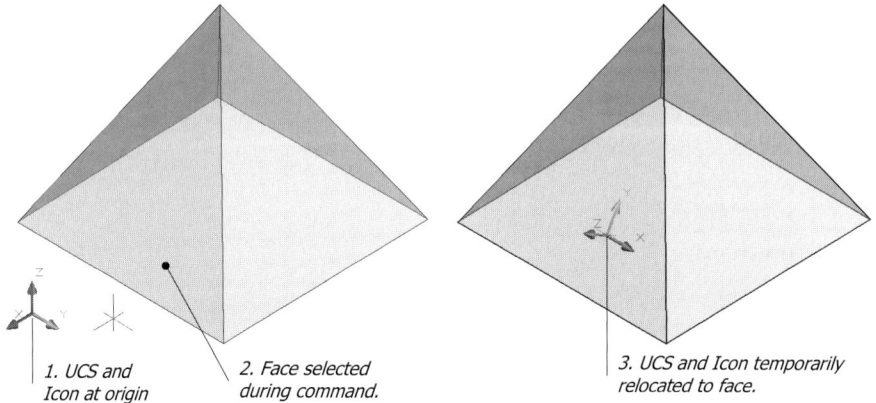

| 1. UCS and Icon at origin | 2. Face selected during command. | 3. UCS and Icon temporarily relocated to face. |

RELATED COMMANDS

UcsMan modifies the UCS via a dialog box.

UcsIcon controls the visibility of the UCS icon.

Plan changes to the plan view of the current UCS.

RELATED SYSTEM VARIABLES

GridDisplay forces the grid to follow the orientation of the dynamic UCS, when set to 8.

UcsAxisAng specifies the default rotation angle when the UCS is rotated around an axis using the **X**, **Y**, or **Z** options of this command.

UcsBase specifies the UCS that defines the origin and orientation of orthographic UCS settings.

UcsDetect toggles dynamic UCS.

UcsFollow aligns the UCS with the view of a newly-activated viewport automatically.

UcsIcon determines the visibility and location of the UCS icon.

UcsOrg specifies the WCS coordinates of UCS icon (default = 0,0,0).

UcsOrtho specifies whether the related UCS is automatically displayed when an orthographic view is restored.

UcsView specifies whether the current UCS is saved when a view is created with the View command.

UcsVp specifies that the UCS reflects the UCS of the currently-active viewport.

UcsXdir specifies the X direction of the current UCS (default = 1,0,0).

UcsYdir specifies the Y direction of the current UCS (default = 0,1,0).

WorldUcs correlates the WCS to the UCS:

0 — Current UCS is WCS.

1 — UCS is the same as WCS (default).

TIPS

- Use the UCS command to draw objects at odd angles in 3D space.

- Although you can create UCSs in paper space, you cannot use 3D viewing commands in paper space.

- Each viewport can have its own UCS.

- UCSs can be aligned with these objects:

Object	UCS Origin	X Axis Passes Through...
Arc	Center	Endpoint closest to the pick point.
Attributes	Insertion point	Extrusion direction.
Blocks	Insertion point	Extrusion direction.
Circle	Center	Pick point.
Dimension	Midpoint of dimtext	Parallel to the X axis of the dimension.
Line	Endpoint nearest to pick	Parallel to the X axis of the line.
Point	Point	Anywhere.
Polyline	Start point	Start point to next vertex.
2D solid	First point	First two points.
Trace	First point	Centerline
3D Face	First point	First two points.
Shape	Insertion point	Extrusion direction.
Text	Insertion point	Extrusion direction.

- UCSs do *not* align with these objects: mlines, rays, xlines, 3D polylines, splines, ellipses, leaders, viewports, 3D solids, 3D meshes, and regions.

- When the GridDisplay system variable is set to 8, the grid matches the orientation of the dynamic UCS.

DEFINITIONS

UCS — user-defined 2D coordinate system oriented in 3D space; sets a working plane, orients 2D objects, defines the extrusion direction, and the axis of rotation.

WCS — world coordinate system; the default 3D x,y,z coordinate system.

UcsIcon

Rel.10 Controls the location and display of the UCS icon.

Command	Alias	Ctrl+	F-key	Alt+	Menu Bar	Dashboard
ucsicon	...	...	...	VLU	View	...
					⤷Display	
					⤷UCS Icon	

Command: ucsicon
Enter an option [ON/OFF/All/Noorigin/ORigin/Properties] <ON>: *(Enter an option.)*

COMMAND LINE OPTIONS

All applies changes of this command to all viewports.

Noorigin displays the UCS icon in the lower-left corner at all times.

OFF turns off the display of the UCS icon.

ON turns on the display of the UCS icon.

ORigin displays the UCS icon at the current UCS origin, or at lower left corner of the viewport when the origin is off-screen.

Properties displays the UCS Icon dialog box.

UCS ICON DIALOG BOX

UCS Icon Style options

  ○**2D** displays flat UCS icon.

  ◉**3D** displays tripod UCS icon.

Cone option (available only with 3D style):

  ☑ Displays arrowheads as 3D cones.

  ☐ Displays plain arrowheads.

Line width changes the line width from 1 to 2 or 3 pixels.

UCS Icon Size options
Slider bar changes the icon size from 5 to 95 pixels.

UCS Icon Color options
Model space icon color selects the icon's color in model space.

Layout tab icon color selects the icon's color in layout (paper space).

RELATED SYSTEM VARIABLE

UcsIcon determines the visibility and location of UCS icon:

    0 — UCS icon not displayed.

    1 — UCS icon displayed in lower-left corner.

    2 — UCS icon displayed at the UCS origin, when possible.

    3 — UCS icon displayed at UCS always (default).

RELATED COMMAND

UCS creates and controls user-defined coordinate systems.

TIPS

- The UCS icon varies, depending on the current viewpoint relative to the active UCS..

- When AutoCAD switches from 2D wireframe mode to one of the VsCurrent command's 3D options, the UCS icon changes to a rendered 3D icon:

- There is generally no need for the UCS icon in 2D drafting, and it can be safely turned off.

- The Options command's 3D Modeling tab has options for toggling the display of the UCS icon:
  - Display in 2D model space.
  - Display in 3D parallel projection.
  - Display in 3D perspective projection.

UcsMan

2000 Displays the UCS dialog box (short for "User-defined Coordinate System MANager").

Commands	Aliases Ctrl+	F-key	Alt+	Menu Bar	Dashboard
ucsman	dducs ...	...	TU	Tools	...
	uc			⌐Named UCS	
+ucsman					

Command: ucsman

Displays dialog box.

UCS DIALOG BOX

Named UCSs tab

Named UCSs lists the names of the AutoCAD-generated and user-defined coordinate systems of the current viewport in the active drawing; the arrowhead points to the current UCS.

Set Current restores the selected UCS.

Details displays the UCS Details dialog box:

Orthographic UCSs tab

Name lists the six standard orthographic UCS views: top, bottom, front, back, left, and right.

Depth specifies the height of the UCS above the x,y plane.

Relative to specifies the orientation of the selected UCS relative to the WCS or to a customized UCS.

Set Current activates the selected UCS.

Details displays the UCS Details dialog box.

Settings tab

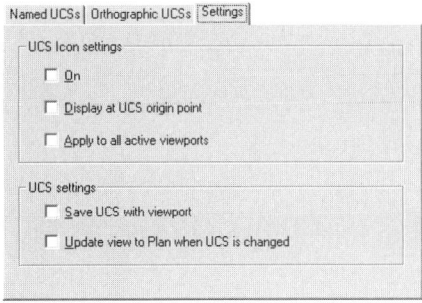

UCS icon settings options

☑ **On** displays the UCS icon in the current viewport; each viewport can display the UCS icon independently.

Display at UCS origin point:

☑ Displays the UCS icon at the origin of the current UCS.

☐ Displays the UCS icon at the lower-left corner of the viewport.

Apply to all active viewports:

☑ Applies these UCS icon settings to all active viewports in the current drawing.

☐ Applies to the current viewport only.

UCS settings options

Save UCS with viewport:

☑ Saves the UCS setting with the viewport.

☐ Determines UCS settings from current viewport.

☑ **Update view to Plan when UCS is changed** restores plan view when the UCS changes.

SHORTCUT MENUS

Right-click the list in the Named UCSs tab:

Set Current sets the selected UCS as active.

Rename renames the selected UCS; you cannot rename the World UCS.

Delete erases the selected UCS; you cannot delete the World UCS.

Details displays the UCS Details dialog box.

Right-click the list in the Orthographic UCS tab:

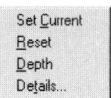

Set Current sets the selected UCS as active.

Reset restores the origin of the selected UCS.

Depth moves the UCS in the z direction; displays dialog box:

Details displays the UCS Details dialog box.

- -

+UCSMAN Command

Command: +ucsman

Tab index <0>: *(Type 1, 2, or 3.)*

COMMAND LINE OPTION

Tab index displays the tab related to the tab number:

- **0** — Named UCS tab.
- **1** — Orthographic UCS tab.
- **2** — Settings tab.

RELATED SYSTEM VARIABLES

See the UCS command.

RELATED COMMANDS

UCS displays the UCS options at the command line.

UcsIcon changes the display of the UCS icon.

Plan displays the plan view of the WCS or a UCS.

TIPS

- Functions of this command were formerly carried out by the DdUcs and DdUcsP commands.

- Unnamed is the first entry, when the current UCS is unnamed. World is the default for new drawings; it cannot be renamed or deleted. Previous is the previous UCS; you can move back through several previous UCSs.

- -

Undefine

<u>Rel. 9</u>   Makes AutoCAD commands unavailable.

Command	Alias	Ctrl+	F-key	Alt+	Menu Bar	Dashboard
undefine	...	...	...	...	...	...

Command: undefine
Enter command name: *(Enter name.)*

Example usage:
Command: undefine
Enter command name: line
Command: line
Unknown command. Type ? for list of commands.
Command: .line
From point:

COMMAND OPTIONS

Enter command name specifies the name of the command to make unavailable.

. *(period)* is the prefix for undefined commands to redefine them temporarily.

RELATED COMMAND

Redefine redefines an AutoCAD command.

TIPS

- Prefixing undefined command with a period bypasses the effect of this command. For example:

    Command: .line

- This command allows AutoLISP and ObjectARX to override native AutoCAD commands.

- Commands created by programs cannot be undefined, including the following programming interfaces:

    AutoLISP and Visual LISP.
    ObjectARx.
    Visual Basic for Applications.
    External commands.
    Aliases.

- In menu macros written with international language versions of AutoCAD, precede command names with an underscore character (_) to translate the command name into English automatically.

Undo

V. 2.5   Undoes the effect of the previous command(s).

Command	Alias	Ctrl+	F-key	Alt+	Menu Bar	Dashboard
undo	...	...	...	...	...	...

Command: undo
Enter the number of operations to undo or [Auto/Control/BEgin/End/Mark/ Back] <1>: *(Enter a number, or an option.)*

COMMAND LINE OPTIONS

Auto treats a menu macro as a single command.

Control limits the options of the **Undo** command.

BEgin groups a sequence of operations (formerly the Group option).

End ends the group option.

Mark sets a marker.

Back undoes back to the marker.

number indicates the number of commands to undo.

Control Options
Enter an UNDO control option [All/None/One] <All>: *(Enter an option.)*

All turns on full undo.

None turns off the undo feature.

One limits the Undo command to a single undo.

RELATED COMMANDS

Oops unerases the most-recently erased object.

Quit leaves the drawing without saving changes.

Redo undoes the most recent undo.

MRedo undoes multiple undoes.

U undoes a single step.

RELATED SYSTEM VARIABLES

UndoCtl determines the state of undo control:
  0 — Undo disabled.
  1 — Undo enabled.
  2 — Undo limited to one command.
  4 — Auto-group mode.
  8 — Group currently active.

UndoMarks specifies the number of undo marks placed in the **Undo** control stream.

TIPS

■ Since the undo mechanism creates a mirror drawing file on disk, disable the Undo command with system variable UndoCtl (set it to 0) when your computer is low on disk space.

■ There are some commands that cannot be undone, such as Save, Plot, and UndoCtl.

Union

2000 Joins two or more solids and regions into a single body.

Command	Alias	Ctrl+	F-key	Alt+	Menu Bar	Dashboard
union	uni	...	...	MNU	Modify ⤷Solids Editing ⤷Union	3D Make

Command: union
Select objects: *(Select one or more solid objects.)*
Select objects: *(Select one or more solid objects.)*
Select objects: *(Press Enter to end command.)*

A box and a cylinder unioned into a single object.

COMMAND LINE OPTION

Select objects selects the objects to join into a single object; you must select at least two solid or region objects.

RELATED COMMANDS

Intersect creates a solid model from the intersection of two objects.

Subtract creates a solid model by subtracting one object from another.

RELATED SYSTEM VARIABLES

ShowHist toggles display of history in solids.

SolidHist toggles the retention of history in solids.

TIPS

- You must select at least two solid or coplanar region objects. The two objects need not overlap for this command to operate.

- History, if enabled, allows you to work with the original objects.

'Units

<u>V. 1.4</u>   Controls the display and format of coordinates and angles, as well as the orientation of angles.

Commands	Aliases	Ctrl+	F-key	Alt+	Menu Bar	Dashboard
units	un	...	...	OU	Format	...
	ddunits				⤷ Units	
-units	-un					

command: units

Displays dialog box:

DRAWING UNITS DIALOG BOX

Length options

Type sets the format for units of linear measure displayed by AutoCAD: Architectural, Decimal, Engineering, Fractional, or Scientific.

Precision specifies the number of decimal places or fractional accuracy.

Angle options

Type sets the current angle format.

Precision sets the precision for the current angle format.

☐ **Clockwise** calculates positive angles in the clockwise direction.

Drawing units for DesignCenter blocks specifies the units when blocks are inserted from the DesignCenter.

Direction displays the Direction Control dialog box.

DIRECTION CONTROL DIALOG BOX

Base Angle options

○ **East** sets the base angle to 0 degrees (default).

○ **North** sets the base angle to 90 degrees.

○ **West** sets the base angle to 180 degrees.

○ **South** sets the base angle to 270 degrees.

⊙ **Other** turns on the Angle option.

Angle sets the base angle to any direction.

Pick an angle dismisses the dialog box temporarily, and allows you to define the base angle by picking two points in the drawing; AutoCAD prompts you 'Pick angle' and 'Specify second point:'.

· ·

-UNITS Command

Command: -units

Report formats:	(Examples)
1. Scientific	1.55E+01
2. Decimal	15.50
3. Engineering	1'-3.50"
4. Architectural	1'-3 1/2"
5. Fractional	15 1/2

With the exception of Engineering and Architectural formats, these formats can be used with any basic unit of measurement. For example, Decimal mode is perfect for metric units as well as decimal English units.

Enter choice, 1 to 5 <2>: *(Enter a value.)*

Enter number of digits to right of decimal point (0 to 8) <4>: *(Enter a value.)*

Systems of angle measure:	(Examples)
1. Decimal degrees	45.0000
2. Degrees/minutes/seconds	45d0'0"
3. Grads	50.0000g
4. Radians	0.7854r
5. Surveyor's units	N 45d0'0" E

Enter choice, 1 to 5 <1>: *(Enter a value.)*

· ·

Enter number of fractional places for display of angles (0 to 8) <0>: *(Enter a value.)*

Direction for angle 0:
  East 3 o'clock = 0
  North 12 o'clock = 90
  West 9 o'clock = 180
  South 6 o'clock = 270
Enter direction for angle 0 <0>: *(Enter a value.)*

Measure angles clockwise? [Yes/No] <N> *(Type Y or N.)*

COMMAND LINE OPTIONS

Report formats selects scientific, decimal, engineering, architectural, or fractional format for length display.

Number of digits to right of decimal point specifies the number of decimal places between 0 and 8.

Systems of angle measure selects decimal degrees, degrees/minutes/seconds, grads, radians, or surveyor's units for angle display.

Denominator of smallest fraction to display specifies the denominator of fraction displays, such as 1/2 or 1/256.

Number of fractional places for display of angles specifies the number of decimal places between 0 and 8.

Direction for angle 0 selects the direction for 0 degrees from east, north, west, or south.

Do you want angles measured clockwise?

Yes measures angles clockwise.

No measures angles counterclockwise.

RELATED SYSTEM VARIABLES

AngBase specifies the direction of zero degrees.

AngDir specifies the direction of angle measurement.

AUnits specifies the units of angles.

AuPrec specifies the displayed precision of angles.

InsUnits specifies the drawing units for blocks dragged from the DesignCenter:

0 — Unitless
1 — Inches
2 — Feet
3 — Miles
4 — Millimeters
5 — Centimeters
6 — Meters
7 — Kilometers
8 — Microinches
9 — Mils
10 — Yards; 3 feet

11 — Angstroms; 0.1 nanometers
12 — Nanometers; 10E-9 meters
13 — Microns; 10E-6 meters
14 — Decimeters; 0.1 meter
15 — Decameters; 10 meters
16 — Hectometers; 100 meters
17 — Gigameters; 10E9 meters
18 — Astronomical Units; 149.597E8 kilometers
19 — Light Years; 9.4605E9 kilometers
20 — Parsecs; 3.26 light years

InsUnitsDefSource specifies source units to be used.

InsUnitsDefTarget specifies target units to be used.

LUnits specifies the units of measurement.

LuPrec specifies the displayed precision of coordinates.

UnitMode toggles the type of display units.

RELATED COMMAND

New sets up drawings with Imperial or metric units.

TIPS

- Because 'Units is a transparent command, you can change units during another command.

- The 'Direction Angle' prompt lets AutoCAD start the angle measurement from any direction.

- AutoCAD accepts the following notations for angle input:

Notation	Meaning
<	Specify an angle based on current units setting.
<<	Bypass angle translation set by Units command to use 0-angle-is-east direction and decimal degrees.
<<<	Bypass angle translation; use angle units set by the Units command and 0-angle-is-east direction.

- The system variable UnitMode forces AutoCAD to display units in the same manner that you enter them.

- At one time, the smallest fraction was 1/2048, but now the smallest is 1/256.

- Do not use a suffix — such as 'r' or 'g' — for angles entered as radians or grads; instead, use the Units command to set angle measurement to radians and grads.

- The Drawing units for DesignCenter blocks option is for inserting blocks from the DesignCenter, and especially when the block was created in other units.

- To prevent blocks from being scaled when dragged from the DesignCenter palette, select Unitless.

UpdateField

<u>**2005**</u>   Forces the update of selected fields.

Command	Alias	Ctrl+	F-key	Alt+	Menu Bar	Dashboard
updatefield	...	...	...	TT	Tools	...
					↳ Update Fields	

Command: updatefield
Select objects: *(Select one or more fields.)*
Select objects: *(Press Enter to end field selection.)*
n **field(s) found**
n **field(s) updated**

COMMAND LINE OPTION

Select objects selects one or more fields; press Ctrl+A to select all objects in the drawing, including field text.

RELATED COMMANDS

Field places field text, which is automatically updated as its value changes.

Find finds fields.

DdEdit edits the properties of field text.

RELATED SYSTEM VARIABLE

FieldEval determines when fields are updated (default = 31, all on):

  0 — Fields are not updated (static).
  1 — Fields are updated when drawing is opened.
  2 — Fields are updated when drawing is saved.
  4 — Fields are updated when plotted.
  8 — Fields are updated with the eTransmit command.
  16 — Fields are updated when the drawing is regenerated.

TIP

- This command forces individual fields to update. Depending on the setting of the FieldEval system variable, all fields are updated automatically.

UpdateThumbsNow

2005   Forces the update of preview images in the Sheet Set Manager palette.

Command	Alias	Ctrl+	F-key	Alt+	Menu Bar	Dashboard
updatethumbsnow	...	...	...	...	...	...

Command: updatethumbsnow

COMMAND LINE OPTIONS
None.

RELATED COMMAND
SheetSet controls sheet sets.

RELATED SYSTEM VARIABLE
UpdateThumbnail determines which thumbnails are updated (default = 15):
   0 — Thumbnail previews not updated.
   1 — Model view thumbnails updated.
   2 — Sheet view thumbnails updated.
   4 — Sheet thumbnails updated.
   8 — Thumbnails updated when sheets and views created, modified, and restored.
   16 — Thumbnails updated when drawing saved.

TIPS
- This command controls when preview images are updated; required when the drawing changes.

- Thumbnails are displayed in the Sheet Set Manager, as illustrated below:

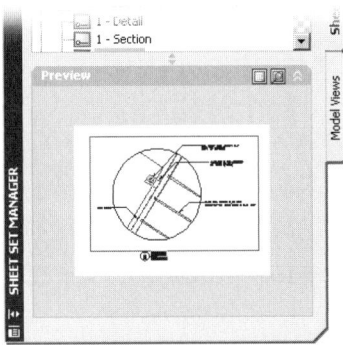

- The SaveAs command's dialog box has the Update sheet and view thumbnails now option.

VbaIde

<u>**2000**</u> Displays the Visual Basic window (short for Visual Basic for Applications Integrated Development Environment).

Command	Alias	Alt+	F-key	Alt+	Menu Bar	Dashboard
vbaide	...	F11	...	TAB	Tools	...
					↳ Macro	
					↳ Visual Basic Editor	

Command: vbaide

Displays window:

MENU BAR

Select Help | Microsoft Visual Basic Help for assistance in using this VBA IDE window.

RELATED COMMANDS

VbaLoad loads a VBA project; displays the Open VBA Project dialog box.

VbaMan displays the VBA Manager dialog box.

VbaRun displays the Macros dialog box.

VbaStmt executes a VBA expression at the command line.

VbaUnload unloads a VBA project.

RELATED SYSTEM VARIABLES

None.

TIPS

- *VBA* is short for "Visual Basic for Applications," a macro programming language common to a number of Windows applications; it is based on the Visual Basic programming language.

- AutoCAD contains sample VBA projects in the *autocad 2008\sample\vba* folder.

- For more information about Visual Basic for Applications, read the online *ActiveX and VBA Developer's Guide* help files included with AutoCAD.

- Because many viruses can be spread via VBA macros, I strongly recommend that your computer have real-time virus protection to prevent infection.

- Because loading a macro from other sources into AutoCAD can also expose your computer to malicious viruses, AutoCAD displays the following warning:

More Info displays AutoCAD's on-line help window.

Always ask before opening projects with macros:

☑ Displays this dialog box.

☐ Prevents this dialog box from being displayed; to turn back on, use the VbaRun command's Options dialog box; see VbaRun command.

Disable Macros loads a VBA project file, but disables macros; you can view, edit, and save the macros. To enable macros, close the project, and then open it again with Enable Macros.

Enable Macros loads the project file with macros enabled.

Do Not Load prevents the project file from being loaded.

VbaLoad, VbaUnload

2000 Loads a VBA project into AutoCAD (short for Visual Basic for Applications LOAD).

Commands	Alias	Ctrl+	F-key	Alt+	Menu Bar	Dashboard
vbaload	...	...	...	TAL	Tools	...
					⌐Macro	
					⌐Load Project	
vbaunload						
-vbaload						

Command: vbaload

Displays the Open VBA Project dialog box; select a .dvd file, and then click Open.
When the project contains macros, the AutoCAD dialog box is displayed; see the VbaIde command.

Command: vbaunload
Unload VBA Project: *(Enter project name.)*

COMMAND LINE OPTIONS

Unload VBA Project specifies the name of the VBA project to unload.

Enter unloads the active global project.

. . . .

-VBALOAD Command
Command: -vbaload
Open VBA project <*projectname*>: *(Enter project name.)*

COMMAND LINE OPTION

Open VBA project specifies the project path and filename.

RELATED COMMANDS

Vbaide displays the Visual Basic for Applications development environment window.

VbaMan displays the VBA Manager dialog box.

VbaRun displays the Macros dialog box.

VbaStmt executes a VBA expression at the command line.

VbaUnload unloads a VBA project.

TIPS

- You may load one or more VBA projects; there is no theoretical limit to the number.

- This command does not load embedded VBA projects; these projects are automatically loaded with the drawing.

- When the project contains macros, AutoCAD displays the AutoCAD dialog box to warn you about protection against macro viruses; see the Vbaide command.

- For more information about Visual Basic for Applications, read the *ActiveX and VBA Developer's Guide* included with AutoCAD.

. . . .

VbaMan

2000 Displays the VBA Manager dialog box (short for Visual Basic for Applications MANager).

Command	Alias	Ctrl+	F-key	Alt+	Menu Bar	Dashboard
vbaman	...	...	...	TAV	Tools	...
					↳Macro	
					↳VBA Manager	

Command: vbaman

Displays dialog box:

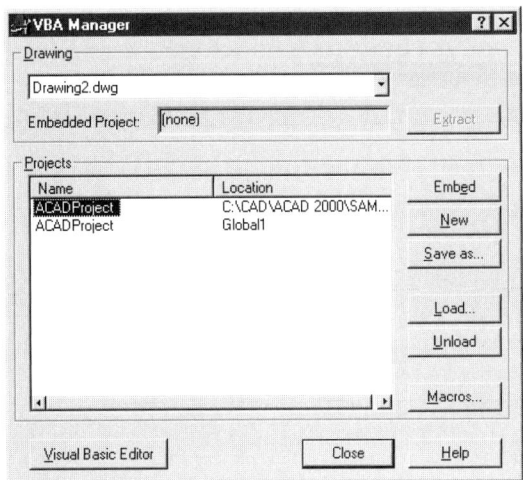

VBA MANAGER DIALOG BOX

Drawing

Drawing lists the names of drawings currently loaded in AutoCAD.

Embedded Project specifies the name of the embedded project.

Extract moves the embedded project from the drawing to a global project file.

Projects

Embed embeds the project in the drawing.

New creates a new project; default name = Global *n*.

Save as saves a global project.

Load displays the Open VBA Project dialog box; see the VbaLoad command.

Unload unloads the global project.

Macros displays the Macros dialog box; see the VbaRun command.

Visual Basic Editor displays the Visual Basic Editor; see the VbaIde command.

VbaRun

2000 Displays the Macros dialog box (short for Visual Basic for Applications RUN).

Command	Alias	Alt+	F-key	Alt+	Menu Bar	Dashboard
vbarun	...	F8	...	TAM	Tools	...
					↳ Macro	
					↳ **Macros**	

-vbarun

Command: vbarun

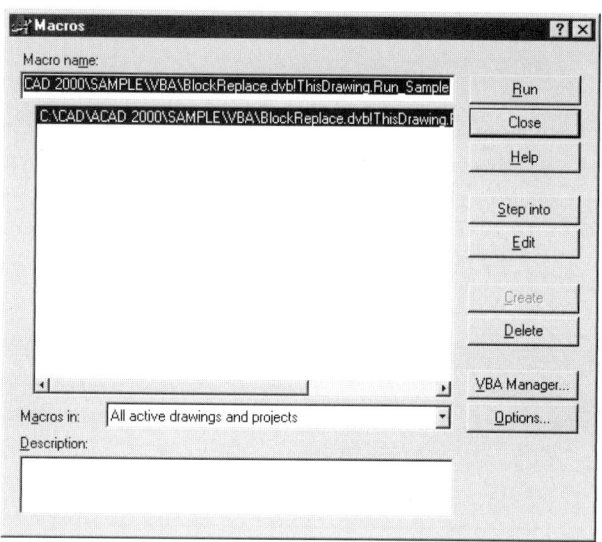

MACROS DIALOG BOX

Macro name specifies the name of the macro; enter a name or select one from the list.

Macros in specifies the projects and drawings containing macros in all active drawings and projects, all active drawings, all active projects, and any single drawing or project currently open in AutoCAD.

Description describes the macro; you may modify the description.

Run runs the macro.

Close closes the dialog box.

Help displays context-sensitive on-line help.

Step into displays the Visual Basic Editor, and executes the macro, pausing at the first executable line of code.

Edit displays the Visual Basic Editor with the macro; see the VbaIde command.

Create displays the Visual Basic Editor with an empty procedure.

Delete erases the selected macro.

VBA Manager displays the VBA Manager dialog box; see the VbaMan command.

Options displays the VBA Options dialog box.

VBA OPTIONS DIALOG BOX

☐ **Enable auto embedding** creates an embedded VBA project for all drawings when you open the drawing:

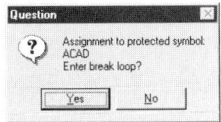

Allow Break on errors:

☑ Stops the macro, and displays the **Visual Basic Editor** with the code, showing the error in the macro.

☐ Displays an error message, and stops the macro.

☑ **Enable macro virus protection** enables virus protection, which displays a dialog box when VBA macros are loaded; see the VbaIde command.

. . . .

-VBARUN Command

Command: -vbarun
Macro name: *(Enter macro name.)*

COMMAND LINE OPTION

Macro name treats a menu macro as a single command.

RELATED COMMANDS

Vbaide displays the Visual Basic for Applications integrated development environment window.

VbaLoad loads a VBA project; displays the Open VBA Project dialog box.

VbaMan displays the VBA Manager dialog box.

VbaStmt executes a VBA expression at the command line.

VbaUnload unloads a VBA project.

TIPS

- A *macro* is an executable subroutine; each project can contain one or more macros.
- When the macro's name is not unique among loaded projects, include the module's project and names in this format: Project.Module.Macro
- When the macro is not yet loaded, include the *.dvb* file name using this format:

    Filenamedvb!Project.Module.Macro

VbaStmt

<u>**2000**</u>   Executes a single line VBA expression at the command prompt (short for Visual Basic for Applications StaTeMenT).

Command	Alias	Ctrl+	F-key	Alt+	Menu Bar	Dashboard
vbastmt	...	...	...	...	...	...

Command: vbastmt
Statement: *(Enter VBA statement.)*

COMMAND LINE OPTION

Statement specifies the VBA statement for AutoCAD to execute.

RELATED COMMANDS

Vbaide displays the Visual Basic for Applications integrated development environment window.

VbaLoad loads a VBA project; displays the Open VBA Project dialog box.

VbaMan displays the VBA Manager dialog box.

VbaRun displays the Macros dialog box.

VbaUnload unloads a VBA project.

RELATED SYSTEM VARIABLES

None.

TIPS

- A VBA *statement* is a complete instruction containing keywords, operators, variables, constants, and expressions.

- A VBA *macro* is an executable subroutine.

- At the 'Statement' prompt, enter a single line of code; use the colon (:) to separate multiple statements on the single line.

View

<u>V. 2.0</u> Saves and displays views by name in the current viewport; creates view categories for sheet sets, live sections, and backgrounds for visual styles; and controls parameters for cameras.

Commands	Aliases Ctrl+	F-key	Alt+	Menu Bar	Dashboard
view	v ...	...	VN	View	Lights
	ddview			↳Named Views	
-view	-v		V3	View	...
			↳3D Views		

Command: view

VIEW DIALOG BOX

Set Current restores the named view; alternatively, double-click the view name.

New displays the New View dialog box.

Update Layers updates view to match the layer visibility settings of the current viewport.

Edit Boundaries highlights the view, by graying out the area outside of the view.

Delete removes the named view from the drawing.

NEW VIEW DIALOG BOX

View name specifies the view name; up to 255 characters long.

View Category specifies the default prefix for named views. In the Sheet Set Manager, select the View List tab to see the names of views, under categories, as illustrated below:

View categories, such as "A-A," listed in the View List tab of the Sheet Set Manager palette.

Boundary options

⊙ **Current display** stores the current viewport as the named view.

○ **Define window** stores a windowed area as the named view.

🔲 **Define View Window** dismisses the dialog box temporarily so that you can pick the two corners that define the view.

AutoCAD clears the dialog box, and prompts:

Specify first corner: *(Pick a point.)*

Specify opposite corner: *(Pick another point.)*

Specify first corner (or press ENTER to accept): *(Press Enter to return to the dialog box.)*

The white rectangle defining the area of the view window.

Settings options

☑ **Save Current Layer Settings with View** toggles whether layer properties (such as freeze/thaw, lock/unlock, plot/no plot, and so on) are stored with the view. You must first use the Layer command to set layer properties.

UCS name specifies the name of a UCS to store with the named view. You must first create named UCSs with the Ucs command.

Live Section specifies the named live section to be displayed when the named view is restored. You must first use the SectionPlane and LiveSection commands to create the named section.

Visual Style specifies the named visual style to be displayed when the view is restored. You must first use the VisualStyles command to create the named visual style.

Background options

▼ selects the background type: default, solid, gradient, image, or sun & sky.

☐ **Save Sun Properties With View** displays simulated sun and sky in the background of the view (*new to AutoCAD 2008*).

... displays the Background dialog box.

BACKGROUND DIALOG BOX

Type selects the type of background:

- **Solid** shows a single color; click the Color bar to display the Select Color dialog box.
- **Gradient** shows two or three colors.
- **Image** shows a picture.

Gradient options

Left: *Two-color gradient at 45 degrees.*
Right: *Three-color gradient at 0 degrees.*

Three color toggles gradients between two and three colors.

Rotation displays the gradient at an angle; range is from 90 (vertical) to 0 (horizontal) to -90 degrees.

Top color, **Middle color**, and **Bottom color** display the Select Color dialog box.

Image options

 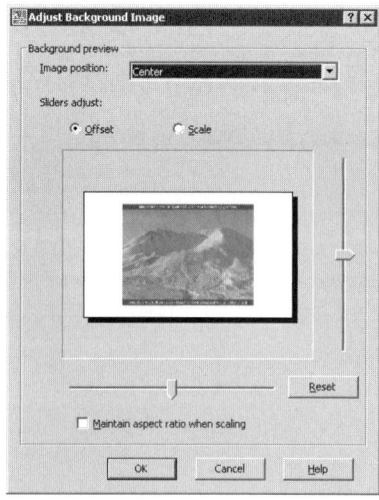

Left: *Selecting the image background.*
Right: *Adjusting the image.*

Browse selects the image from the Select File dialog box. You can choose from Targe, bitmap, TIFF, JPEG, GIF, and PCX formats.

Adjust image displays the Adjust Image dialog box.

ADJUST IMAGE DIALOG BOX

Known as the SetUV command in AutoCAD 2006 and earlier.

Image position:

- **Center** centers the image in the viewport.
- **Stretch** fits the image to the viewport.
- **Tile** repeats the image to fill the viewport.

Left: *Image centered in viewport.*
Center: *Image stretched to fit viewport.*
Right: *Image tiled (repeated) to fit viewport.*
(Image is from Mount St. Helens volcano-cam at www.fs.fed.us/gpnf/volcanocams/msh)

Sliders adjust:

Neither of the following options works in Stretch mode:

- ⊙ **Offset** moves the image relative to the viewport. When image goes outside the viewport boundary, its preview appears dimmed.
- ○ **Scale** changes the size of the image, making it larger or smaller. Negative x scale factors mirror the image; negative y inverts the image.

Left: *Image moved in viewport.*
Center: *Image scaled larger than viewport.*
Right: *Image upside down with negative y-scale factor.*

Maintain aspect ratio when scaling prevents the image from distortion due to unequal scaling.

Reset returns the image to its original size and aspect ratio.

ADJUST SUN & SKY BACKGROUND DIALOG BOX

New to AutoCAD 2008:

The "Sun & Sky" background displays a sun image placed in the sky background, at a position matching the date, time, and geographic location. As the date and time change (sliders in the Dashboard's Lighting control panel), the sun moves through the sky; as it sets, the sky darkens to black. This effect is available only when LightingUnits are set to 1 or 2.

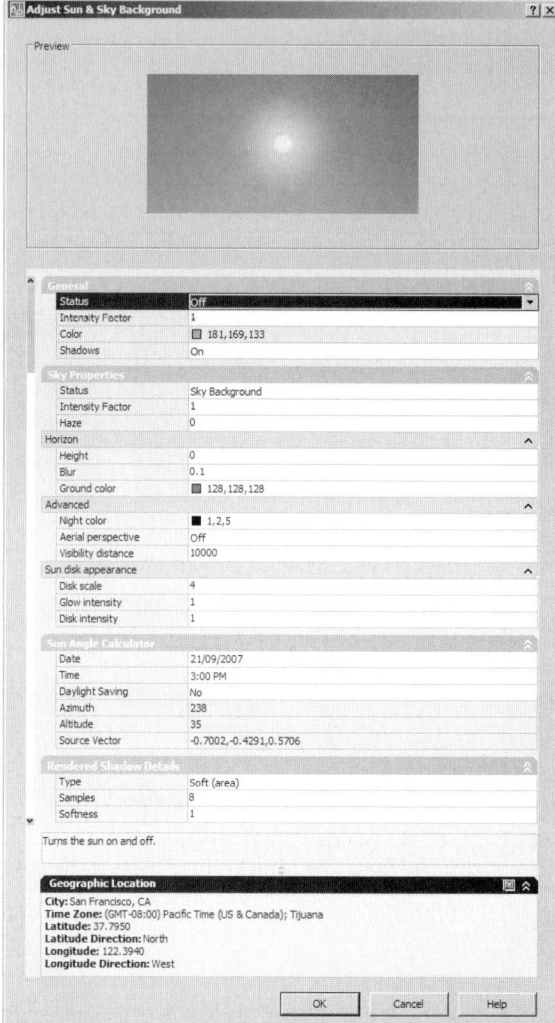

General options

Status toggles the sun effect.

Intensity Factor specifies the brightness of the sun.

Color sets the color of the sun.

Shadows toggles the sun's shadow-casting; leave off for better performance.

Sky Properties options

Status toggles display of the sky effect.

Intensity Factor specifies the brightness of the sky.

Haze controls the scattering effects in the atmosphere.

Horizon options

Height positions the ground plane relative to world zero.

Blur adjusts the blur between the ground plane and the sky.

Ground Color specifies the color of the ground plane.

Advanced options

Night Color specifies the color of the night sky.

Aerial Perspective toggles aerial perspective.

Visibility Distance sets the distance for 10% haze occlusion.

Sun Disk Appearance options

Disk Scale sizes the sun.

Glow Intensity sets the intensity of the sun's glow.

Disk Intensity sets the intensity of the sun's disk.

Sun Angle Calculator options

Date specifies the date.

Time specifies the time.

Daylight Saving toggles daylight savings time.

Rendered Shadow Details options

Type sets the shadow type: soft is available only with photometric lighting.

Samples specifies the number of samples to take on the solar disk.

Softness specifies the edge of shadow appearance.

VIEW PROPERTIES

The properties vary, depending on the type of view selected.

Properties must be set in the View dialog box, because the Properties palette cannot select views.

Properties in gray cannot be edited (read-only).

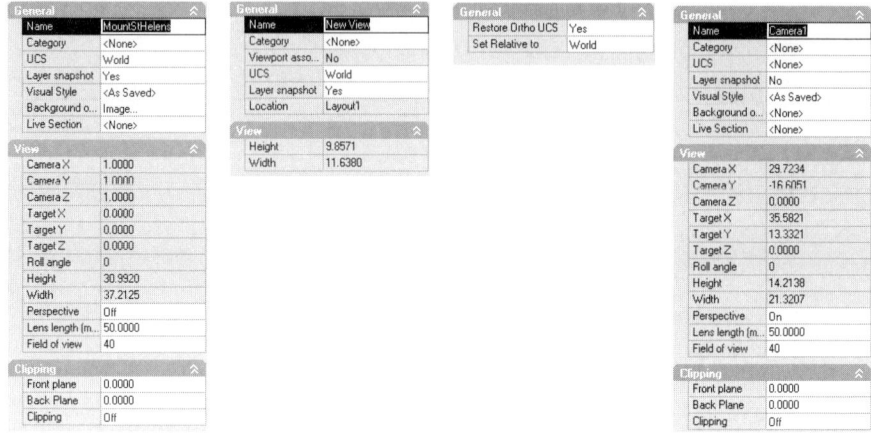

From left to right: *Properties of views in model space, paper space (layout tabs), preset views, and cameras.*

Properties listed in alphabetical order:

Background specifies the type of background color or image: none, Solid, Gradient, Image, or edit.

Camera X, **Y**, and **Z** specifies the x, y, and z coordinates of the view's viewpoint (camera); read-only.

Category lists the names of view categories: none or user-defined.

Clipping selects clipping options: Off, Front On, Back On, or Front and Back On.

Field of View indicates the horizontal field of view angle; coupled with Lens Length.

Front Plane and **Back Plane** indicate the front (and back) clipping plane's offset distance.

Height indicates the view height above or below the x,y-plane; read-only.

Layer Snapshot toggles whether layer parameters are stored with the view: yes or no.

Lens Length (mm) indicates the lens length in millimeters; coupled with Field of View.

Live Section determines whether a live section is displayed when the view is restored: none or user-defined.

Location reports the layout associated with the view; read-only.

Name names the view.

Perspective toggles the projection between perspective and parallel: on or off.

Restore Ortho UCS restores the associated UCS when the orthographic view is current: yes or no.

Roll angle indicates the view's roll angle; read-only.

Set Relative To specifies the base coordinate system for the orthographic view: World or user-defined.

Target X, **Y**, and **Z** specifies the x, y, and z coordinates of the view's look-at point (target); read-only.

UCS names the user-defined coordinate system saved with the view: none, World, or user defined.

Viewport Association toggles the view's association with a sheet set's viewport.

Visual Style specifies the visual style to restore with the view: 2D Wireframe, 3D Wireframe, 3D Hidden, Realistic, Conceptual, or user-defined.

Width indicates the view width (field of view); read-only.

SHORTCUT MENU
Right-click the Views list:

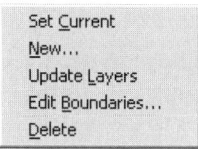

Set Current sets the selected view as active.

New creates a new view; opens the New View dialog box.

Update Layers updates the view to match the layer visibility settings of the current viewport.

Edit Boundaries highlights the view, by graying out the area outside of the view.

Delete erases the selected view; you cannot delete the Current view.

• •

-VIEW Command
Command: -view
Enter an option [?/Delete/Orthographic/Restore/Save/sEttings/Window]: *(Enter an option.)*

COMMAND LINE OPTIONS
? lists the names of views saved in the current drawing.

Delete deletes a named view.

Orthographic restores predefined orthographic views.

Restore restores a named view.

Save saves the current view with a name.

sEttings specifies view settings.

Window saves a windowed view with a name.

Orthographic options
Enter an option [Top/Bottom/Front/BAck/Left/Right] <Top>: *(Enter an option.)*
Select Viewport for view: *(Pick a viewport.)*

Enter an option selects a standard orthographic view for the current viewport: Top, Bottom, Front, BAck, Left, or Right.

Select Viewport for view selects the viewport — in either Model or Layout tab — in which to apply the orthographic view.

sEttings options
Enter an option [Background/Categorize/Layer snapshot/live Section/Ucs/Visual style]: *(Enter an option.)*

Background specifies the type of background for the view.

Categorize selects a sheet set category for the view.

Layer snapshot applies the current layer properties to the view.

live Section assigns a section to the view.

Ucs assigns a user-defined coordinate system to the view.

Visual style selects a visual style for the named view.

STARTUP SWITCH

/v specifies the view name to show when AutoCAD starts up.

RELATED COMMANDS

NewView accesses the New View dialog box directly (*new to AutoCAD 2008*).

Rename changes the names of views via a dialog box.

UCS creates and displays user-defined coordinate systems.

PartialLoad loads portions of drawings based on view names.

Open opens drawings and optionally starts with a named view.

Plot plots named views.

SheetSet uses named views.

RELATED SYSTEM VARIABLES

DefaultViewCategory specifies the default name for categories.

ViewCtr specifies the coordinates of the center of the view.

ViewSize specifies the height of the view.

GeographicLocation specifies the location on earth for positioning the sun.

TIPS

- Name views in your drawing to move quickly from one detail to another.

- The Plot command can plot the named views of drawings; the Open command lets you select a view with which to display the drawing.

- Objects outside of the window created by the Window option may be displayed, but are not plotted.

- As of AutoCAD 2005, this command creates views for sheets.

- As of AutoCAD 2007, it controls the background for visual styles. Backgrounds do not appear in 2D wireframe mode.

Removed Command

The **+View** command was removed from AutoCAD 2007, because the redesigned dialog box no longer has tabs.

ViewPlotDetails

__2005__ Reports plotting errors and successes.

Command	Alias	Alt+	Menu Bar	Dashboard
viewplotdetails	...	FB	File	...
			↳View Plot and Publish Details	

Command: viewplotdetails

Displays dialog box:

PLOT AND PUBLISH DETAILS DIALOG BOX

Green checkmark indicates a successful plot.

Red X warns of plotting error.

View determines whether all messages are displayed, or just errors:

- All messages.
- Errors only.

Copy to Clipboard copies selected text to the Clipboard.

[−] Collapses text under heading.

[+] Expands text under heading.

SHORTCUT MENU

Select text in the dialog box, and then right-click:

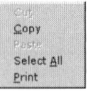

Copy copies selected text to the Clipboard.

Select All selects all text in the dialog box; use the cursor to select portions of text.

Print displays Print dialog box; select a printer and its options, and then click **Print**. *All* text in the dialog box is printed, not just selected text.

The Cut and Paste options are unavailable.

RELATED COMMANDS

Plot plots drawings.

TraySettings toggles the display of icons in the tray.

TIPS

■ When a plot is completed, AutoCAD displays a balloon. Clicking the blue underlined text displays the dialog box.

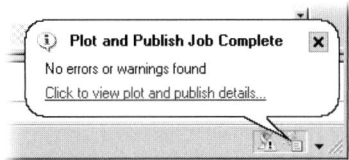

■ You can also access this dialog box by right-clicking the Plotting icon in the tray, and then selecting View Plot Details.

ViewRes

V. 2.5 Controls the roundness of curved objects; determines whether zooms and pans are performed as redraws or regens (short for VIEW RESolution).

Command	Alias	Ctrl+	F-key	Alt+	Menu Bar	Dashboard
viewres	...	...	...	...	...	...

Command: viewres
Do you want fast zooms? [Yes/No] <Y>: *(Type* **Y** *or* **N.***)*
Enter circle zoom percent (1-20000) <1000>: *(Enter a value.)*

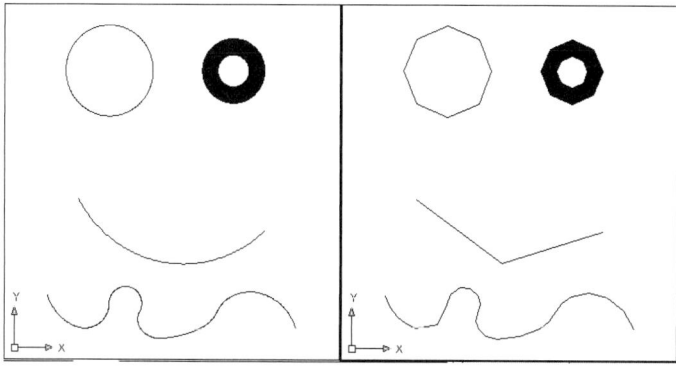

Left: *Circle zoom percent = 1000.*
Right: *Circle zoom percent = 1.*

COMMAND LINE OPTIONS

Do you want fast zooms? does not function; retained only for compatibility with macros and scripts.

Enter circle zoom percent specifies that smaller values display faster, but makes circles look less round (see figure); default = 1000.

RELATED SYSTEM VARIABLE

WhipArc toggles the display of circles and arcs as vectors or as true, rounded objects.

RELATED COMMAND

RegenAuto determines whether AutoCAD uses redraws or regens.

TIPS

- Setting WhipArc to 1 is recommended over increasing the value of ViewRes.

- This command was useful for speeding up the display when computers were much slower than today.

- Back in the days when "fast zooms" could be toggled, every zoom and pan caused a regeneration when the fast zooms were disabled. Today, AutoCAD tries to make every zoom and pan a redraw.

Removed Command

VlConv was removed from AutoCAD Release 14; use 3dsIn instead.

VisualStyles, VisualStylesClose

<u>2007</u>   Creates and edits visual styles.

Command	Alias	Ctrl+	F-key	Alt+	Menu Bar	Dashboard
visualstyles	vsm	...	...	TPV	Tools	Visual Styles
					⌖ Palettes	
					⌖ Visual Styles	

visualstylesclose						

-visualstyles	-vsm					

Command: visualstyles

Displays palette:

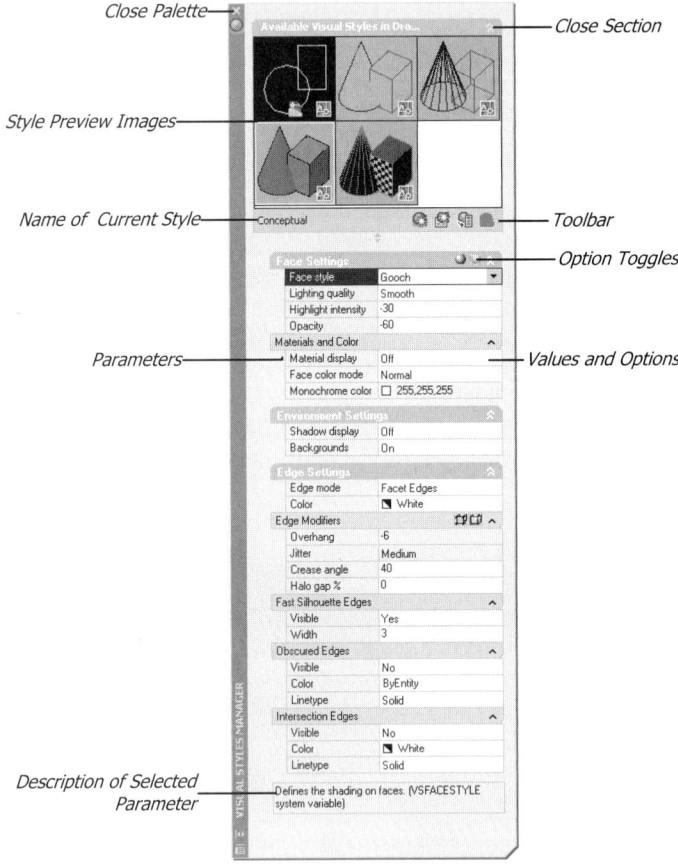

Command: visualstylesclose

Closes palette.

PALETTE OPTIONS

Negative values turn off the option.

Face Settings options

Face Style selects the style of face shading:

- **None** — applies no style to faces, like wireframe mode.
- **Real** — attempts to make the face appear real, like rendered mode; displays materials, if applied.
- **Gooch** — substitutes warm and cool colors for light and dark; allows faces to be seen more easily than with Real.

Lighting Quality toggles faceted and smooth; for curved surfaces only.

Highlight Intensity specifies the size of highlights on faces; for faces without materials. Range 0 (off) to 100. Larger numbers generate larger areas of intensity.

Opacity specifies the transparency of faces; range 0 to 100, where 0 is opaque, 1 is transparent, 2 - 99 are increasing translucent, and 100 (or any negative value) is again opaque.

Materials and Color options

Materials toggles the display of materials and textures, if previously applied with the MaterialAssign command. When other than Off, the Highlight Intensity and Opacity settings do not apply.

- **Off** — materials and textures are not displayed; visual style settings are applied to the model.
- **Materials** — materials are displayed, if assigned.
- **Materials and Textures** — materials and textures are displayed, if applied.

Face Color Mode specifies how colors are displayed:

- **Normal** — face colors displayed normally.
- **Monochrome** — face colors displayed in a monochrome color specified by the Monochrome Color option.
- **Tint** — face colors change their hue and saturation.
- **Desaturate** — face colors reduce their saturation by 30%.

Monochrome Color and **Tint Color** specify the color for monochrome and tint face colors. The color, selected from the Select Color dialog box, covers the model with a single color.

Environment Settings options

No lights need be placed in drawings for visual styles to cast shadows.

Shadow Display specifies the type of shadows to display.

- **Off** — objects cast no shadows. Shadow-casting slows down AutoCAD, so it's best left off unless needed for high-quality images or sun studies.
- **Ground shadows** — objects cast shadows on the shadow plane, but not on each other.
- **Full shadows** — objects cast shadows on the ground and on each other. The graphics board must be capable of full shadow-casting and hardware acceleration must be turned on; see the 3dConfig command.

Backgrounds toggles the display of a background in the viewport; backgrounds are created by the View command.

Edge Settings options

Edge Mode specifies which style of edge to display:

- **None** — facets, isolines, and edges are not displayed; this setting cannot be used when the Face Style set to None.
- **Isolines** — isolines are displayed on curved surfaces, and edges on all objects.
- **Facet Edges** — edges are displayed in a color.

Color specifies the color for edges; one color applies to all edges.

The following parameters appear only when Edge Mode = Isolines:

Number of Lines specifies the number of isolines drawn on curved surfaces; range is 0 to 2047, where the default value of 4 is too low. A good number to use is either 0 (no isolines) or 12; too many isolines completely blacken the curved objects. This parameter does not go into effect until after the Regen command is run.

Always on Top toggles hidden-line removal of isolines:
- **On** — all isolines are displayed.
- **Off** — only foreground isolines are displayed; isolines located "around the back" are hidden.

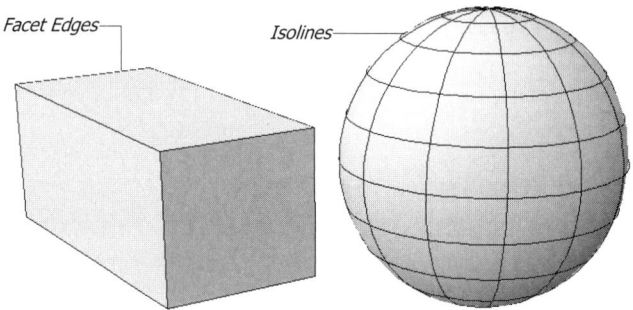

Facet Edges Isolines

Edge Modifiers options

The Edge Modifiers settings apply to isolines and facet edges:

Overhang extends edges beyond their boundaries; range is 0 to 100 pixels.

Jitter skews edge lines to mimic a hand drawn effect by skewing the lines that define edges:
- **Off** — (0) no jitter lines.
- **Low** — (1) few jitters.
- **Medium** — (2) medium jitters.
- **High** — (3) high amount of jitter.

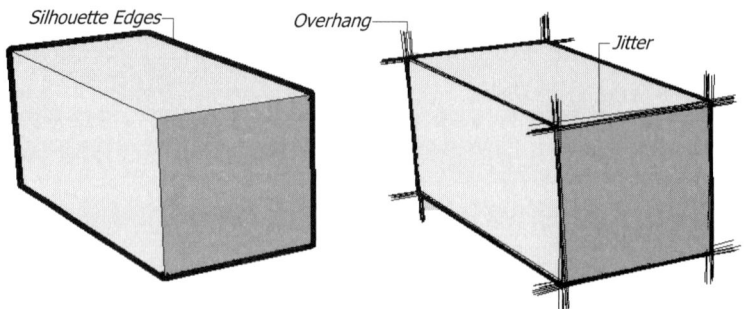

Silhouette Edges Overhang Jitter

The following parameters appear only when Edge Mode = Facet Edges:

Crease Angle specifies the angle beyond which facet edges are not shown on curved surfaces, removing the edge lines; range is from 0 to 180 degrees. A high value, such as 180, turns off the display of facet edges.

Halo Gap % specifies the gap generated between visually overlapping objects; not available in visual styles based on wireframe modes. Range is 0 to 100 pixels; when greater than 0, silhouette edges are not displayed.

Fast Silhouette Edges options

Both of the following settings apply to all objects in the viewport equally:

Visible toggles the display of silhouette edges. AutoCAD does not display edges when the halo gap > 0, opacity > 0 (objects are translucent), or visual styles are wireframe.

Width determines the width of the silhouette edges; range is 1 to 25 pixels.

Obscured Edges options

The following parameters appear only when Edge Mode = Facet Edges:

Visible toggles the visibility of obscured edges and facets:

- **Yes** — hidden lines are shown through the model, although obscured edges are invisible on curved faces when a large crease angle prevents facet lines from showing.
- **No** — hidden lines are not shown.

Color specifies the color for visible obscured edges and facets.

Linetype specifies the line pattern for obscured (hidden) edges and facets. No scale factor can be applied to these linetypes, and you cannot use the linetypes accessed by the Linetype command or stored in the *acad.lin* file. The available linetypes are:

- Off
- Solid
- Dashed
- Dotted
- Short Dash
- Medium Dash
- Long Dash
- Double Short Dash
- Double Medium Dash
- Double Long Dash
- Medium Long Dash
- Sparse Dot

Intersection Edges options

The following parameters appear only when Edge Mode = Facet Edges:

Visible toggles the display of intersection edges. (Autodesk recommends leaving this setting turned off to increase performance.)

Color and Linetype specify the color and linetypes of intersection lines, as described above.

PALETTE TOOLBAR

Name of Current Visual Style — Visual Style

Create New Visual Style —

Delete the Selected Visual Style

Apply the Selected Visual Style to the Current Viewport —

Export the Selected Visual Style to the Tool Palette

-VisualStyle Command

Command: -visualstyle

Enter an option [set Current/Saveas/Rename/Delete/?]: *(Enter an option.)*

set Current prompts for the type of visual style to apply: 2dwireframe, 3dwireframe, 3dHidden, Realistic, Conceptual, Other, or cUrrent. If Other, prompts for its name.

Saveas saves the current visual style; prompts for a name.

Rename renames a visual style; system styles cannot be renamed.

Delete deletes a visual style; system styles cannot be deleted.

? lists the names of visual styles in the current drawing.

RELATED COMMANDS

VsCurrent sets the visual style.

VsSave saves the current visual style by name.

ToolPalettes stores visual styles.

Dashboard accesses the Visual Styles control panel.

Rename renames visual styles.

Purge removes unused visual styles.

View applies visual styles to named views.

RELATED SYSTEM VARIABLES

CMaterial specifies the name of the material.

CShadow specifies the type of shadow cast by objects.

DragVs specifies the default visual style while creating 3D objects.

InterfereObjVs specifies the visual style for interference objects created by Interference.

InterfereVpVs specifies the visual style during interference checking.

Isolines specifies the number of isolines to display on curved solid models.

ShadowPlaneLocation locates the height of the invisible ground plane upon which shadows are cast; can be set to any distance along the z axis, including negative distances.

VsBackgrounds determines whether backgrounds are displayed in visual styles.

VsEdgeColor specifies the edge color; can be any ACI color.

VsEdgeJitter specifies the level of jitter effect.

VsEdgeOverhang extends edge lines beyond intersections.

VsEdges specifies the type of edge to display.

VsEdgeSmooth specifies the crease angle.

VsFaceHighlight specifies the color of highlights; ignored when VsMaterialMode is on.

VsFaceOpacity controls the transparency/opacity of faces.

VsFaceStyle determines how faces are displayed.

VsHaloGap specifies the "halo" gap (gap between intersecting lines).

VsHidePrecision specifies the accuracy of hides and shades.

VsIntersectionColor specifies the color of intersecting polylines.

VsIntersectionEdges toggles the display of intersecting edges.

VsIntersectionLtype specifies the linetype for intersecting polylines.

VsIsoOntop toggles whether isolines are displayed.

VsLightingQuality toggles the quality of lighting.

VsMaterialMode controls the display of material finishes:

VsMonoColor specifies the monochrome tint.

VObscuredColor specifies the color of obscured lines.

VsObscuredEdges toggles the display of obscured edges.

VsObscuredLtype specifies the linetype of obscured lines.

VsShadows determines the quality of shadows.

VsSilhEdges toggles the display of silhouette edges.

VsSilhWidth specifies the width of silhouette edge lines.

VsState reports whether the Visual Styles window is open.

...whew!

TIPS

- The ShadowPlaneLocation system variable allows you to change the elevation of the "ground" (shadow plane) upon which shadows are cast. The default is 0 units, and is measured in the z direction. This plane is independent of anything else.

- You can use the Dashboard's Light control panel to adjust the location and angle of shadow-casting.

Click the Sun Status button, and then move the Date and Time sliders to rotate and lengthen the shadows. The Brightness and Contrast sliders make the objects brighter or more shadowy. They have no effect on the shadows, and work only when Face Style is set to Real.

'VLisp

2000 Opens the VLisp integrated development environment (short for Visual LISP).

Commands	Alias	Ctrl+	F-key	Alt+	Menu Bar	Dashboard
vlisp	vlide	...	...	TIV	Tools	...
					⮡AutoLISP	
					⮡Visual LISP Editor	

Command: vlisp

Displays window:

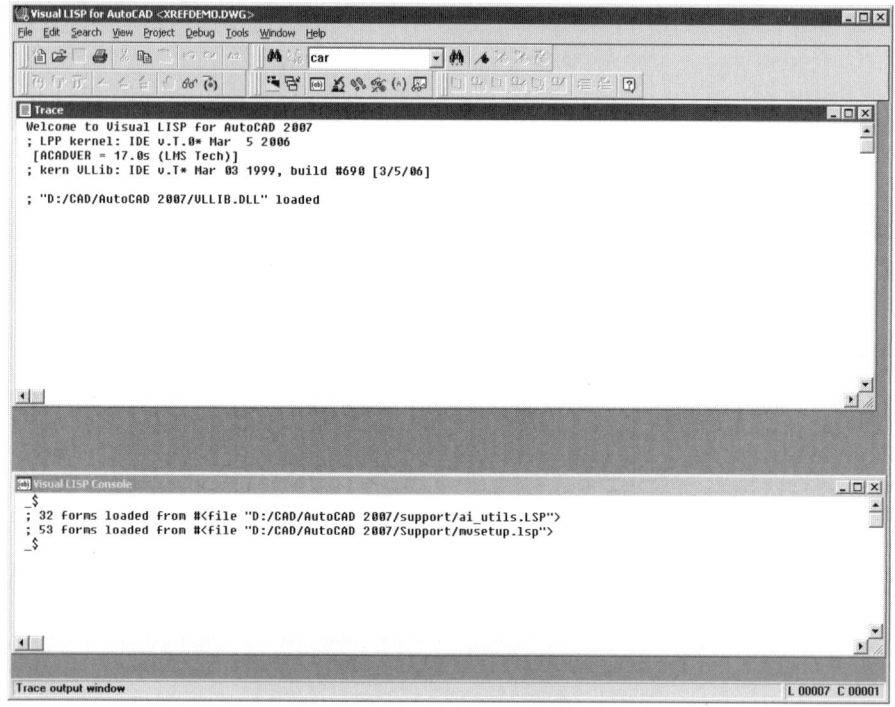

MENU BAR

Select Help | Visual LISP Help Topics for assistance in using this VLISP window.

RELATED COMMAND

AppLoad loads Visual LISP applications, as well as programs written in AutoLISP and other APIs.

TIP

■ Sample VLisp code can be found in the *autocad 2008\sample\vlisp* folder.

VpClip

2000 Clips a layout viewport (short for ViewPort CLIPping).

Command	Alias	Ctrl+	F-key	Alt+	Menu Bar	Dashboard
vpclip	...	...	...	MC♦V	Modify	...
					♦Clip	
					♦Viewport	

Command: vpclip
Select viewport to clip: *(Pick a viewport.)*
Select clipping object or [Polygonal] <Polygonal>: *(Select an object, or type P.)*

The selected viewport disappears, and is replaced by the new clipped viewport.

COMMAND LINE OPTIONS

Select viewport to clip selects the viewport that will be clipped.

Select clipping object selects the object that defines the clipping boundary: closed polyline, circle, ellipse, closed spline, or region.

Polygonal Options
Specify start point: *(Pick a point.)*
Specify next point or [Arc/Close/Length/Undo]: *(Pick a point, or enter an option.)*
Specify next point or [Arc/Close/Length/Undo]: *(Type C to close.)*

Specify start point specifies the starting point for the polygon.

Arc draws an arc segment; see the Arc command.

Close closes the polygon.

Length draws a straight segment of specified length.

Undo undoes the previous polygon segment.

RELATED COMMAND

Mview creates rectangular and polygonal viewports in paper space.

TIPS

- This command does not operate in Model tab.

- The clipping polyline is placed on the current layer. When the layer is frozen, the viewport is not clipped until the layer is thawed. When the layer is off, the polyline is invisible, but still clips.

VpLayer

<u>Rel. 11</u> Controls the visibility of layers in viewports, when a layout tab other than Model is selected (short for ViewPort LAYER).

Command	Alias	Ctrl+	F-key	Alt+	Menu Bar	Dashboard
vplayer	...	...	...	...	...	...

Command: vplayer
Enter an option [?/Freeze/Thaw/Reset/Newfrz/Vpvisdflt]: *(Enter an option.)*
Select a viewport: *(Pick a viewport.)*

COMMAND LINE OPTIONS

Freeze indicates the names of layers to freeze in this viewport.

Newfrz creates new layers that are frozen in all newly-created viewports (short for NEW FReeZe).

Reset resets the state of layers based on the Vpvisdflt settings.

Thaw indicates the names of layers to thaw in this viewport.

Vpvisdflt determines which layers will be frozen in a newly-created viewport and default visibility in existing viewports (short for ViewPort VISibility DeFauLT).

? lists the layers frozen in the current viewport.

RELATED COMMANDS

Layer creates and controls layers in all viewports.

MView creates and joins viewports when tilemode is off.

RELATED SYSTEM VARIABLE

TileMode controls whether viewports are tiled (model) or overlapping (layouts).

VpMax, VpMin

2005 Maximizes or minimizes the selected viewport in the AutoCAD window (short for ViewPortMAXimize).

Command	Alias	Ctrl+	F-key	Alt+	Menu Bar	Dashboard
vpmax	...	...	...	...	...	...
vpmin						

Command: vpmax

When the layout contains more than one viewport, AutoCAD prompts:

Select a viewport to maximize: *(Select a viewport.)*

AutoCAD maximizes the viewport to fill the entire AutoCAD window, and switches to model space for editing.

The red dashed border indicates AutoCAD is in VpMax mode.

COMMAND LINE OPTION

Select a viewport to maximize selects the viewport.

RELATED COMMAND

VPorts creates viewports.

RELATED SYSTEM VARIABLE

VpMaximixedState determines whether the viewport is maximized.

TIPS

- Use the VpMin command to return to the layout tab, and restore the viewport. Alternatively, you can double-click the red border to restore the viewport.

- When the viewport is maximized, AutoCAD displays the Minimize Viewport icon on the status bar. Click the arrows to move from one viewport to the next.

VPoint

V. 2.1   Changes the viewpoint of 3D drawings (short for ViewPOINT).

Command	Alias	Ctrl+	F-key	Alt+	Menu Bar	Dashboard
vpoint	-vp	...	...	V3V	View	...
					⇗3D Views	
					⇗Viewpoint	

Command: vpoint
Current view direction: VIEWDIR=0.0000,0.0000,1.0000
Specify a view point or [Rotate] <display compass and tripod>: *(Enter an option, or press Enter for the compass-tripod.)*

COMMAND LINE OPTIONS

Specify a view point indicates the new 3D viewpoint by coordinates.

Rotate indicates the new 3D viewpoint by angle.

Enter brings up visual guides (see figure below).

RELATED COMMANDS

DdVpoint adjusts the viewpoint via a dialog box.

3dOrbit rotates the viewpoint in real-time.

RELATED SYSTEM VARIABLES

VpointX is the x-coordinate of the current 3D view.

VpointY is the y-coordinate of the current 3D view.

VpointZ is the z-coordinate of the current 3D view.

WorldView determines whether VPoint coordinates are in WCS or UCS.

TIPS

- This command works only in model space, and is largely replaced by the 3dOrbit command

- The *compass* represents the globe, flattened to two dimensions: north pole (0, 0, z) is in the center; equator (x, y, 0), the inner circle; and south pole (0, 0, -z), the outer circle.

- As the cursor is moved on the compass, the *axis tripod* rotates showing the 3D view direction. To select the view direction, pick a location on the globe and press the pick button.

VPorts

Rel. 10 Creates viewports (short for ViewPORTS).

Commands	Alias	Ctrl+	F-key	Alt+	Menu Bar	Dashboard
vports	viewports	...	...	VV	View ↳Viewports	3D Navigate
+vports						
-vports				VV1	View ↳Viewports ↳1 Viewport	

Command: vports

Displays tabbed dialog box.

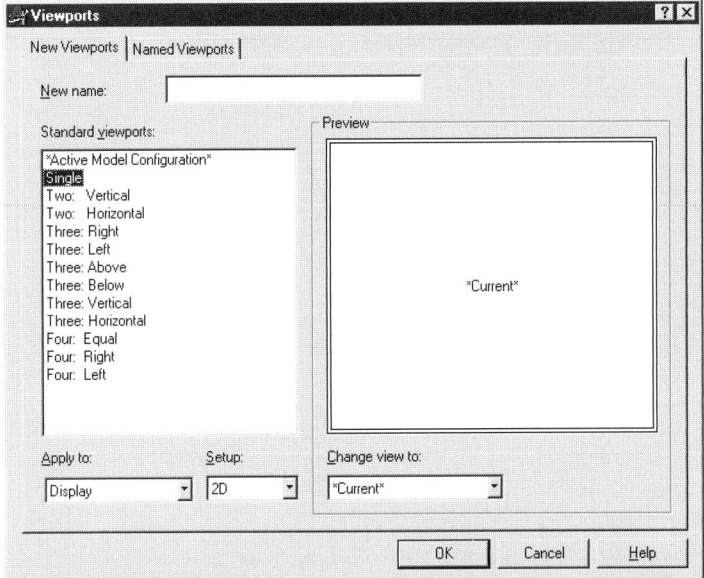

VIEWPORTS DIALOG BOX

In model space

New name specifies the name for the viewport configuration; can be up to 255 characters long.

Standard viewports lists the available viewport configurations.

Preview displays a preview of the viewport configuration.

Apply to applies the viewport configuration to:

- Display.
- Current Viewport.

Setup selects 2D or 3D configuration; the **3D** option applies orthogonal views, such as top, left, and front.

Change view to selects the type of view; in 3D mode, selects a standard orthoganal view.

In paper space

Viewport spacing specifies the spacing between the floating viewports; default = 0 units.

Named Viewports tab

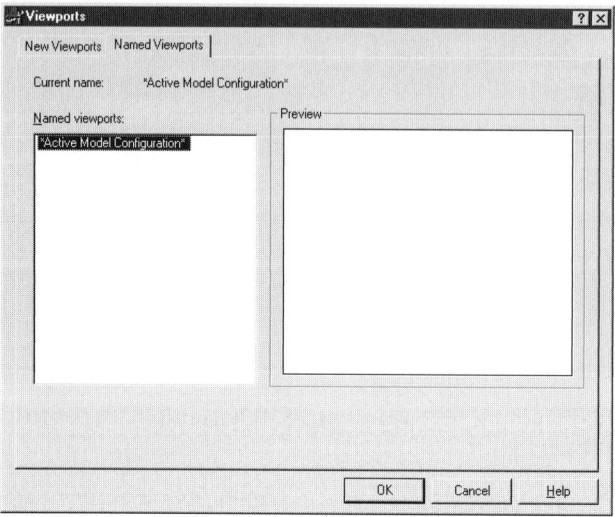

Named viewports lists the names of saved viewport configurations.

. . . .

+VPORTS Command
Command: +vports
Tab index <0>: *(Type 0 or 1.)*

    Tab index specifies the tab to display:
        0 — New Viewports tab (default).
        1 — Named Viewports tab.

. . . .

-VPORTS Command
Command: -vports

    In layout mode, displays MView command prompts. In model space, prompts:

Enter an option [Save/Restore/Delete/Join/SIngle/?/2/3/4] <3>: *(Enter an option.)*

COMMAND LINE OPTIONS

In model space

    Save saves the settings of a viewport by name.

    Restore restores a viewport definition.

    Delete deletes a viewport definition.

    Join joins two viewports together as one when they form a rectangle.

    SIngle joins all viewports into a single viewport.

    ? lists the names of saved viewport configurations.

    4 divides the current viewport into four.

. . . .

2 (Two Viewports) options

Horizontal creates one viewport over another.

Vertical creates one viewport beside another (default).

3 (Three Viewports) options

Horizontal creates three viewports over each other.

Vertical creates three viewports beside each other.

Above creates one viewport over two viewports.

Below creates one viewport below two viewports.

Left creates one viewport left of two viewports.

Right creates one viewport right of two viewports (default).

In paper space
**Specify corner of viewport or [ON/OFF/Fit/Hideplot/Lock/Object/Polygonal/
Restore/LAyer/2/3/4]<Fit>:** *(Enter an option.)*

ON turns on the viewport; the objects in the viewport become visible.

OFF turns off the viewport; the objects in the viewport become invisible.

Fit creates one viewport that fills the display area.

Hideplot removes hidden lines when plotting in layout mode.

Lock locks viewports to prevent editing.

Object converts closed polylines, ellipses, splines, regions, or circles into viewports.

Polygonal creates non-rectangular viewports.

LAyer *(new to AutoCAD 2008)* resets layer property overrides to Bylayer in selected viewport; prompts:

Reset viewport layer property overrides back to global properties [Yes/No]?:
(Type Y or N.)

Select viewports: *(Pick one or more viewports.)*

Other options are identical to those displayed in model tab.

RIGHT CLICK MENU

In paper space, right-click a viewport border for this shortcut menu:

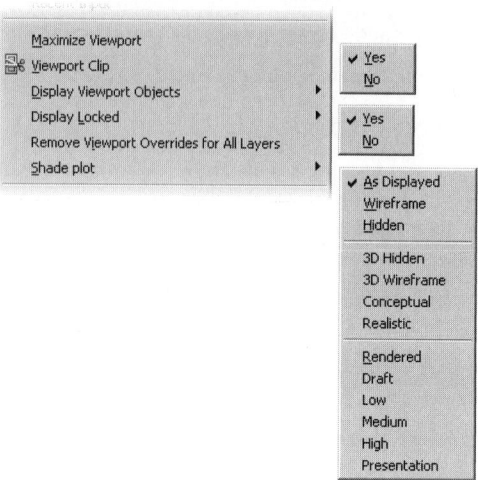

Maximize Viewport maximizes the viewport; see the VpMax command.

Viewport Clip creates, edits, and removes clipping boundaries; see the VpClip command.

Display Viewport Objects toggles visibility of the viewport.

Display Locked toggles locking of the display, which is like locked layers.

Remove Viewport Overrides for All Layers returns properties of objects in this viewpor to ByLayer (*new to AutoCAD 2008*).

Shade Plot selects the type of visual style or rendering to use when plotting this viewport.

RELATED COMMANDS

MView creates viewports in paper space.

RedrawAll redraws all viewports.

RegenAll regenerates all viewports.

VpClip clips a viewport.

RELATED SYSTEM VARIABLES

CvPort identifies the current viewport number.

MaxActVp limits the maximum number of active viewports.

TileMode controls whether viewports can be overlapped or tiled.

VpLayerOverrides reports whether any layer in the current viewport has VP (viewport) property overrides (*new to AutoCAD 2008*).

VpLayerOverridesMode toggles whether viewport property overrides are displayed or plotted.

TIPS

- The Join option joins two viewports only when they form a rectangle in model space.
- You can restore saved viewport arrangements in paper space using the MView command.
- Many display-related commands (such as Redraw and Grid) affect the current viewport only.
- Press Ctrl+R to switch between viewports.
- Use the Layer command to override color, linetype, lineweight, and plot styles on a per-viewport basis.

VsCurrent

2007 Sets the current visual style (short for Visual Style CURRENT).

Command	Aliases	F-key	Alt+	Menu Bar	Dashboard
vscurrent	vs	...	vs	View	Visual Styles
	shademode			⌐Visual Styles	

Command: vscurrent
Enter an option [2dwireframe/3dwireframe/3dHidden/Realistic/Conceptual/ Other]: *(Enter an option, or type O.)*

COMMAND LINE OPTIONS

2dwireframe displays raster and OLE objects; linetypes, and lineweights are visible; this is AutoCAD's default display mode, and is not a visual style.

3dwireframe looks similar to 2D wireframe, but shows the shaded 3D UCS icon and optionally the compass.

3dHidden removes lines hidden by overlapping objects.

Realistic shades objects, smooths edges, and displays materials, if attached to objects.

Conceptual looks like Realistic, but ranges the colors from "cool to warm" to make the model easier to view.

Other prompts for the name of a user-defined visual style: "Enter a visual style name or [?]."

? lists the names of existing visual styles.

RELATED COMMANDS

VisualStyles creates and edits visual styles.

VsSave saves the current visual style by name.

ToolPalettes stores visual styles.

Dashboard accesses the Visual Styles control panel.

TIPS

- Visual styles can be applied through:
  - Using the VsCurrent command's options.
  - Dragging them from the Visual Styles palette into the drawing; see VisualStyles command.
  - Selecting them from the Visual Styles toolbar.
  - Dragging them from the Tools palette into the drawing; see ToolPalettes command.
  - Selecting them from the Visual Styles control panel of the Dashboard; see Dashboard command.

- Visual styles work only in model space.
- Use the View command's Background option to apply a background image or color to visual styles.

VSlide

V. 2.0 Displays slide files in the current viewport (short for View SLIDE).

Command	Alias	Ctrl+	F-key	Alt+	Menu Bar	Dashboard
vslide	...	...	...	...	...	...

Command: vslide

Displays Select Slide File dialog box. Select an .sld file, and then click Open.

COMMAND LINE OPTIONS

None.

RELATED COMMANDS

MSlide creates slide files of the current viewport.

Redraw erases slides from the screen.

RELATED AUTODESK PROGRAM

slidelib.exe creates an SLB-format library file of a group of slide files.

RELATED AUTOCAD FILES

*.sld stores individual slide files.

*.slb stores a library of slide files.

TIP

■ The following applies when FileDia is set to 0, or when this command is used in a script:

• For faster viewing of a series of slides, placing an asterisk before the VSlide command preloads the *.sld* slide file, as in:

> **Command:** *vslide filename

• Use the following format to display a specific slide stored in an SLB slide library file:

> **Command:** vslide
> **Slide file:** acad.slb(slidefilename)

VsSave

2007 Saves the current visual style by name (short for Visual Style SAVE).

Command	Alias	Ctrl+	F-key	Alt+	Menu Bar	Dashboard
vssave	...	...	...	...	...	...

Command: vssave

This command operates only in model space.

Save current visual style as or [?]: *(Enter a name, or type ?.)*

COMMAND LINE OPTIONS

Save current visual style as names the current visual style. If the name already exists, AutoCAD asks, "'*name*' already exists. Do you wish to replace the existing visual style? [Yes/No/Try again]."

? lists the names of existing visual styles, such as:

```
2D Wireframe
3D Hidden
3D Wireframe
Conceptual
Realistic
```

RELATED COMMANDS

VsCurrent sets the current visual style.

VisualStyles creates and edits visual styles.

Rename renames visual styles.

Purge removes unused visual styles from the current drawing.

ToolPalettes stores visual styles and shares them among drawings.

TIPS

- Every drawing contains the following visual styles predefined by Autodesk: 2D Wireframe, 3D Hidden, 3D Wireframe, Conceptual, and Realistic.

- To use a saved visual style in another drawing, drag it from the Visual Styles Manager palette onto the Tools palette. Open the other drawing, and then drag the visual style from the Tool palette into the drawing.

VtOptions

2006 Controls view transitions during pans and zooms.

Command	Alias	Ctrl+	F-key	Alt+	Menu Bar	Dashboard
vtoptions	...	...	...	...	...	...

Command: vtoptions

Displays dialog box:

VIEW TRANSITIONS DIALOG BOX

☑**Enable Animation for Pan and Zoom** smoothes view transitions during pans and zooms.

☑**Enable Animation When View Rotates** smoothes view transitions during view angle changes.

☐**Enable Animation During Scripts** smoothes view transitions during scripts.

Transition Speed specifies the speed of view transitions in milliseconds.

Performance specifies the minimum speed for smooth view transitions in frames per second.

RELATED COMMAND

Zoom utilizes view transitions.

RELATED SYSTEM VARIABLES

VtDuration specifies the duration of view transitions in milliseconds.

VtEnable determines which commands use view transitions.

VtFps specifies the minimum view transition speed in frames per second.

TIPS

- View transitions are particularly useful when viewing in 3D.

- When AutoCAD cannot maintain the transition speed, it switches to an instant transition.

- View transitions make it easier to see where AutoCAD is zooming, but increases the time it takes to complete the zoom.

⚓ **Select Objects** dismisses the dialog box temporarily to select one or more objects.

🖌 **Quick Select** displays the Quick Select dialog box; see the QSelect command.

⊙ **Retain** retains the selected objects in the current drawing after saving them as a drawing file.

○ **Convert to block** converts the selected objects to a block in the drawing, after saving them as a *.dwg* file; names the block under **File name** in the Destination section.

○ **Delete from drawing** deletes selected objects from the drawing, after saving them as a *.dwg* file.

Destination options

File name and path specifies the file name for the block or objects.

▦ ··· displays the Browse for Folder dialog box.

Insert units specifies the units when the *.dwg* file is inserted as a block.

· ·

-WBLOCK Command
Command: -wblock

Displays the Create Drawing File dialog box. Name the file, and then click Save.

Enter name of existing block or

[= (block=output file)/* (whole drawing)] <define new drawing>: *(Enter a name, or use = and * options, or press Enter.)*

COMMAND LINE OPTIONS

Enter name of existing block names a current block in the drawing.

= *(equals)* writes block to a *.dwg* file, using the block's name as the file name.

* *(asterisk)* writes the entire drawing to a *.dwg* file.

ENTER creates a *.dwg* drawing file of the selected objects on disk. (The selected objects are erased from the drawing; use the Oops command to bring them back.)

RELATED COMMANDS

Block creates a block of a group of objects.

Insert inserts a block or another drawing into the drawing.

RELATED SYSTEM VARIABLES

None.

TIPS

- Use the WBlock command to extract blocks from the drawing and store them on a disk drive. This allows the creation of a block library.

- Support for the DesignXML (extended markup language) format was withdrawn from AutoCAD 2004.

· ·

Aliased Command
WebLight is an alias for the Light command.

· ·

Wedge

Rel.11   Draws 3D wedges as solid models.

Command	Alias	Ctrl+	F-key	Alt+	Menu Bar	Dashboard
wedge	we	...	...	DMW	Draw	3D Make
					⤷Modeling	
					⤷Wedge	

Command: wedge
Specify first corner or [Center]: *(Pick a point, or type C.)*
Specify other corner or [Cube/Length]: *(Pick a point, or enter an option.)*
Specify height or [2Point]: *(Pick another point, or enter 2P.)*

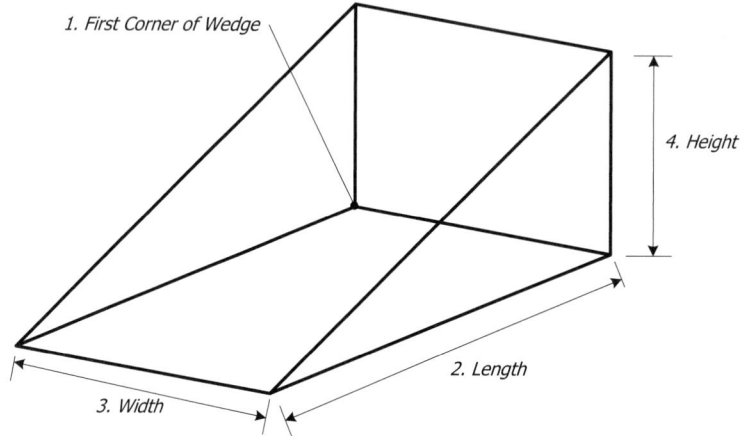

1. First Corner of Wedge
4. Height
2. Length
3. Width

COMMAND LINE OPTIONS
First corner specifies the lower-left corner of the wedge.

Center draws the wedge's base about the center of the sloped face.

Cube draws cubic wedges.

Length specifies the length, width, and height of the wedge.

2Point picks two points to indicate the height.

Center options
Specify center: *(Pick a point.)*
Specify corner or [Cube/Length]: *(Pick a point, or enter an option.)*
Specify height or [2Point]: *(Pick a point.)*

Specify center of wedge indicates the center of the wedge's inclined face.

Specify opposite corner indicates the distance from the midpoint to one corner.

Cube options
Specify length: *(Specify the length.)*

Specify length indicates the length of all three sides.

Length options
Specify length: *(Specify the length.)*
Specify width: *(Specify the width.)*
Specify height: *(Specify the height.)*

Specify length indicates the length parallel to the x-axis.

Specify width indicates the width parallel to the y-axis.

Specify height indicates the height parallel to the z-axis.

RELATED COMMANDS

Ai_Wedge draws wedges as 3D surface models.

Box draws solid boxes.

Cone draws solid cones.

Cylinder draws solid cylinders.

Sphere draws solid spheres.

Torus draws solid tori.

TIPS

- *Length* means size in the x-direction.

- *Width* means size in the y-direction.

- *Height* means size in the z-direction.

- Use negative values for length, width, and height to draw the wedge in the negative x, y, and z directions.

- The IsoLines system variable has no effect on wedges.

- The wedge can be edited using grips:

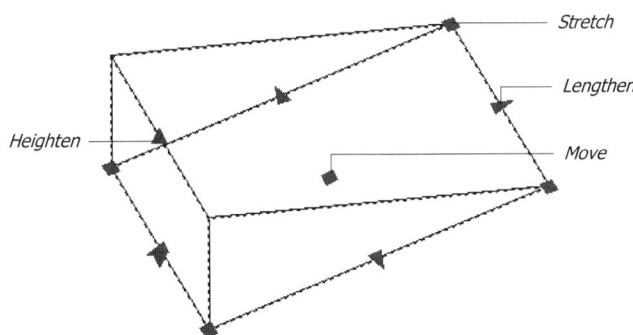

WhoHas

<u>**2000**</u> Determines which computer has drawings open.

Command	Alias	Ctrl+	F-key	Alt+	Menu Bar	Dashboard
whohas	...	...	...	...	...	...

Command: whohas

Displays the Select Drawing to Query dialog box. Select a drawing file, and then click Open.

When the drawing is open, reports:

Owner: ralphg

Computer's Name : HEATHER

Time Accessed : Monday, March 3, 2008 11:27:28 AM

When the drawing is not open, reports:

User: unknown.

COMMAND LINE OPTIONS

None.

RELATED COMMANDS

Open opens drawings.

XAttach attaches drawings that can be opened by other users.

TIP

- This command is meant for use over networks as a convenient way to find out which users are editing specific drawings.

WipeOut

<u>2004</u> Places fills areas of the background color to "wipe out" underlaying portions of drawings.

Command	Alias	Ctrl+	F-key	Alt+	Menu Bar	Dashboard
wipeout	...	...	...	...	...	...

Command: wipeout
Specify first point or [Frames/Polyline] <Polyline>: *(Pick a point, or enter option.)*
Specify next point: *(Pick a point.)*
Specify next point or [Undo]: *(Pick a point, or type U.)*
Specify next point or [Close/Undo]: *(Pick a point, or type C or U.)*
Specify next point or [Close/Undo]: *(Type C to end the command.)*

COMMAND LINE OPTIONS

Specify first point specifies the starting point of the polygon.

Undo undoes the last segment.

Close closes the polygon.

Frames options
Enter mode [ON/OFF] <ON>: *(Type ON or OFF.)*

ON turns on the wipeout boundary polygon.

OFF turns off the boundary polygon.

Polyline options
Select a closed polyline: *(Pick a closed polyline.)*
Erase polyline? [Yes/No] <No>: *(Type Y or N.)*

Select a closed polyline picks a polyline that forms the wipeout boundary.

Erase polyline?

Yes erases the polyline.

No leaves the polyline in place.

RELATED COMMAND

DrawOrder displays overlapping objects in a different order.

TIPS

- The wipeout consists of an image object drawn with the background color. To create rectangular wipeout frames, use the Rectangle command.

- To create a wipeout under text, you may find it easier to use the Background Mask option of the MText command.

- Wipeout boundaries can be edited with grips; wipeouts cannot be edited when the Frames option is turned off. The Frames option applies to all wipeouts in the drawing; you cannot turn frames on and off for individual wipeouts.

- To make text appear above the wipeout, use the DrawOrder command to move the text to the Front — after creating the wipeout.

- Freezing or turning off the layer of the wipeout boundary stops the wipeout action.

WmfIn

Rel.12 Imports *.wmf* and *.clp* files (short for Windows MetaFile IN).

Command	Alias	Ctrl+	F-key	Alt+	Menu Bar	Dashboard
wmfin	...	...	...	IW	Insert	...
					↳Windows Metafile	

Command: wmfin

Displays the Import WMF dialog box. Select a file, and then click Open.

Specify insertion point or [Scale/X/Y/Z/Rotate/PScale/PX/PY/PZ/PRotate]: *(Pick a point, or enter an option.)*

Enter X scale factor, specify opposite corner, or [Corner/XYZ] <1>: *(Specify a value, pick a point, or enter an option.)*

Enter Y scale factor <use X scale factor>: *(Specify a value, or press Enter.)*

Specify rotation angle <0>: *(Specify a value, or press Enter.)*

COMMAND LINE OPTIONS

Insertion point picks the insertion point of the lower-left corner of the WMF image.

X scale factor scales the WMF image in the x direction (default = 1).

Corner scales the WMF image in the x and y directions.

XYZ scales the image in the x, y, and z directions.

Y scale factor scales the image in the y direction (default = x scale).

Rotation angle rotates the image (default = 0).

RELATED COMMANDS

WmfOpts specifies options for importing *.wmf* files.

WmfOut exports selected objects in WMF format.

RELATED FILES

*.clp are Windows Clipboard files.

*.wmf are Windows Metafiles.

TIPS

- The WMF image is placed as a block with the name WMF0; subsequent placements of *.wmf* files increase the number by one: WMF1, WMF2, and so on.

- The *.clp* files are created by the Windows Clipboard. After using Ctrl+C to copy objects to the Clipboard, you can open the Clipboard Viewer, and then save the image as a *.clp* file.

- Exploding the WMF*n* block results in polylines; even circles, arcs, and text are converted to polylines; solid-filled areas are exploded into solid triangles. The *.clp* file is pasted as a block; when exploded, constituent parts are 2D polylines.

- Use WmfOut and WmfIn to explode splines, text, circles, and arc to short line segments. (This is useful for translation to CNC machining files.) Do not zoom or pan between WmfOut and WmfIn.

- WmfIn needs a scale factor of 2 to match the size of objects created by WmfOut.

WmfOpts

<u>Rel.12</u> Specifies options for importing *.wmf* files (short for Windows Meta File OPTionS).

Command	Alias	Ctrl+	F-key	Alt+	Menu Bar	Dashboard
wmfopts	...	...	...	...	...	...

Command: wmfopts

Displays dialog box:

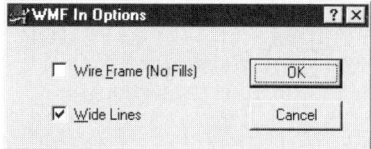

WMF IN OPTIONS DIALOG BOX

Wire Frame

☑ Displays the WMF images with lines only, no filled areas (default).

☐ Displays area fills.

Wide Lines

☑ Displays lines with width (default).

☐ Displays lines with a width of zero.

RELATED COMMANDS

WmfIn imports *.wmf* files.

WmfOut exports selected objects in WMF format.

WmfOut

Rel.12   Exports selected objects in WMF format (short for Windows MetaFile OUTput).

Command	Alias	Ctrl+	F-key	Alt+	Menu Bar	Dashboard
wmfout	...	...	...	FE	File	...
				⮡WMF	⮡Export	
					⮡Metafile	

Command: wmfout
Displays the Create WMF File dialog box. Enter a file name, and then click Save.
Select objects: *(Select one or more objects.)*
Select objects: *(Press Enter to end object selection.)*

COMMAND LINE OPTION
Select objects selects the objects to export. Press Ctrl+A to select all objects visible in the current viewport.

RELATED SYSTEM VARIABLE
WmfBkgnd toggles the background color of exported *.wmf* files:

0 — Transparent background.

1 — AutoCAD background color.

WmfForegnd switches the foreground and background colors of exported *.wmf* files as required:

0 — Foreground is darker than background color.

1 — Background is darker than foreground color.

RELATED COMMANDS
WmfIn imports files in WMF format.

CopyClip copies selected objects to the Clipboard in several formats, including *.wmf*; also called "picture" format.

TIPS
■ The *.wmf* files created by AutoCAD are resolution-dependent; small circles and arcs lose their roundness.

■ The All selection does not select all objects in the drawing; instead, the WmfOut command selects all objects *visible* in the current viewport.

■ Autodesk needs to update its WMF-related commands to handle EMF (enhanced meta format) files.

WorkSpace

Controls workspaces and settings at the command-line.

Command	Alias	Ctrl+	F-key	Alt+	Menu Bar	Dashboard
workspace	...	...	...	**TO**	**Tools**	...
					⇾**Workspaces**	
					⇾*workspace name*	

Command: workspace
Enter workspace option [setCurrent/SAveas/Edit/Rename/Delete/SEttings/ ?] <setCurrent>: *(Enter an option.)*

COMMAND LINE OPTIONS

setCurrent makes a workspace current.

SAveas saves the current interface configuration as a named workspace.

Edit allows modifications to workspaces; displays the Customize User Interface dialog box; see the CUI command.

Rename renames workspaces.

Delete erases workspaces.

SEttings displays the Workspace Settings dialog box; see WsSettings command.

RELATED COMMANDS

Cui creates and modifies workspaces through a dialog box.

WsSave saves the current user interface configuration as a new workspace.

WsSettings specifies options for workspaces.

RELATED SYSTEM VARIABLES

WsCurrent names the current workspace.

STARTUP SWITCH

/w specifies the workspace to show when AutoCAD starts up.

TIPS

- This command is meant for use by scripts and programs.

- The WsCurrent system variable can be used to change workspaces quickly at the command line.

- You can also change workspaces through:
    - The Workspaces item on the Windows menu.
    - The droplist on the Workspaces toolbar.

- The default workspace is named "2D Drafting & Annotation." If it has not been edited, you can select this workspace to return AutoCAD to the way it looked when it was first installed on your computer.

Commands • 849

WsSave

2006 Saves the current user interface as a named workspace (short for WorkSpace Save).

Command	Alias	Ctrl+	F-key	Alt+	Menu Bar	Dashboard
wssave	...	...	...	TO	Tools	...
					⌐Workspaces	
					⌐Save Current As	

-wssave

Command: wssave

Displays dialog box:

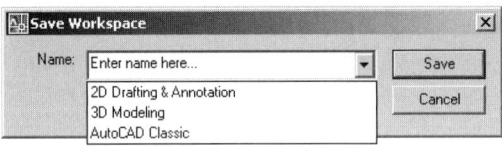

SAVE WORKSPACE DIALOG BOX

Name names the workspace. If the name already exists, AutoCAD warns, "Workspace already exists. Do you wish to replace it?" Click Yes or No.

-WSSAVE Command

Command: -wssave

Save Workspace as <AutoCAD Default>: *(Enter a workspace name.)*

OPTIONS

Save Workspace as names the workspace.

RELATED COMMANDS

Cui creates and modifies workspaces through a dialog box.

Workspace creates, saves, and controls workspaces at the command line.

WsSettings specifies options for workspaces.

RELATED SYSTEM VARIABLES

WsCurrent specifies the name of the current workspace.

TIP

■ Workspaces are saved in *.cui* files, and can be edited with the CUI command.

WsSettings

<u>2006</u>   Specifies which workspaces appear in the Workspace menu.

Command	Alias	Ctrl+	F-key	Alt+	Menu Bar	Dashboard
wssettings	...	...	...	TO	**Tools**	...
					↳**Workspaces**	
					↳**Workspace Settings**	

Command: wssettings

Displays dialog box:

WORKSPACE SETTINGS DIALOG BOX

My Workspace selects the default workspace.

Move Up moves the selected item up the list.

Move Down moves the selected item down the list.

Add Separator adds a gray line above the selected item, which shows up in the Workspaces menu.

⊙ **Do Not Save Changes to Workspace** does not save user interface changes to the current workspace before switching to another workspace.

○ **Automatically Save Workspace Changes** saves changes to the user interface before switching to another workspace.

RELATED COMMANDS

Cui creates and modifies workspaces through a dialog box.

Workspace creates, saves, and controls workspaces at the command line.

WsSave saves the current user interface configuration as a new workspace.

XAttach

<u>**Rel.14**</u> Attaches externally-referenced drawings to the current drawing (short for eXternal reference ATTACH).

Command	Alias	Ctrl+	F-key	Alt+	Menu Bar	Dashboard
xattach	xa	...	...	IR	Insert	...
					⤷ DWG Reference	

Command: xattach

Displays the Select File to Attach dialog box. Select a file, and click Open.

Displays dialog box:

After you click OK, AutoCAD confirms at the command line:

Attach Xref FILENAME: C:\filename.dwg

FILENAME loaded.

EXTERNAL REFERENCE DIALOG BOX

Name specifies the file name of the external *.dwg* file to be attached; thedrop list shows the names of currently-attached xrefs (externally-referenced files).

Browse displays the Select File To Attach dialog box.

Retain Path

☑ Saves the xref's filename and full path in the *.dwg* file.

☐ Saves only the filename of the xref.

Reference Type options

⦿ **Attachment** attaches the xref.

○ **Overlay** overlays the xref.

. . . .

Insertion Point options

☑ **Specify On-screen** specifies the insertion point of the xref in the drawing.

After you click **OK** *to dismiss the dialog box, AutoCAD prompts:*

Specify insertion point or [Scale/X/Y/Z/Rotate/PScale/PX/PY/PZ/PRotate]: *(Pick a point, or enter an option.)*

Scale sets the scale factor for the x, y, and z axes.

X sets the x-scale factor.

Y sets the y-scale factor.

Z sets the z-scale factor.

Rotate specifies the rotation angle.

PScale presets the scale factor for the x, y, and z axes.

PX presets the x-scale factor.

PY presets the y-scale factor.

PZ presets the z-scale factor.

PRotate presets the rotation angle.

Scale options

☐ **Specify On-screen** specifies the scale of the xref in the drawing.

X sets the x-scale factor.

Y sets the y-scale factor.

Z sets the z-scale factor.

☐ **Uniform scale** sets the Y and Z factors equal to X.

After you click OK to dismiss the dialog box, AutoCAD prompts:

Enter X scale factor, specify opposite corner, or [Corner/XYZ] <1>: *(Enter a value, or enter an option.)*

Enter Y scale factor <use X scale factor>: *(Enter a value, or press Enter.)*

X scale factor scales the xref in the x direction.

Corner indicates the x,y scale factor by picking two points of a rectangle.

XYZ specifies the scale factor in the x, y, and z directions.

Y scale factor scales the xref in the y direction.

Rotation Angle options

☐ **Specify On-screen** specifies the rotation of the xref in the drawing.

After you click **OK** *to dismiss the dialog box, AutoCAD prompts:*

Specify rotation angle <0>: *(Enter a value, or press Enter.)*

Rotation angle specifies the rotation angle of the xref.

RELATED SYSTEM VARIABLES

XRefType determines whether xrefs are attached or overlaid, by default.

XEdit determines whether the drawing may be edited in-place, when being referenced by another drawing.

XLoadPath stores the path of temporary copies of demand-loaded xref drawings.

XRefCtl controls whether *.xlg* external reference log files are written.

ProjectName names the project for the current drawing (default = "").

DemandLoad specifies if and when AutoCAD demand-loads a third-party application when a drawing contains custom objects created by the application:

0 — Turns off demand loading.

1 — Loads application when drawings contain proxy objects.

2 — Loads application when the application's commands are invoked.

3 — Loads application when drawings contain proxy objects, or when the application's commands are invoked.

IdxCtl controls the creation of layer and spatial indices:

0 — Creates no indices (default).

1 — Creates layer index.

2 — Creates spatial index.

3 — Creates both layer and spatial indices.

VisRetain specifies how the layer settings — on-off, freeze-thaw, color, and linetype — in xref drawings are defined by the current drawing:

0 — Xref layer definition in the current drawing takes precedence.

1 — Settings for xref-dependent layers take precedence over xref layer definition in the current drawing.

XLoadCtl controls the loading of xref drawings:

0 — Loads the entire xref drawing.

1 — Demand loading; xref is opened.

2 — Demand loading; copy of the xref is opened.

RELATED COMMANDS

XOpen edits xref drawings.

XBind binds portions of xref drawings to the current drawing.

XClip clips the display of xrefs.

ExternalReferences controls xrefs.

TIP

- When AutoCAD cannot find an xref, it searches in the following order:
    - The folder of the current drawing.
    - The project search paths defined in the Options dialog box's Files tab and the ProjectName system variable.
    - The support search paths defined in the Options dialog box's Files tab.
    - The Start In folder specified in the shortcut that launched AutoCAD.

XBind

Rel.11 Binds portions of externally-referenced drawings to the current drawing (short for eXternal BINDing).

Commands	Aliases	Ctrl+	F-key	Alt+	Menu Bar	Dashboard
xbind	xb	...	...	MO⬚EB	Modify ⬚Object 　⬚External Reference 　　⬚Bind	X19
-xbind	-xb					

Command: xbind

Displays dialog box:

XBIND DIALOG BOX

Xrefs lists externally-referenced drawings, along with their bindable objects: blocks, dimension styles, layer names, linetypes, and text styles.

Definitions to Bind lists definitions that will be bound.

Add adds a definition to the binding list.

Remove removes a definition from the binding list.

. . . .

-XBIND Command
Command: -xbind
Enter symbol type to bind [Block/Dimstyle/LAyer/LType/Style]: *(Enter an option.)*
Enter dependent name(s): *(Enter one or more names, separated by commas.)*

. . . .

COMMAND LINE OPTIONS

Block binds blocks to the current drawing.

Dimstyle binds dimension styles to the current drawing.

LAyer binds layer names to the current drawing.

LType binds linetype definitions to the current drawing.

Style binds text styles to the current drawing.

Enter dependent names specifies the named objects to bind.

RELATED COMMANDS

RefEdit edits xref drawings.

ExternalReferences controls xrefs.

TIPS

- The **XBind** command lets you copy named objects from another drawing to the current drawing.

- Before you can use the XBind command, you must first use the XAttach command to attach an xref to the current drawing.

- Blocks, dimension styles, layer names, linetypes, and text styles are known as "dependent symbols."

- When a dependent symbol is part of an xrefed drawing, AutoCAD uses a vertical bar (|) to separate the xref name from the symbol name, as in *filename|layername*.

- After you use the XBind command, AutoCAD replaces the vertical bar with **0**, as in *filename***0***layername*. The second time you bind that layer from that drawing, XBind increases the digit, as in *filename***1***layername*.

- When the XBind command binds a layer with a linetype (other than Continuous), it automatically binds the linetype.

- When the XBind command binds a block — with a nested block, dimension style, layer, linetype, text style, and/or reference to another xref — it automatically binds those objects as well.

XClip

Rel.12 Clips portions of blocks and externally-referenced drawings (short for eXternal CLIP; formerly the XRefClip command).

Command	Alias	Ctrl+	F-key	Alt+	Menu Bar	Dashboard
xclip	xc	...	...	MC⮑X	Modify	X18
					⮑Clip	
					⮑Xref	

Command: xclip
Select objects: *(Select one or more blocks or xrefs.)*
Select objects: *(Press Enter to end object selection.)*
Enter clipping option
[ON/OFF/Clipdepth/Delete/generate Polyline/New boundary] <New>: *(Enter an option.)*

Clipping boundary

Clipping turned off | *Clipping turned on*

COMMAND LINE OPTIONS

Select objects selects the xref or block, *not* the clipping polyline.

ON turns on clipped display.

OFF turns off clipped display; displays all of the xref or block.

Clipdepth sets front and back clipping planes for 3D xrefs and blocks; see 3dClip command.

Delete erases the clipping boundary.

generate Polyline extracts the existing boundary as a polyline.

New boundary places new rectangular or polygonal boundaries, inverts the boundary, or converts polylines into boundaries.

New Boundary options
Delete old boundary? [Yes/No] <Yes>: *(Type Y or N.)*
Current mode: Objects inside/outside boundary will be hidden.
Specify clipping boundary:
[Select polyline/Polygonal/Rectangular/Invert clip] <Rectangular>: *(Enter option.)*

Select Polyline prompts you to select a polyline in the drawing, which becomes the clipping boundary.

Polygonal prompts you to pick points that define the clipping boundary; the prompts are similar to those of the PLine command.

Rectangular prompts you to pick two points to define a rectangular boundary.

Invert Clip inverts the display of objects: objects outside the boundary are displayed, or vice versa (*new to AutoCAD 2008*).

RELATED COMMANDS

XBind binds parts of the xref to the current drawing.

DgnClip clips attached MicroStation V8 DGN files.

DwfClip clips attached DWF files.

ExternalReferences controls xrefs.

RELATED SYSTEM VARIABLE

XClipFrame toggles the display of the clipping boundary.

TIPS

- After drawing the clipping boundary and exiting this command, the boundary will probably be invisible. Use the XClipFrame system variable to show the clipping boundary.

- The properties of the boundary cannot be changed, other than the layer it is on. All other properties are ignored, such as color, linetype, and so on.

- An xref is limited to a single clipping boundary.

- When you select an existing polyline in the drawing, it must follow these rules:
  - Can be open, but AutoCAD closes it.
  - Must be drawn with straight line segments only (no arcs or splines).
  - Cannot intersect itself.

- A spline-fit polyline results in a curved clip boundary, but a curve-fit polyline does not.

- After you draw a polygonal clipping boundary, you can edit it with the PEdit command.

- Inverted clipped areas are displayed only with the 2D Wireframe visual style. As a workaround, use the Wipeout command.

- Use the Off option to reveal the clipped areas temporarily.

- You must use this command to remove clipping boundaries; the Erase command does not work.

XEdges

2007 Extracts wireframes from the edges of 3D solids and surfaces, and 2D regions (short for eXtract EDGES).

Command	Alias	Ctrl+	F-key	Alt+	Menu Bar	Dashboard
xedges	...	...	...	M3E	Modify	...
					⤷**3D Operations**	
					⤷**Extract Edges**	

Command: xedges
Select objects: *(Select one or more solids, surfaces, or regions. Hold down the Ctrl key to select individual edges.)*
Select objects: *(Press Enter.)*

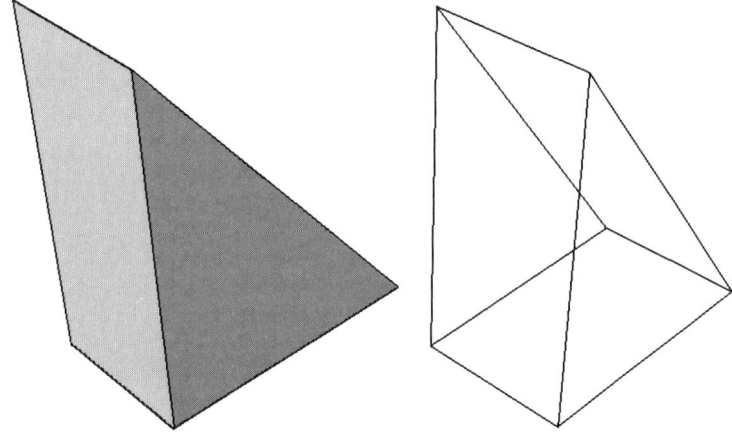

Left: Original 3D solid.
Right: Extracted edges forming 3D wireframe object.

TIPS

- This command does not work with spheres and tori, because they have no edges. Similarly, only the base of cones can be extracted.

- The command leaves edges "on top" of the original solid. Use grips editing to move the solid out of the way.

- Straight edges become lines; curves become arcs, circles, or splines.

- Xedges are created on the current layer.

XLine

Rel.13 Places infinitely long construction lines in drawings.

Command	Alias	Ctrl+	F-key	Alt+	Menu Bar	Dashboard
xline	xl	...	...	DT	Draw	2D Draw
					↳Construction Line	

Command: xline
Specify a point or [Hor/Ver/Ang/Bisect/Offset]: *(Pick a point, or enter an option.)*
Through point: *(Pick a point.)*
Through point: *(Press Enter to end the command.)*

Draws xlines, as illustrated here:

COMMAND LINE OPTIONS

Specify a point picks the midpoint for the xline.

Through point picks another point through which the xline passes.

Ang places the construction line at an angle.

Bisect bisects an angle with the construction line.

From point places the construction line through a point.

Hor places a horizontal construction line.

Offset places the construction line parallel to another object.

Ver places a vertical construction line.

ENTER exits the command.

Angle options
Enter angle of xline <0> or [Reference]: *(Enter an angle, or type R.)*

Enter angle of xline specifies the angle of the xline relative to the x-axis.

Reference specifies the angle relative to two points.

Bisect options
Specify angle vertex point: *(Pick a point.)*
Specify angle start point: *(Pick a point.)*
Specify angle end point: *(Pick a point.)*

Specify angle vertex point specifies the vertex of the angle.

Specify angle start point specifies the angle start point.

Specify angle end point specifies the angle endpoint.

Offset options
Specify offset distance or [Through] <1.0000>: *(Enter a distance, or type T.)*
Select a line object: *(Select a line, xline, ray, or polyline line segment.)*
Specify side to offset: *(Pick a point.)*

Specify offset distance specifies the distance between xlines.

Through picks a point through which the xline should pass.

Select a line object selects the line, xline, ray, or polyline to offset.

Specify side to offset specifies the offset side.

RELATED COMMANDS

Properties modifies characteristics of xline and ray objects.

Ray places semi-infinite construction lines.

RELATED SYSTEM VARIABLE

OffsetDist specifies the current offset distance.

TIPS

- Use xlines to find the bisectors of triangles (using MIDpoint object snap), or to create intersection snap points (using INTersection object snap).

- Ray and xline construction lines are plotted; they do not affect the extents.

- The technical editor asks these mathematical questions:

  - "Which is longer: an xline that is infinitely long in both directions, or a ray that is infinitely long in one direction?"

  - "If you break an xline, it becomes two rays. Now, which is longer?"

XOpen

2004 Opens externally-referenced drawings in new windows (short for eXternal OPEN).

Command	Alias	Ctrl+	F-key	Alt+	Menu Bar	Dashboard
xopen	...	...	...	...	...	...

Command: xopen
Select xref: *(Select an externally-referenced drawing.)*

Left: *original drawing.*
Right: *Opened xref in separate window.*

COMMAND LINE OPTION

Select xrefs selects the xref to open; the xref must be inserted in the current drawing.

TIPS

- This command does not work with blocks.

- When you select a non-xref object, AutoCAD complains, 'Object is not an Xref.'

'Xplode

Rel.12 Explodes complex objects into simpler objects, with user control (short for eXPLODE).

Command	Alias	Ctrl+	F-key	Alt+	Menu Bar	Dashboard
xplode	xp	...	...	...	...	...

Command: xplode
Select objects to XPlode.
Select objects: *(Select one or more objects.)*
Select objects: *(Press Enter to end object selection.)*
Enter an option [Individually/Globally] <Globally>: *(Type I or G.)*
Enter an option [All/Color/LAyer/LType/Inherit from parent block/Explode] <Explode>: *(Enter an option.)*

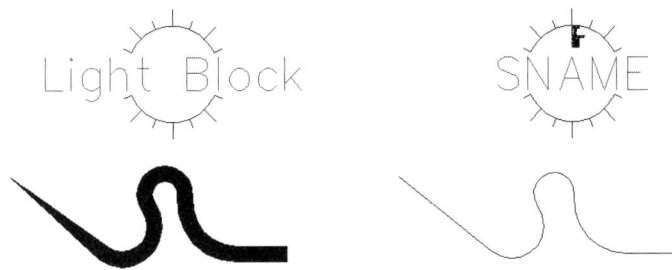

Left*: Block and polyline.*
Right*: Exploded objects lose attribute data and width .*

COMMAND LINE OPTIONS

Select objects selects objects to be exploded.

Individually allows you to specify options for each selected object.

Globally applies options to all selected objects.

All sets the color, layer, linetype and lineweight of exploded objects.

Color sets the color of objects after they are exploded: red, yellow, green, cyan, blue, magenta, white, bylayer, byblock, or any color number.

LWeight specifies a lineweight.

LAyer sets the layer for the exploded objects.

LType specifies any loaded linetype name for the exploded objects.

Inherit from parent block assigns the color, linetype, lineweight, and layer of the exploded objects, based on the original object.

Explode reduces complex objects into their components.

RELATED COMMANDS

Explode explodes the object without options.

U reverses the explosion.

TIPS

- Examples of complex objects include blocks and polylines; examples of simple objects include lines, circles, and arcs.

- Mirrored blocks can be exploded.

- The LWeight option is not displayed when the lineweight is off.

- The 'Enter an option [Individually/Globally]:' prompt appears only when more than one valid object is selected for explosion.

- Specifying BYLayer for the color or linetype means that the exploded objects take on the color or linetype of the object's original layer.

- Specifying BYBlock for the color or linetype means that the exploded objects take on the color or linetype of the original object.

- The default layer is the current layer, not the exploded object's original layer.

- The XPlode command breaks down complex objects as follows:

Object	Exploded into
Attribute	Attribute values are deleted; displays attribute definitions.
Block	Component objects.
Helix	Spline.
Leader	Line segments, splines, mtext, and tolerance objects; arrowheads become solids or blocks.
Mtext	Text.
Multiline	Line and arc segments.
PlaneSurf	Region.
Polyface mesh	Point, line, or 3D faces.
Polysolid	Region.
Region	Lines, arcs, and splines.
Table	Lines.
2D polyline	Line and arc segments; width and tangency are lost.
3D polyline	Line segments.
3D solid	Planar surfaces become regions; nonplanar surfaces become bodies.
3D body	Single-surface body, regions, or curves.

- The Inherit option works only when the parts were originally drawn with color, linetype, and lineweight set to BYBLOCK, and drawn on layer 0.

-XRef

<u>Rel.11</u>   Controls externally-referenced drawings in the current drawing at the command line (short for eXternal REFerence).

Commands	Alias	Ctrl+	F-key	Alt+	Menu Bar	Dashboard
-xref	-xr	...	...	...	...	...

Command: -xref
**Enter an option [?/Bind/Detach/Path/Unload/Reload/Overlay/Attach]
<Attach>:** *(Enter an option.)*

COMMAND LINE OPTIONS
? lists the names of xref files.

Bind makes the xref drawing part of the current drawing.

Detach removes xref files.

Path respecifies paths to xref files.

Unload unloads xref files.

Reload updates the xref files.

Overlay overlays the xref files.

Attach attaches another drawing to the current drawing.

RELATED COMMANDS
ExternalReferences displays a palette of all xrefs, image files, and DWF underlays.

Insert adds another drawing to the current drawing.

RefEdit edits xrefs.

XBind binds parts of xrefs to the current drawing.

XClip clips portions of xrefs.

RELATED SYSTEM VARIABLES
See the XAttach command.

TIPS
- *Warning!* Nested xrefs cannot be unloaded.

- The ExternalReferences command replaces the dialog-based XRef command, as of AutoCAD 2007. Use the undocumented ClassicXref command to access the old dialog box.

'Zoom

V. 1.0   Makes drawings larger or smaller in the current viewport.

Command	Aliases Ctrl+	F-key	Alt+	Menu Bar	Dashboard
zoom	z ...	...	VZ	View	2D Navigate
	rtzoom			↳Zoom	3D Navigate

Command: zoom

Specify corner of window, enter a scale factor (nX or nXP), or [All/Center/Dynamic/Extents/Previous/Scale/Window/Object] <real time>:
(Pick a point, enter an option, or press Enter.)

COMMAND LINE OPTIONS

(pick a point) begins the Window option.

realtime starts real-time zoom.

Enter *or* **Esc** ends real-time zoom.

All displays the drawing limits or extents, whichever is greater.

Dynamic brings up the dynamic zoom view.

Extents displays the current drawing extents.

Previous displays the previous view generated by the Pan, View, or Zoom commands.

Vmax displays the current virtual screen limits (short for Virtual MAXimum; undocumented).

Window indicates the two corners of the new view.

Object zooms to the extents of selected objects.

Center options
Specify center point: *(Pick a point.)*
Enter magnification or height <>: *(Enter a value.)*

Center point indicates the center point of the new view.

Enter magnification or height indicates a magnification value or height of view.

Left options (undocumented)
Lower left corner point: *(Pick a point.)*
Enter magnification or height <>: *(Enter a value.)*

Lower left corner point indicates the lower-left corner of the new view.

Enter magnification or height indicates a magnification value or height of view.

Scale(X/XP) options
*n***X** displays a new view as a factor of the current view.
*n***XP** displays a paper space view as a factor of model space.

RELATED COMMANDS

DsViewer displays the Aerial View palette, which zooms and pans.

Pan moves the view.

View saves zoomed views by name.

VtOptions controls smooth zoom transitions.

3dZoom performs real-time zooms in perspective viewing mode.

RELATED SYSTEM VARIABLES

ViewCtr reports the coordinates of the current view's center point.

ViewSize reports the height of the current view.

VtDuration specifies the duration of view transitions.

VtEnable turns on smooth view transitions.

VtFps specifies the speed of view transitions.

TIPS

- A magnification of 1x leaves the drawing unchanged; a magnification of 2x enlarges objects (zooms in), while 0.5x makes objects smaller (zooms out).

- Transparent zoom is *not* possible during the VPoint, Pan, DView, View and 3dOrbit commands.

- During real-time zoom, right-click in the drawing to see the shortcut menu displayed by the 3dOrbit command. But using the wheel mouse is much faster.

3D

Rel.11 Draws 3D mesh primitives with polygon meshes (short for three Dimensions).

Command	Alias	Ctrl+	F-key	Alt+	Menu Bar	Dashboard
3d	...	...	...	...	...	...

Command: 3d
Enter an option
[Box/Cone/DIsh/DOme/Mesh/Pyramid/Sphere/Torus/Wedge]: *(Enter an option.)*
See the Ai_ commands for details, such as Ai_Box and Ai_Wedge.

RELATED COMMANDS

Ai_Box draws 3D mesh boxes and cubes.

Ai_Cone draws 3D mesh cones.

Ai_Dish draws 3D mesh dishes.

Ai_Dome draws 3D mesh domes.

Ai_Mesh draws 3D meshes.

Ai_Pyramid draws 3D mesh pyramids.

Ai_Sphere draws 3D mesh spheres.

Ai_Torus draws 3D mesh tori.

Ai_Wedge draws 3D mesh wedges..

RELATED SYSTEM VARIABLES

SurfTab1 controls the mesh density in the M direction of mesh objects.

SurfTab2 controls the mesh density in the N direction of mesh objects.

TIPS

- This command creates 3D objects made of 3D polygon meshes, and *not* of 3D solids or surfaces.

- You cannot perform Boolean operations on 3D mesh models.

- To draw cylinders with endcaps, apply thickness to circles.

- Use the Ucs or Align command to place 3D mesh models in space; use the VPoint and 3dOrbit commands to view mesh models from different 3D viewpoints.

- You can apply the Hide and Render commands to 3D mesh models.

3dAlign

Moves and rotates objects in 2D and 3D space.

Command	Alias	Ctrl+	F-key	Alt+	Menu Bar	Dashboard
3dalign	3al	...	...	M3A	Modify	...
					↳3D Operation	
					↳3D Align	

Command: 3dalign
Select objects: *(Select one or more objects.)*
Select objects: *(Press Enter to end object selection.)*

Specify source plane and orientation ...
Specify base point or [Copy]: *(Pick a point, or type C.)*
Specify second point or [Continue] <C>: *(Press Enter, or pick another point.)*
Specify third point or [Continue] <C>: *(Press Enter, or pick a third point.)*

Specify destination plane and orientation ...
Specify first destination point: *(Pck a point.)*
Specify second destination point or [eXit] <X>: *(Press Enter, or pick another point.)*
Specify third destination point or [eXit] <X>: *(Press Enter, or pick a third point.)*

COMMAND LINE OPTIONS

Select objects selects one or more objects to be aligned.

Specify base, **Second**, and **Third point** specify the base, x axis, and y axis, respectively.

Copy copies the objects, instead of moving them.

Continue jumps ahead to the "Specify first destination point" prompt.

Specify First, **Second**, and **Third destination point** specify the basepoint and orientation of the moved/copied/rotated objects.

eXit exits the command.

RELATED COMMANDS

Align moves and rotates objects in 2D space.

3dMove moves objects in 3D space with the assistance of grip tools.

3dRotate rotates objects in 3D space with the assistance of grip tools.

TIP

- Use dynamic UCS to help position the moved/rotated objects more easily.

3dArray

<u>Rel.11</u> Creates 3D rectangular and polar arrays.

Command	Alias	Ctrl+	F-key	Alt+	Menu Bar	Dashboard
3darray	3a	...	...	M33	Modify ↳3D Operation ↳3D Array	...

Command: 3darray
Select objects: *(Select one or more objects.)*
Select objects: *(Press Enter to end object selection.)*
Enter the type of array [Rectangular/Polar] <R>: *(Type R or P.)*

3D rectangular array

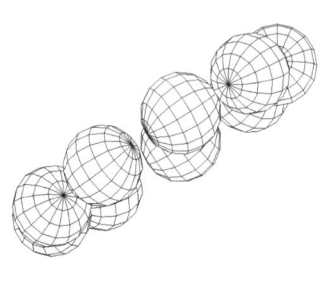

3D Polar array.

COMMAND LINE OPTIONS

Select objects selects the objects to be arrayed.

R creates rectangular 3D arrays.

P creates a polar array in 3D space.

Rectangular Array options
Enter the number of rows (---) <1>: *(Enter a value.)*
Enter the number of columns (|||) <1>: *(Enter a value.)*
Enter the number of levels (...) <1>: *(Enter a value.)*
Specify the distance between rows (---) <1>: *(Enter a value.)*
Specify the distance between columns (|||) <1>: *(Enter a value.)*
Specify the distance between levels (...) <1>: *(Enter a value.)*

Enter the number of rows specifies the number of rows in the x direction.

Enter the number of columns specifies the number of columns in the y direction.

Enter the number of levels specifies the number of levels in the z direction.

Specify the distance between rows specifies the distance between objects in the x direction.

Specify the distance between columns specifies the distance between objects in the y direction.

Specify the distance between levels specifies the distance between objects in the z direction.

Polar Array options
Enter the number of items in the array: *(Enter a number.)*
Specify the angle to fill (+=ccw, -=cw) <360>: *(Enter an angle.)*
Rotate arrayed objects? [Yes/No] <Y>: *(Type Y or N.)*
Specify center point of array: *(Pick a point.)*
Specify second point on axis of rotation: *(Pick a point.)*

Enter the number of items specifies the number of objects to array.

Specify the angle to fill specifies the distance along the circumference that objects are arrayed (default = 360 degrees).

Rotate arrayed objects?

Yes objects rotate so that they face the central axis (default).

No objects do not rotate.

Specify center point of array specifies the center point of the array and one end of the axis.

Specify second point on axis of rotation specifies the other end of the array axis.

ESC interrupts the drawing of arrays.

RELATED COMMANDS

Array creates a rectangular or polar array in 2D space.

Copy creates one or more copies of the selected object.

MInsert creates a rectangular block-array of blocks.

TIP

- This command does not operate in paper space.

'3dClip

__2000__ Performs real-time front and back clipping (short for three
Dimensional CLIPping).

Command	Alias	Ctrl+	F-key	Alt+	Menu Bar	Dashboard
3dclip	...	...	...	...	...	...

Command: 3dclip

Displays window:

SHORTCUT MENU

Right-click Adjust Clipping Planes window:

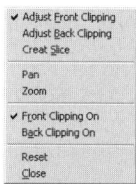

Adjust Front Clipping switches to front clipping mode.

Adjust Back Clipping switches to back clipping mode.

Create Slice switches to slicing — ganged front and back clipping — mode.

Pan moves the model within the window.

Zoom enlarges and reduces the view of the model within the window.

Front Clipping On toggles front clipping on and off.

Back Clipping On toggles back clipping on and off.

Reset removes clipping planes, and returns view to original state.

Close closes the Adjust Clipping Planes window.

TOOLBAR

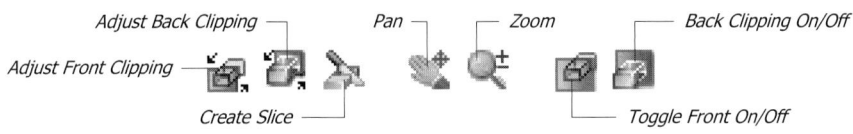

RELATED COMMANDS

3dOrbit provides real-time 3D viewing of the drawing.

3dPan performs real-time 3D sideways panning.

3dZoom performs real-time 3D zooming.

RELATED SYSTEM VARIABLES

PerspectiveClip determines the location of eyepoint clipping (*new to AutoCAD 2008*).

TIPS

- Use this command to hide objects in the front of 3D scenes, or to expose the interior of 3D models.

- Follow these steps to use this command effectively:
    1. Click Toggle Front On/Off to turn on front clipping.
    2. Drag the front clipping line over the model to cut off its front.
    3. Repeat, if necessary, with the back clipping plane.

- When you exit the Adjust Clipping Planes window, AutoCAD remains in 3dOrbit mode, and retains the clipped view. Press Esc to return to the 'Command:' prompt.

- Clipping planes are relative to the viewport. Clipping planes do not change when using 3dOrbit or other commands to rotate the view.

- The DWF Viewer is also capable of real-time clipping of 3D models exported in 3D DWF format by the 3dDwf command.

'3dConfig

Configures display characteristics of your computer's graphics board.

Command	Alias	Ctrl+	F-key	Alt+	Menu Bar	Dashboard
3dconfig	...	...	...	...	...	Visual Styles
-3dconfig						

Command: 3dconfig

Displays dialog box.

DIALOG BOX OPTIONS

☑ **Degrade When FPS Is Below** determines how AutoCAD degrades the screen image when motion falls below the specified FPS (frames per second), such as during the 3dOrbit command. Default is 5; range is 1 to 60 FPS.

Degradation Order specifies which visual effects are removed from display to prevent further degradation. The first item is degraded or removed first.

Move Up/Down changes the degradation order.

View Tune Log shows the results of AutoCAD's automatic evaluation of your computer's graphics board; displays the Performance Tuner Log dialog box.

. . . .

Manual Tune allows you to set the graphics board's capabilities; displays the Manual Performance Tuning dialog box.

Check for Updates launches Internet Explorer to access Autodesk's certification site.

Manual Performance Turning dialog box

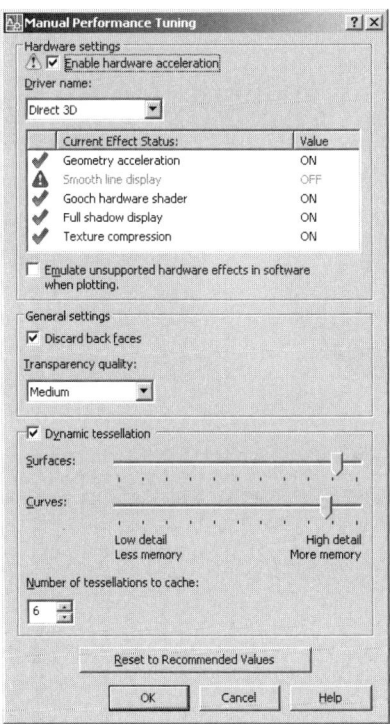

☑ **Enable Hardware Acceleration** toggles graphics generated by AutoCAD (software) or the graphics board (hardware). Usually hardware is faster, but if the graphics board lacks the capabilities listed under Effect, then it is better to turn off this option.

Warning! When Geometry Acceleration is turned off, full shadows cannot be displayed.

✔ (green) — Graphics board is capable of displaying all forms of graphics required by AutoCAD; hardware acceleration should be turned on.

⚠ (yellow) — Graphics card will run most but not all graphics; turn on hardware acceleration at the risk of AutoCAD locking up.

⚠ (red) — Graphics card is not capable; hardware acceleration is unavailable.

Driver Name selects the device driver for the graphics board. AutoCAD provides two drivers, and graphics board manufacturers may include their own:

- **OpenGL** — uses *wopengl9.hdi* driver file.
- **Direct 3D** — uses *direct3d9.hdi* driver file; required for Windows Vista.

☐ **Emulate supported hardware effects in software when plotting** emulates shadows and other effects not supported by the graphics board when plotting (*new to AutoCAD 2008*).

☑ **Discard Back Faces** does not draw the backs of faces, which are not seen anyhow.

Transparency Quality:

- **Low (Faster)** — dithers the image to simulate transparency.
- **Medium** — blends the image for improved quality.
- **High (Slower)** — highest quality but slower speed (default for hardware acceleration).

Left: *Transparency simulated by dithering.*
Right: *Hardware-accelerated transparency.*

☑ **Dynamic Tessellation** determines the smoothness of curved 3D objects.

Surface Tessellation specifies the level of surface details; less tessellation uses less memory.

Curve Tessellation specifies the level of curve details.

Number of Tessellations to Cache:

- **1** — tessellation is the same for all viewports; some objects may be regenerated during zooming.
- **2 or more** — required when drawings have two or more viewports with different views.

Reset to Recommended Values returns all options to values selected by Autodesk.

Command: -3dconfig
Configure: 3DCONFIG
Enter option [Adaptive degradation/Dynamic tessellation/General options/ acceLeration/Plot emulation/eXit] <Adaptive degradation>: *(Enter an option, or type V for undocumented adVanced options.)*

COMMAND LINE OPTIONS

Adaptive degradation allows AutoCAD to switch to lower-quality rendering to maintain display speed.

Dynamic tessellation specifies the smoothness of faceted objects in 3D drawings.

General options handles options not dependent on the graphics board.

Geometry specifies the display of isolines and back faces in 3D drawings.

acceLeration specifies hardware or software acceleration.

Plot emulation plots effects not supported by the graphics board, such as shadows *(new to AutoCAD 2008)*.

adVanced lists undocumented options.

eXit exits the command.

Adaptive degradation options
Enter mode [ON/OFF] <ON>: *(Type ON or OFF.)*

ON or **OFF** turns on or off adaptive degradation.

When on, the following options become available:
Enter speed to maintain (0-60fps) <5>: *(Enter a value.)*

Maintain speed fps specifies the speed at which to display, in frames per second; range is 5fps to 60fps.

Dynamic tessellation options
Enter option [Surface tessellation/Curve tessellation/Tessellations to cache/ eXit] <Surface tessellations>: *(Enter an option.)*

Surface tessellation specifies the number of tessellation lines to display on surfaces; ranges from 0 to 100 (default = 92).

Curve tessellation specifies the number of tessellation lines to display on curved surfaces; ranges from 0 to 100 (default = 87).

Tessellations to cache specifies the number of tessellations to cache; ranges from 1 to 4 (default = 4).

General options
Formerly named Geometry options.
Enter option [Discard backfaces/Transparency quality/eXit] <Discard backfaces>: *(Enter an option.)*

Discard backfaces:

On discards back faces to enhance performance.

Off displays back faces.

Configure: Transparency - Enter mode sets the quality of transparency to low, medium, or high.

acceLeration options

Enter option [Hardware/Software/eXit] <Hardware>: *(Enter an option.)*

Hardware configures graphics card to perform 3D display rendering, the faster option.

Software configures 3D display through software, if the graphics board is unable to.

Enter option [Driver name/Geometry acceleration/Antialias lines/Shadows enabled/Texture compression/eXit] <Driver name>:

Driver name selects a specific driver for the graphics board.

Geometry acceleration allows a more precise display, if turned on and supported by the graphics board.

Antialias lines uses anti-aliasing to drawn smoother lines.

Shadows enabled enables shadow-casting, if the graphics board is capable. (This option is not displayed when Software is selected.)

Texture compression toggles image compression (*new to AutoCAD 2008*).

adVanced

Redraw on window expose toggles redrawing windows when they become visible.

RELATED COMMAND LINE SWITCH

/nohardware turns off hardware acceleration when AutoCAD starts up (*new to AutoCAD 2008*).

TIPS

- Turning on Hardware Accerleration can improve the speed and quality of displays in realistic and conceptual visual styles. However, some effects in AutoCAD no longer work, such as transparent Tools palettes.

- Turning on Hardware Acceleration can cause AutoCAD to crash when starting a 3D drawing.

- When Texture Compression is on, the graphics board uses less memory for storing image-based materials and images; however, opening drawings may take longer and image quality may be reduced.

'3dDistance/Zoom/Swivel/Pan

2000   This collection of commands is a subset of 3dOrbit.

Command	Alias	Ctrl+	F-key	Alt+	Menu Bar
3ddistance	...	...	...	VC⌦A	View ⌦Camera ⌦Adjust Distance
3dzoom	...	...	...	...	...
3dswivel	...	...	...	VC⌦S	View ⌦Camera ⌦Swivel
3dpan	...	...	...	...	...

 Command: 3ddistance

Interactively changes the 3D viewing distance, closer to and further from the target.

 Command: 3dzoom

Interactively zooms the 3D view, just like the 3dDistance command.

 Command: 3dswivel

Interactively twists the 3D view about the target; the effect is like a rotating pan.

 Command: 3dpan

Interactively pans the 3D view side to side; the effect is flat movement.

COMMAND LINE OPTIONS

Enter *or* Esc exits the command.

RELATED COMMANDS

3dClip performs real-time 3D front and back clipping.

3dOrbit performs real-time 3D viewing of the drawing.

3dOrbitCtr specifies the target point for 3D orbiting views.

TIPS

■ You can use the 3dPan and 3dZoom commands when the current drawing is in perspective mode; when you try to use the regular Pan and Zoom commands, AutoCAD complains, '** That command may not be invoked in a perspective view **.'

■ Autodesk notes that the following commands exist but are not documented:
  • 3dZoomTransparent
  • 3dSwivelTransparent
  • 3dPanTransparent

These appear to be used by the 3dOrbit command's shortcut menu.

3dDwf

2006 Saves drawings in 3D DWF format.

Command	Alias	F-key	Alt+	Menu Bar	Dashboard
3ddwf	3ddwfpublish	...	...	...	...

Command: 3ddwf

This command operates only in model tab.

Displays Export 3D DWF dialog box. Enter a file name, and then click Save.

To specify options, select Tools | Options from the file dialog box.

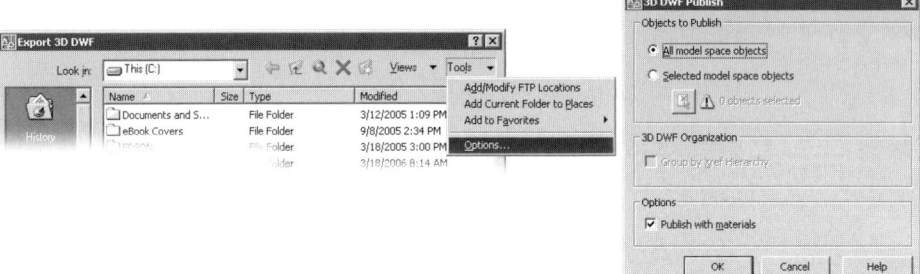

3D DWF PUBLISH DIALOG BOX

Objects to Publish options

⊙ **All Model Space Objects** saves the entire drawing.

○ **Select Model Space Objects** saves selected objects.

removes dialog box for selecting objects. AutoCAD prompts, "Select objects."

3D DWF Organization options

☐ **Group by Xref Hierarchy** sorts objects by external-reference hierarchy in the DWF Viewer. This option is grayed out when the drawing has links to externally-referenced drawings.

Options options

☑ **Publish with materials** includes material definitions for objects.

RELATED COMMANDS

Plot exports drawings in 2D DWF format.

Publish exports drawing sets in 2D DWF format.

AutoPublish exports drawings in DWF format when saved or closed (*new to AutoCAD 2008*).

RELATED SYSTEM VARIABLE

3dDwfPrec specifies the precision of 3D *.dwf* files:

1 — 1.000 units.	4 — 0.100
2 — 0.500	5 — 0.010
3 — 0.200	6 — 0.001

TIPS

- This command works only in model space; it cannot create *.dwf* files of layouts (paper space).

- This command was renamed from 3dDwfPublish in AutoCAD 2007.

- The level of precision defined by the 3dDwfPrec system variable affects file size and visual accuracy. For example, as precision is increased from level 1 to 6, the *steering.dwf* file size increases 11x, from 45KB to 513KB, while the visual accuracy of curves improves significantly:

Left: 3D DWF file created with precision set to 1.
Right: Same drawing exported with precision set to 6.

- When done saving the drawing in 3D DWF format, AutoCAD asks, "Filename.dwf was published successfully. Would you like to see it now?" Click Yes to view in DWF Viewer, which is included free with AutoCAD:

- To markup DWF files, download the free DWF Design Review software from Autodesk at www.autodesk.com/designreview (after registration).

Renamed Command

3dDwfPublish was renamed 3dDwf in AutoCAD 2007.

3dFace

V. 2.6 Draws 3D faces with three or four corners.

Command	Alias	Ctrl+	F-key	Alt+	Menu Bar	Dashboard
3dface	3f	...	...	DMMF	Draw	...
					⌐Surfaces	
					⌐3D Face	

Command: 3dface
Specify first point or [Invisible]: *(Pick a point, or type I followed by the point.)*
Specify second point or [Invisible]: *(Pick a point, or type I followed by the point.)*
Specify third point or [Invisible] <exit>: *(Pick a point, or type I followed by the point.)*
Specify fourth point or [Invisible] <create three-sided face>: *(Pick a point, or type I followed by the point.)*
Specify third point or [Invisible] <exit>: *(Press Enter to exit the command.)*

COMMAND LINE OPTIONS

First point picks the first corner of the face.

Second point picks the second corner of the face.

Third point picks the third corner of the face.

Fourth point picks the fourth corner of the face; or press Enter to create a triangular face.

Invisible makes the edge invisible.

RELATED COMMANDS

3D draws a 3D object: box, cone, dome, dish, pyramid, sphere, torus, or wedge.

Properties modifies the 3d face, including visibility of edges.

Edge changes the visibility of the edges of 3D faces.

EdgeSurf draws 3D surfaces made of 3D meshes.

PEdit edits 3D meshes.

PFace draws generalized 3D meshes.

RELATED SYSTEM VARIABLE

SplFrame controls the visibility of edges.

TIPS

- A 3D face is the same as a 2D solid, except that each corner can have a different z coordinate.

- Unlike the procedure for Solid, corner coordinates are entered in natural order.

- The **i** (short for invisible) suffix must be entered before osnap modes, point filters, and corner coordinates.

- *Invisible* 3D faces, where all four edges are invisible, do not appear in wireframe views; they hide objects behind them in hidden-line mode, and are rendered in shaded views.

- You can use the Properties and Edge commands to change the visibility of 3D face edges.

- 3D faces cannot be extruded.

3dMesh

Rel.10 Draws polygon 3D meshes.

Command	Alias	Ctrl+	F-key	Alt+	Menu Bar	Dashboard
3dmesh	...	...	...	DMMM⮑	Draw	...
					⮑Modeling	
					⮑3D Mesh	

Command: 3dmesh
Enter size of mesh in M direction: *(Enter a value.)*
Enter size of mesh in N direction: *(Enter a value.)*
Specify location for vertex (0, 0): *(Pick a point.)*
Specify location for vertex (0, 1): *(Pick a point.)*
Specify location for vertex (1, 0): *(Pick a point.)*
Specify location for vertex (1, 1): *(Pick a point.)*

COMMAND LINE OPTIONS

Enter size of mesh in M direction specifies the m-direction mesh size (between 2 and 256).

Enter size of mesh in N direction specifies the n-direction mesh size (between 2 and 256).

Specify location for vertex (*m*,*n*) specifies a 2D or 3D coordinate for each vertex.

RELATED COMMANDS

3D draws a variety of 3D objects.

3dFace draws a 3D face with three or four corners.

Explode explodes a 3D mesh into individual 3D faces.

PEdit edits a 3D mesh.

PFace draws a generalized 3D face.

Xplode explodes a group of 3D meshes.

TIPS

- It is more convenient to use the EdgeSurf, RevSurf, RuleSurf, and TabSurf commands than the 3dMesh command. The 3dMesh command is meant for use by AutoLISP and other programs.

- The range of values for the m- and n-mesh size is 2 to 256.

- The number of vertices = **m** x **n**.

- The first vertex is (0,0). The vertices can be at any point in space.

- The coordinates for each vertex in row **m** must be entered before starting on vertices in row **m+1**.

- Use the PEdit command to close the mesh, since it is always created open.

- The SurfU and SurfV system variables do not affect the 3D mesh object.

3dMove

<u>2007</u> Moves objects along the x, y, or z axis.

Command	Alias	Ctrl+	F-key	Alt+	Menu Bar	Dashboard
3dmove	3m	...	...	M3M	Modify	3D Make
					⬑3D Operations	
					⬑3D Move	

Command: 3dmove
Current positive angle in UCS: ANGDIR=counterclockwise ANGBASE=0
Select objects: *(Pick one or more objects.)*
Select objects: *(Press Enter.)*
Specify base point or [Displacement] <Displacement>: *(Pick a point, or type D.)*
Specify second point or <use first point as displacement>: *(Pick another point.)*

COMMAND LINE OPTIONS

Select objects selects the objects to be rotated.

Specify base point indicates the starting point for the move.

Displacement specifies relative x,y,z displacement when you press Enter at the next prompt.

Specify second point of displacement indicates the distance to move.

3D Move grip tool

Z Axis (blue)
Base Point
X Axis (red)
Y Axis (green)

RELATED COMMANDS

Move moves objects in 2D (current UCS) and 3D, but without the grip tools.

3dRotate rotates objects in 3D space.

RELATED SYSTEM VARIABLES

GtAuto toggles the display of grips tools.

GtDefault determines whether the 3dMove or Move command are activated automatically in 3D space when you enter the Move command.

GtLocation locates the grip tool at the UCS icon or on the last selected object.

TIPS

- You can drag the base point (in the middle of the grip tool) to relocate the base point.

- When dynamic UCS is turned on, the grip tool aligns itself with the selected face. (Press Ctrl+D or click DUCS to toggle dynamic UCS).

'3dOrbit, '3dCOrbit, '3dFOrbit

2000 Provides interactive 3D viewing of drawings.

Command	Alias	Ctrl+	F-key	Alt+	Menu Bar	Dashboard
3dorbit	orbit	...	...	VB	View	3D Navigate
	3do				�バ Orbit	
					�バ Constrained	
3dcorbit	...	...	...	VB	View	3D Navigate
					�バ Orbit	
					�バ Continuous	
3dforbit	...	...	...	VBF	View	3D Navigate
					�バ Orbit	
					�バ Free Orbit	

Command: 3dorbit *or* 3dcorbit *or* 3dforbit
Press ESC or ENTER to exit, or right-click to display shortcut-menu. *(Move the cursor to change the viewpoint, and then press Enter to exit the command.)*

SHORTCUT MENU
Right-click the drawing:

Exit exits the command.

Current Mode: Constrained Orbit reports the current mode; not a command option.

Other Navigation Modes displays a submenu that accesses other 3D viewing commands.

Enable Orbit Auto Target maintains the view on a target point instead of on the center of the viewport.

Animation Settings displays the Animation Settings dialog box.

Zoom Window zooms into a rectangular area defined by two pick points; see the Zoom command.

Zoom Extents zooms to the extents of the drawing to display all visible objects.

Zoom Previous returns to the prior view.

Parallel displays the view in parallel projection; more accurate.

Perspective displays the view in perspective projection; more realistic.

Reset View resets to the view to when you first began the command.

Preset Views displays a submenu of standard views, such as top and side.

Named Views displays a submenu of user-named views, if any are defined in the drawing (see the **View** command).

Visual Styles displays a submenu for selecting the type of visual style (formerly named Shading Modes); see the VisualStyles command.

Visual Aids displays a submenu of visual aids for navigating in 3D space.

Other Navigation Modes submenu

To switch between navigation modes, you can press the number keys 1 - 9 at any time during this command, without having to access the following shortcut submenu.

Constrained Orbit constrains movement to the x,y plane or the z axis; see the 3dOrbit command.

Free Orbit freely rotates the 3D view (3dFOrbit).

Continuous Orbit keeps the view spinning in the direction determined by dragging the mouse across the model (3dCOrbit).

Adjust Distance zooms the viewpoint nearer and farther away (3dDistance).

Swivel pans the viewpoint from side to side (3dSwivelTransparent).

Walk moves through the 3D model at a fixed height (3dWalk).

Fly moves through the 3D model at any height (3dFly).

Zoom zooms in and out (3dZoomTransparent).

Pan pans around (3dPanTransparent).

Animation Settings dialog box

Visual Style selects the visual style by which to save the animation.

Resolution specifies the image quality; ranges from 160x120 (low quality, small file) to 1024x768 (high quality, large size).

Frame rate (FPS) specifies the frames per second, ranging from 1 to 60fps.

Format selects the movie format in which to save the file: Microsoft AVI, QuickTime MOV, generic MPG, or Windows WMV.

*To record an animation to file during the 3dOrbit command, click the **Record** button in the Dashboard's Navigation control panel:*

Start Recording Animation ————

Playback Animation File ————

Stop Recording and Save to File

Pause Recording

*1. Click **Start Recording Animation**.*

2. Move cursor to animate the view.

*3. Click **Stop Recording**, and then name the file.*

Visual Aids options

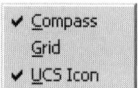

✔ Compass
Grid
✔ UCS Icon

Compass toggles the display of the compass to help you navigate in 3D space.

Grid toggles the display of the grid as lines, to help you see the x,y-plane; see the Grid command.

UCS Icon toggles the display of the 3D UCS icon; see the UcsIcon command.

RELATED COMMANDS

Vpoint creates static 3D viewpoints.

3dClip performs real-time 3D front and back clipping.

3dDistance performs real-time 3D forward and backward panning.

3dPan and **3dZoom** perform real-time 3D panning and zooming.

3dWalk and **3dFly** move through 3D views.

AniPath saves flythroughs to movie files.

RELATED SYSTEM VARIABLE

Perspective reports whether the viewport is displaying a parallel or perspective projection.

TIPS

- If you select one or more objects before invoking this command, they will appear while orbiting.

- To activate this command transparently during other commands, hold down the Shift key while pressing the mouse wheel.

- No transparent commands can be entered while this command is active.

- This command is meant to replace the DView command.

'3dOrbitCtr

2004   Specifies the target point for 3D orbiting views.

Command	Alias	Ctrl+	F-key	Alt+	Menu Bar	Dashboard
3dorbitctr	...	...	...	...	...	...

Command: 3dorbitctr
Specify orbit center point: *(Pick a point.)*

The command then enters 3dOrbit mode.

COMMAND LINE OPTION

Specify orbit center point specifies the center of view rotation.

RELATED COMMANDS

3dCOrbit places the drawing in real-time, 3D, continuous orbit mode.

3dFOrbit performs real-time 3D viewing.

3dOrbit performs real-time 3D viewing fixed on the x,y-plane.

3dPoly

Rel.10 Draws 3D polylines (short for 3D POLYline).

Command	Alias	Ctrl+	F-key	Alt+	Menu Bar	Dashboard
3dpoly	3p	...	...	D3	Draw	...
					↳**3D Polyline**	

Command: 3dpoly
Specify start point of polyline: *(Pick a point.)*
Specify endpoint of line or [Undo]: *(Pick a point, or type U.)*
Specify endpoint of line or [Undo]: *(Pick a point, or type U.)*
Specify endpoint of line or [Close/Undo]: *(Pick a point, type U or C, or press Enter to end the command.)*

COMMAND LINE OPTIONS

Specify start point indicates the starting point of the 3D polyline.

Close joins the last endpoint with the start point.

Undo erases the last-drawn segment.

Specify endpoint of line indicates the endpoint of the current segment.

ENTER ends the command.

RELATED COMMANDS

Explode reduces a 3D polyline into lines and arcs.

PEdit edits 3D polylines.

PLine draws 2D polylines.

TIPS

- Because 3D polylines are made of straight lines, you can use the PEdit command to spline-fit the polyline.

- 3D polylines do not support linetypes and widths.

- You may use lineweights to fatten up a 3D polyline.

- Use the .xy point filter to place pick points with object snaps, and then specify the z coordinate.

3dRotate

2007 Rotates objects about the x, y, or z axis.

Command	Alias	Ctrl+	F-key	Alt+	Menu Bar	Dashboard
3drotate	3r	...	...	M3R	Modify	3D Draw
					⇘3D Operations	
					⇘3D Rotate	

Command: 3rotate
Current positive angle in UCS: ANGDIR=counterclockwise ANGBASE=0
Select objects: *(Pick one or more objects.)*
Select objects: *(Press Enter.)*
Specify base point: *(Pick a point; locates the grip tool.)*
Pick a rotation axis: *(Pick one of the three axes.)*
Specify angle start point: *(Pick a point.)*
Specify angle end point: *(Pick another point.)*

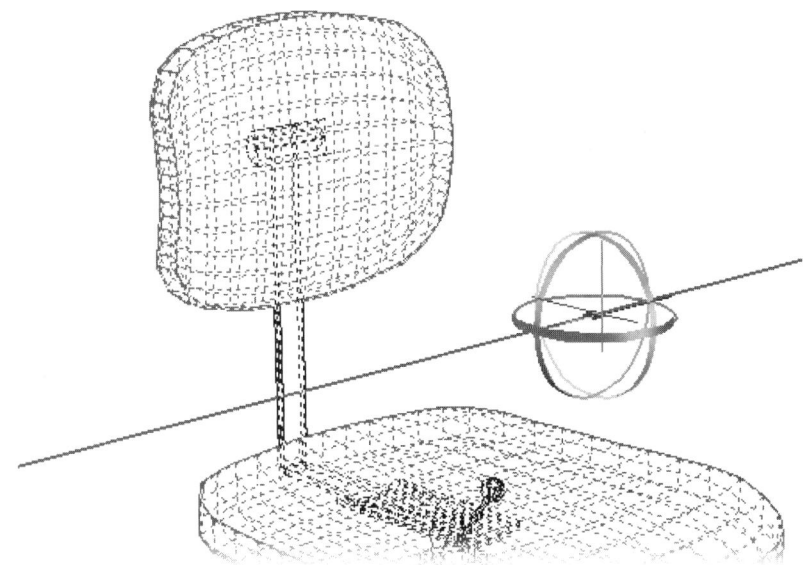

COMMAND LINE OPTIONS

Select objects selects the objects to be rotated.

Specify base point picks the point about which the objects will be rotated.

Pick a rotation axis selects the x (red), y (green), or z (blue) axis about which to rotate; the selected axis turns yellow.

Specify angle start point picks the starting point for rotation.

Specify angle end point picks the ending point for rotation.

3D Rotate grip tool

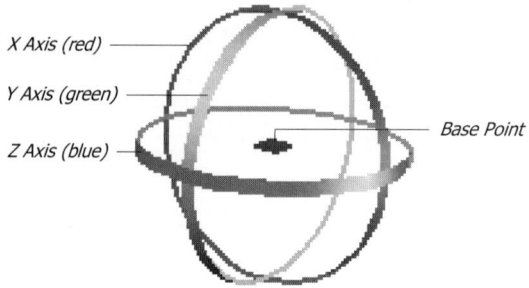

X Axis (red)
Y Axis (green)
Z Axis (blue)
Base Point

RELATED COMMANDS

Rotate rotates objects in 2D space (current UCS).

Rotate3d rotates objects in 3D space without the grip tool.

3dMove moves objects in 3D space.

RELATED SYSTEM VARIABLES

GtAuto toggles the display of grips tools.

GtDefault determines whether the 3dRotate or Rotate command is activated automatically in 3D space when you enter the Rotate command.

GtLocation locates the grip tool at the UCS icon or on the last selected object.

AngDir specifies the direction of positive angles (default =counterclockwise).

AngBase specifies the 0-angle relative to the current UCS (default = 0).

TIPS

- After starting the 3dRotate command, press the Spacebar to switch to the 3dMove command.

- When dynamic UCS is turned on, the grip tool aligns itself with the selected face. (Press Ctrl+D or click DUCS to toggle dynamic UCS.)

3dsIn

Rel.13   Imports *.3ds* files created by 3D Studio and other applications (short for 3D Studio IN).

Command	Alias	Ctrl+	F-key	Alt+	Menu Bar	Dashboard
3dsin	...	...	...	I3	Insert ⤷3D Studio	...

Command: 3dsin

Displays the 3D Studio File Import dialog box. Select a .3ds file, and then click Open.

AutoCAD displays dialog box:

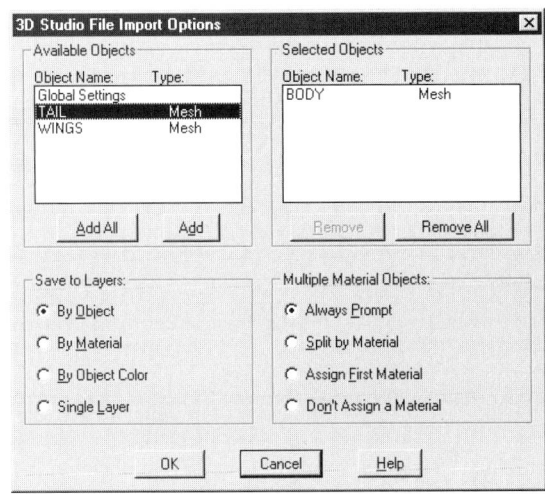

3D STUDIO FILE IMPORT OPTIONS DIALOG BOX

Available and Selected Objects options

Object Name names the object.

Type specifies the type of object.

Add adds the object to the Selected Objects list.

Add All adds all objects to the Selected Objects list.

Remove removes the object from the Selected Objects list.

Remove All removes all objects from the Selected Objects list.

Save to Layers options

⊙ **By Object** places each object on its own layer.

○ **By Material** places the objects on layers named after materials.

○ **By Object Color** places the objects on layers named "Color*nn*."

○ **Single Layer** places all objects on layer "AvLayer."

Multiple Material Objects options

⊙ **Always Prompt** prompts for each material.

○ **Split by Material** splits objects with more than one material into multiple objects, each with one material.

○ **Assign First Material** assigns the first material to the entire object.

○ **Don't Assign to a Material** removes all 3D Studio material definitions.

RELATED COMMANDS

None.

RELATED FILES

.3ds are 3D Studio files.

.tga are converted bitmap and animation files.

TIPS

- You are limited to selecting a maximum of 70 3D Studio objects.

- Conflicting object names are truncated and given a sequence number.

- The By Object option gives the AutoCAD layer the name of the object.

- The By Object Color option places all objects on layer "ColorNone" when no colors are defined in the 3DS file.

- 3D Studio assigns materials to faces, elements, and objects; AutoCAD assigns materials only to objects, colors, and layers.

- 3D Studio bitmaps are converted to *.tga* (Targa format) bitmaps. Converted *.tga* files are saved to the *.3ds* file folder.

- Only the first frame of animation files (CEL, CLI, FLC, and IFL) is converted to Targa bitmap files.

- 3D Studio "ambient lights" lose their color; 3D Studio "omni lights" become point lights in AutoCAD; 3D Studio "cameras" become named views in AutoCAD.

Removed Command

3dsOut was removed from AutoCAD 2007.

'3dWalk, '3dFly

2007 Moves through 3D models interactively.

Command	Alias	Ctrl+	F-key	Alt+	Menu Bar	Dashboard
3dwalk	**3dw**	...	...	**VKK**	**View**	**3D Navigate**
	3dnavigate				⮡**Walk and Fly**	
					⮡**Walk**	
3dfly	...	...	...	**VKF**	**View**	**3D Navigate**
					⮡**Walk and Fly**	
					⮡**Fly**	

Command: 3dwalk *or* 3dfly

Press ESC or ENTER to exit, or right-click to display shortcut-menu. *(Drag mouse to move around, or enter these keystrokes:)*

Movement	Left Side	Right Side of Keyboard
Move forward	W	Up-arrow
Move back	S	Down-arrow
Move left	A	Left-arrow
Move right	D	Right-arrow
Toggle Walk-Fly mode	F	...
Display mappings	Tab	...
Exit walk/fly mode	Esc	Enter

Both commands display the same palette:

Close Palette — Toolbar

Preview

Target Indicator

Position Indicator

Close Options

Preview Options

Palette Options — Resize Palette

POSITION LOCATOR

General

Position indicator color	■ Red
Position indicator size	Large
Position indicator blink	Off
Target indicator	On
Target indicator color	□ Green
Preview transparency	N/A
Preview visual style	Realistic

PALETTE OPTIONS

Position Indicator Color specifies the color of the large dot that shows your current position in the 3D model.

Position Indicator Size sets the size of the indicator between Small (the default), Medium, and Large.

Position Indicator Blink toggles blinking of the position indicator.

Target Indicator toggles the display of the target.

Target Indicator Color specifies the color of the target indicator.

Preview Transparency changes the translucency of this palette, if supported by the graphics board; range is 0 (opaque) to 95 (nearly transparent).

Preview Visual Style specifies the visual style of the preview image.

PALETTE TOOLBAR

The zoom and pan buttons apply to the preview image in the palette:

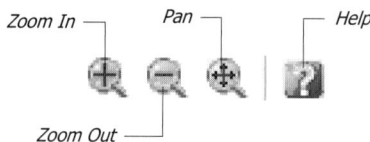

Zoom In — Pan — Help

Zoom Out —

SHORTCUT MENU

The shortcut menu is nearly-identical to that of the 3dOrbit command:

Unique to the 3dWalk and 3dFly commands are these options:

Display Position Locator toggles the display of the Position Locator palette.

Walk and Fly Settings displays the Walk and Fly Settings dialog box (WalkFlySettings command).

RELATED COMMANDS

WalkFlySettings presets the step size and rate.

AniPath saves flythroughs to movie files.

3dOrbit circles 3D models.

RELATED SYSTEM VARIABLES

StepsPerSec specifies the number of steps per second.

StepSize specifies the distance per step.

TIPS

- The 3dWalk command keeps the elevation fixed, while 3dFly is independent of elevation.

- If AutoCAD is not in perspective mode, this command displays the following dialog box:

- In addition to moving the cursor in the drawing, you can also click-and-drag the view in the Preview area of the Position Locator palette. Notice the hand cursor in the figure below:

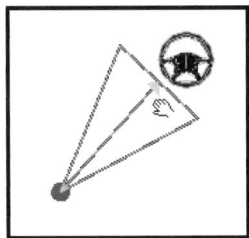

- Press Tab to show the window displaying walk'n fly keystrokes; they cannot be customized.

- You can switch between viewing modes using these keystrokes:
    - **1** — constrained orbit (same as the 3dOrbit command).
    - **2** — free orbiting (3dFOrbit).
    - **3** — continuous orbiting (3dCOrbit).
    - **4** — adjust distance (3dDistance).
    - **5** — swivel (3dSwivel).
    - **6** — walk (3dWalk).
    - **7** — fly (3dFly).
    - **8** — zoom (3dZoom).
    - **9** — pan (3dPan).

Express Tools

The following commands are included with the Express Tools package.

Command	Description
AcadInfo	Reports on the status of AutoCAD, and stores the data in *acadinfo.txt*.
AliasEdit	Edits the aliases stored in *acad.pgp*.
AlignSpace	Aligns model space objects, whether in different viewports or with objects in paper space.
ArcText	Places text along an arc.
AttIn / AttOut	Imports and extracts attributes in tab-delimited format.
BCount	Counts inserted blocks, and then generates a report.
BlockReplace	Replaces all inserts of one block with another.
BlockToXref	Converts blocks to xrefs.
Block?	Lists objects stored in blocks.
BreakLine	Creates the break-line symbol.
BScale	Scales blocks from their insertion points.
Burst	Explodes blocks, and converts attributes to text.
CdOrder	Sets the display order by color.
ChUrls	Invokes DdEdit-like editor for editing hyperlinks (URL addresses).
ClipIt	Adds arcs, circles, and polylines to the XClip command.
CopyM	Runs the Copy command with repeat, divide, measure, and array options.
DimEx / DimIm	Exports and imports dimension styles to/from ASCII files.
DimReassoc	Restores measurement to overridden text.
▣ DumpShx	Converts SHX files to SHP.
▣ DwgLog	Keeps a log file for every drawing.
EditTime	Pauses the timer when not editing.
ExpressMenu	Loads the menu for Express Tools.
ExpressTools	Loads the express tools collection and its menu.
ExOffset	Adds options to the Offset command.
ExTrim	Trims all objects at the specified cutting line.
ExPlan	Adds options to the Plan command.
Flatten	Reduces 3D drawings to 2D.
FS	Selects objects that touch the selected object (renamed from FastSelect).
GAttE	Changes attributes globally.
GetSel	Selects objects based on layer and type.
ImageEdit	Launches the image editor to edit selected images.
▣ Julian	Provides AutoLISP routines for converting Julian dates.
LayoutMerge	Combines layouts into a single layout.
Lsp	Provides AutoLISP function searching utilities.
LspSurf	Provides a LISP file viewer.
MkLtype	Creates linetypes from selected objects.
MkShape	Creates shapes from selected objects.

▣ New to AutoCAD 2008.

. . . .

Command	Description
MoCoRo	Moves, copies, rotates, and scales objects.
MoveBak	Moves *.bak* files to specified folders.
MPEdit	Acts exactly like the PEdit command.
MStretch	Stretches with multiple selection windows.
NCopy	Copies objects nested inside blocks and xrefs.
OverKill	Removes overlapping duplicate objects.
Plt2Dwg	Imports HPGL files into the drawings.
Propulate	Updates, lists, and clears drawing properties.
PsBScale	Sets and updates the scale of blocks relative to paper space.
PsTScale	Sets text height relative to paper space.
QlAttach	Associates leaders with annotation objects.
QlAttachSet	Associates leaders with annotations.
QlDetachSet	Disassociates leaders from annotations.
QQuit	Closes all drawings, and then exits AutoCAD.
ReDirMode	Sets options for the ReDir command.
ReDir	Changes paths for xrefs, images, shapes, and fonts.
RepUrls	Replaces hyperlinks.
Revert	Closes the drawing, and re-opens the original.
RtEdit	Edits remote text objects.
RText	Inserts and edits remote text objects.
RtUcs	Changes UCSs in real time.
SaveAll	Saves all drawings.
ShowUrls	Lists URLs in a dialog box.
Shp2blk	Converts a shape definition to a block definition.
Ssx	Creates selection sets.
SuperHatch	Uses images, blocks, external references, or wipeouts as hatch patterns.
SysvDlg	Launches an editor for system variables.
TCase	Changes text between Sentence, lower, UPPER, Title, and tOGGLE cASE.
TCircle	Surrounds text and multiline text with circles, slots, and rectangles.
TCount	Prefixes text with sequential numbers.
TextFit	Fits text between points.
TextMask	Places masks behind selected text.
TextUnmask	Removes masks behind selected text.
TFrames	Toggles the frames surrounding images and wipeouts.
TJust	Justifies text created with the MText and AttDef commands.
TOrient	Re-orients text, multiline text, and block attributes.
TrEx	Works like the Trim and Extend commands.
TScale	Scales text, multiline text, attributes, and attribute definitions.
Txt2Mtxt	Converts single-line to multiline text.
TxtExp	Explodes selected text into polylines.
VpScale	Lists the scale of the selected viewports.
VpSynch	Synchronizes viewports with a master viewport.
XData	Attaches xdata to objects.
XdList	Lists xdata attached to objects.
XList	Displays properties of objects nested in blocks and xref.

Obsolete & Removed Commands

The following commands have been removed or replaced since AutoCAD v 2.5. As of AutoCAD 2007, some operate only in 2D wireframe mode, such as Hide.

Command	Introduced	Removed	Replacement	Reaction
AcadBlockDialog	2000	2004	Block	Displays the Block dialog box
AcadWBlockDialog	2000	2004	WBlock	Displays the Write Block dialog box
AmeLite	R11	R12	Region	"Unknown command"
AscText	R11	R13	MText	"Unknown command"
Ase...	R12	R13	ASE...	"Unknown command"
				(Most R12 ASE commands were combined into fewer ASE commands with R13)
Ase...	R13	2000	dbConnect	"Unknown command"
AseUnload	R12	R14	Arx Unload	"Unknown command"
▣ Assist	2000i	2008	Infobar	"Unknown command"
▣ AssistClose	2000i	2008	*none*	"Unknown command"
Axis	v1.4	R12	*none*	"Discontinued command"
Background	R14	2007	View	"Unknown command"
BHatch	R12	2006	Hatch	Displays Hatch&Gradient dialog box
BMake	R12	2000	Block	Displays Block Definition dialog
BPoly	R12	2004	Boundary	Displays the Boundary dialog box
CConfig	R13	2000	PlotStyle	"Discontinued command"
Config	R12	R14	Options	Displays Options dialog
Content	2000	2004	AdCenter	Displays DesignCenter window
DdAttDef	R12	2000	AttDef	Displays Attribute Def dialog
DdAttE	R9	2000	AttEdit	Displays Edit Attributes dialog
DdAttExt	R12	2000	AttExt	Displays Attribute Ext dialog
DdChProp	R12	2000	Properties	Displays Properties window
DdColor	R13	2000	Color	Displays Select Color dialog
DdEModes	R9	R14	Object Properties	Dialog box explains change
DdGrips	R12	2000	Options	Displays Options dialog
DDim	R12	2000	DimStyle	Displays Dim Style Mgr dialog
DdInsert	R12	2000	Insert	Displays Insert dialog
DdLModes	R9	R14	Layer	Displays Layer Manager dialog
DdLType	R9	R14	Linetype	Displays Linetype Mgr dialog
DdModify	R12	2000	Properties	Displays Properties window
DdOSnap	R12	2000	DSettings	Displays Drafting Settings dialog
DdPlotStamp	2000i	2004	PlotStamp	Displays Plot Stamp dialog box
DdRename	R12	2000	Rename	"Unknown command"
DdRModes	R9	2000	DSettings	Displays Drafting Settings dialog

▣ Removed in AutoCAD 2008

Command	Introduced	Removed	Replacement	Reaction
DdSelect	R12	2000	Options	Displays Options dialog
DdUcs	R10	2000	UcsMan	Displays UCS dialog
DdUcsP	R12	2000	UcsMan	Displays UCS dialog
DdUnits	R12	2000	Units	Displays Units dialog
DdView	R12	2000	View	Displays View dialog
DText	v2.5	2000	Text	Executes Text command
DLine	R11	R13	MLine	"Unknown command"
DwfOut	R14	2004	Publish	Executes Plot command.
DwfOutD	R14	2000	DwfOut	"Unknown command"
🗒 EAttExt	2002	2008	DataExtraction	Runs DataExtraction command.
End	R11	R13	Quit	"Discontinued command"
EndRep	v1.0	v2.5	Minsert	"Discontinued command"
EndSv	v2.0	v2.5	End	"Discontinued command"
EndToday	2000i	2004	*none*	"Unknown command"
ExpressTools	2000	2002	ExpressTools	Restored in AutoCAD 2004
Files	v1.4	R14	*Explorer*	"Discontinued command"
FilmRoll	v2.6	R13	*none*	"Unknown command"
FlatLand	R10	R11	*none*	Accepts only 0 as a value
Fog	R14	2007	RenderEnvironment	
GifIn	R12	R14	ImageAttach	"No longer supported"
HpConfig	R12	2000	PlotStyle	"Discontinued command"
IgesIn	v2.5	R13	*none*	"Discontinued command"
IgesOut	v2.5	R13	*none*	"Discontinued command"
Image	R14	2007	ClassicImage	Displays External References palette
InetCfg	R14	2000	*none*	"Unknown command"
InetHelp	R14	2000	Help	"Unknown command"
InsertUrl	R14	2000	Insert	Displays Insert dialog
ListUrl	R14	2000	QSelect	"Unknown command"
MakePreview	R13	R14	RasterPreview	"Discontinued command"
MenuLoad	R13	2006	CuiLoad	Load Customizations dialog box
MenuUnload	R13	2006	CuiUnload	Unload Customizations dialog box
MeetNow	2000i	2004	*none*	"Unknown command"
OceConfig	R13	2000	PlotStyle	"Discontinued command"
OpenUrl	R14	2000	Open	Displays Select File dialog
OSnap	v2.0	2000	DSettings	Displays Drafting Settings dialog
Painter	R14	2004	MatchProp	Runs MatchProp command
PcxIn	R12	R14	ImageAttach	"No longer supported"
PsDrag	R12	2000i	*none*	"Unknown command"
PsIn	R12	2000i	*none*	"Unknown command"

Command	Introduced	Removed	Replacement	Reaction
Preferences	R11	2000	Options	Displays Options dialog box
PrPlot	v2.1	R12	Plot	"Discontinued command"
QPlot	v1.1	v2.0	SaveImg	"Unknown command"
RConfig	R12	R14	*none*	"Unknown command"
RendScr	R12	2007	RenderWin	Displays Render window
RenderUnload	R12	R14	Arx Unload	"Unknown command"
Repeat	v1.0	v2.5	Minsert	"Discontinued command"
Replay	R12	2007	*none*	"Unknown command"
RMat	R13	2007	Materials	Displays Materials palette
RmlIn	2000i	2006	Markup	"Unknown command"
SaveAsR12	R13	R14	SaveAs	"Unknown command"
SaveUrl	R14	2000	SaveAs	Displays Save Drawing As dialog
Scene	R12	2007	*none*	"Unknown command"
SetUV	R14	2007	MaterialMap	Runs MaterialMap command
ShadeMode	2000	2007	-ShadeMode	Runs VsCurrent command
ShowMat	R13	2007	List	Runs List Command
Snapshot	v2.0	v2.1	Saveimg	"Unknown command"
Sol...	R11	R13	*(AME commands lost their SOL-prefix)*	
Stats	R12	2007	*none*	"Unknown command"
TbConfig	R12	2000i	CUI	Displays CUI dialog box
TiffIn	R12	R14	ImageAttach	"No longer supported"
Today	2000i	2004	*none*	"Unknown command"
Toolbar	R13	2000i	CUI	Displays CUI dialog box
+View	...	2007	View	"Unknown command"
VlConv	R13	R14	3dsIn & 3dsOut	"Unknown command"
XRef	R14	2007	ClassicXref	Displays External References palette
3dDwfPublish	2006	2007	3dDwf	Displays Export 3D DWF dialog box
3dLine	R9	R11	Line	"Unknown command"
3dsOut	R13	2007	*none*	"Unknown command"

APPENDIX C
System Variables

AutoCAD stores information about its current state, the drawing, and the operating system in nearly 600 system variables. These variables help users and programmers determine the state of the AutoCAD system.

CONVENTIONS

The following pages list all documented system variables, plus several more not documented by Autodesk. The list uses the following conventions:

Bold   System variable is documented by Autodesk in AutoCAD 2008.

Italicized   System variable is not listed by the SetVar command nor in Autodesk's documentation.

~~*Strikethru Italic*~~   System variable was removed from AutoCAD in this release or earlier.

▭   System variable must be accessed through the SetVar command. Otherwise, the equivalent command is executed.

◪   System variable is new to AutoCAD 2008.

Default   Indicates the default value, as set in the *acad.dwt* template drawing file. Other template files may have different default values.

TIPS

- The SetVar command lets you change the value of all system variables, except those marked read-only (R/O).

- You can get a list of most system variables at the 'Command:' prompt with the ? option of the SetVar command, as follows:

        Command: setvar
        Variable name or ?: ?
        Variable(s) to list <*>: *(Press Enter.)*

- In this appendix, "Toggle" means that the system variable has one of two values: 0 (off, closed, disabled) or 1 (on, open, enabled).

- "ACI color" means the system variable takes on one of AutoCAD's 255 basic colors, not including ByLayer or ByBlock. ACI is short for "AutoCAD color index."

Variable	Default	Meaning
~~LInfo~~		*Removed from AutoCAD 2004.*
⌨ _AuthorPalettePath	*varies*	*Path to the customized Tool palette folder.*
_PkSer (R/O)	*varies*	*Software serial number, such as "117-69999999".*
_Server (R/O)	0	Network authorization code.
⌨ _ToolPalettePath	*varies*	*Path to the Tool palette folder.*
_VerNum (R/O)	*varies*	*Internal program build number, such as "Z.54.01".*

. . . .

A

Variable	Default	Meaning

AcadLspAsDoc	0	Controls whether *acad.lsp* is loaded into:
		  0 The first drawing only.
		  1 Every drawing.

AcadPrefix (R/O)	*varies*	Specifies paths used by AutoCAD search in Options | Files dialog box.
AcadVer (R/O)	"17.1"	Specifies the AutoCAD version number.
AcisOutVer	70	Controls the ACIS version number; values are 15, 16, 17, 18, 20, 21, 30, 40, or 70.
AcGiDumpMode	*0*	*Value of 0 or 1.*
AdcState (R/O)	0	Toggle: reports if DesignCenter is active.

AFlags	0	Controls the default attribute display mode:
		  0 No mode specified.
		  1 Invisible.
		  2 Constant.
		  4 Verify.
		  8 Preset.
		  16 Lock position in block.
		  32 Multiple-line attributes. ▣

AngBase	0	Controls the direction of zero degrees relative to the UCS.

AngDir	0	Controls the rotation of positive angles:
		  0 Clockwise.
		  1 Counterclockwise.

▣ AnnoAllVisible	1	Toggles display of annotative objects at the current scale:
		  0 Displays only objects matching VP scale.
		  1 Displays all annotative objects (default).

▣ AnnoAutoScale	4	Updates annotative objects when the annotation scale is changed:
		  1 Adds the new annotation scale to annotative objects (except on layers that are off, frozen, locked, or have viewports set to freeze).
		  2 As above, but includes objects on locked layers (excludes objects on layers that are off, frozen, or have viewports set to freeze).

. . . .

		3 The opposite of 2: applies to annotative objects on all layers, except objects on locked layers. **4** Applies to all objects regardless of layer status. **-1** Same as 1, but turned off. **-2** Same as 2, but turned off. **-3** Same as 3, but turned off. **-4** Same as 4, but turned off.
▣ **AnnotativeDwg**	0	Toggles whether the drawing acts like an annotative block when inserted into other drawings: **0** Nonannotative block. **1** Annotative block. (This sysvar is read-only when drawing contains annotative text.)
ApBox	0	Toggles the display of the AutoSnap aperture box cursor.
▦ **Aperture**	10	Controls the object snap aperture in pixels: **1** Minimum size. **50** Maximum size.
ApState	*0*	*Toggle: reports the state of the ApBox variable.*
~~*AssistState*~~ *(R/O)*	*0*	*Removed from AutoCAD 2008.*
▦ **Area** (R/O)	0.0	Reports the area measured by the last Area command.
AttDia	0	Controls the user interface for entering attributes: **0** Command-line prompts. **1** Dialog box.
▣ **AttIpe**	0	Toggles display of in-place attribute text editor's toolbar: **0** Reduced-function toolbar. **1** Full-function toolbar.
AttMode	1	Controls the display of attributes: **0** Off. **1** Normal. **2** On; displays invisible attributes.
▣ **AttMulti**	1	Toggles creation of multi-line attributes: **0** Attributes can be single-line only. **1** Attributes can be multi-line
AttReq	1	Toggles attribute values during insertion: **0** Uses default values. **1** Prompts user for values.
AuditCtl	0	Toggles creation of *.adt* audit log files: **0** File not created. **1** File created.

Variable	Default	Meaning
AUnits	0	Controls the type of angular units:
		0 Decimal degrees.
		1 Degrees-minutes-seconds.
		2 Grads.
		3 Radians.
		4 Surveyor's units.
AUPrec	0	Controls the number of decimal places displayed by angles; range is 0 to 8.
AutoDwfPublish	1	Toggles automatic publishing of DWF Files when drawings are saved or closed:
		0 DWF are not published.
		1 DWF are published.
AutoSnap	55	Controls the AutoSnap display (sum):
		0 Turns off all AutoSnap features.
		1 Turns on marker.
		2 Turns on SnapTip.
		4 Turns on magnetic cursor.
		8 Turns on polar tracking.
		16 Turns on object snap tracking.
		32 Turns on tooltips for polar tracking and object snap tracking.
AuxStat	*0*	*-32768 Minimum value.*
		32767 Maximum value.
~~*AxisMode*~~	*0*	*Removed from AutoCAD 2002.*
AxisUnit	*0.0*	*Obsolete system variable.*

. . . .

B

Variable	Default	Meaning
BackgroundPlot	2	Controls background plotting and publishing (ignored during scripts):
		0 Plot foreground; publish foreground.
		1 Plot background; publish foreground.
		2 Plot foreground; publish background.
		3 Plot background; publish background.
BackZ	0.0	Controls the location of the back clipping plane offset from the target plane.
BAction Color	"7"	Specifies ACI text color for actions in Block Editor.
BDependencyHighlight	1	Toggles highlighting of dependent objects when parameters, actions, or grips selected in Block Editor:
		0 Not highlighted.
		1 Highlighted.
BGripObjColor	"141"	Specifies ACI color of grips in Block Editor.
BGripObjSize	8	Controls the size of grips in Block Editor; range is 1 to 255.
BgrdPlotTimeout	*20*	*Controls the timeout for failed background plots; ranges from 0 to 300 seconds.*

. . . .

Variable	Default	Meaning
BindType	0	Controls how xref names are converted when being bound or edited: **0** From *xref|name* to *xref0name*. **1** From *xref|name* to *name*.
▧ BlipMode	0	Toggles the display of blip marks.
BlockEditLock	0	Toggles the locking of dynamic blocks being edited: **0** Unlocked. **1** Locked.
BlockEditor (R/O)	0	Toggle: reports whether Block Editor is open.
BParameterColor	"7"	Specifies ACI color of parameters in the Block Editor.
BParameterFont	"Simplex.shx"	Controls the font used for parameter and action text in the Block Editor.
BParameterSize	12	Controls the size of parameter text and features in Block Editor: **1** Minimum. **255** Maximum.
BTMarkDisplay	1	Controls the display of value set markers: **0** Unlocked. **1** Locked.
BVMode	0	Controls the display of invisible objects in the Block Editor: **0** Invisible. **1** Visible and dimmed.

. . . .

C

Variable	Default	Meaning
CalcInput	1	Controls how formulas and global constants are evaluated in dialog boxes: **0** Not evaluated. **1** Evaluated after pressing Alt+Enter.
CameraDisplay	0	Toggles the display of camera glyphs.
CameraHeight	0	Specifies the default height of cameras.
▣ CAnnoScale	"1:1"	Names the current annotative scale for the current viewport.
▣ CAnnoScaleValue(R/O)	1.0	Reports the current annotation scale.
CDate (R/O)	*varies*	Specifies the current date and time in the format YyyyMmDd.HhMmSsDd, such as 20080503.18082328
CeColor	"BYLAYER"	Controls the current color.
CeLtScale	1.0	Controls the current linetype scaling factor.
CeLType	"BYLAYER"	Controls the current linetype.

. . . .

Variable	Default	Meaning
CeLWeight	-1	Controls the current lineweight in millimeters; valid values are 0, 5, 9, 13, 15, 18, 20, 25, 30, 35, 40, 50, 53, 60, 70, 80, 90, 100, 106, 120, 140, 158, 200, and 211, plus the following: **-1** BYLAYER. **-2** BYBLOCK. **-3** Default, as defined by LwDdefault.
CenterMT	0	Controls how corner grips stretch uncentered multiline text: **0** Center grip moves in same direction; opposite grip stays in place. **1** Center grip stays in place; both side grips move in direction of stretch.
ChamferA	0.0	Specifies the current value of the first chamfer distance.
ChamferB	0.0	Specifies the current value of the second chamfer distance.
ChamferC	0.0	Specifies the current value of the chamfer length.
ChamferD	0	Specifies the current value of the chamfer angle.
ChamMode	0	Toggles the chamfer input mode: **0** Chamfer by two lengths. **1** Chamfer by length and angle.
⊞ *CipMode*	*0*	*Toggles Customer Involvement Program.*
CircleRad	0.0	Specifies the most-recent circle radius.
CLayer	"0"	Specifies name of current layer.
CleanScreenState (R/O)	0	Toggle: reports whether cleanscreen mode is active.
CliState (R/O)	1	Reports the command line palette.
CMaterial	"ByLayer"	Sets the name of the current material.
CmdActive (R/O)	1	Reports the type of command currently active (used by programs): **1** Regular command. **2** Transparent command. **4** Script file. **8** Dialog box. **16** Dynamic data exchange. **32** AutoLISP command. **64** ARX command.
CmdDia	1	Toggles QLeader's inplace text editor.
CmdEcho	1	Toggles AutoLISP command display: **0** No command echoing. **1** Command echoing.
CmdInputHistoryMax	20	Controls the maximum command input items stored; works with InputHistoryMode.

Variable	Default	Meaning
▣ **CMleaderStyle**	"Standard"	Reports name of current mleader style.
CmdNames (R/O)	*varies*	Reports the name of the command currently active, such as "SETVAR".

CMLJust	0	Controls the multiline justification mode:
		  0 Top.
		  1 Middle.
		  2 Bottom.

CMLScale	1.0	Controls the scale of overall multiline width:
		  -*n* Flips offsets of multiline.
		  0 Collapses to single line.
		  n Scales by a factor of *n*.

CMLStyle	"STANDARD"	Specifies the current multiline style name.
Compass	0	Toggles the display of the 3D compass.

Coords	1	Controls the coordinate display style:
		  0 Updated by screen picks.
		  1 Continuous display.
		  2 Polar display upon request.

▣ **CopyMode**	0	Toggles whether the Copy command repeats itself:
		  0 Command repeats copying.
		  1 Command makes one copy, then exits.

CPlotStyle	"ByColor"	Specifies the current plot style; options for named plot styles are: ByLayer, ByBlock, Normal, and User Defined.
CProfile (R/O)	"<<Unnamed Profile>>"	Specifies the name of the current profile.
CpuTicks (R/O)	*592020023071334.1*	*Reports the number of CPU ticks.*
CrossingColor	3	ACI color of crossing rectangle.

CShadow	0	Shadows cast by 3D objects, if graphics board is capable (see 3dConfig command):
		  0 Casts and receives shadows.
		  1 Casts shadows.
		  2 Receives shadows.
		  3 Ignores shadows.

CTab	"Model"	Specifies the name of the current tab.
CTableStyle	"Standard"	Specifies the name of the current table style name.
~~CurrentProfile~~	"*<<Unnamed Profile>>*"	*Removed from AutoCAD 2000; replaced by CProfile.*

CursorSize	5	Controls the cursor size as a percent of the viewport size.
		  1 Minimum size.
		  100 Full viewport.

CVPort	2	Specifies the current viewport number.

. . . .

D

DashboardState	0	Toggle: reports whether Dashboard is open.

. . . .

Variable	Default	Meaning
DataLinkNotify	2	Control reporting on data links: **0** All notifications disabled. **1** Displays data link icon in tray. **2** Displays the icon and a warning balloon in the tray.
Date (R/O)	*varies*	Reports the current date in Julian format, such as 2448860.54043252
DbcState (R/O)	0	Toggles: specifies whether dbConnect Manager is active.
~~DBGListAll~~	*0*	*Removed from AutoCAD 2002.*
DblClkEdit	1	Toggles editing by double-clicking objects.
DBMod (R/O)	4	Reports how the drawing has been modified: **0** No modification since last save. **1** Object database modified. **2** Symbol table modified. **4** Database variable modified. **8** Window modified. **16** View modified. **32** Field modified.
DctCust	"sample.cus"	Specifies the name of custom spelling dictionary.
DctMain	"enu"	Controls the code for spelling dictionary: **enu** American English **eng** British English (ise) **enc** Canadian English **cat** Catalan **csy** Czech **dan** Danish **nld** Dutch (primary) **fin** Finnish **fra** French (accented capitals) **frc** French (unaccented capitals) **deu** German (post-reform) **deo** German (pre-reform) **ita** Italian **nor** Norwegian (Bokmal) **ptb** Portuguese (Brazilian) **ptg** Portuguese (Iberian) **rus** Russian **esp** Spanish **sve** Swedish
DefaultLighting	1	Toggles distant lighting.
DefaultLightingType	0	Toggles between new (1) and old (0) type of lights.
DefaultViewCategory	*""*	*Specifies the default name for View Category in the View command's New View dialog box*
DefLPlStyle	"ByColor"	Reports the default plot style for layer 0.
DefPlStyle	"ByColor"	Reports the default plot style for new objects.

Variable	Default	Meaning
DelObj	1	Toggles the deletion of source objects: **-2** Users are prompted whether to erase all defining objects. **-1** Users are prompted whether to erase profiles and cross sections. **0** Objects retained. **1** Profiles and cross sections erased. **2** All defining objects erased.
DemandLoad	3	Controls application loading when drawing contains proxy objects: **0** Apps not demand-loaded. **1** Apps loaded when drawing opened. **2** Apps loaded at first command. **3** Apps loaded when drawing is opened or at first command.
ⓐ **DgnFrame**	0	Controls display of DGN underlay frame: **0** Does not display frame. **1** Displays and plots frame. **2** Displays but does not plot frame.
ⓐ **DgnOsnap**	1	Toggles object snapping to DGN elements: **0** Does not osnap. **1** Osnaps.
DiaStat (R/O)	0	Reports whether user exited dialog box by clicking: **0** **Cancel** button. **1** **OK** button.

· ·

Dimension Variables

Variable	Default	Meaning
DimADec	0	Controls angular dimension precision: **-1** Use DimDec setting (default). **0** Zero decimal places (minimum). **8** Eight decimal places (maximum).
DimAlt	Off	Toggles alternate units: **On** Enabled. **Off** Disabled.
DimAltD	2	Controls alternate unit decimal places.
DimAltF	25.4	Controls alternate unit scale factor.
DimAltRnd	0.0	Controls rounding factor of alternate units.
DimAltTD	2	Controls decimal places of tolerance alternate units; range is 0 to 8.
DimAltTZ	0	Controls display of zeros in alternate tolerance units: **0** Zeros not suppressed. **1** All zeros suppressed. **2** Include 0 feet, but suppress 0 inches. **3** Includes 0 inches, but suppress 0 feet. **4** Suppresses leading zeros. **8** Suppresses trailing zeros.

· ·

| --- | --- | --- |
DimAltU	2	Controls display of alternate units:
		1 Scientific.
		2 Decimal.
		3 Engineering.
		4 Architectural; stacked.
		5 Fractional; stacked.
		6 Architectural.
		7 Fractional.
		8 Windows desktop units setting.
DimAltZ	0	Controls the display of zeros in alternate units:
		0 Suppresses 0 ft and 0 in.
		1 Includes 0 ft and 0 in.
		2 Includes 0 ft; suppress 0 in.
		3 Suppresses 0 ft; include 0 in.
		4 Suppresses leading 0 in dec dims.
		8 Suppresses trailing 0 in dec dims.
		12 Suppresses leading and trailing zeroes.
DimAnno (R/O)	0	Reports whether current dimstyle is:
		0 Not annotative.
		1 Annotative.
DimAPost	""	Specifies the prefix and suffix for alternate text.
DimArcSym	0	Specifies the location of the arc symbol:
		0 Before dimension text.
		1 Above the dimension text.
		2 Not displayed.
DimAso	On	Toggles associative dimensions:
		On Dimensions are created associative.
		Off Dimensions are not associative.
DimAssoc	2	Controls how dimensions are created:
		0 Dimension elements are exploded.
		1 Single dimension object, attached to defpoints.
		2 Single dimension object, attached to geometric objects.
DimASz	0.18	Controls the default arrowhead length.
DimAtFit	3	Controls how text and arrows are fitted when there is insufficient space between extension lines (leader is added when DimTMove = 1):
		0 Text and arrows outside extension lines.
		1 Arrows first outside, then text.
		2 Text first outside, then arrows.
		3 Either text or arrows, whichever fits better.
DimAUnit	0	Controls the format of angular dimensions:
		0 Decimal degrees.
		1 Degrees.Minutes.Seconds.
		2 Grads.
		3 Radians.
		4 Surveyor units.

Variable	Default	Meaning
DimAZin	0	Controls the display of zeros in angular dimensions: **0** Displays all leading and trailing zeros. **1** Suppresses 0 in front of decimal. **2** Suppresses trailing zeros behind decimal. **3** Suppresses zeros in front and behind the decimal.
DimBlk	""	Specifies the name of the arrowhead block:

➤ Closed filled
▷ Closed blank
⇒ Closed
● Dot
✒ Architectural tick
╱ Oblique
⇒ Open
⊙ Origin indicator
⊙ Origin indicator 2
→ Right angle
⇉ Open 30
◆ Dot small
○ Dot blank
◇ Dot small blank
⊣□ Box
⊣■ Box filled
◁ Datum triangle
◀ Datum triangle filled
✗ Integral
  None

Architectural tick: "Archtick"
Box filled: "Boxfilled"
Box: "Boxblank"
Closed blank: "Closedblank"
Closed filled: "" (default)
Closed: "Closed"
Datum triangle filled: "Datumfilled"
Datum triangle: "Datumblank"
Dot blanked: "Dotblank"
Dot small: "Dotsmall"
Dot: "Dot"
Integral: "Integral"
None: "None"
Oblique: "Oblique"
Open 30: "Open30"
Open: "Open"
Origin indication: "Origin"
Right-angle: "Open90"

Variable	Default	Meaning
DimBlk1	""	Specifies the name of first arrowhead's block; uses same list of names as under DimBlk. **.** No arrowhead.
DimBlk2	""	Specifies name of second arrowhead block.
DimCen	0.09	Controls how center marks are drawn: *-n* Draws center lines. **0** No center mark or lines drawn. *+n* Draws center marks of length *n*.
DimClrD	0	ACI color of dimension lines: **0** BYBLOCK (default). **256** BYLAYER.
DimClrE	0	Specifies ACI color of extension lines and leaders.
DimClrT	0	Specifies ACI color of dimension text.
DimDec	4	Controls the number of decimal places for the primary tolerance; range is 0 to 8.
DimDLE	0.0	Controls the length of the dimension line extension.
DimDLI	0.38	Controls the increment of the continued dimension lines.
DimDSep	"."	Specifies the decimal separator (must be a single character).

Variable	Default	Meaning
DimExe	0.18	Controls the extension above the dimension line.
DimExO	0.0625	Specifies the extension line origin offset.
~~DimFit~~	*3*	*Obsolete: Autodesk recommends use of DimATfit and DimTMove instead.*

DimFrac	0	Controls the fraction format when DimLUnit set to 4 or 5:
		  0 Horizontal.
		  1 Diagonal.
		  2 Not stacked.

DimFXL	1	Default length of fixed extension lines.
DimFxlOn	0	Toggles fixed extension lines.
DimGap	0.09	Controls the gap between text and the dimension line.
DimJogAngle	45	Specifies default angle for jogged dimension lines.

DimJust	0	Controls the positioning of horizontal text:
		  0 Center justify.
		  1 Next to first extension line.
		  2 Next to second extension line.
		  3 Above first extension line.
		  4 Above second extension line.

DimLdrBlk	""	Specifies the name of the block used for leader arrowheads; same as DimBlock. . Suppresses display of arrowhead.
DimLFac	1.0	Controls the linear unit scale factor.
DimLim	Off	Toggles the display of dimension limits.
DimLtEx1	""	Specifies linetype for the first extension line.
DimLtEx2	""	Specifies linetype for second extension line.
DimLtype	""	Specifies linetype name for dimension line.

DimLUnit	2	Controls dimension units (except angular); replaces DimUnit:
		  1 Scientific.
		  2 Decimal.
		  3 Engineering.
		  4 Architectural.
		  5 Fractional.
		  6 Windows desktop.

DimLwD	-2	Controls the dimension line lineweight; valid values are BYLAYER, BYBLOCK, or integer multiples of 0.01mm.
DimLwE	-2	Controls the extension lineweight.
DimPost	""	Specifies the default prefix or suffix for dimension text (maximum 13 characters): "" No suffix.
DimRnd	0.0	Controls the rounding value for dimension distances.

Variable	Default	Meaning
DimSAh	Off	Toggles separate arrowhead blocks: **Off** Use arrowheads defined by DimBlk. **On** Use arrowheads defined by DimBlk1 and DimBlk2.
DimScale	1.0	Controls the overall dimension scale factor: **0** Value is computed from the scale between current model space viewport and paper space. **>0** Scales text and arrowheads.
DimSD1	Off	Toggles display of the first dimension line: **On** First dimension line is suppressed. **Off** Not suppressed.
DimSD2	Off	Toggles display of the second dimension line: **On** Second dimension line is suppressed. **Off** Not suppressed.
DimSE1	Off	Toggles display of the first extension line: **On** First extension line is suppressed. **Off** Not suppressed.
DimSE2	Off	Toggles display of the second extension line: **On** Second extension line is suppressed. **Off** Not suppressed.
DimSho	On	Toggles dimension updates while dragging: **On** Dimensions are updated during drag. **Off** Dimensions are updated after drag.
DimSOXD	Off	Toggles display of dimension lines outside of extension lines: **On** Dimension lines not drawn outside extension lines. **Off** Are drawn outside extension lines.
⌨ **DimStyle** (R/O)	"STANDARD"	Reports the current dimension style.
DimTAD	0	Controls the vertical position of text: **0** Centered between extension lines. **1** Above dimension line, except when dimension line not horizontal and DimTIH = 1. **2** On side of dimension line farthest from the defining points. **3** Conforms to JIS.
DimTDec	4	Controls the number of decimal places for primary tolerances; range is 0 to 8.
DimTFac	1.0	Controls the scale factor for tolerance text height.
DimTFill	0	Toggles background fill color for dimension text.
DimTFillClr	0	Specifies background color for dimension text.
DimTIH	On	Toggles alignment of text placed inside extension lines:

Variable	Default	Meaning
		Off Text aligned with dimension line.
		On Text is horizontal.
DimTIX	Off	Toggles placement of text inside extension lines:
		Off Text placed inside extension lines, if room.
		On Text forced between the extension lines.
DimTM	0.0	Controls the value of the minus tolerance.
DimTMove	0	Controls how dimension text is moved:
		0 Dimension line moves with text.
		1 Adds a leader when text is moved.
		2 Text moves anywhere; no leader.
DimTOFL	Off	Toggles placement of dimension lines:
		Off Dimension lines not drawn when arrowheads are outside.
		On Dimension lines drawn, even when arrowheads are outside.
DimTOH	On	Toggles text alignment when outside of extension lines:
		Off Text aligned with dimension line.
		On Text is horizontal.
DimTol	Off	Toggles generation of dimension tolerances:
		Off Tolerances not drawn.
		On Tolerances are drawn.
DimTolJ	1	Controls vertical justification of tolerances:
		0 Bottom.
		1 Middle.
		2 Top.
DimTP	0.0	Specifies the value of the plus tolerance.
DimTSz	0.0	Controls the size of oblique tick strokes:
		0 Arrowheads.
		>0 Oblique strokes.
DimTVP	0.0	Controls the vertical position of text when **DimTAD** = 0:
		1 Turns on DimTAD (=1).
		>-0.7 *or* **<0.7** Dimension line is split for text.
DimTxSty	"STANDARD"	Specifies the dimension text style.
DimTxt	0.18	Controls the text height.
DimTZin	0	Controls the display of zeros in tolerances:
		0 Suppresses 0 ft and 0 in.
		1 Includes 0 ft and 0 in.
		2 Includes 0 ft; suppress 0 in.
		3 Suppresses 0 ft; include 0 in.
		4 Suppresses leading 0 in decimal dim.
		8 Suppresses trailing 0 in decimal dim.
		12 Suppresses leading and trailing zeroes.
~~DimUnit~~	*2*	*Obsolete; replaced by DimLUnit and DimFrac.*

Variable	Default	Meaning
DimUPT	Off	Controls user-positioned text:
		Off Cursor positions dimension line.
		On Cursor also positions text.
DimZIN	0	Controls the display of zero in feet-inches units:
		0 Suppresses 0 ft and 0 in.
		1 Includes 0 ft and 0 in.
		2 Includes 0 ft; suppress 0 in.
		3 Suppresses 0 ft; include 0 in.
		4 Suppresses leading 0 in decimal dim.
		8 Suppresses trailing 0 in decimal dim.
		12 Suppresses leading and trailing zeroes.
DispSilh	0	Toggles the silhouette display of 3D solids.
Distance (R/O)	0.0	Reports the distance last measured by the Dist command.
~~Dither~~		*Removed from Release 14.*
DonutId	0.5	Controls the inside diameter of donuts.
DonutOd	1.0	Controls the outside diameter of donuts.
▥ **DragMode**	2	Controls the drag mode:
		0 No drag.
		1 On if requested.
		2 Automatic.
DragP1	10	Controls the regen drag display; range is 0 to 32767.
DragP2	25	Controls the fast drag display; range is 0 to 32767.
DragVs	""	Specifies the default visual style when 3D solids are created by dragging the cursor; disabled when visual style is 2D wireframe.
DrawOrderCtrl	3	Controls the behavior of draw order:
		0 Draw order not restored until next regen or drawing reopened.
		1 Normal draw order behavior.
		2 Draw order inheritance.
		3 Combines options 1 and 2.
DrState	0	Toggles the Drawing Recovery palette.
DTextEd	2	Controls the user interface of DText/Text command:
		0 In-place text editor.
		1 Edit text dialog box.
		2 Can click elsewhere in drawing to start new text string.
DwfFrame	2	Display of the frame around DWF overlays:
		0 Frame is turned off.
		1 Frame is displayed and plotted.
		2 Frame is displayed but not plotted
DwfOsnap	1	Toggles osnapping of the DWF frame.

Variable	Default	Meaning

DwgCheck	0	Toggles checking of whether drawing was edited by software other than AutoCAD:
		  0 Suppresses dialog box.
		  1 Displays warning dialog box.
		  2 Warning appears on command line.

DwgCodePage (R/O)	*varies*	Same value as SysCodePage.
DwgName (R/O)	*varies*	Reports the current drawing file name, such as "drawing1.dwg".
DwgPrefix (R/O)	*varies*	Reports the drawing's drive and folder, such as "d:\acad 2008\".

DwgTitled (R/O)	0	Reports whether the drawing file name is:
		  0 "drawing1.dwg".
		  1 User-assigned name.

~~*DwgWrite*~~		*Removed from AutoCAD Release 14.*

🔲 **DxEval**	12	Controls which commands cause AutoCAD to check for changed external data files (sum):
		  0 Never.
		  1 Open.
		  2 Save.
		  4 Plot.
		  8 Publish.
		  16 eTransmit and Archive.
		  32 Save with automatic update.
		  64 Plot with automatic update.
		  128 Publish with automatic update.
		  256 eTransmit/Archive with automatic update.

DynDiGrip	31	Controls the dynamic dimensions displayed during grip stretch editing (DynDiVis =2):
		  0 None.
		  1 Resulting dimension.
		  2 Length change dimension.
		  4 Absolute angle dimension.
		  8 Angle change dimension.
		  16 Arc radius dimension.

DynDiVis	1	Controls the dynamic dimensions displayed during grip stretch editing:
		  0 First (in the cycle order).
		  1 First two (in the cycle order).
		  2 All (as specified by DynDiGrip).

DynMode	0	Controls dynamic input features. (Click DYN on status bar to turn on hidden modes.)
		  -3 Both on hidden.
		  -2 Dimensional input on hidden.
		  -1 Pointer input on hidden.
		  0 Off.
		  1 Pointer input on.
		  2 Dimensional input on.
		  3 Both on.

Variable	Default	Meaning
DynPiCoords	0	Toggles pointer input coordinates: **0** Relative. **1** Absolute.
DynPiFormat	0	Toggles pointer input coordinates: **0** Polar. **1** Cartesian.
DynPiVis	1	Controls when pointer input is displayed: **0** When user types at prompts for points. **1** When prompted for points. **2** Always.
DynPrompt	1	Toggles display of prompts in Dynamic Input tooltips.
DynToolTips	1	Toggles which tooltips are affected by tooltip appearance settings: **0** Only Dynamic Input value fields. **1** All drafting tooltips.

. . . .

E

Variable	Default	Meaning
EdgeMode	0	Toggles edge mode for the Trim and Extend commands: **0** No extension. **1** Extends cutting edge.
Elevation	0.0	Specifies the current elevation, relative to current UCS.
EnterpriseMenu (R/O)	"."	Reports the path and .*cui* file name.
EntExts	*1*	*Controls how drawing extents are calculated:* *0 Extents calculated every time; slows down AutoCAD, but uses less memory.* *1 Extents of every object are cached as a two-byte value (default).* *2 Extents of every object are cached as a four-byte value (fastest but uses more memory).*
EntMods (R/O)	*0*	*Increments by one each time an object is modified since the drawing was opened; ranges from 0 to 4.29497E9.*
ErrNo (R/O)	0	Reports error numbers from AutoLISP, ADS, & Arx.
ErState (R/O)	0	Tggole: reports display of the External References palette.
~~*ExtDir*~~		*Removed from Release 14.*
Expert	0	Controls the display of prompts: **0** Normal prompts. **1** 'About to regen, proceed?' and 'Really want to turn the current layer off?' **2** 'Block already defined. Redefine it?' and 'A drawing with this name already exists. Overwrite it?' **3** Linetype command messages.

. . . .

Variable	Default	Meaning
		4 UCS Save and VPorts Save.
		5 DimStyle Save and DimOverride.
ExplMode	1	Toggles whether the Explode and Xplode commands explode non-uniformly scaled blocks:
		0 Does not explode.
		1 Explodes.
ExtMax (R/O)		Upper-right coordinate of drawing extents.
	-1.0E+20, -1.0E+20, -1.0E+20	
ExtMin (R/O)		Lower-left coordinate of drawing extents.
	1.0E+20, 1.0E+20, 1.0E+20	
ExtNames	1	Controls the format of named objects:
		0 Names are limited to 31 characters, and can include A - Z, 0 - 9, dollar ($), underscore (_), and hyphen (-).
		1 Names are limited to 255 characters, and can include A - Z, 0 - 9, spaces, and any characters not used by Windows or AutoCAD for special purposes.

· ·

F

Variable	Default	Meaning
FaceTRatio	0	Controls the aspect ratio of facets on rounded 3D bodies:
		0 Creates an *n* by 1 mesh.
		1 Creates an *n* by *m* mesh.
FaceTRres	0.5000	Controls the smoothness of shaded and hidden-line objects; range is 0.01 to 10.
~~FfLimit~~	...	*Removed from AutoCAD Release 14.*
FieldDisplay	1	Toggles background to field text:
		0 No background.
		1 Gray background.
FieldEval	31	Controls how fields are updated:
		0 Not updated
		1 Updated with Open.
		2 Updated with Save.
		4 Updated with Plot.
		8 Updated with eTransmit.
		16 Updated with regeneration.
FileDia	1	Toggles the user interface for file-access commands, such as Open and Save:
		0 Displays command-line prompts.
		1 Displays file dialog boxes.
FilletRad	0.0	Specifies the current fillet radius.
FillMode	1	Toggles the fill of solid objects, wide polylines, fills, and hatches.
Flatland (R/O)	*0*	*Obsolete system variable.*
FontAlt	"simplex.shx"	Specifies the font used for missing fonts.
FontMap	"acad.fmp"	Specifies the name of the font mapping file.

· ·

Variable	Default	Meaning
Force_Paging	*0*	*Ranges from 0 to 4.29497E9.*
FrontZ (R/O)	0.0	Reports the front clipping plane offset.
FullOpen (R/O)	1	Reports whether the drawing is: **0** Partially loaded. **1** Fully open.
FullPlotPath	1	Specifies the format of file name sent to plot spooler: **0** Drawing file name only. **1** Full path and drawing file.

. . . .

G

Variable	Default	Meaning
GfAng	*0*	*Controls the angle of gradient fill; 0 to 360 degrees.*
GfClr1	*"RGB 000,000,255"*	*Specifies the first gradient color in RGB format.*
GfClr2	*"RGB 255,255,153"*	*Specifies the second gradient color in RGB format.*
GfClrLum	*1.0*	*Controls the level of gray in one-color gradients:* **0** *Black.* **1** *White.*
GfClrState	*1*	*Specifies the type of gradient fill:* **0** *Two-color.* **1** *One-color.*
GfName	*1*	*Specifies the style of gradient fill:* **1** *Linear.* **2** *Cylindrical.* **3** *Inverted cylindrical.* **4** *Spherical.* **5** *Inverted spherical.* **6** *Hemispherical.* **7** *Inverted hemispherical.* **8** *Curved.* **9** *Inverted curved.*
GfAShift	*0*	*Controls the origin of the gradient fill:* **0** *Centered.* **1** *Shifted up and left.*
GlobCheck	*0*	*Controls reporting on dialog boxes:* **-1** *Turn off local language.* **0** *Turn off.* **1** *Warns if larger than 640x400.* **2** *Also reports size in pixels.* **3** *Additional information.*
GridDisplay	2	Determines grid display (sum of bitcodes): **0** Grid restricted to area specified by the Limits command. **1** Grid is infinite. **2** Adaptive grid, with fewer grid lines when zoomed out. **4** Generates more grid lines when zoomed in. **8** Grid follows the x,y plane of the dynamic UCS.

. . . .

Variable	Default	Meaning
GridMajor	5	Specifies number of minor grid lines per major line.
GridMode	0	Toggles the display of the grid.
GridUnit	0.5,0.5	Controls the x, y spacing of the grid.
GripBlock	0	Toggles the display of grips in blocks: **0** At block insertion point. **1** Of all objects within block.
GripColor	160	Specifies ACI color of unselected grips.
GripDynColor	140	Specifies ACI color of grips in dynamic blocks.
GripHot	1	Specifies ACI color of selected grips.
GripHover	3	Specifies ACI grip color when cursor hovers.
GripLegacy	*0*	*Toggle of unknown meaning.*
GripObjLimit	100	Controls the maximum number of grips displayed; range is 1 to 32767; 0 = grips never suppressed.
Grips	1	Toggles the display of grips.
GripSize	5	Controls the size of grip; range 1 - 255 pixels.
GripTips	1	Toggles the display of grip tips when cursor hovers over custom objects.
GtAuto	1	Toggles display of grip tools.
GtDefault	0	Toggles which commands are the default commands in 3D views: **0** Move and Rotate. **1** 3dMove and 3dRotate
GtLocation	0	Controls location of grip tools: **0** Aligns grip tool with UCS icon. **1** Aligns with the last selected object.

. . . .

H

HaloGap	0	Controls the distance by which haloed lines are shortened in 2D wireframe visual style; a percentage of 1 unit.
Handles (R/O)	1	Reports whether object handles (not grips) can be accessed by applications; obsolete.
HidePrecision	0	Controls the precision of hide calculations in 2D wireframe visual style: **0** Single precision, less accurate, faster. **1** Double precision, more accurate, but slower (recommended).
HideText	On	Controls the display of text during Hide: **On** Text is neither hidden nor hides other objects, unless text object has thickness. **Off** Text is hidden and hides other objects.
Highlight	1	Toggles object selection highlighting.
HPAng	0	Specifies current hatch pattern angle.

. . . .

Variable	Default	Meaning
HpAssoc	1	Toggles associativity of hatches: **0** Not associative. **1** Associative.
HpBound	1	Controls the object created by the Hatch and Boundary commands: **0** Region. **1** Polyline.
HpDouble	0	Toggles double hatching.
HpDrawOrder	3	Controls draw order of hatches and fills: **0** None. **1** Behind all other objects. **2** In front of all other objects. **3** Behind the hatch boundary. **4** In front of the hatch boundary.
HpGapTol	0	Controls largest gap allowed in hatch boundaries; ranges from 0 to 5000 units.
HpInherit	0	Toggles how MatchProp copies the hatch origin from source object to destination objects: **0** As specified by HpOrigin. **1** As specified by the source hatch object.
🗚 **HpMaxlines**	1000000	Controls the maximum number of hatch lines; range is 100 - 10,000,000.
HpName	"ANSI31"	Specifies default hatch name.
HpObjWarning	10000	Specifies the maximum number of hatch boundaries that can be selected before AutoCAD flashes warning message; range is 1 to 1073741823.
HpOrigin	0,0	Specifies the default origin for hatch objects.
HpOrigMode	0	Controls the default hatch origin point:. **0** Specified by HpOrigin. **1** Bottom-left corner of hatch's rectangular extents. **2** Bottom-right corner of hatch's rectangular extents. **3** Top-right corner of hatch's rectangular extents. **4** Top-left corner of hatch's rectangular extents. **5** Center of hatch's rectangular extents.
HpScale	1.0	Specifies the current hatch scale factor; cannot be zero.
HpSeparate	0	Controls the number of hatch objects made from multiple boundaries: **0** Single hatch object created. **1** Separate hatch object created.
HpSpace	1.0	Controls the default spacing of user-defined hatches; cannot be zero.

Variable	Default	Meaning
HyperlinkBase	""	Specifies the path for relative hyperlinks.

. . . .

I

Variable	Default	Meaning
ImageHlt	0	Toggles image frame highlighting when raster images are selected.
ImpliedFace	1	Toggles detection of implied faces.
IndexCtl	0	Controls creation of layer and spatial indices: **0** No indices created. **1** Layer index created. **2** Spatial index created. **3** Both indices created.
InetLocation	"http://www.autodesk.com"	Specifies the default URL for Browser.
InputHistoryMode	15	Controls the content and location of user input history (bitcode sum of): **0** No history displayed. **1** Displayed at the command line, and in dynamic prompt tooltips accessed with Up and Down arrow keys. **2** Current command displayed in the shortcut menu. **4** All commands in the shortcut menu. **8** Blipmark for recent input displayed in the drawing.
InsBase	0.0,0.0,0.0	Controls the default insertion base point relative to the current UCS for Insert and XRef commands.
InsName	""	Specifies the default block name: . Set to no default.
InsUnits	1	Specifies drawing units of blocks dragged into drawings: **0** Unitless. **1** Inches. **2** Feet. **3** Miles. **4** Millimeters. **5** Centimeters. **6** Meters. **7** Kilometers. **8** Microinches. **9** Mils. **10** Yards. **11** Angstroms. **12** Nanometers. **13** Microns. **14** Decimeters. **15** Decameters. **16** Hectometers. **17** Gigameters.

. . . .

Variable	Default	Meaning
		18 Astronomical Units.
		19 Light Years.
		20 Parsecs.
InsUnitsDefSource	4	Controls source drawing units value; ranges from 0 to 20; see above.
InsUnitsDefTarget	4	Controls target drawing units; see list above.
IntelligentUpdate	20	Controls graphics refresh rate in frames per second; range is 0 (off) to 100 fps.
InterfereColor	"1"	Specifies color of interference objects.
InterfereObjVs	"Realistic"	Specifies visual style of interference objects.
InterfereVpVs	"3d wireframe"	Specifies visual style during interference checking.
IntersectionColor	257	Specifies ACI color of intersection polylines in 2D wireframe visual style:
		0 ByBlock.
		256 ByLayer.
		257 ByEntity.
IntersectionDisplay	Off	Toggles display of 3D surface intersections during Hide command in 2D wireframe:
		Off Does not draw intersections.
		On Draws polylines at intersections.
ISaveBak	1	Toggles creation of *.bak* backup files.
ISavePercent	50	Controls the percentage of waste in saved *.dwg* file before cleanup occurs:
		0 Slower full saves.
		>0 Faster partial saves.
		100 Maximum.
IsoLines	4	Controls the number of contour lines on 3D solids; range is 0 to 2047.

. . . .

L

Variable	Default	Meaning
LastAngle (R/O)	0	Reports the end angle of last-drawn arc.
LastPoint	0,0,0	Reports the x,y,z coordinates of the last-entered point.
LastPrompt (R/O)	"*varies*"	Reports the last string on the command line.
Latitude	"37.7950"	Specifies last-used angle of latitude.
🅰 **LayerEval**	1	Controls when newly-added (unreconciled) layers are detected:
		0 Never.
		1 When xref layers are added to the drawing
		2 When new layers are added to drawings and xrefs.
LayerFilterAlert	2	Controls the deletion of layer filters in excess of 99 filters *and* the number of layers:
		0 Off.
		1 Deletes all filters without warning, when layer dialog box opened.

. . . .

Variable	Default	Meaning
		2 Recommends deleting all filters when layer dialog box opened.
		3 Displays dialog box for selecting filters to erase, upon opening the drawing.
⊡ **LayerNotify**	15	Controls when AutoCAD displays alerts about unreconciled layers (bitcode):
		0 Off.
		1 Plotting.
		2 Opening drawings.
		4 Loading, reloading, and attaching xrefs.
		8 Restoring layer states.
		16 Saving drawings.
		32 Inserting blocks, etc.
⊡ **LayLockFadeCtl**	90	Controls the fading of locked layers during the LayIso command:
		0 Not faded.
		>0 Faded up to 90 percent.
		<0 Not faded, but the value is saved.
LayoutRegenCtl	2	Controls display list for layouts:
		0 Display-list regen'ed with each tab change.
		1 Display-list is saved for model tab and last layout tab.
		2 Display list is saved for all tabs.
LazyLoad	*0*	*Toggle: unknown purpose.*
LegacyCtrlPick	0	Toggles function of Ctrl+pick:
		0 Selects faces, edges, and vertices of 3D solids.
		1 Cycles through overlapping objects.
LensLength (R/O)	50.0	Reports perspective view lens length, in mm.
LightGlyphDisplay	1	Toggles display of light glyph.
⊡ **LightingUnits**	2	Controls the type of lighting used:
		0 Generic lighting.
		1 International units of photometric lighting.
		2 American units of photometric lighting.
LightListState	0	Toggles display of Light List palette.
⊡ **LightsInBlocks**	1	Toggles use of lights in blocks:
		0 Off.
		1 On.
LimCheck	0	Toggles drawing limits checking.
LimMax	12.0,9.0	Controls the upper right drawing limits.
LimMin	0.0,0.0	Controls the lower left drawing limits.
⊡ **LinearBrightness**	0	Controls the overall brightness of generic lighting in renderings; range is -10 to 10.
⊡ **LinearContrast**	0	Controls the overall contrast of generic lighting in renderings; range is -10 to 10.

Variable	Default	Meaning

LispInit	1	Toggles AutoLISP functions and variables:
		  0 Preserved from drawing to drawing.
		  1 Valid in current drawing only.

Locale (R/O)	"enc"	Reports ISO language code; see DctMain.
LocalRootPrefix (R/O)	"d:\docume..."	Reports the path to folder holding local customizable files.

LockUi	0	Controls the position and size of toolbars and palettes; hold down Ctrl key to unlock temporarily (bitcode sum):
		  0 Toolbars and palettes unlocked.
		  1 Docked toolbars locked.
		  2 Docked palettes locked.
		  4 Floating toolbars locked.
		  8 Floating palettes locked.

LoftAng1	90	Angle of loft to first cross-section; range is 0 to 359.9 degrees.
LoftAng2	90	Specifies angle of loft to second cross-section.
LoftMag1	0.0	Specifies magnitude of loft at first cross-section; range is 1 to 10.
LoftMag2	0.0	Specifies magnitude of loft at last cross-section.

LoftNormals	1	Specifies location of loft normals:
		  0 Ruled.
		  1 Smooth.
		  2 First normal.
		  3 Last normal.
		  4 Ends normal.
		  5 All normal.
		  6 Use draft angle and magnitude.

LoftParam	7	Specifies the loft shape:
		  1 Minimizes twists between cross-sections.
		  2 Aligns start-to-end direction of each cross-section.
		  4 Generates simple solids and surfaces, instead of spline solids and surfaces.
		  8 Closes the surface or solid between the first and last cross-sections.

LogExpBrightness	65.0	Controls the overall brightness of photometric lighting in renderings; range is 0 to 200.
LogExpContrast	50.0	Controls the overall contrast of photometric lighting in renderings; range is 0 to 100.

LogExpDaylight	2	Controls how daylight is displayed photometric renderings:
		  0 Off.
		  1 On.
		  2 Same as sun status.

LogExpMidtones	1.00	Controls the overall midtones of photometric lighting in renderings; range is 0.01 to 20.

Variable	Default	Meaning
🔲 *LogExpPhysicalScale*	*1500.000*	*Physically scales photometric lights.*
LogFileMode	0	Toggles writing command prompts to *.log* file.
LogFileName (R/O)	"...\Drawing1.log"	Reports file name and path for *.log* file.
LogFilePath	"d:\acad 2008\"	Specifies path to the *.log* file.
LogInName (R/O)	"*username*"	Reports user's login name; truncated after 30 characters.
~~*LongFName*~~		*Removed from AutoCAD Release 14.*
Longitude	-122.3940	Current angle of longitude.
🔲 **LTScale**	1.0	Controls linetype scale factor; cannot be 0.

LUnits	2	Controls linear units display:
		  1 Scientific.
		  2 Decimal.
		  3 Engineering.
		  4 Architectural.
		  5 Fractional.

LUPrec	4	Controls decimal places (or inverse of smallest fraction) of linear units; range is 0 to 8.
LwDefault	25	Controls the default lineweight, in millimeters; must be one of the following values: 0, 5, 9, 13, 15, 18, 20, 25, 30, 35, 40, 50, 53, 60, 70, 80, 90, 100, 106, 120, 140, 158, 200, or 211.
LwDisplay	0	Toggles whether lineweights are displayed; setting saved separately for Model space and each layout tab.

LwUnits	1	Toggles units used for lineweights:
		  0 Inches.
		  1 Millimeters.

. . . .

Variable	Default	Meaning
M		
MacroTrace	*0*	*Toggles diesel debug mode.*
MatState	0	Toggles display of Materials palette.
MaxActVP	64	Controls the maximum number of viewports to display; range is 2 to 64.
MaxObjMem	*0*	*Controls the maximum number of objects in memory; object pager is turned off when value = 0, <0, or 2,147,483,647.*
MaxSort	1000	Controls the maximum names sorted alphabetically; range is 0 to 32767.

MButtonPan	1	Toggles the behavior of the wheel mouse:
		  0 As defined by AutoCAD *.cui* file.
		  1 Pans when dragging with wheel.

MeasureInit	0	Toggles drawing units for default drawings:
		  0 English.
		  1 Metric.

. . . .

Variable	Default	Meaning
Measurement	0	Toggles current drawing units: **0** English. **1** Metric.
MenuCtl	1	Toggles the display of submenus in side menu: **0** Only with menu picks. **1** Also with keyboard entry.
MenuEcho	0 …	Controls menu and prompt echoing (sum): **0** Displays all prompts. **1** Suppresses menu echoing. **2** Suppresses system prompts. **4** Disables ^P toggle. **8** Displays all input-output strings.
MenuName (R/O)	"acad"	Reports path and file name of *.cui* file.
Millisecs (R/O)	*248206921*	*Reports number of milliseconds since timing began.*
MirrText	0	Toggles text handling by Mirror command: **0** Retains text orientation. **1** Mirrors text.
ModeMacro	""	Invokes Diesel macros.
MsLtScale	1	Controls how linetypes are scaled: **0** Not scaled by the annotation scale. **1** Scaled by the annotation scale.
MsmState (R/O)	0	Specifies if Markup Set Manager is active: **0** No. **1** Yes.
MsOleScale	1.0	Controls the size of text-containing OLE objects when pasted in model space: **-1** Scales by value of PlotScale. **0** Scales by value of DimScale. **>0** Scale factor.
MTextEd	"Internal"	Controls the name of the MText editor: **.** Uses default editor. **0** Cancels the editing operation. **-1** Uses the secondary editor. **"blank"** MTEXT internal editor. **"Internal"** MTEXT internal editor. **"oldeditor"** Previous internal editor. **"Notepad"** Windows Notepad editor. **":lisped"** Built-in AutoLISP function. *string* Name of editor fewer than 256 characters long using this syntax: *:AutoLISPtextEditorFunction#TextEditor.*
MTextFixed	2	Controls the mtext editor appearance: **0** Mtext editor is used. **1** Mtext editor remembers its location. **2** Difficult-to-read text is displayed horizontally at a larger size.
MTJigString	"abc"	Specifies the sample text displayed by mtext editor; maximum 10 letters; enter . for no text.

Variable	Default	Meaning
MyDocumentsPrefix (R/O)		Reports path to the *my documents* folder of the currently logged-in user.
		"C:\\Documents and Settings*username*\\My Documents"

N

Variable	Default	Meaning
NodeName (R/O)	*"AC$"*	*Reports the name of the network node; range is one to three characters.*
NoMutt	0	Toggles display of messages (a.k.a. muttering) during scripts, LISP, macros:
		0 Displays prompt, as normal.
		1 Suppresses muttering.
NorthDirection	0	Specifies angle of sun relative to positive y axis.
NwfState	*1*	*Reports whether New Features Workshop displays when AutoCAD starts.*

O

Variable	Default	Meaning
ObscuredColor	257	Specifies ACI color of objects obscured by Hide in 2D wireframe visual style.
ObscuredLtype	0	Specifies linetype of objects obscured by Hide in 2D wireframe visual mode:
		0 Invisible.
		1 Solid.
		2 Dashed.
		3 Dotted.
		4 Short dash.
		5 Medium dash.
		6 Long dash.
		7 Double short dash.
		8 Double medium dash.
		9 Double long dash.
		10 Medium long dash.
		11 Sparse dot.
OffsetDist	-1.0	Controls current offset distance:
		<0 Offsets through a specified point.
		>0 Default offset distance.
OffsetGapType	0	Controls how polylines reconnect when segments are offset:
		0 Extends segments to fill gap.
		1 Fills gap with fillet (arc segment).
		2 Fills gap with chamfer (line segment).
OleFrame	2	Controls the visibility of the frame around OLE objects:
		0 Frame is not displayed and not plotted.
		1 Frame is displayed and is plotted.
		2 Frame is displayed but is not plotted.
OleHide	0	Controls display and plotting of OLE objects:
		0 All OLE objects visible.
		1 Visible in paper space only.

		2 Visible in model space only.
		3 Not visible.
OleQuality	3	Controls the quality of display and plotting of embedded OLE objects:
		0 Monochrome.
		1 Low quality graphics.
		2 High quality graphics.
		3 Automatically selected mode.
OleStartup	0	Toggles loading of OLE source applications to improve plot quality:
		0 Does not load OLE source application.
		1 Loads OLE source app when plotting.
OpmState	0	Toggles whether Properties palette is active.
OrthoMode	0	Toggles orthographic mode.
OsMode	20517	Controls current object snap mode (sum):
		0 NONe.
		1 ENDpoint.
		2 MIDpoint.
		4 CENter.
		8 NODe.
		16 QUAdrant.
		32 INTersection.
		64 INSertion.
		128 PERpendicular.
		256 TANgent.
		512 NEARest.
		1024 QUIck.
		2048 APPint.
		4096 EXTension.
		8192 PARallel.
		16383 All modes on.
		16384 Object snap turned off via OSNAP on the status bar.
OSnapCoord	2	Controls keyboard overrides object snap:
		0 Object snap overrides keyboard.
		1 Keyboard overrides object snap.
		2 Keyboard overrides object snap, except in scripts.
OSnapHatch	0	Toggles whether hatches are snapped:
		0 Osnaps ignore hatches.
		1 Hatches are snapped.
OSnapNodeLegacy	1	Toggles whether osnap snaps to mtext insertion points.
OsnapZ	0	Toggles osnap behavior in z direction:
		0 Uses the z-coordinate.
		1 Uses the current elevation setting.

Variable	Default	Meaning
OsOptions	3	Determines when objects with negative z values are osnaped: **0** Uses the actual z coordinate. **1** Substitutes z coordinate with the elevation of the current UCS.

. . . .

P

Variable	Default	Meaning
PaletteOpaque	2	Controls transparency of palettes: **0** Turned off by user. **1** Turned on by user. **2** Unavailable, but turned on by user. **3** Unavailable, and turned off by user.
PaperUpdate	0	Toggles how AutoCAD plots layouts with paper size different from plotter's default: **0** Displays a warning dialog box. **1** Changes paper size to that of the plotter configuration file.
PDMode	0	Controls point display style (sum): **0** Dot. **1** No display. **2** +-symbol. **3** x-symbol. **4** Short line. **32** Circle. **64** Square.

0	1	2	3	4
·		+	×	'

32	33	34	35	36
⊙	○	⊕	⊠	○

64	65	66	67	68
□	□	⊞	⊠	⊡

96	97	98	99	100
○	○	⊕	⊠	○

Variable	Default	Meaning
PDSize	0.0	Controls point display size: **>0** Absolute size, in pixels. **0** 5% of drawing area height. **<0** Percentage of viewport size.
PEditAccept	0	Toggles display of the PEdit command's 'Object selected is not a polyline. Do you want to turn it into one? <Y>:' prompt.
PEllipse	0	Toggles object used to create ellipses: **0** True ellipse. **1** Polyline arcs.
Perimeter (R/O)	0.0	Reports perimeter calculated by the last Area, DbList, and List commands.
Perspective	0	Toggles perspective mode; not available in 2D wireframe visual mode.
PerpectiveClip	5.0	Controls position of eyepoint clipping, as a percentge; range is 0.01% - 10.0%.
PFaceVMax (R/O)	4	Reports the maximum vertices per 3D face.
PHandle	*0*	*Ranges from 0 to 4.29497E9; unknown usage.*

. . . .

Variable	Default	Meaning
PickAdd	1	Toggles meaning of Shift key on selection sets: **0** Adds to selection set. **1** Removes from selection set.
PickAuto	1	Toggles selection set mode: **0** Single pick mode. **1** Automatic windowing and crossing.
PickBox	3	Controls selection pickbox size; range is 0 to 50 pixels..
PickDrag	0	Toggles selection window mode: **0** Pick two corners. **1** Pick a corner; drag to second corner.
PickFirst	1	Toggles command-selection mode: **0** Enter command first. **1** Select objects first.
PickStyle	1	Controls how groups and associative hatches are selected: **0** Includes neither. **1** Includes groups. **2** Includes associative hatches. **3** Includes both.
Platform (R/O)	*"varies"*	Reports the name of the operating system.
PLineGen	0	Toggles polyline linetype generation: **0** From vertex to vertex. **1** From end to end.
PLineType	2	Controls the automatic conversion and creation of 2D polylines by PLine: **0** Not converted; creates old-format polylines. **1** Not converted; creates optimized lwpolylines. **2** Polylines in older drawings are converted on open; PLine creates optimized lwpolyline objects.
PLineWid	0.0	Controls current polyline width.
~~PlotId~~	""	*Obsolete; has no effect in AutoCAD.*
PlotOffset	0	Toggles the plot offset measurement: **0** Relative to edge of margins. **1** Relative to edge of paper.
PlotRotMode	2	Controls the orientation of plots: **0** Lower left = 0,0. **1** Lower left plotter area = lower left of media. **2** X, y-origin offsets calculated relative to the rotated origin position.
~~Plotter~~	*0*	*Obsolete; has no effect in AutoCAD.*

Variable	Default	Meaning
PlQuiet	0	Toggles display during batch plotting and scripts (replaces CmdDia): **0** Plot dialog boxes and nonfatal errors are displayed. **1** Logs nonfatal errors; plot dialog boxes are not displayed.
PolarAddAng	""	Holds a list of up to 10 user-defined polar angles; each angle can be up to 25 characters long, each separated with a semicolon (;). For example: 0;15;22.5;45.
PolarAng	90	Controls the increment of polar angle; contrary to Autodesk documentation, you may specify any angle.
PolarDist	0.0	Controls the polar snap increment when SnapStyl is set to 1 (isometric).
PolarMode	0	Controls polar and object snap tracking: **0** Measure polar angles based on current UCS (absolute), track orthogonally; don't use additional polar tracking angles; and acquire object tracking points automatically. **1** Measure polar angles from selected objects (relative). **2** Use polar tracking settings in object snap tracking. **4** Use additional polar tracking angles (via PolarAng). **8** Press Shift to acquire object snap tracking points.
PolySides	4	Controls the default number of polygon sides; range is 3 to 1024.
Popups (R/O)	1	Reports display driver support of AUI: **0** Not available. **1** Available.
PreviewEffect	2	Controls the visual effect for previewing selected objects: **0** Dashed lines. **1** Thick lines. **2** Thick dashed lines.
PreviewFilter	7	Controls the exclusion of objects from selection previewing (bitcode sum): **0** No objects excluded. **1** Objects on locked layers. **2** Objects in xrefs. **4** Tables. **8** Multiline text. **16** Hatch patterns. **32** Groups.
Product (R/O)	"AutoCAD"	Reports the name of the software.
Program (R/O)	"acad"	Reports the name of the software's executable file.

Variable	Default	Meaning
ProjectName	""	Controls the project name of the current drawing; searches for xref and image files.
ProjMode	1	Controls the projection mode for Trim and Extend commands: **0** Does not project. **1** Projects to x,y-plane of current UCS. **2** Projects to view plane.
ProxyGraphics	1	Toggles saving of proxy images in drawings: **0** Not saved; displays bounding box. **1** Image saved with drawing.
ProxyNotice	0	Toggles warning message displayed when drawing contains proxy objects.
ProxyShow	1	Controls the display of proxy objects: **0** Not displayed. **1** All displayed. **2** Bounding box displayed.
ProxyWebSearch	0	Toggles checking for object enablers: **0** Does not check for object enablers. **1** Checks for object enablers when an Internet connection is present.
PsLtScale	1	Toggles paper space linetype scaling: **0** Uses model space scale factor. **1** Uses viewport scale factor.
PSolHeight	4.0	Specifies default height of polysolid objects.
PSolWidth	0.25	Specifies default width of polysolid objects.
PsProlog	""	Specifies the PostScript prologue file name.
PsQuality	75	Controls resolution of PostScript display, in pixels: **<0** Display as outlines; no fill. **0** Displays no fills. **>0** Displays filled.
PStyleMode	1	Toggles the plot color matching mode of the drawing: **0** Uses named plot style tables. **1** Uses color-dependent plot style tables.
PStylePolicy (R/O)	1	Reports whether the object color is associated with its plot style: **0** Not associated. **1** Associated.
PsVpScale	0.0	Controls the view scale factor (ratio of units in paper space to units in newly-created model space viewports) 0 = scaled to fit.
PublishAllSheets	1	Determines which sheets (model space and layouts) are loaded automatically into the Publish command's list: **0** Current drawing only. **1** All open drawings.

Variable	Default	Meaning
ⓐ **PublishCollate**	1	Controls how sheets are published: **0** Sheet sets are processed one sheet at a time; a separate plot is created for each sheet. **1** Sheets sets are processed as a single job and single plot file.
PUcsBase (R/O)	""	Reports name of UCS defining the origin and orientation of orthographic UCS settings; in paper space only.

. . . .

Q

Variable	Default	Meaning
QAFlags	*0*	*Controls the quality assurance flags:* *0 Turned off.* *1 The ^C metacharacters in a menu macro cancels grips, just as if user pressed **ESC**.* *2 Long text screen listings do not pause.* *4 Error and warning messages are displayed at the command line, instead of in dialog boxes.* *128 Screen picks are accepted via the AutoLISP (command) function.*
QaUcsLock	*0*	*Toggles; purpose unknown.*
QcState	0	Toggles whether QuickCalc palette is open.
QTextMode	0	Toggles quick text mode.
QueuedRegenMax	*2147483647*	*Ranges between very large and very small numbers.*

. . . .

R

Variable	Default	Meaning
R14RasterPlot	*0*	*Toggle; purpose unknown.*
RasterDpi	300	Controls the conversion of millimeters or inches to pixels, and vice versa; range is 100 to 32767.
ⓐ *RasterPercent*	*20*	*Percentage of system memory to allocate to plotting raster images; range is 0 - 100%.*
RasterPreview (R/O)	1	Toggles creation of BMP preview image.
ⓐ *RasterThreshold*	*20*	*Amount of RAM to allocate to plotting raster images; range is 0 - 2000MB.*
RecoveryMode	2	Controls recording of drawing recovery information after software failure: **0** Note recorded. **1** Recorded; Drawing Recovery palette not displayed automatically. **2** Recorded, and Drawing Recovery palette displays automatically.
RefEditName	""	Specifies the reference file name when in reference-editing mode.
RegenMode	1	Toggles regeneration mode: **0** Regens with each view change. **1** Regens only when required.

. . . .

Variable	Default	Meaning
Re-Init	0	Controls the reinitialization of I/O devices: **1** Digitizer port. *2 Plotter port; obsolete.* **4** Digitizer. *8 Plotter; obsolete.* **16** Reloads PGP file.
RememberFolders	1	Toggles the path search method: **0** Path specified in desktop AutoCAD icon is default for file dialog boxes. **1** Last path specified by each file dialog box is remembered.
RenderPrefsState	0	Toggles display of the Render Preferences palette.
🄰 **RenderUserLights**	1	Determines which lights are rendered: **0** Default *or* user-defined lights are rendered. **1** Default *and* user-defined lights are rendered.
ReportError	1	Determines if AutoCAD sends an error report to Autodesk: **0** No error report created. **1** Error report is generated and sent to Autodesk.
RoamableRootPrefix (R/O)		Reports the path to the root folder where roamable customized files are located.
		"d:\documents and settings*username*\application data\aut..."
~~RIAspect~~		*Removed from AutoCAD Release 14.*
~~RIBackG~~		*Removed from AutoCAD Release 14.*
~~RIEdge~~		*Removed from AutoCAD Release 14.*
~~RIGamut~~		*Removed from AutoCAD Release 14.*
~~RIGrey~~		*Removed from AutoCAD Release 14.*
~~RIThresh~~		*Removed from AutoCAD Release 14.*
RTDisplay	1	Toggles raster display during real-time zoom and pan: **0** Displays the entire raster image. **1** Displays raster outline only.

. . . .

S

Variable	Default	Meaning
🄰 **SaveFidelity**	0	Toggles how annotative objects are translated to earlier releases (bitcode): **0** No changes made. **1** Each scale representation is saved to a separate layer.
SaveFile (R/O)	""	Reports the automatic save file name.
SaveFilePath	"...\temp\"	Specifies the path for automatic save files.
SaveName (R/O)	""	Reports the drawing's save-as file name.

. . . .

Variable	Default	Meaning
SaveTime	10	Controls the automatic save interval, in minutes; 0 = disable auto save.
ScreenBoxes (R/O)	0	Reports the maximum number of menu items supported by display; 0 = screen menu turned off.
ScreenMode (R/O)	3	Reports the state of AutoCAD display: **0** Text screen. **1** Graphics screen. **2** Dual-screen display.
ScreenSize (R/O)	*varies*	Reports the current viewport size, in pixels, such as 719.0000,381.0000.
SDI	*0*	*Controls the multiple-document interface (SDI is "single document interface"):* ***0** Turns on MDI.* ***1** Turns off MDI (only one drawing may be loaded into AutoCAD).* ***2** Disables MDI for apps that cannot support MDI; read-only.* ***3** (R/O) Disables MDI for apps that cannot support MDI, even when SDI= 1.*
SelectionAnnoDisplay	1	Toggles how alternate scale representations are displayed when annotative objects are selected: **0** Not displayed. **1** Displayed according to the value of XFadeCtl.
SelectionArea	1	Toggles use of colored selection areas.
SelectionAreaOpacity	25	Controls the opacity of color selection areas; range is 0 (transparent) to 100 (opaque).
SelectionPreview	3	Controls selection preview: **0** Off. **1** On when commands are inactive. **2** On when commands prompt for object selection.
SetBylayerMode	127	Controls which properties are affected by the SetByLayer command: **0** None. **1** Colors. **2** Linetypes. **4** Lineweights. **8** Materials. **16** Plot styles. **32** ByBlock is changed to Bylayer. **64** Blocks are changed from ByBlock to ByLayer.
ShadEdge	3	Controls shading by Shade command: **0** Faces only shaded. **1** Faces shaded, edges in background color. **2** Only edges, in object color.

Variable	Default	Meaning
		3 Faces in object color, edges in background color.
ShadeDif	70	Controls percentage of diffuse to ambient light; range is 0 to 100 percent.
ShadowPlaneLocation	0.0	Default height of the shadow plane.
ShortcutMenu	11	Controls display of shortcut menus (sum):
		0 Disables all.
		1 Default shortcut menus.
		2 Edit shortcut menus.
		4 Command shortcut menus when commands are active.
		8 Command shortcut menus only when options available at command line.
		16 Shortcut menus when the right button held down longer.
ShowHist	1	Toggles display of history in solids:
		0 Original solids are not displayed.
		1 Display of original solids depends on Show History property settings.
		2 Displays all original solids.
ShowLayerUsage	1	Toggles layer-usage icons in Layers dialog box.
ShpName	""	Specifies the default shape name:
		. Set to no default.
SigWarn	1	Toggles display of dialog box when drawings with digital signatures are opened:
		0 Only when signature is invalid.
		1 Always.
SketchInc	0.1	Controls the Sketch command's recording increment.
SkPoly	0	Toggles sketch line mode:
		0 Record as lines.
		1 Record as a polyline.
SnapAng	0	Controls rotation angle for snap and grid; when not 0, grid lines are not displayed.
SnapBase	0.0,0.0	Controls current origin for snap and grid.
SnapIsoPair	0	Controls current isometric drawing plane:
		0 Left isoplane.
		1 Top isoplane.
		2 Right isoplane.
SnapMode	0	Toggles snap mode.
SnapStyl	0	Toggles snap style:
		0 Normal.
		1 Isometric.
SnapType	0	Toggles snap for the current viewport:
		0 Standard snap.
		1 Polar snap.
SnapUnit	0.5,0.5	Controls x,y spacing for snap distances.
SolidCheck	1	Toggles solid validation.

Variable	Default	Meaning
SolidHist	1	Toggles retention of history in solids.
SortEnts	127	Controls object display sort order:
		0 Off.
		1 Object selection.
		2 Object snap.
		4 Redraw.
		8 Slide generation.
		16 Regeneration.
		32 Plot.
		64 PostScript output.
SpaceSwitch	*1*	*Either 1 or 9; purpose unknown.*
SplFrame	0	Toggles polyline and mesh display:
		0 Polyline control frame not displayed; display polygon fit mesh; 3D faces invisible edges not displayed.
		1 Polyline control frame displayed; display polygon defining mesh; 3D faces invisible edges displayed.
SplineSegs	8	Controls number of line segments that define splined polylines; range is -32768 to 32767.
		<0 Drawn with fit-curve arcs.
		>0 Drawn with line segments.
SplineType	6	Controls type of spline curve:
		5 Quadratic Bezier spline.
		6 Cubic Bezier spline.
SsFound	""	Specifies path and file name of sheet sets.
SsLocate	1	Toggles whether sheet set files are opened with drawing:
		0 Not opened.
		1 Opened automatically.
SsmAutoOpen	1	Toggles whether the Sheet Set Manager is opened with drawing (SsLocate must be 1):
		0 Not opened.
		1 Opened automatically.
SsmPollTime	60	Controls time interval between automatic refreshes of status data in sheet sets; range is 20 to 600 seconds (SsmSheetStatus = 2).
SsmSheetStatus	2	Controls refresh of status data in sheet sets:
		0 Not automatically refreshed.
		1 Refresh when sheet set is loaded or updated.
		2 Also refresh as specified by SsmPollTime.
SsmState (R/O)	0	Toggles whether Sheet Set Manager is open.
StandardsViolation	2	Controls whether alerts are displayed when CAD standards are violated:
		0 No alerts.
		1 Alert displayed when CAD standard violated.

Variable	Default	Meaning
		2 Displays icon on status bar when file is opened with CAD standards, and when non-standard objects are created.
Startup	0	Controls which dialog box is displayed by the New and QNew commands: **0** Displays Select Template dialog box. **1** Displays Startup and Create New Drawing dialog box.
StartupToday		*Removed from AutoCAD 2004.*
StepSize	6.0	Specifies length of steps in walk mode; range is 1E-6 to 1E+6.
StepsPerSec	2	Specifies speed of steps in walk mode; range is 1-30.
SunPropertiesState	0	Toggles display of Sun Properties palette.
SunStatus	0	Toggles display of light by the sun.
SurfTab1	6	Controls density of m-direction surfaces and meshes; range is 5 to 32766.
SurfTab2	6	Density of n-direction surfaces and meshes; range is 2 to 32766.
SurfType	6	Controls smoothing of surface by PEdit: **5** Quadratic Bezier spline. **6** Cubic Bezier spline. **8** Bezier surface.
SurfU	6	Controls surface density in m-direction; range is 2 to 200.
SurfV	6	Specifies surface density in n-direction; range is 2 to 200.
SysCodePage (R/O)	"ANSI_1252"	Reports the system code page; set by operating system.

. . . .

T		
TableIndicator	1	Toggles display of column letters and row numbers during table editing.
▦ **TableToolbar**	1	Toggles display of cell editing toolbar: **0** Not displayed. **1** Displayed.
TabMode	0	Toggles tablet mode.
Target (R/O)	0.0,0.0,0.0	Reports target coordinates in the current viewport.
Tbaskbar	*1*	*Toggles whether each drawing appears as a button on the Windows taskbar.*
TbCustomize	1	Toggles whether toolbars can be customized.
TDCreate (R/O)	*varies*	Reports the date and time that the drawing was created, such as 2448860.54014699.
TDInDwg (R/O)	*varies*	Reports the duration since the drawing was loaded, such as 0.00040625.

. . . .

Variable	Default	Meaning
TDuCreate (R/O)	*varies*	Reports the universal date and time when the drawing was created, such as 2451318.67772165.
TDUpdate (R/O)	*varies*	Reports the date and time of last update, such as 2448860.54014699.
TDUsrTimer (R/O)	*varies*	Reports the decimal time elapsed by user-timer, such as 0.00040694.
TDuUpdate (R/O)	*varies*	Reports the universal date and time of the last save, such as 2451318.67772165.
TempOverrides	1	Toggles temporary overrides.
TempPrefix (R/O)	"d:\temp"	Reports the path for temporary files set by Temp variable.
TextEval	0	Toggles the interpretation of text input during the -Text command: **0** Literal text. **1** Read (and ! as AutoLISP code.
TextFill	1	Toggles the fill of TrueType fonts when plotted: **0** Outline text. **1** Filled text.
☐ TextOutputFileFormat	0	Controls Unicodes for plot and text window log files: **0** ANSI format. **1** UTF-8 (Unicode). **2** UTF-16LE (Unicode). **3** UTF-16BE (Unicode).
TextQlty	50	Controls the resolution of TrueType fonts when plotted; range is 0 to 100.
TextSize	0.2	Controls the default height of text (2.5 in metric units).
TextStyle	"Standard"	Specifies the default name of text style.
Thickness	0.0	Controls the default object thickness.
TileMode	1	Toggles the view mode: **0** Displays layout tab. **1** Displays model tab.
TimeZone	-80000	Specifies current time zone.
ToolTipMerge	0	Toggles the merging of tooltips during dynamic display.
ToolTips	1	Toggles the display of tooltips.
TpState (R/O)	0	Reports if Tool Palettes palette is open.
TraceWid	0.0500	Specifies current width of traces.
TrackPath	0	Controls the display of polar and object snap tracking alignment paths: **0** Displays object snap tracking path across the entire viewport.

Variable	Default	Meaning
		1 Displays object snap tracking path between the alignment point and "From point" to cursor location.
		2 Turns off polar tracking path.
		3 Turns off polar and object snap tracking paths.
TrayIcons	1	Toggles the display of the tray on status bar.
TrayNotify	1	Toggles service notifications displayed by the tray.
TrayTimeout	0	Controls length of time that tray notifications are displayed; range is 0 to 10 seconds.
TreeDepth	3020	Controls the maximum branch depth (in $xxyy$ format):
		xx Model-space nodes.
		yy Paper-space nodes.
		>0 3D drawing.
		<0 2D drawing.
TreeMax	10000000	Controls the memory consumption during drawing regeneration.
TrimMode	1	Toggles trims during Chamfer and Fillet:
		0 Leaves selected edges in place.
		1 Trims selected edges.
TSpaceFac	1.0	Controls the mtext line spacing distance measured as a factor of "normal" text spacing; ranges from 0.25 to 4.0.
TSpaceType	1	Controls the type of mtext line spacing:
		1 At Least: adjusts line spacing based on the height of the tallest character in a line of mtext.
		2 Exactly: uses the specified line spacing; ignores character height.
TStackAlign	1	Controls vertical alignment of stacked text:
		0 Bottom aligned.
		1 Center aligned.
		2 Top aligned.
TStackSize	70	Controls size of stacked text as a percentage of the current text height; range is 1 to 127%.

. . . .

U

Variable	Default	Meaning
UcsAxisAng	90	Controls the default angle for rotating the UCS around an axis (via the UCS command using the X, Y, or Z options); valid values limited to: 5, 10, 15, 18, 22.5, 30, 45, 90, or 180.
UcsBase	""	Specifies name of UCS that defines the origin and orientation of orthographic UCS settings.
UcsDetect	0	Toggles dynamic UCS mode.

. . . .

Variable	Default	Meaning
UcsFollow	0	Toggles view displayed with new UCSs: **0** No change. **1** Automatically aligns UCS with new view.
⌨ **UcsIcon**	3	Controls display of the UCS icon: **0** Off. **1** On. **2** Displays at UCS origin, if possible. **3** On, and displayed at origin.
UcsName (R/O)	""	Reports the name of current UCS view: **""** Current UCS is unnamed.
UcsOrg (R/O)	0.0,0.0,0.0	Reports the origin of current UCS relative to WCS.
UcsOrtho	1	Controls whether the related orthographic UCS settings are restored automatically: **0** UCS setting remains unchanged when orthographic view is restored. **1** Related ortho UCS is restored automatically when an ortho view is restored.
UcsView	1	Toggles whether the current UCS is saved with a named view.
UcsVp	1	Toggles whether the UCS in active viewports remains fixed (locked) or changes (unlocked) to match the UCS of the current viewport.
UcsXDir (R/O)	1.0,0.0,0.0	Reports the x-direction of current UCS relative to WCS.
UcsYDir (R/O)	0.0,1.0,0.0	Reports the y-direction of current UCS relative to WCS.
UndoCtl (R/O)	21	Reports the status of undo: **0** Undo disabled. **1** Undo enabled. **2** Undo limited to one command. **4** Auto-group mode. **8** Group currently active. **16** Combines zooms and pans.
UndoMarks (R/O)	0	Reports the number of undo marks.
UnitMode	0	Toggles the type of units display: **0** As set by Units command. **1** As entered by user.
UpdateThumbnail	15	Controls how thumbnails are updated (sum): **0** Thumbnail previews not updated. **1** Sheet views updated. **2** Model views updated. **4** Sheets updated. **8** Updated when sheets or views are created, modified, or restored. **16** Updated when the drawing is saved.
UseAcis	*0*	*Toggle involving ACIS.*

Variable	Default	Meaning
UserI1 *thru* **UserI5**	0	Five user-definable integer variables.
UserR1 *thru* **UserR5**	0.0	Five user-definable real variables.
UserS1 *thru* **UserS5**	""	Five user-definable string variables; values are not saved.

. . . .

V

Variable	Default	Meaning
ViewCtr (R/O)	*varies*	Reports x,y,z-coordinate of center of current view, such as 15,9,56.
ViewDir (R/O)	*varies*	Reports current view direction relative to UCS (0.0,0.0,1.0 = plan view).
ViewMode (R/O)	0	Reports the current view mode: **0** Normal view. **1** Perspective mode on. **2** Front clipping on. **4** Back clipping on. **8** UCS-follow on. **16** Front clip not at eye.
ViewSize (R/O)	*varies*	Reports the height of current view in drawing units.
ViewTwist (R/O)	0	Reports the twist angle of current view.
VisRetain	1	Controls xref drawing's layer settings: **0** Xref-dependent layer settings are not saved in the current drawing. **1** Xref-dependent layer settings are saved in the current drawing, and take precedence over settings in the xref'ed drawing the next time the current drawing is loaded.
VpLayerOverrides (R/O)	0	Reports whether VP layer properties are overridden in the current viewport: **0** None. **1** At least one layer.
VpLayerOverridesMode	1	Controls display and plot of VP layer property overrides: **0** Not displayed or plotted. **1** Displayed and plotted.
VpMaximizedState (R/O)	0	Toggle: reports whether viewport has been maximized by VpMax command.
VsBackgrounds	1	Toggles display of backgrounds in visual styles.
VsEdgeColor	"ByEntity"	Specifies the edge color.
VsEdgeJitter	-2	Specifies the level of pencil effect: **0** *or* -*n* None. **1** Low. **2** Medium. **3** High.

. . . .

Variable	Default	Meaning
VsEdgeOverhang	-6	Specifies extension of pencil lines beyond edges; range is 1 to 100 pixels; *-n* = none.

VsEdges	1	Specifies types of edges to display:
		  0 No edges displayed.
		  1 Isolines displayed.
		  2 Facets and edges displayed.

VsEdgeSmooth	1	Specifies crease angle; range is 0 to 180 degrees.
VsFaceColorMode	0	Determines color of faces.
VsFaceHighlight	-30	Specifies the size of highlights; range is -100 to 100. Ignored when VsMaterialMode is 1 or 2 and objects have materials attached.
VsFaceOpacity	-60	Controls the transparency/opacity of faces; range is -100 to 100 (fully opaque).

VsFaceStyle	0	Determines how faces are displayed:
		  0 None.
		  1 Real.
		  2 Gooch.

VsHaloGap	0	Specifies halo gap; range is 0 to 100 pixels.
VsHidePrecision	0	Toggles accuracy of hides and shades.
VsIntersectionColor	"7 (white)"	Specifies the color of intersecting polylines.
VsIntersectionEdges	0	Toggles the display of intersecting edges.

VsIntersectionLtype	1	Specifies the linetype for intersecting polylines:
		  0 Off.
		  1 Solid.
		  2 Dashed.
		  3 Dotted.
		  4 Short dash.
		  5 Medium dash.
		  6 Long dash.
		  7 Double-short dash.
		  8 Double-medium dash.
		  9 Double-long dash.
		  10 Medium-long dash.
		  11 Sparse dot.

VsIsoOntop	0	Toggles whether isolines are displayed.

VsLightingQuality	1	Toggles the quality of lighting:
		  0 Facets displayed.
		  1 Facets smoothed.

VsMaterialMode	0	Controls the display of material finishes:
		  0 No materials displayed.
		  1 Materials displayed.
		  2 Materials and textures displayed.

VSMax (R/O)	*varies*	Reports the upper-right corner of virtual screen, such as 37.46,27.00,0.00.
VSMin (R/O)	*varies*	Reports the lower-left corner of virtual screen, such as -24.97,-18.00,0.0.
VsMonoColor	"RGB:255,255,255"	Specifies the monochrome tint.

Variable	Default	Meaning
VsObscuredColor	"ByEntity"	Specifies the color of obscured lines.
VsObscuredEdges	1	Toggles the display of obscured edges.
VsObscuredLtype	1	Specifies the linetype of obscured lines.
VsShadows	0	Determines the quality of shadows:
		0 No shadows displayed.
		1 Ground shadows displayed.
		2 Full shadows displayed.
VsSilhEdges	0	Toggles the display of silhouette edges.
VsSilhWidth	5	Specifies the width of silhouette edge lines; range is 1 to 25 pixels.
VsState	0	Toggles the Visual Styles palette.
VtDuration	750	Controls the duration of smooth view transition; range is 0 to 5000 seconds.
VtEnable	3	Controls smooth view transitions for pans, zooms, view rotations, and scripts:
		0 Turned off.
		1 Pan and zoom.
		2 View rotation.
		3 Pan, zoom, and view rotation.
		4 During scripts only.
		5 Pan and zoom during scripts.
		6 View rotation during scripts.
		7 Pan, zoom, and view rotation during scripts.
VtFps	7	Controls minimum speed for smooth view transitions; range is 1 to 30 frames per second.

. . . .

W

Variable	Default	Meaning
WhipArc	0	Toggles display of circular objects:
		0 Displays as connected vectors.
		1 Displays as true circles and arcs.
WhipThread	1	Controls multithreaded processing on two CPUs (if present) during redraws and regens:
		0 Single-threaded calculations.
		1 Regenerations multi-threaded.
		2 Redraws multi-threaded.
		3 Regens and redraws multi-threaded.
WindowAreaColor	5	Specifies ACI color of windowed selection area.
WmfBkgnd	Off	Toggles background of *.wmf* files:
		Off Background is transparent.
		On Background is same as AutoCAD's background color.
WmfForegnd	Off	Toggles foreground colors of exported WMF images:
		Off Foreground is darker than background.
		On Foreground is lighter than background.
WorldUcs (R/O)	1	Toggles matching of WCS with UCS:
		0 Current UCS does not match WCS.
		1 UCS matches WCS.

. . . .

Variable	Default	Meaning
WorldView	1	Toggles view during 3dOrbit, DView, and VPoint commands: **0** Current UCS. **1** WCS.
WriteStat (R/O)	1	Toggle: reports whether *.dwg* file is read-only: **0** Drawing file cannot be written to. **1** Drawing file can be written to.
WsCurrent	*"varies"*	Controls name of current workspace.

. . . .

X

Variable	Default	Meaning
XClipFrame	0	Toggles visibility of xref clipping boundary.
XEdit	1	Toggles editing of xrefs: **0** Cannot in-place refedit. **1** Can in-place refedit.
XFadeCtl	50	Controls faded display of objects not being edited in-place: **0** No fading; minimum value. **90** 90% fading; maximum value.
XLoadCtl	2	Controls demand loading: **0** Demand loading turned off; entire drawing is loaded. **1** Demand loading turned on; xref file opened. **2** Demand loading turned on; a *copy* of the xref file is opened.
XLoadPath	"...\temp"	Specifies path for storing temporary copies of demand-loaded xref files.
XRefCtl	0	Toggles creation of *.xlg* xref log files.
XrefNotify	2	Controls notification of updated and missing xrefs: **0** No alert displayed. **1** Icon indicates xrefs are attached; a yellow alert indicates missing xrefs. **2** Also displays balloon messages when an xref is modified.
XrefType	0	Toggles xrefs: **0** Attached. **1** Overlaid.

. . . .

Z

Variable	Default	Meaning
ZoomFactor	60	Controls the zoom level via mouse wheel; range from 3 to 100.
ZoomWheel	0	Switches the zoom direction when mouse wheel is rotated forward: **0** Zooms in. **1** Zooms out.

. . . .

Variable	Default	Meaning

....

3

▣ **3dConversionMode** 1 Controls how material and light definitions are converted when pre-AutoCAD 2008 drawings are opened:
  0 None converted.
  1 Converted automatically.
  2 Converted after prompted.

3dDwfPrc 2 Level of precision in drawings exported as *.dwf* files:
  1 1
  2 0.5
  3 0.2
  4 0.1
  5 0.01
  6 0.001

▣ **3dSelectionMode** 1 Toggles how visually overlapping objects are selected (other than 2D and 3D wireframe):
  0 Uses traditional 3D selection.
  1 Uses line-of-sight selection.

....

AutoCAD 2008 Keystrokes & Shortcuts

Cursor Icons

Crosshair cursor with object snap aperture.

Cursor is orthographic.

Object is annotative object 🔲.

Object has two or more annotative scales 🔲.

Object or layer is locked.

Table is linked to external data file 🔲.

Drawing is being panned.

Drawing has panned to its extents.

Drawing is in realtime zoom.

Camera distance is adjusted.

Camera is swiveled.

3D view is in constrainted orbit.

3D view is in continuous orbit.

3D view is in free orbit.

Selection Set Options

Ctrl+*pick*	Selects faces.
Pick	Selects one object.
ALL	Selects all objects.
AU	AUtomatic: *(pick)* or BOX.
BOX	Left to right = Crossing; right to left = Window.
C	Crossing.
CP	Crossing polygon.
F	Fence.
G	Group.
L	Last.
M	Multiple (no highlighting).
P	Previous.
SI	Single selection.
SU	Selects faces.
W	Window.
WP	Window polygon.

Selection modes:

A	Adds to selection set (default).
R	Removes from selection set.
Shift	Remove sfrom selection set.
U	Undoes change to selection set.

UCS Icons

Old-style UCS Icon

New-style UCS Icon

Cannot-Edit Icon

3D UCS Icon

Paperspace Icon